SULFUR NUTRITION AND ASSIMILATION IN HIGHER PLANTS

REGULATORY AGRICULTURAL AND ENVIRONMENTAL ASPECTS

Edited by

L.J. De Kok, I. Stulen, H. Rennenberg, C. Brunold and W.E. Rauser

SPB Academic Publishing bv, The Hague, 1993

CIP-DATA KONINKLIJKE BIBLIOTHEEK, DEN HAAG

Sulfur

Sulfur nutrition and assimilation in higher plants :
regulatory agricultural and environmental aspects / ed.
by L.J. De Kok ... [et al.]. – The Hague : SPB Academic
Publishing. – Ill.
With index, ref.
ISBN 90-5103-084-3 bound
Subject headings: plant physiology.

ISBN 90-5103-084-3

CONTENTS

PREFACE

This book contains the invited papers of the second Workshop on Sulfur Metabolism in Higher Plants, which was held in Garmisch-Partenkirchen, F.R.G., from April 21 to 25, 1992 and organized by the Fraunhofer Institute for Atmospheric Environmental Research, and the Universities of Bern, Guelph and Groningen.

The book is the second volume reviewing sulfur metabolism in higher plants. The first volume: *Sulfur Nutrition and Sulfur Assimilation in Higher Plants*, edited by H. Rennenberg, Ch. Brunold, L.J. De Kok and I. Stulen, published in 1990 by SPB Academic Publishing, The Hague, dealt mainly with the fundamental, environmental and agricultural aspects. Although the present volume also covers various environmental and agricultural aspects of sulfur metabolism, the emphasis is set on the regulatory aspects of sulfur uptake and assimilation and the physiological function of the various sulfur compounds in plants. In addition, the present knowledge on the molecular aspects of sulfur assimilation as well as the interaction of sulfur metabolism with nitrogen and selenium metabolism is discussed.

A selection of papers derived from the posters shown at the workshop are published in a special issue of Phyton (Austria) entitled: "Progress in Sulfur Metabolism of Higher Plants", Phyton (1992) 32(3): 1-159. The content of this issue is listed in this book.

We are pleased to dedicate this book to Prof. dr. L. Bergmann, University of Köln, F.R.G. for his outstanding contribution to a better understanding of the synthesis and function of glutathione in plants.

Luit J. De Kok
Ineke Stulen
Heinz Rennenberg
Christian Brunold
Wilfried E. Rauser

editors

L. Bergmann.

FOREWORD

Exploring Glutathione and Homoglutathione Metabolism

Ludwig Bergmann

First, I want to thank the editors for having done me the great honor of dedicating this book to me. These proceedings will certainly be stimulating to all interested in the field of sulfur nutrition and metabolism, as they contain a wealth of new information and the formulation of principal questions that should be investigated.

Having worked mainly on metabolism and development of plant cell cultures, I came to the field of sulfur metabolism as an outsider more by chance than by careful planning. At a time where much science is done in form of carefully planned projects and where research very often is confined by the policy of advisory committees and regulations of granting agencies, it may be illuminating to trace the way that led to our interest in γ-glutamyl-cysteinyl-tripeptides in higher plants.

Our original interest in this group of thiols arose from studies on substances released into the media of suspension cultures of *Nicotiana tabacum*. Some of these substances are required to induce cell division and growth in single cells plated in low population densities on agar, and can be replaced by adding of kinetine and glutamine to the basic media. We became interested in the nature of these substances when a student of mine, Heike Logemann, observed that conditioned media of cultures grown photoheterotrophically were more effective in sustaining growth of single tobacco cells than those of heterotrophic cultures (Logemann & Bergmann 1974). We were surprised by this result as roots rather than green leaves are thought to be the main sources of cytokinins and so we tried to quantify the cytokinin in the conditioned media. Using the famous soja test of C.O. Miller we found that the growth of the soja calli was severely inhibited by the conditioned media derived from tobacco cells.

Since the conditioned media induced cell divisions in tobacco but not in soja, I asked Heinz Rennenberg, who was then looking for a subject for his thesis work, to isolate and identify the inhibitory substances. This was difficult because there was a vast array of carbohydrates, amino acids and phenylpropane derivatives in the media which all had to be separated and tested. Nevertheless, Rennenberg was successful and one day came up with a peptide fraction that inhibited the growth of soja calli at high dilutions. Fortunately, there was only one peptide in this fraction and this turned out to be glutathione (Rennenberg 1976). We soon found that the green tobacco cultures released nearly 30 times as much glutathione into the cultured media as chloroplast-free heterotrophic suspension cultures. We also observed that glutathione was not only released into the medium but was also taken up by the cells and used as sulfur source after depletion of the sulfate provided in the medium (Bergmann & Rennenberg 1978). This led to the suggestion that glutathione in plants may function as a storage and transport form of reduced sulfur, which was subsequently proved by transport experiments with whole plants (Rennenberg *et al.* 1979; Bonas *et al.* 1982) and has been elegantly demonstrated more recently in translocation studies in maize seedlings by Rauser *et al.* (1991).

Sulfur Nutrition and Assimilation in Higher Plants
edited by L.J. De Kok et al., pp. ix–xii

The overproduction of glutathione in photoheterotrophically grown tobacco cultures also led to the idea that glutathione was probably synthesized in plastids and that a high rate of glutathione synthesis in green cells might be associated with the differentiation of chloroplasts. Although the first report on high glutathione contents of chloroplasts by Foyer & Halliwell (1976) seemed to support this idea, it was shown to be wrong. Glutathione in chlorenchyma cells can be synthesized in chloroplasts as well as in the cytosol. To demonstrate this, we had to study the enzymes of glutathione synthesis, γ-glutamyl-cysteine synthetase and glutathione synthetase, and to localize them inside the cell. This proved to be much more difficult than we had expected. Despite several attempts we were unable to repeat the experiments of Webster & Varner (1955) who claimed to have measured the activities of both enzymes in plants.

Our efforts were only successful with the development of methods for derivating thiols with monobromobimanes and the publication of the pilot paper by Newton *et al.* (1981) on the separation of bimane derivatives by reverse phase HPLC. These methods allow the determination of γ-glutamyl-cysteine synthetase and glutathione synthetase activity in crude extracts by measuring simultaneously the formation of γ-glutamyl-cysteine and glutathione, respectively, and the consumption of cysteine and/or γ-glutamyl-cysteine as substrates. In his thesis work, Rüdiger Hell characterized both enzymes with this technique and demonstrated that the activities of both enzymes can be found in isolated chloroplasts of several plant species as well as in cytosol (Hell & Bergmann 1988, 1990). These results provide evidence for the synthesis of glutathione in the chloroplasts and in the cytosol. However, both compartments may be independent, because all experiments to demonstrate the uptake of glutathione or glutathione disulfide by isolated chloroplasts have failed so far, but further studies are necessary to establish this point.

In addition to glutathione we were also attracted by homoglutathione, the γ-glutamyl-cysteinyl-tripeptide, which is characteristic for several members of the *Fabales*. There were two reasons for this interest. First, to support our idea of the function of glutathione as a transport form of reduced sulfur we wanted to know if homoglutathione was also used in legumes for the long distance transport of reduced sulfur. This indeed proved to be the case. Experiments by Peter Macnicol (Macnicol & Bergmann 1984) demonstrated that in *Vigna radiata* homoglutathione serves as a major transport form of organic sulfur from the leaves to the developing seeds, which fits nicely into the scheme.

The second reason we became interested in homoglutathione was the question of its synthesis. Like glutathione synthetase from mammalian cells, we found that the glutathione synthetase from tobacco and from pea leaves exhibited high specificities for glycine and did not accept β-alanine as a substrate. We were therefore puzzled by the finding of Webster (1953) who reported, in his first paper on glutathione synthesis, high rates of glutathione synthesis by incubating enzyme preparations of *Phaseolus vulgaris* with ^{14}C-glycine. As homoglutathione is the main thiol present in *Phaseolus vulgaris* the data of Webster (1953) either are incorrect or indicate the existence of a glutathione synthetase in *Phaseolus* which has a different specificity than those of tobacco or pea.

To elucidate this question we first had to elaborate an efficient method for the isolation of homoglutathione, since homoglutathione required as HPLC standard was not on the market. This was done by Sigrid Klapheck (1988) who purified homoglutathione from seeds of *Phaseolus coccineus* by using anion-exchange chromatography and CuO_2-precipitation. In this way she achieved yields which were nearly 100 times higher than those obtained by the classical method of Carnegie (1963).

Having chromatographically pure homoglutathione as a standard, it was then possible to characterize the tripeptide synthetase of *Phaseolus coccineus*. This characterization revealed a high affinity of the enzyme for β-alanine and a poor rate of glutathione synthesis even at 100 μM glycine, which led us to designate the enzyme as homoglutathione synthetase (Klapheck *et al.* 1988). By comparing the rates of homoglutathione and glutathione synthesis in leaf extracts of different species of *Phaseolus*, it became apparent that the occurrence of a β-alanine specific homoglutathione synthetase is a general feature of the *Phaseoleae*.

Like glutathione, oxidized homoglutathione is reduced by NADPH in the presence of glutathione reductase and it has been tacitly assumed that homoglutathione is used as the reductant in photo-scavenging of hydrogen peroxide produced during photosynthesis instead of glutathione. However, the concentrations of homoglutathione we found in chloroplasts of *Phaseolus coccineus* and of *Glycine max.* were much lower than the concentrations of glutathione in chloroplasts of *Spinacia oleracea* and *Pisum sativum*, where the ascorbate-glutathione-reduction-pathway has been established. These low concentrations led us to question the function of homoglutathione in photo-scavenging hydrogen peroxide and to look for the enzymes involved in the ascorbate-glutathione-reduction-pathway in chloroplasts of plants containing homoglutathione. In contrast to the activities of ascorbate peroxidase and monodehydroascorbate reductase, that were comparable to the activities measured in isolated chloroplasts of *Spinacia oleracea* and *Pisum sativum*, no or only very small activities of glutathione dehydrogenase and of glutathione disulfide reductase could be detected in chloroplasts of *Phaseolus occineus* and *Glycine max*. This finding strongly suggests that the regeneration of ascorbate in these chloroplasts mainly proceeds by monodehydroascorbate reductase activity, whereas homoglutathione only plays a secondary role in H_2O_2-scavenging (Klapheck *et al.* 1992). To me, this result is a fine example that we should be very careful with generalizations.

There is another point which made the low homoglutathione content in the chloroplasts of *Phaseolus coccineus* interesting to us. Several experiments have shown that the rate of glutathione synthesis in plant cells can be modulated by substrate availability and may be regulated by feedback inhibition of γ-glutamyl-cysteine synthetase, as glutathione *in vitro* produces substantial inhibition of γ-glutamyl-cysteine synthetase activity at concentrations found in plant cells. Thus, the question arises, how is the homoglutathione concentration in the chloroplasts of *Phaseolus* kept so low. The enzymes needed for homoglutathione synthesis, γ-glutamyl-cysteine synthetase and homoglutathione synthetase, are both found in isolated chloroplasts of *Phaseolus coccineus* and in the cytosol. Assessing the amino acid pools, Heinrich Zopes (Zopes *et al.* 1992) recently found very low β-alanine concentrations in the chloroplasts. The concentrations were one order of magnitude smaller than the apparent K_m value for β-alanine of the homoglutathione synthetase and probably could limit the rate of homoglutathione synthesis. Zopes *et al.* (1992) further demonstrated that the chloroplasts apparently are unable to synthesize β-alanine because one of the key enzymes of β-alanine synthesis in *Phaseolus coccineus*, β-ureidopropionase is lacking in the chloroplasts. β-Ureidopropionase cleaves N-carbamoyl-β-alanine, a product in the uracil degradation pathway, into CO_2, NH_3 and β-alanine. Its absence in the chloroplasts leads to the interesting question of how pyrimidine nucleic acids are degraded in chloroplasts.

Finally, I would like to mention one more result that came out of our engagement with glutathione synthesis. Several of the experiments of Webster & Varner (1955) on glutathione synthesis were done with wheat germ as the enzyme source. Therefore, we have worked also with wheat and by careful analysis of the thiol composition of *Triticum aestivum* detected a new homologue of glutathione, γ-glutamyl-cysteinyl-serine (Klapheck *et al.* 1992). This new thiol is widespread within the family of *Poaceae* and its functions seem to be similar to those of glutathione (Bergmann & Rennenberg, this volume).

References

Bergmann, L. & Rennenberg, H. 1978. Efflux and production of glutathione in suspension cultures of *Nicotiana tabacum*. Z. Pflanzenphysiol. 88: 175-185.

Bonas, U., Schmitz, K., Rennenberg, H. & Bergmann, L. 1982. Phloem transport of sulfur in *Ricinus*. Planta 155: 82-88.

Carnegie, P.R. 1963. Isolation of a homologue of glutathione and other acidic peptides from seedlings of *Phaseolus aureus*. Biochem. J. 89: 459-470.

Foyer, C.H. & Halliwell, B. 1976. The presence of glutathione and glutathione reductase in chloroplasts: A proposed role in ascorbic acid metabolism. Planta 133: 21-25.

Hell, R. & Bergmann, L. 1988. Glutathione synthetase in tobacco suspension cultures: catalytic properties and localization. Physiol. Plant. 72: 70-76.

Hell, R. & Bergmann, L. 1990. γ-Glutamylcysteine synthetase in higher plants: catalytic properties and and subcellular localization. Planta 180: 603-612.

Klapheck, S. 1988. Homoglutathione: isolation, quantification and occurrence in legumes. Physiol. Plant. 74: 733-739.

Klapheck, S., Zopes, H., Levels, H.-G. & Bergmann, L. 1988. Properties and localization of the homoglutathione synthetase from *Phaseolus coccineus* leaves. Physiol. Plant. 74: 733-739.

Klapheck, S., Chrost, B., Starke, J. & Zimmermann, H. 1992. γ-Glutamylcysteinylserine – a new homologue of glutathione in plants of the family *Poaceae*. Botanica Acta 105: 174-179.

Logemann, H. & Bergmann, L. 1974. Influence of light and medium on the plating efficiency of isolated cells from callus cultures of *Nicotiana tabacum* var. "Samsun". Planta 121: 283-292.

Macnicol, P.K. & Bergmann, L. 1984. A role for homoglutathione in organic sulphur transport to the developing mung bean seed. Plant Sci. Let. 36: 219-223.

Newton, G.L., Dorian, R. & Fahey R.C. 1981. Analysis of biological thiols: Dervatisation with monobromobimane and separation by reverse-phase high-performance liquid chromatography. Anal. Biochem. 114: 383-387.

Rauser, W.E., Schupp, R. & Rennenberg, H. 1991. Cysteine, γ-glutamyl-cysteine and glutathione levels in maize seedlings. Distribution and translocation in normal and cadmium-exposed plants. Plant Physiol. 97: 128-138.

Rennenberg, H. 1976. Glutathione in conditioned media of tobacco suspension cultures. Phytochem. 15: 1433-1434.

Rennenberg, H., Schmitz, K. & Bergmann, L. 1979. Long-distance transport of sulfur in *Nicotiana tabacum*. Planta 147: 57-62.

Webster, G.C. 1953. Peptide-bond synthesis in higher plants. III. The synthesis of glutathione. Arch. Biochem. Biophys. 47: 241-250.

Webster, G.C. & Varner, J.E. 1955. Peptide-bond synthesis in higher plants. III. The formation of glutathione from γ-glutamylcysteine. Arch. Biochem. Biophys. 55: 95-103.

Zopes, H., Klapheck, S. & Bergmann, L. 1992. The functions of homoglutathione and hydroxymethylglutathione for scavenging of hydrogen peroxide. Plant Cell Physiol. *submitted.*

Section 1.
Regulatory aspects of sulfur uptake and metabolism

Contents

MEMBRANE AND LONG-DISTANCE TRANSPORT OF SULFATE

David T. Clarkson[1], Malcolm J. Hawkesford[1] and Jean-Claude Davidian[2]
[1]*Department of Agricultural Sciences, University of Bristol, AFRC Institute of Arable Crops Research, Long Ashton Research Station, Bristol, BS18 9AF, UK* and [2]*Biochimie et Physiologie Végétales, ENSA-INRA, 34060 Montpellier Cedex 01, France*

New opportunities

It is not easy to add very significantly to the excellent review of many topics related to sulfate transport written by Cram (1990) in the proceedings of the first Sulfur-Workshop volume. Progress has not been spectacular in any aspect of sulfate transport strictly related to higher plants, which may reflect the limited number of workers in the field. In the last few years, however, two quite unrelated developments have increased our interest in the properties of the sulfate transporter of crop plants and have given hope that we may soon learn much more about molecular aspects of its structure and the regulation of its expression.

Firstly, legislation has enforced a clean-up of the atmosphere. For many years fortuitous additions of sulfur from pollutants and from compound fertilizers, such as triple superphosphate, provided sulfur inputs which exceeded the growth requirements of crops and natural vegetation. Both of these sources are in sharp decline so that, many crops now respond to sulfur fertilization and there is a need to understand plant factors which ensure efficient acquisition and utilization of sulfur.

Secondly, success in cloning the gene for the sulfate permease in *Neurospora* has given us a glimpse of what the higher plant transporter may look like (Ketter *et al.* 1991). Another success has been the complementation of yeast mutants with genes from higher plant cDNA libraries; this approach has yielded a gene for a potassium channel from *Arabidopsis* (Sentenac *et al.* 1992). Molecular genetics quickens the pace of research and in a short while we will probably learn more than we have ever known.

There is, then, both the chance and the practical need to have a more complete picture about the uptake of sulfate and the internal management of the sulfur economy of plants. This review takes stock of what is known as we prepare to leave the darkness and welcome the new dawn.

The conceptual model of sulfate co-transport at the plasma membrane

The energization of the thermodynamically "uphill" transport of anions into cells is widely believed to be driven by the proton electrochemical potential gradient across the plasma membrane (PM) (Fig. 1). This idea is so firmly fixed that it seems heretical to point out that actual evidence on this point is hard to find and usually incomplete, especially with regard to SO_4^{2-}. The observations of Lass & Ullrich-Eberius (1984)

Sulfur Nutrition and Assimilation in Higher Plants
edited by L.J. De Kok et al., pp. 3–19

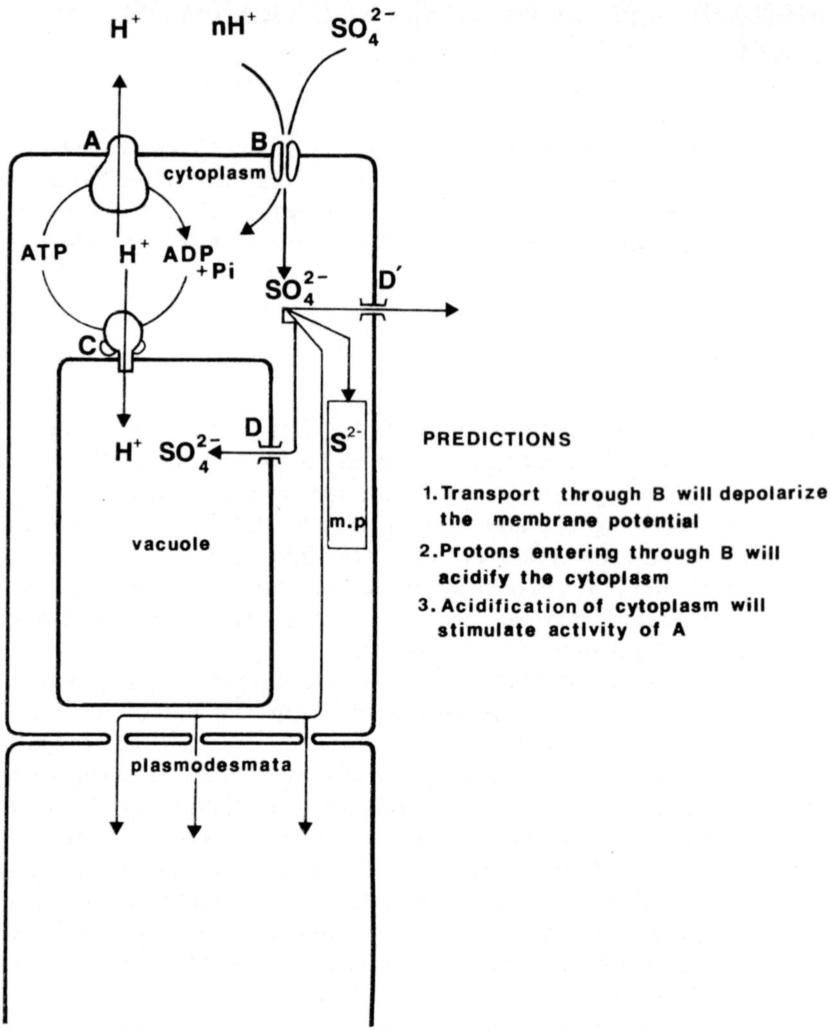

Fig. 1. A hypothetical scheme for the early events in SO_4^{2-} transport in a root cell. The mechanism depends on a proton motive force built up by the ATPase (A) in the plasma membrane (PM). Protons are shown as being co-transported with SO_4^{2-} through a permease (B) with a stoichiochemistry of $3H^+$: $1SO_4^{2-}$. The small pool of cytoplasmic SO_4^{2-} can be depleted by a "downhill" (*i.e.* diffusive) movement into the vacuole through an unknown mechanism D, the driving force being the inside positive membrane potential created by the tonoplast (TP) proton pump (C). Transport via the symplast towards the root xylem involves movement through plasmodesmata. In most circumstances the flux by this pathway is greater than the flux into the vacuole (Bell 1991). Some of the SO_4^{2-} is reduced by enzymes of the assimilatory pathway thus entering the metabolised pool (m.p.). This flux varies among species and must involve transport across the membranes of plastids where the ATP sulfurylase is located. SO_4^{2-} efflux at the PM is through an unknown mechanism D'. Some consequences of proton/SO_4^{2-} co-transport are predicted.

with *Lemna* confirmed the first of the predictions in Fig. 1. The initial influx of SO_4^{2-} after a period of deprivation was accompanied by a depolarization of the membrane potential, the magnitude of which depended on the external $[SO_4^{2-}]$ and

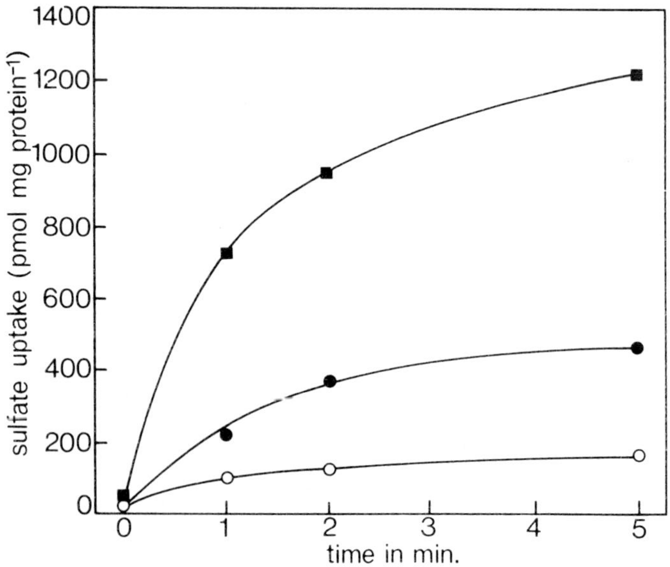

Fig. 2. Dependence of sulfate uptake into isolated plasma membrane vesicles of *Brassica napus* roots on the pH gradient. Membrane vesicles, isolated by two phase partitioning were equilibrated at either pH 8 (squares and open circles) or pH 6 (solid circles) were then transferred to pH 6 (squares and solid circles) or pH 8 (open circles) in the presence of 100 μM sulfate. Uptake was estimated at the times indicated using radioactive tracer and a filtration method for vesicle separation (Hawkesford & Davidian, *in press*).

its influx. However, neither the external pH nor the cytoplasmic pH were measured at that time, so the other expectations of the scheme were not verified. Given the probable ΔpH across the PM and the observed depolarization of the membrane potential (Δy), energetic calculations demand a stoichiometry of $3H^+$ per SO_2^{2-} transported. In cells with a large PM influx some perturbation of cytoplasmic pH might occur. More recently, Ullrich & Novacky (1990) have shown that, for Cl^- and $H_2PO_4^-$, depolarization of the membrane potential was, indeed, accompanied by cytoplasmic acidification in the roots hairs of *Limnobium*. In this species, unlike *Lemna*, electrodes measuring Δy and pH can be specifically located in the cytoplasm, close to the PM surface. Sulfate was not examined in these experiments. The much larger influx of NO_3^-, which might have caused an even more marked acidification of the cytoplasm than either Cl^- or $H_2PO_4^-$, did exactly the reverse. Alkalinization of the cytoplasm was thought to be due to nitrate reduction and the contributions of K^+ transport to the ''strong ion difference''. Ullrich & Novacky (1990) point out that whether or not changes in the internal or external pH are measured during co-transport depends, not on H^+ movements themselves, but on the movement of strong counter ions (*e.g.* K^+ or Cl^-). Thus the pH of a compartment is determined by the concentration differences of the strong ions present (see Stewart 1983), and we should not expect pH changes to report directly the numbers of protons co-transported with a substrate.

An alternative approach to the role of ΔpH in SO_4^{2-} transport has been to explore the energization of SO_4^{2-} accumulation in right-side-out PM vesicles using pH

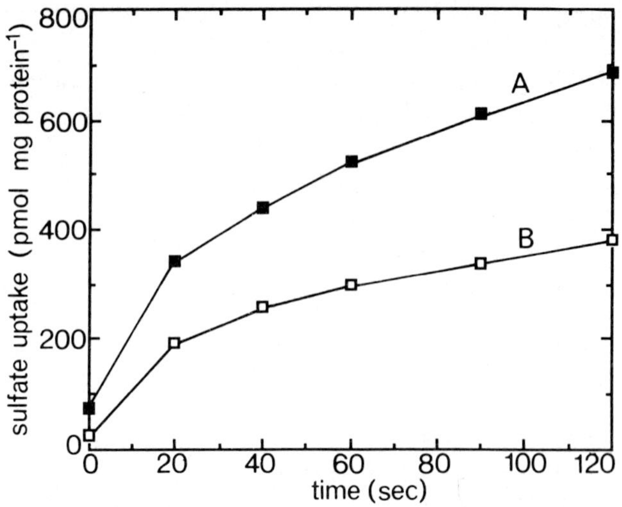

Fig. 3. Uptake of sulfate into plasma membrane vesicles isolated from sulfate-deficient *Brassica napus* roots (A) or from sulfate replete material (B). Membranes isolated by two phase partitioning were equilibrated at pH 8 prior to transfer to 100 μM sulfate at pH 6, and uptake was estimated at the times indicated using radioactive tracer and filtration method for vesicle separation (Hawkesford & Davidian, *in press*).

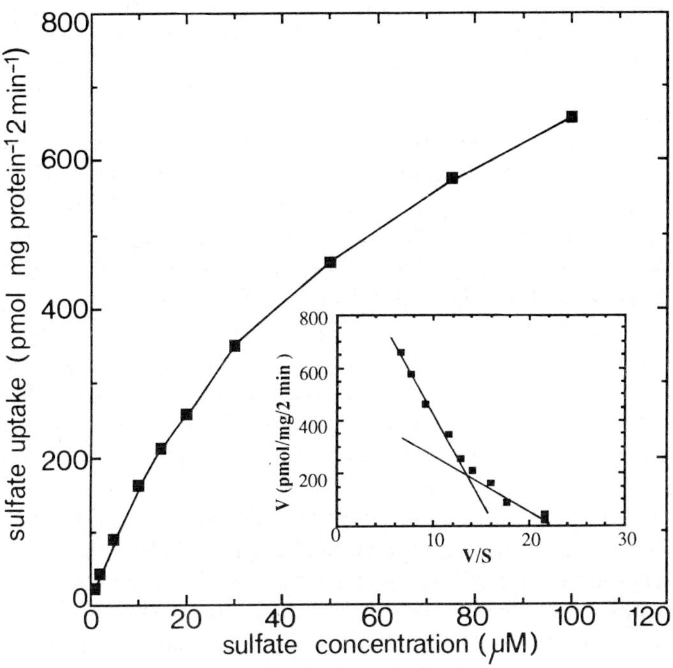

Fig. 4. Concentration dependence of sulfate uptake into isolated plasma membrane vesicles of *Brassica napus* roots. Membranes were equilibrated at pH 8 and then transferred to 100 μM sulfate at pH 6. Vesicles were filtered either immediately, or after two minutes. The data as plotted are the results of two minute labeling minus the results for time zero (non-specific binding). The inset shows an Eadie-Hofstee transformation of this data and two possible components which may be derived from it.

jumps (this report, and Hawkesford & Davidian, *in press*). With no other form of energization, a switch of the external pH from 8 to 6 brings about $^{35}SO_4^{2-}$ uptake into the vesicles (Fig. 2). SO_4^{2-} transport by this system depended critically on the magnitude of the ΔpH across the PM and on the sulfur status of the roots from which the PM preparations were made (Figs. 2, 3). Kinetic analysis of the concentration dependence of SO_4^{2-} transport indicates discontinuities of the absorption isotherm (Fig. 4), suggesting "multiphasic" behaviour described for intact roots and cells (Nissen 1974, 1991). This matter will be discussed below. In the lower range of $[SO_4^{2-}]$ the K_m for SO_4^{2-} transport was dependent on external pH, being least (12 μM) at pH 5 and greater (K_m 67 μM) at pH 6.5 or above. The K_m for H^+ was also affected by $[SO_4^{2-}]$ being lowest (0.5 μM) at high $[SO_4^{2-})$. A detailed report of this work is in preparation; to our knowledge this is the first report of SO_4^{2-} transport from isolated, higher plant membranes. It suggests a method for studying the stoichiometry of H^+/SO_4^{2-} co-transport.

Multiple forms or states of the sulfate carrier

The strenuous debate about the interpretation of uptake kinetics in terms of a single transport system existing in a number of discrete states which are determined by external substrate concentration (Nissen 1971, 1991), or of several kinds of transporter system operating over different substrate concentration ranges (*e.g.* Epstein 1973, 1976; Borstlap 1983), shows no signs of being over. The genetic separation of constitutive, low affinity, and facultative, high affinity transport in micro-organisms, *e.g.* in *Neurospora* (Marzluf 1970), has been advanced to support the latter point of view. However, it has been questioned, particularly with respect to SO_4^{2-}, $H_2PO_4^-$ and K^+ whether higher plants really do have separate facultative transport systems switched on by nutrient starvation (Clarkson & Lüttge 1991). This makes the "argument-by-analogy" with microbes less potent. In addition, many authors have pointed out that major shifts in K_m are not a feature of higher plant response to SO_4^{2-} deprivation (Lee 1982; Clarkson *et al.* 1983) and it is, therefore, probable that higher plants depend on the tight regulation of a constitutive system. Over the "environmental" range of $[SO_4^{2-}]$ encountered by plants in natural and agricultural conditions the evidence is compatible with transport by a single mechanism. It is only at unusually large external $[SO_4^{2-}]$ that uptake becomes apparently less dependent on metabolism and may display linear kinetics. The real point of dispute becomes whether the discontinuities in absorption isotherms represent discrete states of the transport system, which must be explained ultimately in molecular terms, or whether they are manifestations of other changes which occur as the concentration of an ion changes in the external medium. A theoretical explanation of departures for normal Michaelian kinetics is available (Sanders *et al.* 1984; Sanders 1986). A random ligand-binding model (Sanders 1986) for a co-transport system, such as the nH^+/SO_4^{2-} discussed above, predicts both single and multiple kinetics in response to substrate concentration. If, for example, $[H^+]$ is saturating, single Michaelian functions always result and increasing internal $[SO_4^{2-}]$ could cause non-competitive inhibition of transport. Small relative differences in such factors as $[H^+]$ or membrane potential (see also Gerson & Poole 1971) or in the rate constants of the reactions of the carrier cycle, can be readily shown to generate dual isotherms. An idea, which is more compatible with the most recent paper by Nissen (1991), in which he advances some ideas on the molecular aspects of multiphasic behaviour, is that there can be "slip" in a carrier mechanism which gives it the versatility to transport the substrate with or without driver-ion (Eddy 1980). Transport without the driver ion, *e.g.* H^+, is necessarily "downhill" thermodynamically and is typically a low-affinity system, while the high affinity transport depends on H^+/substrate transport. This resembles the idea advanced by Nissen (1991) where it is postulated that, at high external concentrations, the transporter is transformed into a channel (*sic*). It is clear that "channel" is not used in the conventional way since Nissen excludes the possibility of diffusive movement through it. In the case of SO_4^{2-} uptake, however, there is no way in which the transport could be downhill given its double negative charge and membrane potential of -100 to -150 mV in most plant roots, unless the cytoplasmic [SO_4^{2-}] is assigned very low values. While there is no direct evidence on this latter point, compartmental analyses suggest that cytoplasmic [SO_4^{2-}] is not negligible (see below). We conclude that there is likely to be only one transporter protein (or permease) in higher plant membranes operating over a wide range of sulfate concentrations. If true, this simplifies the process of searching for mutants, defective in sulfate transport, and their phenotypic complementation using cDNA libraries.

Molecular aspects of a plasma membrane sulfate transporter

The genes encoding enzymes of the sulfate uptake and assimilatory pathway in fungi and yeast show increased expression when cellular levels of sulfur become limiting (*e.g.* Jarai & Marzluf 1991). In *Neurospora crassa* a number of regulatory genes, *cys-3* (Fu & Marzluf 1990) and *scon*1 and 2 (Paietta 1990), control expression of the structural genes, and are, in turn, regulated via intracellular pools of available sulfur compounds, possibly cysteine. An analogous complex regulatory pathway for sulfate uptake and assimilation exists in *Saccharomyces cerevisiae* (see *e.g.* Thomas *et al.* 1990; Cherest & Surdin-Kerjan 1992). *Neurospora crassa* possesses two distinct sulfate transporters, encoded by separate genes, whose expression is regulated by sulfur repression and which show a specific developmental regulation of expression. Sulfate permease I (*CYS*13) occurs in the conidiospores and permease II (*CYS*14) in the mycelia (Marzluf 1970). The structural gene for the *Neurospora crassa* sulfate permease II was recently isolated, sequenced and shown to encode for a plasma membrane polypeptide of 781 amino acids, of molecular weight 87 kDa (Ketter *et al.* 1991). In common with many secondary transporters including two other fungal H^+/anion co-transporters, *e.g.* the H^+/phosphate transporter (Mann *et al.* 1989), twelve transmembrane helices may be postulated (Ketter *et al.* 1991; Fig. 5). No significant stretches of primary sequence similarity between any of these transporters have been observed. Of some note however is the large extracellular hydrophilic loop between helices 11 and 12, and the preponderance of basic residues in both this loop and in the loop between helices 8 and 9. Similarly, in the model for the phosphate transporter (Mann *et al.* 1989) a large number of positively charged residues are located on one side of the membrane (possible the extracellular side) and a large hydrophilic loop is predicted (on the postulated intracellular side). The significance

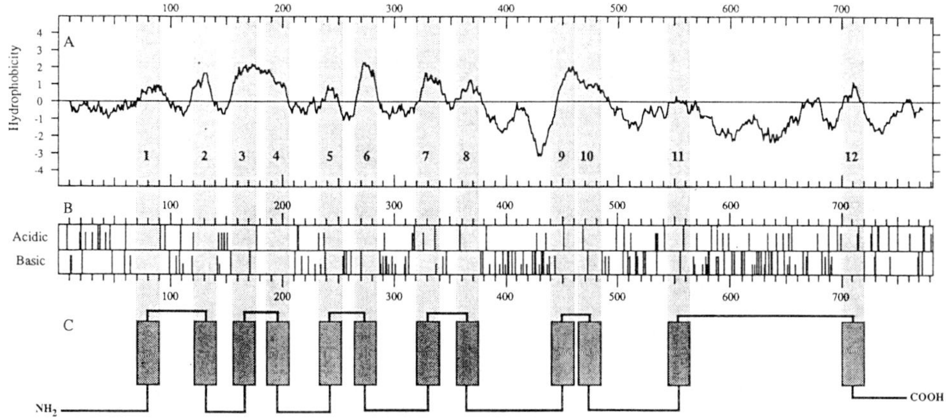

Fig. 5. Secondary structure model predicted for the *Neurospora crassa* sulfate permease II. The sequence of Ketter *et al.* (1991) was used to predict the hydropathy profile of the predicted protein (panel A) using the algorithm of Kyte & Doolittle (1982) with a window of 19 amino acids. Positioning of transmembrane regions is based upon the analysis of Ketter *et al.* (1991) with some modifications. Panel B shows the distribution of charged residues: in the acidic section, intermediate bars are aspartic acid and full sized bars are glutamic acid, whilst in the basic section the small size bars indicate histidine, intermediate bars lysine and full bars, arginine. Panel C shows the hypothetical arrangement of transmembrane helices. The derivation of these helices (numbered 1-12) from the hydropathy profile is shown by the shaded area, and directly corresponds with both panels A and B. Residue number is also indicated.

of these charge clusters, or of buried charged residues, in the transport function is unknown, but it may be speculated that they provide a binding site for sulfate ion at the vestibule to the transport channel.

The activity of the sulfate transporter in barley roots was particularly sensitive to inhibition by the arginyl-binding reagents, phenylglyoxal and the more polar (and less membrane permeable) hydroxy-phenylglyoxal (Clarkson *et al.* 1992). This suggests that the arrays of arginine residues, seen in the loop between putative helices 11 and 12 in the *Neurospora* sulfate permease II, may have counterparts on the outer face of the plant SO_4^{2-} transporter.

The model of the sulfate transporter (Fig. 5) by placing clusters of positively charged groups on the outer face of the plasma membrane, does not conform to the "rules" for the location of charged residues predicted by von Heijne (1986). Such models are always tentative and other interpretations are possible, for instance, the final two transmembrane segments might be eliminated so that a long, hydrophilic C-terminal region would be present in the cytoplasm.

Attempts to identify the plant plasma membrane sulfate transporter

The commonly observed increase of sulfate uptake capacity following a period of sulfur deprivation, suggests several strategies for the identification of the genes and gene products responsible for this behaviour, and specifically for the transporter. Differential protein synthesis in response to sulfate and phosphate deprivation has

Table 1. A hypothetical calculation of the rate at which the cytoplasm will fill with SO_4^{2-} when starved cells are returned to SO_4^{2-} medium.

Consider	a tissue with a volume of 1cm³ in which the volume fraction of cytoplasm (0.1)	= 0.1 cm³
Assume	$[SO_4^{2-}]$ in cytoplasm pool at t_0	= 0
	$[SO_4^{2-}]$ in cytoplasm when pool is full	= 5×10^{-6} mol cm^{-3}
	$[SO_4^{2-}]$ influx	= 2×10^{-6} mol cm^{-3} h^{-1}
Calculate	SO_4^{2-} content of cytoplasm	= 0.5×10^{-6} moles
	time for delivery	= 0.25 h
Compare	$t^{1/2}$ for repression of transporter 1.5 – 2.5 h (see text)	
	$t^{1/2}$ for cytoplasmic $^{35}SO_4^{2-}$ exchange ≈ 10 min	

been observed for cultured roots of *Lycopersicon esculentum* (Hawkesford & Belcher 1991) and for *Brassica napus* (Davidian, in preparation). In *L. esculentum*, synthesis of a 28-kDa soluble and 38-, 43- and 47-kDa plasma membrane polypeptides were observed to have increased synthesis after 4 days sulfur deprivation. Longer periods resulted in enhanced synthesis of additional polypeptides. Analysis of *in vitro* translation products has confirmed increased translation of the corresponding mRNAs, or more likely an increased abundance of the specific mRNAs, and the induction of a 52 kDa translation product, probably ATP sulfurylase (Hawkesford, unpublished results). Confirmation that one or more of the induced polypeptides is the transporter will require additional evidence, preferably derived from the isolation and reconstitution of transporter activity in membrane vesicle.

Regulation of sulfate transport observed at the physiological level

Higher plant cells display a remarkable degree of control over the activity of their sulfate transport system. It appears to be a constitutive transport system, which shows no sign of requiring a sulfate "signal" for induction as does the nitrate transporter. Its expression can be strongly repressed in cells with adequate/supra-optimal SO_4^{2-} supply. Net uptake of SO_4^{2-} by *Lemna paucicostata* was 500-fold more rapid in SO_4^{2-}-starved fronds than in sulfur-replete controls without any significant change in K_m (Datko & Mudd, 1984). The stability of the K_m after large shifts in the V_{max} of the system has also been seen in roots of intact barley plants (Lee 1982; Clarkson *et al.* 1983), cultured haploid carrot cells (Furner & Sung 1982) and in intact seedlings of *Zea mays* genotypes (Renosto & Ferrari 1975).

The high transport rates generated by SO_4^{2-} deprivation are rapidly repressed when SO_4^{2-} is restored to the external medium. The provision of other sulfur sources, such as reduced glutathione (GSH) (Rennenberg *et al.* 1988, 1989) and cysteine (Datko & Mudd 1984; Herschbach & Rennenberg 1991) also represses SO_4^{2-} uptake rapidly. Some authors regard SO_4^{2-} itself as a primary regulator but, at least in barley roots, its direct interaction with the transporter in some kind of allosteric regulation does not appear to be compatible with the kinetics of repression. In Table 1 there is a hypothetical calculation showing that the cytoplasmic pool of SO_4^{2-}

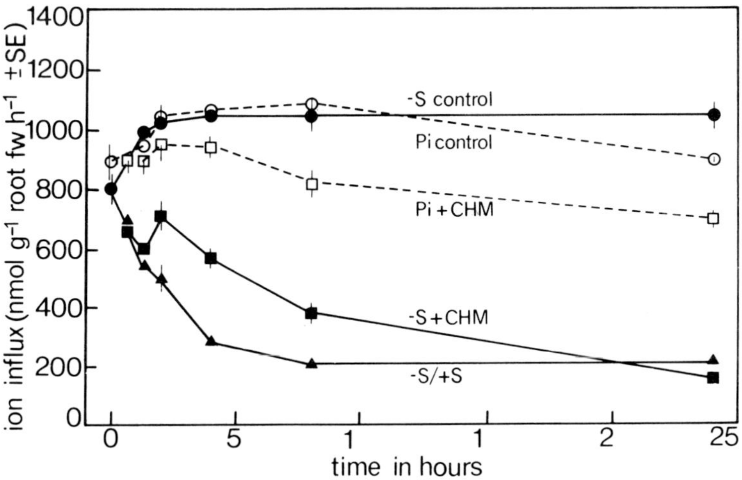

Fig. 6. A comparison of the effects of CHM (50 mmol m^{-3}) and SO_4^{2-} (150 mmol m^{-3}) on repressing the uptake (measured over 10 min) of $^{35}SO_4^{2-}$. The barley roots used had been deprived of SO_4^{2-} for 2 d before the experiment began. Data for effects of CHM on P_i uptake are also shown. The roots were doubly labeled with ^{35}S and ^{32}P. Standard errors are shown when larger than symbols (from Clarkson *et al.* 1992).

would go from zero to being "full" in a period of 0.25 h when sulfate supply is restored to a starved cell or organ. Export from the cytoplasmic pool to the vacuole of an intact root would be small (≈ 10% of influx) over such a period and would not, therefore, substantially increase the time taken to refill the cytoplasmic pool. If we compare this with the data in Fig. 6 it becomes clear that repression is much slower than the estimated refilling of the sulfate pool. The similarity between the collapse of the SO_4^{2-} transport system when cycloheximide (CHM) is added to sulfur-starved roots and repression due to SO_4^{2-} in the absence of CHM, may be fortuitous but it is clear that continued SO_4^{2-} transport depends on protein synthesis (Clarkson *et al.* 1992). This point has been made several times before using different species (Renosto & Ferrari 1975; Jensén & König 1982; Rennenberg *et al.* 1989). All these reports lead to the conclusion that the SO_4^{2-} transporter turns over rapidly ($t^{1/2}$ 1.5 – 2.5 h). It is by no means clear how SO_4^{2-} interacts with this process. Control could be exercised over the transcription of the gene for the transporter or the translation of the mRNA. The relatively slow effects of cordycepin (which blocks polyadenylation of mRNA) and α-amanitin (which blocks transcription) on sulfate uptake in tobacco cells led to the suggestion that the effects of GSH and SO_4^{2-} in regulating transporter activity might be at the translational level, the implication being that the mRNA might turn over more slowly than the transport protein (Rennenberg *et al.* 1989). The resolution of this question will come when the gene for the transporter has been identified so that its control can be investigated. Meanwhile, we have seen that cordycepin acts rather slowly in the inhibition of *in vivo* synthesis of most proteins in cultured tomato roots. The effectiveness of the cordycepin treatment was indicated by marked inhibition of some PM proteins, while others, including the 28 kD polypeptide induced by sulfate-starvation (Hawkesford & Belcher 1991) was unaffected. Similar results were obtained with *in vitro* translation experiments on mRNAs isolated from identical treatments (Hawkesford, Belcher & Clarkson, unpublished data).

In *Neurospora* the sulfate permease II protein (*CYS*-14) turns over with a half-life of approximately 2 h (Jarai & Marzluf 1991) while the half-life of the *cys*-14 mRNA is much faster ($t^{1/2}$ / 15 min) (Ketter *et al.* 1991). While the principal regulation of transport activity is assigned to the transcription of the *cys*-14 gene via the positively acting DNA binding protein *CYS*-3, it seems that protein turnover provides an additional level of control in the operation of the sulfur regulatory circuit in *Neurospora*.

Transport at the tonoplast

Proton pumping mechanisms in the tonoplast, transport H^+ from the cytoplasm into the vacuole. This electrogenic transport, which can be driven by the hydrolysis of ATP or of PP_i by different mechanisms, sets up a membrane potential across the tonoplast which is inside positive (characteristically between +10 to +30 mV) and a ΔpH which is inside acid relative to the cytoplasm. The transport of SO_4^{2-} into the vacuole may be "downhill" thermodynamically, a membrane potential of +29 mV being enough to maintain at equilibrium a [SO_4^{2-}] ten-fold greater than that of the cytoplasm. The kinetics of tracer exchange between the cytoplasm and the vacuole were reviewed by Cram (1990) and there is little that can be added here. When tissues are in a steady state with respect to their SO_4^{2-} content it can be shown that the rate constant for the exchange or turnover of vacuolar SO_4^{2-} (k_v) is slow relative to that in the cytoplasm (k_c). The k_v for intact roots of *Macroptilium atropurpureum* was 9.5 x 10^{-2} h^{-1} while k_c was 4.03 h^{-1} (Bell 1991). Tracer exchange kinetics revealed that the quantities of SO_4^{2-} in the cytoplasm and vacuole were 0.6 and 4.2 μmol g^{-1} root fresh weight respectively. Assuming volume fractions of 0.1 for the cytoplasm and 0.8 for the vacuole and 0.1 for the apoplast, the concentrations would be 6 mol m^{-3} and 5.25 mol m^{-3} in the cytoplasm and vacuole respectively. These values are comparable to those measured in the organelles and the vacuoles of barley mesophyll protoplasts by Kaiser *et al.* (1989). These authors showed that any substantial increase in SO_4^{2-} content, brought about by greater provision of external SO_4^{2-}, was found in vacuole rather than the cytoplasm. Isolated vacuoles retained their SO_4^{2-} for extended periods (> 20 min) as long as a membrane potential ($\Delta\psi$) was maintained across the tonoplast. Collapse of the $\Delta\psi$ in the presence of gramicidin resulted in the loss of SO_4^{2-} to the external medium, but collapse of ΔpH by nigericin did not. Labeled $^{35}SO_4^{2-}$ from the external medium entered isolated vacuoles without direct energization of the membrane, but the addition of Mg ATP to the external solution increased the rate of $^{35}SO_4^{2-}$ entry 1.7-fold. This effect was due to the maintenance of the membrane potential. Chemical analysis of the SO_4^{2-} content of the vacuoles treated with 4 mol m^{-3} SO_4^{2-} over a 40 min period showed that only in the presence of Mg-ATP was there any net uptake of SO_4^{2-}. In the absence of Mg-ATP, only tracer exchange occurred.

The above results have implications for the net removal of SO_4^{2-} from vacuoles. Given the membrane potentials expected between the vacuole and cytoplasm, the cytoplasm [SO_4^{2-}] would have to fall to a value <20% of that in the vacuole before export from the vacuole becomes a "downhill" process, thus the diffusion of SO_4^{2-}

out of vacuoles may occur only when the cytoplasmic SO_4^{2-} has been substantially reduced. For this reason, and because of the relatively slow rate constant for vacuolar exchange (k_v), the cytoplasm may not be effectively buffered by the vacuole against changes in concentration. This may contribute to the highly responsive behaviour of the SO_4^{2-} transport system at the PM to variations in the external SO_4^{2-} supply.

There are some well-documented cases where the leaf vacuoles of stress-tolerant species are found to contain [SO_4^{2-}] of 100 mol m^{-3} or greater. Clearly such accumulation serves an important osmotic role, maintaining turgor in the face of drought or salinity-stress. Kaiser *et al.* (1989) speculate that tonoplast SO_4^{2-} transport may need to be energized in such circumstances, perhaps via H^+/SO_4^{2-} antiport as has been suggested by Zhen *et al.* (1991) for NO_3^- uptake into vacuoles.

Transport across plastid envelope membranes

ATP sulfurylase, the first enzyme of the assimilation pathway, is located in plastids. Hence in Fig. 1, entry of SO_4^{2-} into the metabolized pool is shown to involve transport into a membrane-bound compartment. The exact nature of this process is unknown, particularly in root plastids, but in chloroplasts there are some clues that the phosphate translocator (Heldt & Rapley 1970) may be involved. Firstly, it was noted that when spinach chloroplasts were suspended in 20 mol m^{-3} SO_4^{2-}, there was complete inhibition of CO_2-dependent O_2 evolution (Baldry *et al.* 1968), which may be explained by depletion of stomatal P_i by exchange with SO_4^{2-} across the envelope membrane (Mourioux & Douce 1979). In intact spinach chloroplasts, SO_4^{2-} uptake occurred at a slow rate (14 n mol mg chlorophyll^{-1} min^{-1}) via a non-saturable mechanism (Hampp & Zeigler 1977). Rates of P_i exchange across the envelope are more that ten times greater. Far from competing with SO_4^{2-} uptake, provision of 5 mol m^{-3} P_i doubled the rate of SO_4^{2-} uptake from solutions containing 1-4 mol m^{-3} SO_4^{2-}. This result is hard to interpret because there may be interactions between P_i and energization of the membrane and the synthesis of metabolites which can exchange with SO_4^{2-}. If chloroplasts were preloaded with $^{35}SO_4^{2-}$ or $^{35}SO_3^{2-}$ exchange occurred when phosphoglyceric acid (PGA) and dihydroxyacetone phosphate (DHAP) were put in the bathing medium. P_i also promoted exchange, but malate and glucose-6-phosphate did not (Hampp & Zeigler 1977). Another study with barley showed that SO_4^{2-} inhibited $^{35}SO_4^{2}$ uptake into chloroplasts and that pyridoxal-phosphate (a potent inhibitor of the phosphate translocator) eliminated SO_3^{2-}uptake (Pfanz *et al.* 1987).

There are some attractions in the idea that SO_4^{2-} and P_i influxes into the stroma are linked. Exchange of SO_4^{2-} for photosynthetically synthesized PGA and DHAP (both divalent anions at stromal pH) would ensure that SO_4^{2-} entry into the chloroplast occurs in conditions where ATP will be available for sulfate activation ATP sulfurylase. It is hard to see, however, that such a mechanism can operate in non-green plastids so some other pathway needs to be looked for.

Analysis of the chloroplast genome in the liverwort *Marchantia* (Umesono *et al.* 1988) has revealed two genes (mbpX and mbpY) with 39% and 45% identity respectively to genes of the cyanobacterial sulfate permease. In *Synechococcus* the genes encoded an ATP-binding protein (*cys* A1) and one of the channel-forming proteins (*cys* A2) (Laudenbach & Grossman 1991). There is no similarity between these genes and those of the phosphate translocator (Flügge *et al.* 1989). An energized SO_4^{2-} transport system may be needed to get enough SO_4^{2-} into the plastid if cytosolic $[SO_4^{2-}]$ is low (K_m for SO_4^{2-} of the ATP sulfurylase is reported as 0.5-3 mol m^{-3}; see Stulen & De Kok, this volume), or if, as in non-green plastids, no metabolite is available for exchange. Root plastids might be an interesting place to look for some ATP-binding transport mechanism.

Release of sulfate into xylem vessels

It is probable that sulfate crosses the root by radial movement through the symplast. Entry of SO_4^{2-} into the root apoplast is repelled by the arrays of fixed negative charges in the walls. In at least one species, *Allium cepa*, it has been proposed that there is, in addition, a physical barrier to sulfate movement in the apoplasm in the walls of root hypodermis (Peterson 1987). Passage from cell to cell probably occurs through plasmodesmata. Since these structures are generally undisturbed by secondary and tertiary wall development of the endodermis, it seems probable that a large proportion of the root surface will contribute to the flow of sulfate passing to the shoot in the xylem. The innermost region of the root symplasm is comprised of xylem parenchyma cells. Across the plasma membranes of these interesting, but inaccessible cells, solutes pass into the xylem vessels. There is, therefore, a net efflux from these cells. The management of this process is not well understood and has long been the subject of speculation. Recent explanations are attempting to incorporate what is known about ion channels, and their regulation by such factors as voltage, calcium and turgor, in a scheme for solute deposition in the xylem (*e.g.* Schwencke & Wagner 1992). It is clear that these processes of ion release must interact with the rate at which water moves across the root and up the xylem (*e.g.* Pettersson 1966; Smith & Lang 1988). For example, when water fluxes are great, solutes released from the xylem parenchyma will be diluted to a greater extent than where flows are less, (*e.g.* in an excised root). Dilution will steepen the electrochemical potential gradient down which SO_4^{2-} can diffuse into the xylem. High transpiration will also have more far-reaching effects in reducing $[K^+]$; the membrane potential of many cells can collapse quite abruptly when $[K^+]$ increases in the range 1-10 mol m^{-3} and changes over this range could quite easily occur when transpiration is shut down at night or for any other reason. Channel opening may occur as cells depolarize. It is hard to find direct ways of verifying that such events occur but the behaviour of guard cells, which also preside over net salt efflux, can be used as a conceptual model.

Excised roots of tobacco have been used in a study to compare the effects of GSH and cysteine on sulfate uptake and movement into the xylem (Herschbach & Rennenberg 1991). Using a partitioned treatment cell, both uptake and exudation of tracer ($^{35}SO_4^{2-}$) from the cut end could be measured. In these experiments, exogenous GSH (0.1 mol m^{-3}) inhibited SO_4^{2-} fluxes by 70% in the first hour of treatment (a more rapid response than noted earlier with cultured tobacco cells (Rennenberg *et al.* 1988). The effect on $^{35}SO_4^{2-}$ movement into the xylem was inhibited to a greater extent than uptake, but the method used precluded any measurement of volume flow through the system. It is hard to decide, therefore, if there is some additional flow-related effect on the release of SO_4^{2-} from the symplasm. These experiments were also of interest because both uptake and loading of the xylem were very sensitive to the provision of exogenous cysteine. Within 3 h, the provision of 0.1 m mol m^{-3} L-cysteine had diminished sulfate influx by 60% and transport to the xylem by 80%. Again the loading step appeared more sensitive, but as the authors point out, the proportion of SO_4^{2-} entering the xylem became appreciably greater as the influx increased. This might explain why roots which had been grown for 2 d in the absence of SO_4^{2-} loaded a far higher proportion of the uptake into the xylem over a 1 h period than SO_4^{2-}-sufficient roots. As calculated in Table 1 of the present paper, cytoplasmic SO_4^{2-} may have risen greatly above normal steady state levels due to the derepressed transport system.

Using a similar experimental system with excised roots of *M. atropurpureum*, Bell (1991) found that decreasing the external [SO_4^{2-}] from 0.25 mol m^{-3} to zero increased the normally slow loss of sulfate from root vacuoles to the xylem. This may well have been due to rapid lowering of cytoplasmic SO_4^{2-}, reversing the direction of the gradient for SO_4^{2-} diffusion across the tonoplast, as discussed above.

Translocation in the phloem and cycling in the plant

Pale green or yellow-green young leaves on a plant, where older leaves remain dark green, is a well-recognized symptom of sulfur deficiency which points to some curious features about the compartmentation and translocation of sulfur. At first sight, it seems that there may be a failure to translocate sulfur in the phloem but several lines of evidence show that this is not the case. Working with tobacco plants, in which an incision has been made in mature leaves, Rennenberg *et al.* (1979) showed that $^{35}SO_4^{2-}$ introduced into the cut was translocated out of the leaf and recovered shortly after in the stem, as SO_4^{2-}. Normally, such mature leaves appear to export little of the SO_4^{2-} they contain if the SO_4^{2-} supply to the plant is withdrawn, so that young leaves quickly develop the sulfur stress symptoms referred to above. Thus, we have to conclude that it is not that mature leaves cannot translocate SO_4^{2-} but that they do not. Where leaves are cut, SO_4^{2-} can enter the leaf apoplast and gain direct access to the veins. In the intact leaf, most of the sulfate will be present in the mesophyll vacuoles: it seems probable that net export of this sulfate in times of sulfur stress is slow (Clarkson *et al.* 1983; Smith & Lang 1988; Dietz 1989). A further curiosity of behaviour was found in soybean plants by Smith and Lang (1988). They found that, despite transpiring at a rate equal to or greater than that of young leaves, mature leaves retained very little of the daily delivery of $^{35}SO_4^{2-}$ they would have received in the transpiration stream; >95% of the label ending up in the developing leaves. The data suggested that scarcely any of the $^{35}SO_4^{2-}$ delivered to mature leaves escaped from the veins to mix with the SO_4^{2} in the mesophyll cells. Such results indicate a very effective xylem to phloem transfer. The rapid internal cycling of $^{35}SO_4^{2-}$, absorbed by donor roots, but then quickly translocated from leaves to unlabeled receiver roots in wheat plants, points to the same conclusion (Larsson *et al.* 1991). In similar cycling experiments with *Ricinus*, phloem sap contained ^{35}S (largely SO_4^{2-}) with a specific activity >20-fold greater than that in the leaves from which it originated, again suggesting that labeled sulfate entered and left the mature leaves without mixing with the bulk (mesophyll) sulfate pool (Saker *et al.* 1989; Saker, unpublished results).

It seems that, once their growth has been completed, mesophyll cells become cut off from events influencing the sulfur nutrition of other parts of the plant. This is curious behaviour, quite unlike that of P_i or N which can be drawn readily from mature leaves to mitigate deficiencies in other parts of the plant. There is some net export of SO_4^{2-} during prolonged sulfur stress but the rate is too slow to support new growth and does not set in motion the leaf senescence program. This apparent isolation is not found in all cell types within a plant. In *M. atropurpureum*, for example, the SO_4^{2-} in the vacuoles of cells in the stem and in the roots is quickly exported (Clarkson *et al.* 1983). Bell (1991) and Bell *et al.* (1990) have attempted to analyze this situation by comparing rates of tracer exchange and fluxes across the tonoplast in root and leaf tissue of *M. atropurpureum*. The results conform to expectations by showing the rate constant for vacuolar exchange, k_v, was 3 to 5-fold greater in roots than in mature leaves under steady sulfur nutrition and 8 to 10-fold greater when plants were deprived of SO_4^{2-}. To study tracer exchange in leaves they must first be cut into narrow strips so as to expose the mesophyll cells to the bathing solution. In view of the restrictions on SO_4^{2-} movement into the leaf apoplasm which seem to be imposed by rapid xylem/phloem exchange in leaves, it is probable that the results from leaf strips may overestimate the exchanges and fluxes which occur in the intact leaf.

The phloem also contains GSH and usually very much smaller amounts of methionine and cysteine; these can, therefore, be delivered to roots from the shoot. It has been proposed (Rennenberg *et al.* 1989; Herschbach & Rennenberg 1991) that GSH, because of its strong interactions with the transport of SO_4^{2-} by the root, may modulate transport activity and thus act as a quantitative message which informs the root about the sulfur status of the shoot: in many species, excessive sulfur levels are reflected by large increases in GSH in leaves (De Kok 1990). Sulfate is also translocated to roots and, arguably, could perform a similar function. In cycling experiments with wheat (Larsson *et al.* 1991) and *Ricinus* (Saker, unpublished results) only labeled sulfate appeared to be re-exported from receiver roots in the xylem. It is not entirely clear what the functional significance of this difference might be.

Concluding remarks

The most immediate task is to isolate the genes for the plasma membrane sulfate transporter. It will then be possible to resolve quickly the question whether the physiological regulation, which has been evident for years, is primarily a matter of gene transcription, mRNA translation or allosteric regulation. It will also be possible to see whether the regulation of transporter activity is part of a regulatory circuit in which the expression of a number of structural genes of sulfur metabolism can be co-ordinated.

The unknown nature of the sulfate transport processes at the tonoplast and plastid envelope would benefit from some closer analysis at the physiological level. Both storage and assimilation of SO_4^{2-} can be influenced by these transports and there is a need to see if they are regulated in parallel with what is going on at the PM.

Studying control over the release of SO_4^{2-} into xylem vessels in roots and of xylem/phloem exchange poses major technical difficulties due to inaccessibility and the delicacy of the tissue involved. Much ingenuity will be required to understand such matters.

Acknowledgements

The authors acknowledge financial support from AFRC and INRA which has made our collaborative research possible.

References

Baldry, C.W., Cockburn, W. & Walker, D.A. 1968. Inhibition by sulphate of oxygen evolution associated with photosynthetic carbon assimilation. Biochim. Biophys. Acta 153: 476-483.

Bell, C.I. 1991. The efficiency of sulfur use and its regulation of transport in *Macroptilium atropurpureum*, cv. Siratro. Ph.D. Thesis, University of Newcastle upon Tyne, 207 pp.

Bell, C.I., Cram, W.J. & Clarkson, D.T. 1990. Turnover of sulfate in leaf vacuoles limits retranslocation under sulfur stress. In: Rennenberg, H, Brunold, C., De Kok, L.J. & Stulen, I. (eds.), Sulfur Nutrition and Assimilation in Higher Plants, SPB Academic Publishing, The Hague, pp. 163-165.

Borstlap, A.C. 1983. The use of model-fitting in the interpretation of "dual" uptake isotherms. Plant Cell & Environ. 6: 407-416.

Cherest, H. & Surdin-Kerjan, Y. 1992. Genetic analysis of a new mutation conferring cysteine auxotrophy in *Saccharomyces cerevisiae*: Updating of the sulfur metabolism pathway. Genetics 130: 51-58.

Clarkson, D.T. & Lüttge, U. 1991. Mineral nutrition: inducible and repressible nutrient transport systems. Prog. Bot. 2: 61-83.

Clarkson, D.T., Smith, F.W. & Vanden Berg, P.J. 1983. Regulation of sulfate transport in a tropical legume, *Macroptilium atropurpureum*, cv. Siratro. J. Exp. Bot. 34: 1463-1483.

Clarkson, D.T., Hawkesford, M.J., Davidian, J.-C. & Grignon, C. 1992. Contrasting responses of sulfate and phosphate transport in barley (*Hordeum vulgare* L.) roots to protein-modifying reagents and inhibition of protein synthesis. Planta 187: 306-314.

Cram, W.J. 1990. Uptake and transport of sulfate. In: Rennenberg, H, Brunold, C., De Kok, L.J. & Stulen, I. (eds.), Sulfur Nutrition and Sulfur Assimilation in Higher Plants, SPB Academic Publishing, The Hague, pp. 3-11.

Datko, A.H. & Mudd, S.H. 1984. Sulfate uptake and its regulation in *Lemna paucicostata* Hegelm 6746. Plant Physiol. 75: 466-473.

De Kok, L.J. 1990. Sulfur metabolism in plants exposed to atmospheric sulfur. In: Rennenberg, H., Brunold, C., De Kok, L.J., Stulen (eds.), Sulfur Nutrition and Assimilation in Higher Plants, SPB Academic Publishing, The Hague, pp. 111-130.

Dietz, K.-J. 1989. Recovery of spinach leaves from sulfate and phosphate deficiency. J. Plant Physiol. 134: 551-557.

Eddy, A.A. 1980. Slip and leak models of gradient-coupled transport. Trans. Biochem. Soc. 8: 271-273.

Epstein, E. 1973. Mechanism of ion transport through plant cell membranes. Int. Rev. Cytol. 34: 123-168.

Epstein, E. 1976. Kinetics of ion transport and the carrier concept. In: Lüttge, U. & Pitman, M.G. (eds.), Encyclopedia of Plant Physiology, Vol. 2. part B, Springer Verlag, Berlin, Heidelberg, New York, pp. 70-94.

Flügge, U.I., Fischer, K., Gross, A., Sebald, A., Lottspeich, F. & Eckerskorn, C. 1989. The triose phosphate-3-phosphoglycerate-phosphate translocator from spinach chloroplasts: nucleotide sequence of a full-length cDNA clone and import of the *in vitro* synthesized precursor protein into chloroplasts. EMBO J. 8: 39-46.

Fu, Y.-H. & Marzluf, G.A. 1990. *Cys*3, the positive-acting sulfur regulatory gene of *Neurospora crassa*, encodes a sequence-specific DNA-binding protein. J. Biol. Chem. 265: 11942-11947.

Furner, I.J. & Sung, Z.R. 1982. Regulation of sulfate uptake in carrot cells: properties of a hypercon-

trolled variant. Proc. Natl. Acad. Sci. USA 179: 1149-1153.
Gerson, D.F. & Poole, R.J. 1971. Anion absorption by plants. A unary interpretation of dual mechanisms. Plant Physiol. 48: 509-511.
Hampp, R. & Zeigler, I. 1977. Sulfate and sulfite translocation *via* the phosphate translocator of the inner envelope membrane of chloroplasts. Planta 137: 309-312.
Hawkesford, M.J. & Belcher, A.R. 1991. Differential protein synthesis in response to sulfate and phosphate deprivation: identification of possible components of plasma membrane transport system in cultured tomato roots. Planta 185: 323-329.
Hawkesford, M.J., Davidian, J.-C. & Grignon, C. 1993. Sulphate/H^+ cotransport in plasma membrane vesicles isolated from roots of *Brassica napus* L.: increased transport in membranes isolated from sulfur-starved plants. Planta *in press*.
Heldt, E.W. & Rapley, L. 1970. Specific transport of inorganic phosphate, 3-phosphoglycerate and dihydroxyacetone phosphate, and of dicarboxylates across the inner membrane of spinach chloroplasts. FEBS Lett. 10: 143-148.
Herschbach, C. & Rennenberg, H. 1991. Influence of glutathione (GSH) sulfate influx, xylem loading and exudation in excised tobacco roots. J. Exp. Bot. 42: 1021-1029.
Jarai, G. & Marzluf, G.A. 1991. Sulfate transport in *Neurospora crassa*: regulation, turnover and cellular localization of the *CYS*-14 protein. Biochemistry 30: 4768-4773.
Jensén, P. & König, T. 1982. Development of regulation mechanisms for SO_4^{2-} – influx in spring wheat. Physiol. Plant. 55: 459-464.
Kaiser, G., Martinoia, E., Schroppel-Meier, G. & Heber, V. 1989. Active transport of sulfate into the vacuole of plant cells provides halotolerance and can detoxify SO_2. J. Plant Physiol. 133: 756-763.
Ketter, J.S., Jarai, G., Fu, Y.-H. & Marzluf, G.A. 1991. Nucleotide sequence, messenger RNA stability, and DNA recognition elements of cis-14, the structural gene for sulfate permease II in *Neurospora crassa*. Biochemistry 30: 1780-1787.
Kyte, J. & Doolittle, R.F. 1982. A simple method for displaying the hydropathic character of a protein. J. Mol. Biol. 157: 105-132.
Larsson, C.-M., Larsson, M., Purves, J.V. & Clarkson, D.T. 1991. Translocation and cycling through roots of recently absorbed nitrogen and sulfur in wheat (*Triticum aestivum*) during vegetative and generative growth. Physiol. Plant. 82: 345-352.
Lass, B. & Ullrich-Eberius, C.I. 1984. Evidence for proton/sulfate co-transport and its kinetics in *Lemna gibba* G1. Planta 161: 53-60.
Laudenbach, D.E. & Grossman, A.R. 1991. Characterization and mutagenesis of sulfur-regulated genes in a cyanobacterium: evidence for function in sulfate transport. J. Bacterial. 173: 2739-2750.
Lee, R.B. 1982. Selectivity and kinetics of ion uptake by barley plants following nutrient deficiency. Ann. Bot. 50: 429-449.
Mann, B.J., Bowman, B.J., Grotelueschen, J. & Metzenberg, R.L. 1989. Nucleotide sequence of pho-4$^+$, encoding a phosphate-repressible phosphate permease of *Neurospora* crassa. Gene 83: 281-289.
Marzluf, G.A. 1970. Genetic and biochemical studies of distinct sulfate permease species in different developmental stages of *Neurospora crassa*. Arch. Biochem. Biophys. 138: 254-263.
Mourioux, D. & Douce, R. 1979. Transport du sulfate à travers la double membrane tumitante, ou enveloppe, des chloroplasts d'épinard. Biochimie 61: 1283-92.
Nissen, P. 1971. Uptake of sulfate by roots and leaf slices of barley mediated by single, multiphasic mechanisms. Physiol. Plant. 24: 315-324.
Nissen, P. 1974. Uptake mechanisms: inorganic and organic. Annu. Rev. Plant Physiol. 25: 53-79.
Nissen, P. 1991. Multiphasic uptake mechanisms in plants. Int. Rev. Cytol. 126: 89-134.
Paietta, J.V. 1990. Molecular cloning and analysis of the *scon*-2 negative regulatory gene of *Neurospora crassa*. Mol. & Cell. Biol. 10: 5207-5214.
Peterson, C.A. 1987. The exodermal casparian band of onion roots blocks the apoplastic movement of sulfate ions. J. Exp. Bot. 38: 2068-2081.
Pettersson, S. 1966. Artificially induced water and sulfate transport through sunflower roots. Physiol. Plant. 19: 581-601.
Pfanz, H., Martinoia, E., Lange, O.-L. & Heber, U. 1987. Mesophyll resistances to SO_2 fluxes into leaves. Plant Physiol. 85: 922-927.
Rennenberg, H., Schmitz, K. & Bergmann, L. 1979. Long distance transport of sulfur in *Nicotiana tabacum*. Planta 147: 56-62.

Rennenberg, H., Polle, A., Martini, N. & Thoene, B. 1988. Interaction of sulfate and glutathione transport in cultured tobacco cells. Planta 176: 68-74.

Rennenberg, H., Kemper, O. & Thoene, B. 1989. Recovery of sulfate transport into heterotrophic tobacco cells from inhibitors by reduced glutathione. Physiol. Plant. 76: 271-276.

Renosto, F. & Ferrari, G. 1975. Mechanism of sulfate transport inhibition by cycloheximide in plant tissues. Plant Physiol. 56: 478-480.

Saccomani, M. & Ferrari, G. 1989. Sulfate influx and efflux in seedlings of maize genotypes of different productivity. Maydica 34: 171-177.

Saker, L.R., Clarkson, D.T. & Purves, J.V. 1989. Internal cycling of nitrogen and sulfur in *Ricinus* (Castor Bean). Long Ashton Res. Stn. Ann. Rep. 1987, pp. 75-76.

Sanders, D. 1986. Generalized kinetic analysis of ion-driven co-transport systems: II. Random ligand binding as a simple explanation for non-Michaelian kinetics. J. Memb. Biol. 90: 67-87.

Sanders, D., Hansen, U.-P., Gradmann, D. & Slayman, C.L. 1984. Generalized kinetic analysis of ion-driven co-transport systems: a unified interpretation of selective ionic effects on Michaelis parameters. J. Memb. Biol. 73: 123-152.

Schwenke, H. & Wagner, E. 1992. A new concept of root exudation, Plant Cell Environ. 15: 289-299.

Sentenac, H., Bonneand, N., Minet, M., Lacroute, F., Salmon, J.-M., Gaymard, F. & Grignon, C. 1992. Cloning and expression in yeast of a plant K^+ transport system. Science 256: 663-665.

Smith, I.K. & Cheema, H.K. 1985. Sulfate transport into plants and excised roots of soybean (*Glycine max* L.). Ann. Bot. 56: 219-224.

Smith, I.K. & Lang, A.L. 1988. Translocation of sulfate in soybean (*Glycine max* L. Merr.). Plant Physiol. 86: 798-802.

Stewart, P.A. 1983. Modern quantitative acid-base chemistry. Can. J. Physiol. Pharmacol. 61: 1444-1461.

Thomas, D., Barbey, R. & Surdin-Kerjan, Y. 1990. Gene-enzyme relationship in the sulfate assimilation pathway of *Saccharomyces cerevisiae*. Study of the 3-phosphoadenylsulfate reductase structural gene. J. Biol. Chem. 265: 15518-15524.

Ullrich, C.I. & Novacky, A.J. 1990. Extra- and intracellular pH and membrane potential changes induced by K^+, Cl^-, $H_2PO_4^-$, and NO_3^- uptake and fusicoccin in root hairs of *Limnobium stoloniferum*. Plant Physiol. 94: 1561-1567.

Umesono, K., Inokuchi, H., Shiki, Y., Takeuchi, M., Chang, Z., Fukuzawa, H., Kohchi, T., Shirai, K., Ohyama, K. & Ozeki, H. 1988. Structure and organization of *Marchantia polymorpha* chloroplast genome. II. Gene organization of the large single copy region from *rps-12* to *atpB*. J. Mol. Biol. 203: 299-331.

Von Heijne, G. 1986. The distribution of positively charged residues in bacterial inner membrane proteins correlates with trans-membrane topology. EMBO J. 5: 3021-3027.

Zhen, R.-G., Koyro, H.-W., Leigh, R.A., Tomos, A.D. & Miller, A.J. 1991. Compartmental nitrate concentrations in barley root cells neasured with nitrate-selective microelectrodes and by single-cell sap sampling. Planta 185: 356-361.

SOME MOLECULAR ASPECTS OF SULFATE METABOLISM IN PHOTOSYNTHETIC ORGANISMS

J.A. Schiff[1], A.I. Stern[2], T. Saidha[1] and J. Li[3]
[1]*Photobiology Group, Biology Department, Brandeis University, Waltham, MA 02254, U.S.A.;* [2]*Botany Department, University of Massachusetts, Amherst, MA 01003, U.S.A.;* [3]*Boyce Thompson Institute, Cornell University, Ithaca, NY 14853, U.S.A.*

Introduction

Our aim in what follows is to try to tie together what is known about the various reactions of sulfate metabolism insofar as present information permits and to indicate where new research directions might lead. We will be mainly concerned with persistent questions which continue to be asked and still demand answers. Since these areas have, for the most part, been recently reviewed (Rennenberg *et al.* 1990), we will build upon this background and endeavor to add some new literature and ideas. In this way, we hope to comprehensively examine some long-standing questions in the light of newer information.

The nutritional, ecological and evolutionary background of sulfate metabolism has previously been reviewed (Schiff 1983; Brunold 1990). Sulfate is the most common and ubiquitous form of sulfur in the environment. Cycles exist, involving diverse organisms, which reduce sulfate and oxidize the reduction products back to sulfate. Only the oxidation of sulfur compounds such as hydrogen sulfide by the chemosynthetic bacteria has been observed to be energetically coupled to the reduction of carbon dioxide, but these organisms have recently achieved prominence as the primary producers of biological communities near volcanic vents deep in the ocean where chemosynthesis traps energy and makes it available to the other members of the community (Jannasch & Taylor 1984).

Assimilatory sulfate reduction appears to be present in all photosynthetic organisms and their close relatives, in the fungi, and in most procaryotic organisms (Schiff 1983). Sulfate reduction, along with nitrate reduction and carbon reduction fades out in the more animal-like groups and is generally absent in protozoa and in multicellular animals. In photosynthetic eucaryotes and a number of photosynthetic procaryotes, the pathway of assimilatory sulfate reduction is thought to begin with adenosine 5'-phosphosulfate (APS) as the sulfo donor (Schiff 1983; Brunold 1990) without the participation of thioredoxin. Although a thioredoxin-requiring PAPS reductase has been reported from spinach leaves, its significance remains to be ascertained (Schwenn 1989). The PAPS-thioredoxin system of sulfate reduction is characteristic of certain fungi such as yeast and bacteria such as *Escherichia coli* and *Salmonella* (Schiff 1983; Brunold 1990). Various other combinations of APS or PAPS with or without thioredoxin have been noted among the photosynthetic procaryotes (Schmidt 1982).

The most primitive sulfate reducers appear to be the anaerobic sulfate reducing bacteria such as *Desulfovibrio* that use sulfate as the electron acceptor for respiration

Sulfur Nutrition and Assimilation in Higher Plants
edited by L.J. De Kok et al., pp. 21–35

rather than oxygen. APS serves as the sulfo donor for this dissimilatory sulfate reduction and ATP is formed during the process of oxidative phosphorylation (Schiff 1983). These organisms may have arisen during the anaerobic phase of the origin of life. When oxygen entered the atmosphere through the action of oxygenic photosynthesizers, it became available as an electron acceptor in respiration. It appears that dissimilatory sulfate reduction which served both as a source of reduced sulfur compounds and ATP gave way to organisms carrying out assimilatory sulfate reduction (for formation of reduced sulfur compounds) and aerobic respiration (for ATP formation) as separate processes (Schiff 1983). This would lead to the evolution of most of the organisms we know today that have assimilatory sulfate reduction including photosynthetic aerobic procaryotes and eucaryotes. Sulfate reduction, as already mentioned, disappeared in the evolution of the primitive animals.

Our concern in what follows is the situation in the modern photosynthetic organisms, chiefly the photosynthetic eucaryotes such as the algae and multicellular plants.

Sulfate uptake

Although a fair amount of information has been gathered about sulfate uptake into eucaryotic cells (Cram 1990), little is known of the actual molecular mechanisms that underlie sulfate transport. Even in procaryotes where the powerful aid of genetics and related techniques has provided a more detailed description of the components of the sulfate transport system (which would also presumably be applicable to organelles such as the chloroplast and mitochondrion) (Silver & Walderhaug 1992), the molecular mechanisms underlying sulfate transport remain elusive.

Multiple forms of enzymes of sulfate reduction and their localization

ATP sulfurylase (ATPS)

ATP sulfurylase (ATP: sulfate adenylyltransferase (EC 2.7.7.4)) catalyzes the reaction of ATP and sulfate to form adenosine 5'-phosphosulfate (APS) and pyrophosphate (PPi):

$$ATP + SO_4^{2-} = APS + PPi$$

This enzyme catalyzes the first reaction in sulfate utilization in all cells, as far as is known, and is, therefore, the first enzyme of sulfate reduction and of the formation of adenosine 3'-phosphate 5'-phosphosulfate (PAPS), the universal donor of the sulfo group in the formation of sulfate esters. The enzyme has been detected in a wide variety of organisms, and has been purified from several sources (Li *et al.* 1991). The enzyme varies widely in molecular weight and number of subunits depending on the source.

The reaction as written favors ATP and sulfate formation because the free energy of hydrolysis of the phosphate-phosphate anhydride bond is less than the free energy of hydrolysis of the phosphate-sulfate anhydride bond. However, this free energy deficit is offset by the hydrolysis of PPi catalyzed by inorganic pyrophosphatase and by removal of APS by enzymatic reactions having a low K_m for this compound; this enables the ATP sulfurylase reaction to proceed in the direction of formation of APS, although APS is frequently an inhibitor of ATP sulfurylase, apart from kinetic or equilibrium considerations. PAPS has also been found to be an inhibitor of ATPS from certain filamentous fungi but is not a strong inhibitor of ATPS from spinach leaves, cabbage leaves, *Saccharomyces cerevisae* or rat liver (Renosto *et al.* 1990). It was suggested that allosteric regulation of the ATPS reaction by PAPS may be characteristic of those organisms that use PAPS for two divergent sets of reactions, assimilatory sulfate reduction and sulfate ester formation.

Earlier work indicated that multiple forms of ATPS may be present in cells. Thus two forms separating on DEAE-cellulose chromatography or on electrophoresis were obtained from Furth mouse mastocytoma cells but no differences were found in enzymatic properties (Shoyab & Marx 1970). Earlier work with higher plants indicated that ATPS was primarily in the chloroplasts (Burnell 1984; Schmidt 1986; Gerwick *et al.* 1980) with two electrophoretic forms in the bundle sheath chloroplasts of *Panicum miliaceum* (Gerwick *et al.* 1980) and multiple forms in hypocotyls of beet (Paynter & Anderson 1974). A more recent study (Lunn *et al.* 1990) has shown that there are two forms of ATPS in spinach leaves that are separable by anion-exchange chromatography; activity in the major peak is localized in the chloroplasts and the minor peak is found in the cytosol. No ATPS activity was attributable to the mitochondria.

ATPS appears to be absent from the chloroplasts of *Euglena gracilis* var. *bacillaris* (Saidha *et al.* 1988). A mutant of this organism W_{10}BSmL lacking plastids yields two forms of ATPS which have been purified to homogeneity (Li *et al.* 1991); ATPSm is located mainly in the mitochondria as part of the sulfate metabolizing center on the outside of the inner membrane while ATPSc is mainly in the soluble fraction of the cells. Although they are similar in molecular weight, the two forms yield different CNBr cleavage patterns indicating differences in amino acid sequence, are different in pH optima and isoelectric points and show different sensitivities to inhibition by APS. Much higher concentrations of inorganic pyrophosphate and Mg^{2+} are required to saturate the reverse reaction of ATPSc compared with ATPSm; therefore one might expect that pyrophosphate would more readily prevent the forward reaction of ATPSm compared to ATPSc. The situation in *Euglena* between mitochondrion and cytosol may be similar to that in higher plants between chloroplast and cytosol. In higher plants, the chloroplast contains inorganic pyrophosphatase and correspondingly low levels of pyrophosphate while the cytoplasm has little inorganic pyrophosphatase and accumulates pyrophosphate to serve as substrate in various reactions of carbohydrate metabolism (Dancer *et al.* 1990). In *Euglena* ATPSm which is sensitive to pyrophosphate is located on the outside of the inner mitochondrial membrane along with inorganic pyrophosphatase (Saidha *et al.* 1985) which would keep the pyrophosphate levels low. If the *Euglena* cytoplasm is similar to that of higher plants, inorganic pyrophosphate would accumulate but ATPSc is, appropriately, relatively insensitive to pyrophosphate.

The role of ATPS in higher plant plastids and in *Euglena* mitochondria seems clear; the enzyme is present to provide APS for sulfate reduction and for PAPS formation (Schiff 1983; Saidha *et al.* 1988; Brunold 1990). The function of ATPS in the cytoplasm, however, is far from clear and requires further study. For example, it is not known whether the cytoplasmic enzyme is active in the forward or reverse reaction (or both) *in vivo*.

APS sulfotransferase (APSST) and the putative sulfo-carrier

APS sulfotransferase is thought to be the first step in assimilatory sulfate reduction in all photosynthetic eucaryotes and in some photosynthetic procaryotes (Schiff 1983; Brunold 1990). The enzyme has been known for some time and a number of its properties were studied in partially purified preparations from *Chlorella* and higher plants (Schmidt 1976; Tsang & Schiff 1976) because it was difficult to purify the enzyme extensively in active form. Recently, it has been possible to purify an active APS sulfotransferase from *Euglena* to apparent homogeneity; this was facilitated by the unusual resistance of the *Euglena* enzyme to irreversible denaturation by sodium dodecylsulfate (SDS); in fact, low concentrations of SDS stimulate enzyme activity (Li & Schiff 1991).

APS sulfotransferase is usually assayed in a reaction in which $AP^{35}S$ and a thiol participate to form acid-volatile radioactivity assayed as $^{35}SO_2$ (Brunold 1990; Li & Schiff 1991). Almost any thiol is active in this reaction and the product formed is sulfite ($^{35}SO_3^{2-}$), with monothiols and ring-forming dithiols (such as dithiothreitol), and thiosulfate ($^-S\text{-}^{35}SO_3^-$) with vicinal dithiols such as BAL (2,3-dithiopropan-1-ol). In all cases the acid-volatile SO_3^- portions originate from APS; the ^-S portion of thiosulfate when it is formed appears to come from the thiol (Tsang & Schiff 1976). Although the acid-volatile assay using $AP^{35}S$ and a thiol (usually dithiothreitol or mercaptoethanol) has become commonplace, the reaction *in vivo* is thought to involve a carrier as the acceptor for the sulfo group of APS (Schiff 1983; Brunold 1990).

APSST is found in higher plant chloroplasts (Fankhauser & Brunold 1978) and on the outside of the inner mitochondrial membrane of *Euglena* along with the sulfate activating enzymes and the rest of the assimilatory sulfate-reducing system (Saidha *et al.* 1985, 1988). The highly purified active enzyme from *Euglena* is a 102 kDa tetramer and exhibits multiple forms with pI values from 5.0 to 5.5 on isoelectrofocusing. The tetramer is held together by disulfide bonds; treatment with thiols yields inactive monomers of 24 kDa (Li & Schiff 1991). The presence of AMP (a competitive inhibitor of the enzyme from *Euglena* (Li & Schiff 1992) and other sources (Schmidt 1975) and adenosine 5'-phosphoramidate (APA), an uncompetitive inhibitor of the *Euglena* enzyme (Li & Schiff 1992), protect the *Euglena* enzyme from thiol inactivation (Li & Schiff 1991), as does adenosine 5'-monosulfate (Li & Schiff 1991).

As already noted, APSST will use almost any thiol as an acceptor for the sulfo group of APS in the formation of acid-volatile radioactivity. Fractionation of extracts of *Chlorella* led to the identification of glutathione as the endogenous thiol that was most active in the APSST reaction (Tsang & Schiff 1978). Phytochelatin, a thiol-containing metal-binding polypeptide related to glutathione, was active as the thiol in the APSST reaction from maize (Brunold, personal communication). However, due to the lack of specificity of the APSST for thiols, this information does not help to identify the endogenous acceptor(s) for the sulfo group of APS, which presumably include the postulated sulfo-carrier.

Some years ago, it was found that the protein of fairly crude or partially purified extracts of *Chlorella* containing APSST would bind the radioactivity from $AP^{35}S$ in the absence of thiols (Abrams & Schiff 1973); the binding activity was at least an order of magnitude lower in extracts from *Chlorella* mutants lacking APSST activity. The main findings in this work have been confirmed and extended using more highly purified APSST preparations from *Euglena* (Li & Schiff 1992). Incubation of these APSST preparations with $AP^{35}S$, but no thiol, yields labeled APSST which moves to the position of the active enzyme tetramer on SDS-PAGE in the absence of thiols and heat. Since most of the label is not exchangeable with APS, most of the radioactivity is not in the form of bound substrate. Radioactivity is associated with at least two sites on the enzyme. Pronase digestion of the labeled enzyme followed by paper electrophoresis yields ^{35}S-labeled S-sulfocysteine indicating that the sulfo group of $AP^{35}S$ is transferred to the thiol group of at least one cysteine residue of the APSST protein resulting in covalently-bound radioactivity in the form of a Bunte salt (E-S-SO_3^-). Radioactivity not covalently bound to the protein can be extracted from the enzyme using acidic reagents such as formic-acetic acid, trifluoroacetic acid or trichloroacetic acid. This extract yields one major spot on paper electrophoresis at pH 2.0, and one or two other spots including sulfate (sulfate may arise from acidic decomposition of residual APS). The major spot called compound A, is stable in acid and from its mobility on paper electrophoresis at various pH values appears to have a high negative charge to mass ratio (since it moves quite rapidly towards the positive pole), is a strong acid (it moves rapidly at pH 2 and very likely contains at least one SO_3^- group) and probably lacks carboxyl and amino groups since its mobility does not change with pH in the range of 2.0–5.8. Although stable in acid, compound A decomposes to form other compounds including sulfate and thiosulfate as the pH is raised. This decomposition can be slowed or eliminated by treatment with iodoacetamide before the pH is raised indicating the presence of at least one reactive thiol group in the compound. Presumably a free thiol group is necessary for decomposition in base and this group is, or becomes, a thiol group of the thiosulfate that is formed when the pH of compound A (untreated with iodoacetamide) is raised. Since the non-covalently bound material becomes labeled with $AP^{35}S$, it is an excellent candidate for the sulfo-carrier. It should be pointed out, however, that it is not certain whether compound A itself is the native form of the non-covalently bound material of APSST. Extraction in acid may preserve its native configuration or might modify it somewhat during detachment from the protein. However, compound A is an acid-stable, isolatable form of the non-covalently bound radioactivity associated with the APSST after incubation with $AP^{35}S$ and may, therefore, represent either the sulfo-carrier or a derivative of it.

The properties of compound A do not appear to accord with those of known compounds. Its rather high mobility on paper electrophoresis suggests a compound of rather low molecular weight. However, acid stability and decomposition with increasing pH rule out a large number of compounds including polythionates and sulfane sulfonates of various chain lengths. Barring some unusual oxidation state of inorganic sulfur, the usual compounds we are familiar with do not seem to fit the known properties of compound A in all particulars. It might be pointed out, for example, that free thiosulfate and sulfite are unstable in acid and yield SO_2, plus elemental sulfur in the case of thiosulfate, under conditions where compound A is stable. One possibility that should be considered seriously is a metal complex involving a positively-charged metal with ligands to negatively-charged groups on the APSST and to negatively-charged sulfur anions, as is found in ferredoxin and other non-heme iron proteins for example. Metal complexes of the phytochelatins offer another example; some of these complexes are reported to contain acid-volatile sulfur in the form of sulfite and sulfide (Rauser 1990; Steffens 1990). It would appear that certain positively-charged metal cations bound to polypeptides are very versatile in binding available sulfur anions. The ferredoxin and the phytochelatin complexes are unstable in acid, however, and release volatile sulfur. Thus they do not presently seem to be closely related to the APSST-carrier complex (compound A is stable in acid) but they may serve as models for how the APSST-carrier complex is constructed. Using the information already presented, we might visualize a divalent metal cation (M^{2+}) which is bound through, say two ligands to negatively-charged (perhaps thiol) groups on the APSST apoprotein. Other ligands of the metal may then bind sulfur anions including sulfate and thiosulfate forming an enzyme-metal-sulfur anion complex. This complex may be stable enough to be released intact when the APSST protein is denatured with acid. However, treatment with increasing pH would provide hydroxyl ions (OH^-) which might displace the sulfur anions (including sulfate and thiosulfate) leading to their release and the observed decomposition of compound A. Presumably the thiosulfate in the metal complex would be protected from acid decomposition by being bound to the metal. Similarly, the base-catalyzed decomposition of compound A would be prevented in the intact enzyme (which operates at high pH (Li & Schiff 1991) by complex formation with the enzyme. Iodoacetamide might stabilize compound A and prevent base decomposition by substituting the thiol group of the thiosulfate thereby stabilizing the complex.

It might be pointed out that several other sulfur-labeled compounds besides sulfate and thiosulfate are released during base-catalyzed decomposition of compound A, so other sulfur-containing anions may be involved. Sulfite and sulfide, if present together in the complex, would have the potential to form a large variety of sulfur compounds of intermediate oxidation state, a complex mixture long known as Wackenroeder's liquid. However, the information already gathered indicates that if sufficient amounts of APSST can be isolated, physical techniques may be useful in determining whether metal complexes are present and, if they are, more about their composition and structure. Experiments with APSST are presently limited by the small amounts of purified enzyme available. Perhaps physiological conditions can be utilized which lead to an increase in the formation of APSST or molecular biological techniques might be employed to manufacture large amounts of the enzyme. This would have to be done under appropriate physiological conditions, however, if the correctly configured enzyme containing the non-covalently-bound fraction is to be obtained.

After removal of the non-covalently bound radioactivity by treatment of the labeled APSST with TFA, the precipitated protein still contains radioactive S-sulfocysteine indicating that the S-sulfocysteine remains covalently bound to the protein (as a modified cysteine residue) when the non-covalently bound radioactivity is extracted with TFA (Li & Schiff 1992). The radioactivity in the intact, untreated labeled *Euglena* APSST is exchangeable with sulfite and, even more actively, with thiosulfate (Li & Schiff 1992). The exchange of the SO_3^- group of the S-sulfocysteine of the

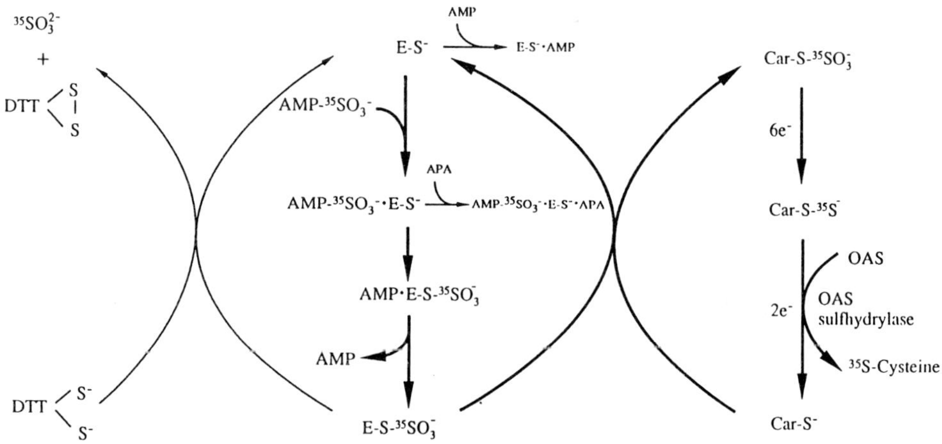

Fig. 1. Possible intermediates in the adenosine 5'-phosphosulfate (APS) sulfotransferase (APSST) reaction from *Euglena* (Li & Schiff 1992). ·, non-covalent bond; -, covalent bond; AMP-$^{35}SO_3^-$ (= AP^{35}S), ^{35}S-labeled adenosine 5'-phosphosulfate; Car-S$^-$ represents a sulfo-carrier (perhaps compound A) bound non-covalently to the APSST (E-S$^-$, where S$^-$ represents the thiol group of a cysteine residue); OAS, *O*-acetyl-L-serine; APA, adenosine 5'-phosphoramidate. Starting top center, AP^{35}S binds to the enzyme and the sulfo group is transferred to E-S$^-$ to yield E-S-$^{35}SO_3^-$. Reactions of E-S$^-$ and the enzyme-substrate complex with the inhibitors AMP and APA, respectively, are shown as side reactions. In the usual assay (shown to the left) E-S-$^{35}SO_3^-$ probably reacts with the added thiol (in this case dithiothreitol, DTT) to yield ^{35}S-sulfite and, ultimately, acid-volatile radioactivity as $^{35}SO_2$. However, *in vivo* the reaction may involve transfer of the sulfo group of E-S-$^{35}SO_3^-$ to a carrier (heavy arrows to the right). This carrier may be identical with the non-covalently bound material of the APSST ("compound A"). Further reduction would then occur on the carrier (perhaps via organic thiosulfate reductase) and reaction with OAS would yield cysteine. Alternatively, free sulfite might be released reductively from the carrier and be reduced via sulfite reductase to sulfide which could react with OAS to form cysteine (Schiff 1983; Brunold 1990).

APSST protein with sulfite is expected since Bunte salts (in this case E-S-SO_3^-) are known to exchange readily with sulfite (Saidha & Schiff 1989). However, exchange of E-S-SO_3^- with thiosulfate would not be expected to take place. Thus the high exchange of labeled APSST with thiosulfate (and, perhaps some of the exchange with sulfite) probably represents exchange with the non-covalently bound radioactivity and is yet another reason to suspect that the non-covalently bound material (compound A, the putative sulfo-carrier) contains a thiosulfate moiety.

These results with *Euglena* APSST can be assembled in the form of a hypothetical scheme which suggests a mechanism for the APSST reaction (Fig. 1) (Li & Schiff 1992). Starting top center, the APSST bearing at least one free thiol group (E-S$^-$) forms an enzyme-substrate complex with AP^{35}S (= AMP-$^{35}SO_3^-$). The labeled sulfo group of AP^{35}S would then be transferred to the thiol group of a cysteine residue of the enzyme forming S-sulfocysteine (E-S-$^{35}SO_3^-$). The usual acid-volatile assay is shown to the left where E-S-$^{35}SO_3^-$ reacts with an added thiol (in this case dithiothreitol) to yield sulfite and, ultimately, SO_2 when acidified. However, the reaction of APSST *in vivo* probably involves a carrier (perhaps compound A) as shown to the right (heavy arrows). E-S-$^{35}SO_3^-$ could transfer the sulfo group to the carrier yielding Car-S-$^{35}SO_3^-$ and reduction of the sulfo group could take place on the carrier to yield the persulfide (Car-S-$^{35}S^-$) and, ultimately, cysteine. If the carrier is a metal ion, the two sulfurs indicated could constitute a thiosulfate moiety attached to the metal. Because free sulfite and sulfide usually are not found during assimilatory sulfate reduction in the APS system, a pathway involving bound intermediates and reduction via an organic thiosulfate reductase has been favored (Schiff 1983; Brunold 1990). However, more recently, free sulfite has been found to accumulate during assimilatory sulfate reduction in *Euglena* mitochondria (Saidha *et al.* 1988) and a path involving reductive release of free sulfite and reduction via sulfite reductase to free sulfide is not ruled out. Although the scheme (Fig. 1) suggests a sequential transfer of the sulfo group of APS to the enzyme and then to the non-covalently bound material as the probable sequence, a parallel route in which the sulfo group is transferred independently to the enzyme and to the non-covalently bound material is not ruled out.

O-acetyl serine sulfhydrylase

O-acetyl serine sulfhydrylase (*O*-acetylserine (thiol)lyase, EC 4.2.99.8) which catalyzes the last reaction in the assimilatory sulfate reducing pathway forming cysteine (see Fig. 1) has also been reported to exist in multiple forms within the cell. Enzyme activity has been found both inside and outside the chloroplast in leaves (Schmidt 1976) and multiple forms have been reported by various authors (Masada *et al.* 1975; Bertognelli & Wedding 1977; Nakamura & Tamara 1989). Fankhauser and Brunold (1979) found that one form was particularly associated with chloroplasts in spinach leaves. More recently (Lunn *et al.* 1990) three forms of the enzyme separable by anion-exchange chromatography were found in spinach leaves; each of the forms is found in a different compartment, either the cytoplasm, the chloroplasts or the mitochondria. The authors suggest that cysteine may not be able to move between each of the cellular compartments capable of protein synthesis and that the enzyme is present in each of these compartments to supply cysteine for protein formation. Forms of *O*-acetyl serine sulfhydrylase separable by anion-exchange chromatography were assignable, respectively, to proplastids, mitochondria and to cytoplasm in cauliflower (*Brassica oleracea*) inflorescence (Rolland *et al.* 1992).

Euglena mutant W_{10}BSmL lacking plastids (Osafune & Schiff 1983) yields two forms of *O*-acetyl serine sulfhydrylase which are separable by ion-exchange chromatography on Reactive Blue Agarose (type 3000); one form is associated with mitochondria, the other appears to be soluble (Li & Schiff, unpublished results), in agreement with previous work using separation of organelles on gradients (Brunold & Schiff 1976).

Sulfonolipids

The thylakoid sulfolipid (diacylsulfoquinovosylglycerol)

Since its discovery by Benson many years ago, the thylakoid sulfolipid has been found in a wide variety of photosynthetic organisms, both eucaryotic and procaryotic (Benson *et al.* 1959), but its biosynthesis has not yet yielded to investigation

(Kleppinger-Sparace *et al.* 1990). The sulfonic acid group of the 6-sulfoquinovose moiety of the sulfolipid appears to originate from sulfite, or a compound at the oxidation-reduction level of sulfite (Kleppinger-Sparace *et al.* 1990); in *Euglena* chloroplasts the sulfur of sulfite or cysteine are precursors (Saidha & Schiff 1989). Thus an important source of the sulfonic acid group is probably the sulfate-reducing pathway of each organism synthesizing the sulfolipid. In photosynthetic eucaryotes, APS appears to be the major sulfo donor for sulfate reduction (Schiff 1983; Brunold 1990) and would be expected to also be the source of the sulfo group of the sulfolipid in these organisms. In procaryotes (Schmidt 1982) some organisms use APS and others PAPS as the sulfo donor for sulfate reduction and, therefore, the sulfo donor for sulfolipid formation should correspond to the donor used for sulfate reduction.

In higher plants, sulfate reduction (and sulfolipid formation) is localized in the plastids (Kleppinger-Sparace *et al.* 1990). Thus the pathway would be expected to proceed via APS and sulfite to sulfolipid in this compartment. Indeed, exogenous APS is preferred over PAPS for sulfolipid formation in spinach chloroplasts (Kleppinger-Sparace & Mudd 1990). In *Euglena*, sulfate activation and reduction are present on the outside of the inner mitochondrial membrane and the sulfite and cysteine formed are released from the organelle and serve as precursors of the sulfonic acid group of the sulfolipid formed in the chloroplasts. Sulfate activation and sulfate utilization are absent from *Euglena* chloroplasts (Saidha *et al.* 1988). Carbon 3 and sulfur of cysteic acid were found to be incorporated into the 6-sulfoquinovose of the sulfolipid by *Euglena* cells (Davies *et al.* 1966). However, added non-radioactive cysteic acid failed to interfere with the labeling of chloroplast sulfolipid from $^{35}SO_4^{2-}$ when mitochondria and chloroplasts were incubated together or when the labeled mitochondrial products (including sulfite and cysteine) were incubated with *Euglena* chloroplasts (Saidha & Schiff 1989), indicating that cysteic acid itself does not participate in either the mitochondrial or chloroplast portions of sulfolipid biosynthesis. Perhaps cysteic acid is converted to other sulfolipid precursors in intact cells before reaching the plastids. Two excellent inhibitors of sulfur labeling of sulfolipid in *Euglena* chloroplasts incubated with $^{35}SO_4^{2-}$ and mitochondria are S-sulfocysteine and cysteine sulfinic acid. Both compounds act on the mitochondrial side of the process and probably dilute the label in the sulfite pool through exchange with sulfite or by releasing non-radioactive sulfite. Both compounds are also highly effective in preventing labeling of the sulfolipid from $^{35}SO_4^{2-}$ in spinach chloroplasts (Saidha & Schiff 1989).

Although some sense can be made of the metabolic source of sulfur for sulfolipid formation, the steps between sulfite (or a compound at the level of sulfite) and the formation of the sulfolipid remain obscure. Many investigators have considered the problem and the several reviews on the subject present a number of interesting ideas. However, in the absence of data, none can be favored at present. An interesting lead is the finding that UDP- (and to a lesser extent GDP-) sulfoquinovose stimulates sulfolipid formation in osmotically-shocked chloroplasts (Heinz *et al.* 1989). Another direction is indicated by the isolation of mutants of the photosynthetic purple nonsulfur bacterium *Rhodobacter sphaeroides* which are deficient in sulfolipid accumulation; although one gene was isolated having a putative gene product of 33.6 kDa, the function of the protein is unknown (Benning & Somerville 1992).

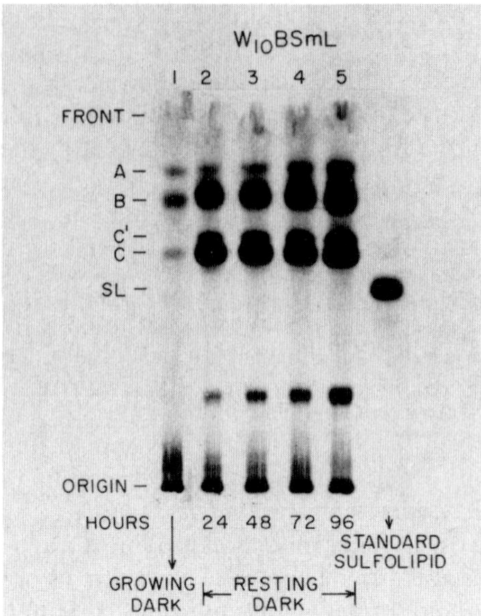

Fig. 2. Taurolipids in *Euglena* (Saidha *et al.* 199O). Dark-grown cultures of *Euglena gracilis* var. *bacillaris* aplastidic mutant W_{10}BSmL were used to inoculate cultures grown on Hutner's pH 3.5 glutamate/malate medium (Saidha & Schiff 1989) in darkness with added $^{35}SO_4^{2-}$. Alternatively, the cultures growing in darkness were mixed with resting medium (Stern *et al.* 1964) (containing KH_2PO_4, $MgCl_2$ and mannitol (a non-utilizable osmotic agent)) and when division ceased (1-2 days) these cultures received $^{35}SO_4^{2-}$ (zero time of resting) and were allowed to incubate in darkness for the indicated time. The cells were then harvested by centrifugation, extracted with chloroform/methanol (2: 1 v/v) and the extracts were subjected to chromatography on silica gel thin layers along with ^{35}S-labeled thylakoid sulfolipid extracted from labeled wild-type cells grown in the light as a chromatographic standard (Saidha & Schiff 1989). After development in chloroform: methanol: acetic acid: water (65: 35: 8: 4 by volume) the thin layer plate was placed on film and the developed autoradiogram is shown. Lane 1, extract of growing cells incubated with $^{35}SO_4^{2-}$; lanes 2-5, extracts of resting cells incubated with $^{35}SO_4^{2-}$ for 24, 48, 72 and 96 hours, respectively. A, B and C indicate lipid spots which yield ^{35}S-taurine on mild acid hydrolysis (C' is probably similar, but was not tested). SL indicates thylakoid sulfolipid standard. (Since W_{10}BSmL lacks plastids, no thylakoid sulfolipid is formed in the various incubations (Saidha & Schiff 1989.) Note that only very small amounts of labeled taurolipids are labeled in growing cells; the amounts increase greatly under resting conditions.

Taurolipids

A group of sulfonic acid-containing lipids different from the thylakoid sulfolipid has become of interest recently (Saidha *et al.* 1990; Saidha, Stern & Schiff, unpublished results). Unlike growing cells, resting (non-dividing) cells of *Euglena gracilis* var. *bacillaris* incubated with $^{35}SO_4^{2-}$ accumulate a series of ^{35}S-containing lipids which can be separated on silica-gel thin layers (Fig. 2); each of these lipids yields ^{35}S-taurine on mild acid hydrolysis. Treatment of the ^{35}S-lipids with dinitrofluorobenzene (DNFB) (which forms dinitrophenyl (DNP) derivatives of free amino groups) followed by hydrolysis yields only taurine; treatment of the lipids after hydrolysis with DNFB yields DNP-taurine. This shows that the amino group of taurine is unavailable for reaction with DNFB in the intact lipid, and indicates that the taurine is bound to the lipid through at least the amino group. The taurolipids of dark-grown resting wild-type *Euglena* (which forms a normal chloroplast on light exposure) and mutant W_3BUL (in which a proplastid remnant undergoes only limited light-induced development) both show an increase in taurolipid labeling from $^{35}SO_4^{2-}$ on light exposure. Dark-grown resting cells of mutant W_{10}BSmL (which lack plastids completely) show much higher labeling than wild type or W_3 and the labeling is not very sensitive to light. This indicates that the presence of a plastid exerts an inhibitory effect on taurolipid formation which is relieved by light. The same group of taurolipids found in whole cells of W_{10}BSmL is also found in purified mitochondria from W_{10} indicating that the mitochondrion is a significant depot for taurolipids in resting *Euglena* cells. The increased labeling of taurolipids in W_{10} would be consistent with the fact that elimination of plastids leads to formation of increased mitochondrial material in *Euglena* (Lefort 1964; Buetow 1989). Similar taurolipids have been described from cells and mitochondria of *Tetrahymena* (Kaya & Sano 1991) and since *Euglena* is closely related to the flagellated protozoa, this group of lipids might be found in other protozoa as well. A different sort of lipid in which taurine residues are linked by carbon-carbon bonds to long carbon chains are found in the diatom *Nitzschia alba* (Anderson *et al.* 1978, 1979) and the bacterium *Capnocytophaga* (Godchaux & Leadbetter 1980).

Taurine has been implicated indirectly in the preservation of membrane integrity (Hayes *et al.* 1975; Wright *et al.* 1986). The sulfonic acid groups of the taurolipids (and perhaps other lipids such as the thylakoid sulfolipid) might be present to provide negative charges which would repel negatively charged deleterious oxygen species from the vicinity of membranes. The sulfonic acid group is appropriate for such a function since it has a low pK and, therefore, is ionized at physiological pH, is very unreactive, is resistant to hydrolysis from the carbon to which it is attached and would be expected to provide stable negative charges to a membrane with which it is associated. Thus the formation of taurolipids in resting *Euglena* cells might represent a protective stress response to nutritional deprivation.

As already noted, *Euglena* mitochondria convert sulfate to sulfite which is then used for chloroplast sulfolipid biosynthesis; the sulfite formed by the mitochondria might also be used for taurolipid formation within the mitochondrion itself.

Sulfotransfer from PAPS and formation of tyrosine-*O*-sulfate

There are numerous enzymatic reactions known in which the sulfo group of adenosine 3'-phosphate 5'-phosphosulfate (PAPS) is transferred to form a sulfo derivative, usually a sulfate ester, of a phenol, a polysaccharide or a steroid, for example. In animals, the enzymatic transfer of the sulfo group of PAPS to phenols to form the corresponding phenol sulfates is a well known detoxification reaction leading to the excretion of the phenol as the sulfate ester (DeMeio 1975; Roy 198?; Ramaswamy & Jakoby 1987). This reaction is catalyzed by phenol sulfotransferases of broad specificity for phenols but high specificity for PAPS; APS is not used as a substrate.

Although the naturally-occurring phenol, tyrosine, is excreted by animals as the sulfate ester (tyrosine *O*-sulfate), none of the known phenol sulfotransferases appear to use free tyrosine as a substrate (DeMeio 1975; Roy 1981). It was, therefore, of some interest to find that *Euglena* mitochondria and chloroplasts excrete ^{35}S-tyrosine-*O*-sulfate into the surrounding medium when supplied with $PAP^{35}S$ (Saidha *et al.* 1989); no tyrosine need be added indicating that the organelles supply the tyrosine from internal sources. This raises the interesting possibility that sulfation of tyrosine might not only serve as a detoxifying mechanism but may also serve as a mechanism for transporting tyrosine out of the organelles.

The phenol sulfotransferase from *Euglena* has been purified to homogeneity (Saidha & Schiff 1991; Saidha & Schiff, unpublished results). The enzyme is a single polypeptide of 25 kDa and accepts free tyrosine as a substrate forming tyrosine-*O*-sulfate. This appears to be the only phenol sulfotransferase present in *Euglena*. However, although the *Euglena* enzyme will use free tyrosine as a substrate, like other phenol sulfotransferases it is comparatively non-specific and will use other phenols besides tyrosine. Like other phenol sulfotransferases it is specific for PAPS as the sulfo donor and will not use APS. Since this enzyme may be membrane-bound in *Euglena* mitochondria and chloroplasts, it is strategically-placed to participate both in tyrosine transport and detoxification.

Concluding remarks

Like any area of fundamental research, sulfur metabolism continues to generate new findings that suggest still further experiments. Certainly the nature of the reactions in the pathway of assimilatory sulfate reduction in photosynthetic organisms between APS and reduced intermediates forming cysteine continues to be an important and intriguing problem requiring much further work. In addition, branches off this pathway, particularly in the direction of sulfolipid formation, for example, still offer challenging questions for future work. The function of the sulfonolipids and other compounds containing oxidized sulfur is still obscure and requires further exploration. The existence of multiple forms of various enzymes of sulfur metabolism and their cellular localization raise questions concerning compartmentation of sulfur biochemistry. Indeed, because we are fortunate to have a useful isotope of sulfur (^{35}S) available and because sulfur metabolism is somewhat more specialized than the global carbon metabolism of the cell, a study of the movement of sulfur compounds between cellular compartments can offer insights into intracellular metabolic traffic and organelle nutrition, particularly in relation to the development of the cell and its organelles. There is clearly much to be done and interesting explorations and findings await us.

Acknowledgements

Supported by grants from the National Science Foundation. JAS is the Abraham and Etta Goodman Professor of Biology. J.L. was a Goodman Graduate Fellow. We thank Ann Blanchard and Nancy O'Donoghue for the preparation, respectively, of the manuscript and the illustrations.

References

Abrams, W.R. & Schiff, J.A. 1973. Studies of sulfate utilization by algae. 11. An enzyme-bound intermediate in the reduction of adenosine 5'-phosphosulfate (APS) by cell-free extracts of wild-type *Chlorella* and mutants blocked for sulfate reduction. Arch. Mikrobiol. 94: 1-10.

Anderson, R., Kates, M. & Volcani, B.E. 1978. Identification of the sulfolipids in the non-photosynthetic diatom *Nitzschia alba*. Biochim. Biophys. Acta 528: 89-106.

Anderson, R., Kates, M. & Volcani, B.E. 1979. Studies on the biosynthesis of the sulfolipids in the diatom *Nitzschia alba*. Biochim. Biophys. Acta 573: 557-561.

Benning, C. & Somerville, C.R. 1992. Isolation and genetic complementation of a sulfolipid-deficient mutant of *Rhodobacter sphaeroides*. J. Bacteriol. 174: 2352-2360.

Benson, A.A., Daniel, H. & Wiser, R. 1959. A sulfolipid in plants. Proc. Natl. Acad. Sci. USA 45: 1582-1597.

Bertognelli, B.L. & Wedding, R. 1977. Purification and initial kinetic characterization of different forms of *O*-acetylserine sulfhydrylase from seedlings of two species of *Phaseolus*. Plant Physiol. 60: 115-121.

Brunold, C. & Schiff, J.A. 1976. Studies of sulfate utilization by algae. 15. Enzymes of assimilatory sulfate reduction in *Euglena* and their cellular localization. Plant Physiol. 57: 430-436.

Brunold, C. 1990. Reduction of sulfate to sulfide. In: Rennenberg, H., Brunold, C., De Kok, L.J. & Stulen, I. (eds.), Sulfur Nutrition and Sulfur Assimilation in Higher Plants. SPB Academic Publishing, The Hague, pp. 13-31.

Buetow, D.E. 1989. The mitochondrion. In: Buetow, D. E. (ed.), The Biology of *Euglena*, vol. IV. Academic Press, New York, pp. 247-314.

Burnell, J.N. 1984. Sulfate assimilation in C_4 plants: Intercellular and intracellular location of ATP sulfurylase, cysteine synthase and cystathionine beta lyase in maize leaves. Plant Physiol. 75: 873-875.

Cram, W.J. 1990. Uptake and transport of sulfate. In: Rennenberg, H., Brunold, C., De Kok, L. J. & Stulen, I. (eds.), Sulfur Nutrition and Sulfur Assimilation in Higher Plants. SPB Academic Publishing, The Hague, pp. 3-11.

Dancer, J., Veith, R., Feil, R., Komor, E. & Stitt, M. 1990. Independent changes of inorganic pyrophosphate and the ATP/ADP or UTP/UDP ratios in plant cell suspension cultures. Plant Sci. 66: 59-63.

Davies, W.H., Mercer, E.I. & Goodwin, T.W. 1966. Some observations on the biosynthesis of the plant sulfolipid by *Euglena gracilis*. Biochem. J. 98: 369-373.

DeMeio, R.H. 1975. Sulfate activation and transfer. In: Greenberg, D. M. (ed.), Metabolism of Sulfur Compounds. Academic Press, New York, pp. 287-358.

Fankhauser, H. & Brunold, C. 1978. Localization of adenosine 5'-phosphosulfate sulfotransferase in spinach leaves. Planta 143: 285-289.

Fankhauser, H. & Brunold, C. 1979. Localization of *O*-acetylserine sulfhydrylase in *Spinacia oleracea* L. Plant Sci. Lett. 14: 185-192.

Gerwick, B.C., Ku, S.B. & Black, C.C. 1980. Initiation of sulfate activation: A variation in C_4 photosynthesis plants. Science 209: 513-515.

Godchaux III, W. & Leadbetter, E.R. 1980. *Capnocytophaga* ssp. contain sulfonolipids that are novel in procaryotes. J. Bacteriol. 144: 592-602.

Hayes, K.C., Carey, R.E. & Schmidt. S.Y. 1975. Retinal degeneration associated with taurine deficiency in the cat. Science 188: 949-951.

Heinz, E., Schmidt, H., Hoch, M., Jung, K-H., Binder, H. & Schmidt, R.R. 1989. Synthesis of different nucleoside 5'-diphosphosulfoquinovoses and their use for studies on sulfolipid biosynthesis in chloroplasts. Eur. J. Biochem. 184: 445-454.

Jannasch, H.W. & Taylor, C.D. 1984. Deep-sea microbiology. Annu. Rev. Microbiol. 38: 487-514.

Kaya, K. & Sano, T. 1991. Definition of total biosynthesis pathway of taurolipids in *Tetrahymena* cells. Biochim. Biophys. Acta 1084: 101-104.

Kleppinger-Sparace, K.F., Sparace, S.A. & Mudd, J.B. 1990. Plant sulfolipids. In: Rennenberg, H., Brunold, C., De Kok, L.J. & Stulen, I. (eds.), Sulfur Nutrition and Sulfur Assimilation in Higher Plants. SPB Academic Publisher, The Hague, pp. 77-88.

Kleppinger-Sparace, K.F. & Mudd, J.B. 1990. Biosynthesis of sulfoquinovosyldiacylglycerol in higher plants: Use of adenosine-5'-phosphosulfate and adenosine-3'-phosphate-5'-phosphosulfate as precursors. Plant Physiol. 93: 256-263.

Lefort, M. 1964. Modifications du chondriome dans les cellules etiolees de l'*Euglena gracilis* (Klebs).

Comptes Rendus de L'Academie Des Sciences, Paris 258: 4318-4321.
Li, J. & Schiff, J.A. 1991. Purification and properties of adenosine 5'-phosphosulphate sulphotransferase from *Euglena*. Biochem. J. 274: 355-360.
Li, J., Saidha, T. & Schiff, J.A. 1991. Purification and properties of two forms of ATP sulfurylase from *Euglena*. Biochim. Biophys. Acta 1078: 68-76.
Li, J. & Schiff, J.A. 1992. Adenosine 5'-phosphosulfate sulfotransferase from *Euglena*: Enzyme-bound intermediates. Plant Cell Physiol. 33: 63-72.
Lunn, J.E., Droux, M., Martin, J. & Douce, R. 1990. Localization of ATP sulfurylase and *O*-acetyl serine (thiol)lyase in spinach leaves. Plant Physiol. 94: 1345-1352.
Masada, M., Fukushima, K. & Tamura, G. 1975. Cysteine synthase from rape leaves. J. Biochem. 77: 1107-1115.
Nakamura, K. & Tamara, G. 1989. Five isoforms of cysteine synthase in rape leaves. Agric. Biol. Chem. 53: 2537-2536.
Osafune, T. & Schiff, J. A. 1983. W_{10}BSmL, a mutant of *Euglena gracilis* var. *bacillaris* lacking plastids. Exp. Cell Res. 148: 530-535.
Paynter, D.I. & Anderson, J.W. 1974. Comparative enzymology of the adenosine triphosphate sulfurylases from leaf and swollen hypocotyl tissue of *Beta vulgaris* – multiple enzyme forms in hypocotyl tissue. Plant Physiol. 53: 180-186.
Ramaswamy, S.G. & Jakoby, W.B. 1987. Sulfotransferase assays. In: Jakoby, W.B. & Griffith, O.W. (eds.), Methods in Enzymology, vol. 143. Academic Press, New York, pp. 201-207.
Rauser, W. E. 1990. Phytochelatins. Annu. Rev. Biochem. 59: 61-86.
Rennenberg, H., Brunold, C., De Kok, L.J. & Stulen, I. 1990. Sulfur Nutrition and Sulfur Assimilation in Higher Plants. SPB Academic Publishing, The Hague.
Renosto, F., Martin, R.L., Wailes, L.M., Daley, L.A. & Segal, I.H. 1990. Regulation of inorganic sulfate activation in filamentous fungi: Allosteric inhibition of ATP sulfurylase by 3'-phosphoadenosine 5'-phosphosulfate. J. Biol. Chem. 265: 10300-10308.
Rolland, N., Droux, M. & Douce, R. 1992. Subcellular distribution of *O*-acetylserine (thiol)lyase in cauliflower (*Brassica oleracea* L.) inflorescence. Plant Physiol. 98: 927-935.
Roy, A.B. 1981. Sulfotransferases. In: Mulder, G. J. (ed.), Sulfation of Drugs and Related Compounds. CRC Press, Boca Raton, Florida, pp. 83-130.
Saidha, T., Stern, A.I., Lee, D.-H. & Schiff, J.A. 1985. Localization of a sulphate-activating system within *Euglena* mitochondria. Biochem. J. 232: 357-365.
Saidha, T., Na, S.-Q., Li, J. & Schiff, J.A. 1988. A sulphate metabolizing centre in *Euglena* mitochondria. Biochem. J. 253: 533-539.
Saidha, T. & Schiff, J.A. 1989. The role of mitochondria in sulfolipid biosynthesis by *Euglena* chloroplasts. Biochim. Biophys. Acta 1001: 268-273.
Saidha, T., Hanfstingl, U. & Schiff, J.A. 1989. Formation of tyrosine-*O*-sulfate by mitochondria and chloroplasts of *Euglena*. Arch. Biochem. Biophys. 272: 237-244.
Saidha, T., Stern, A.I. & Schiff, J.A. 1990. Taurine conjugates in the lipid fraction of *Euglena*. Plant Physiol. 93 (Suppl.): 145.
Saidha, T. & Schiff, J.A. 1991. A purified PAPS: Phenol sulfotransferase from *Euglena gracilis* using free tyrosine as substrate. Plant Physiol. 96 (Suppl.): 155.
Schiff, J.A. 1983. Reduction and other metabolic reactions of sulfate. In: Lauchli, A. & Bieleski, R. L. (eds.), Encyclopedia Plant Physiol. New Series 15A, Inorganic Plant Nutrition. Springer, Berlin, pp. 402-421.
Schmidt, A. 1975. Inhibition of the adenosine 5'-phosphosulfate sulfotransferase activity from spinach, maize and *Chlorella* by adenosine 5'-monophosphate. Planta 127: 93-95.
Schmidt, A. 1976. The adenosine 5'-phosphosulfate sulfotransferase from spinach (*Spinacia oleracea* L.). Stabilization, partial purification and properties. Planta 130: 257-263.
Schmidt, A. 1982. Assimilation of sulfur. In: Schiff, J. A. & Lyman, H. (eds.), On the Origins of Chloroplasts. Elsevier/North Holland, New York, pp. 179-191.
Schmidt, A. 1986. Regulation of sulfur metabolism in plants. Progr. Bot. 48: 133-150.
Schwenn, J.D. 1989. Sulfate assimilation in higher plants: A thioredoxin-dependent PAPS-reductase from spinach leaves. Z. Naturforsch. 44C: 504-508.
Shoyab, M. & Marx, W. 1970. Two forms of ATP sulfurylase in Furth mouse mastocytoma. Life Sci. 9: 1151-1158.
Silver, S. & Walderhaug, M. 1992. Gene regulation of plasmid and chromosome-determined inorganic

ion transport in bacteria. Microbiol. Rev. 56: 195-228.

Steffens, J.C. 1990. The heavy metal-binding peptides of plants. Annu. Rev. Plant Physiol. Plant Mol. Biol. 41: 553-575.

Stern, A.I., Schiff, J.A. & Epstein, H.T. 1964. Studies of chloroplast development in *Euglena*. V. Pigment biosynthesis, photosynthetic oxygen evolution and carbon dioxide fixation during chloroplast development. Plant Physiol. 39: 220-226.

Tsang, M.L.-S. & Schiff, J.A. 1976. Studies of sulfate utilization by algae. 17. Reactions of the adenosine 5'-phosphosulfate (APS) sulfotransferase from *Chlorella* and studies of model reactions which explain the diversity of side products with thiols. Plant Cell Physiol. 17: 1209-1220.

Tsang, M.L.-S. & Schiff, J.A. 1978. Studies of sulfate utilization by algae. 18. Identification of glutathione as a physiological carrier in assimilatory sulfate reduction by *Chlorella*. Plant Sci. Lett. 11: 177-183.

Wright, C.E., Tallan, H.H. & Lin, Y.Y. 1986. Taurine: Biological update. Annu. Rev. Biochem. 55: 427-453.

GENE REGULATION OF SULFUR ASSIMILATION

Nicholas M. Kredich
Departments of Medicine and Biochemistry, Duke University Medical Center, Durham, North Carolina 27710, U.S.A.

Introduction

Sulfur assimilation is a fundamental biologic process in which oxidized forms of sulfur are reduced and then incorporated into organic molecules, usually cysteine or homocysteine. These two amino acids then provide sulfur for most if not all other cellular organic compounds containing reduced sulfur with a valency of -2. Given the global importance of such compounds, one might predict the regulation of sulfur assimilation to be as complex and interesting as the comparable systems for nitrogen and carbon assimilation. Studies of the gene regulatory system known as the cysteine regulon in *S. typhimurium* and *E. coli* are confirming that expectation, and it seems likely that further elucidation of this process in plants and other organisms will be equally rewarding.

S. typhimurium and *E. coli* provide well-recognized advantages for studying regulatory systems owing to their relatively simple genome and the ease with which genetic manipulations can be accomplished. Cysteine auxotrophs were among the first isolated and characterized in *S. typhimurium* in pioneering studies by Clowes (1958a, 1958b) and by Demerec and his co-workers (Demerec *et al.* 1955; Mizobuchi *et al.* 1962), who had identified most of the genes involved in cysteine biosynthesis by 1962. The availability of these well-defined mutant strains allowed Monty and his co-workers (Dreyfuss & Monty 1963; Leinweber & Monty 1963) to delineate most of the steps in sulfur assimilation in *S. typhimurium*, which were subsequently confirmed in *E. coli* (Jones-Mortimer 1968a, 1973). More recently, the genes for these activities have been cloned and factors regulating their expression have been studied *in vitro*. The focus of this article will be on the regulation of this pathway in *S. typhimurium* and *E. coli* with special emphasis on gene regulation.

The cysteine biosynthetic pathway and its overall regulation

The pathway for assimilatory sulfate reduction in *S. typhimurium* and *E. coli* (Fig. 1) differs from that of plants by requiring synthesis of PAPS and it's reduction to sulfite, which is then reduced to sulfide by NADPH-sulfite reductase (Dreyfuss & Monty 1963; Kredich 1987). Serine acetyltransferase catalyzes the synthesis of *O*-acetylserine, which reacts with sulfide to give cysteine in a reaction that can be catalyzed by either of two isozymes, *O*-acetylserine (thiol)-lyase-A and -B (Becker *et al.* 1969; Becker & Tomkins 1969). The -B isozyme can also use thiosulfate instead of sulfide in this reaction to give cysteine thiosulfonate as a product (Nakamura *et al.* 1984). Reduction of cysteine thiosulfonate to cysteine provides an alternative pathway for cysteine biosynthesis that shares some features with the glutathione

Sulfur Nutrition and Assimilation in Higher Plants
edited by L.J. De Kok et al., pp. 37–47

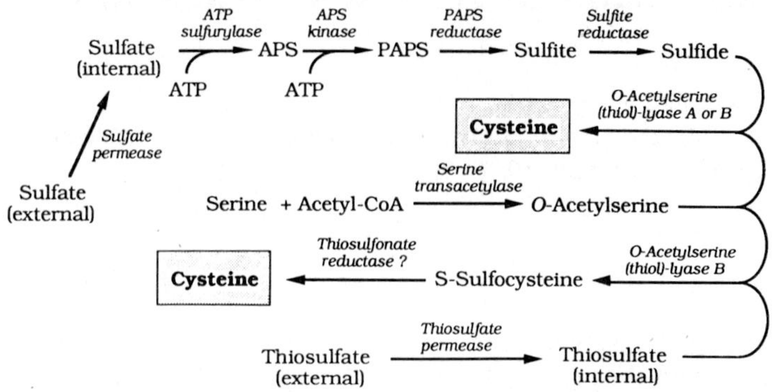

Fig. 1. Pathway of cysteine biosynthesis in *S. typhimurium* and *E. coli* (Kredich 1987). There are three branches to this pathway: synthesis of *O*-acetylserine from serine and acetyl-CoA; sulfate uptake and reduction to sulfide; and thiosulfate uptake. The reduction of *S*-sulfocysteine has not yet been attributed to a specific enzyme.

thiosulfonate reduction pathway proposed for sulfate assimilation in *Chlorella* (Tsang & Schiff 1978) and in higher plants (see Schmidt 1986 for a review).

Cysteine biosynthesis is regulated in *S. typhimurium* and *E. coli* through an interplay between feedback inhibition of catalytic activity and a complicated system of gene regulation termed the cysteine regulon, in which metabolites in the pathway act as inducers or anti-inducers. The end-product cysteine plays a major role by feedback inhibiting serine acetyltransferase (Kredich & Tomkins 1966; Kredich *et al.* 1969), the enzyme that synthesizes *O*-acetylserine. In turn, *O*-acetylserine or its derivative *N*-acetylserine acts as an inducer that is required for expression of genes required for sulfate reduction (Jones-Mortimer *et al.* 1968; Kredich 1971; Ostrowski & Kredich 1989). There is no evidence for significant feedback inhibition of enzymes of the sulfate assimilation pathway by sulfide or other metabolites, but sulfide and thiosulfate do act as anti-inducers of the pathway at the level of gene regulation (Ostrowski & Kredich 1990; Hryniewicz & Kredich 1991). Thus, synthesis of the carbon precursor of cysteine is sensitive to the end product cysteine; the reductive branch of the pathway is sensitive to its end product sulfide; and the third branch, which consists only of a permease (Hryniewicz *et al.* 1990), is sensitive to its end product thiosulfate. Furthermore, cysteine indirectly regulates the overall pathway by inhibiting synthesis of inducer (Fig. 2).

General aspects of gene regulation

Approximately 20 genes function to provide *S. typhimurium* and *E. coli* with cysteine, either through *de novo* synthesis from inorganic sulfur or by the uptake of extracellular cystine (Baptist & Kredich 1977). These genes are arranged in nine or more groups on the chromosome (Table 1). Most are expressed at very low to unmeasurable levels when cells are grown on cysteine or on sulfide and at much higher levels when sulfur is limiting. High level expression (''derepression'') requires a transcriptional activator protein, which is the product of *cysB*, and an inducer, which serves

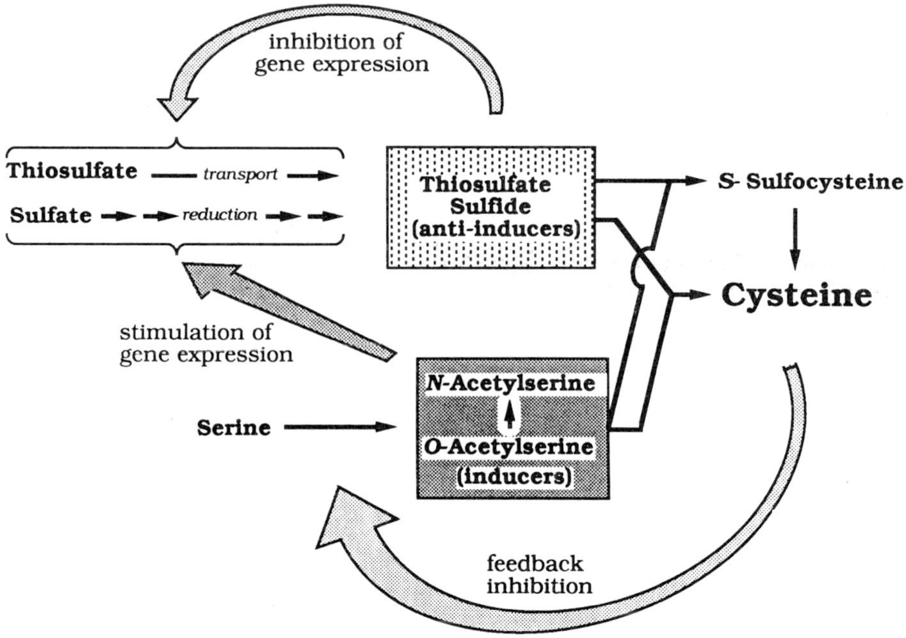

Fig. 2. Regulatory mechanisms in the cysteine biosynthetic pathway. Cysteine feedback inhibits serine acetyltransferase with a K_i of approximately 1 μM (Kredich & Tomkins 1966; Kredich *et al.* 1969). The product of this reaction, *O*-acetylserine, is not only a direct precursor of cysteine, but also serves as an inducer of genes required for sulfate uptake and reduction and for thiosulfate uptake. The end products of these two pathway branches, sulfide and thiosulfate, also act at the level of gene expression as anti-inducers.

as a signal for sulfur limitation (Jones-Mortimer 1968b; Kredich 1971). Genes that are regulated by CysB protein and its inducer comprise the cysteine regulon (Kredich 1987). *O*-Acetylserine was considered for many years to be the inducer, but recent findings have attributed that function to *N*-acetylserine (Ostrowski & Kredich 1989). *N*-Acetylserine is derived from *O*-acetylserine by an intramolecular *O*- to *N*-acyl migration that occurs non-enzymatically at a rate of about 1% per min at neutral pH (Flavin & Slaughter 1965) (Fig. 3). Comparison of *in vivo* inducer activities of the two compounds suggests that the activity of *O*-acetylserine is in large part or perhaps entirely due to its inevitable contamination by *N*-acetylserine (Ostrowski & Kredich 1989). Although the conversion of *O*-acetylserine to *N*-acetylserine occurs spontaneously, it is interesting to speculate whether an enzyme might be required to speed up the reaction *in vivo*.

Derepression of the cysteine regulon requires sulfur limitation, which is due to both inhibition of *O*-acetylserine synthesis by cysteine and the anti-inducer activities of sulfide and thiosulfate. Cysteine itself is not an anti-inducer *in vitro*, but behaves as one *in vivo* when added to a cell culture together with inducer (Jones-Mortimer *et al.* 1968; Kredich 1971) because it is degraded to sulfide by the inducible enzyme cysteine desulfhydrase (Kredich *et al.* 1972).

Table 1. Chromosomal organization of genes of the cysteine biosynthetic pathway. [a] Chromosomal locations are from the maps of Sanderson & Roth (1988) for *S. typhimurium* and Bachmann (1990) for *E. coli*; [b] *sbp* was cloned and sequenced by Hellinga & Evans (1985); [c] CTS-1 is an L-cystine transport system that is regulated as part of the cysteine regulon (Baptist & Kredich, 1977). It is probably encoded for by four genes, which have not been identified.

Gene (Cluster)	Location *S. typh.*	(min)[a] *E. coli*	Regulation by CysB protein	Activity
(*cysPTWAM*)	49	52	Positive	
cysP				Thiosulfate binding protein
cysTWA				Sulfate/thiosulfate permease
cysM				*O*-Acetylserine (thiol)-lyase B
cysK	49	52	Positive	*O*-Acetylserine (thiol)-lyase A
(*cysJIH*)	60	59	Positive	
cysJ				Sulfite reductase flavoprotein
cysI				Sulfite reductase hemoprotein
cysH				PAPS sulfotransferase
(*cysDNC*)	60	59	Positive	
cysDN				ATP sulfurylase
cysC				APS kinase
cysB	33	28	Negative	CysB activator protein
cysE	79	80	None	Serine transacetylase
cysG	72	73	None	Siroheme synthesis
sbp	?	89	Positive	Sulfate binding protein
CTS-1 genes[c]	?	?	Positive	L-Cystine transport

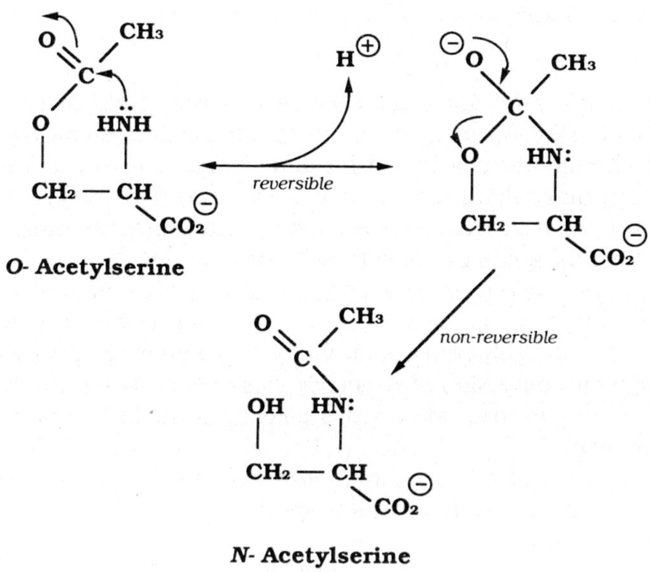

Fig. 3. Conversion of *O*-acetylserine to *N*-acetylserine by an intramolecular acyl shift. The reaction occurs spontaneously at a rate of 1% per min at pH 7.6 (Flavin & Slaughter 1965).

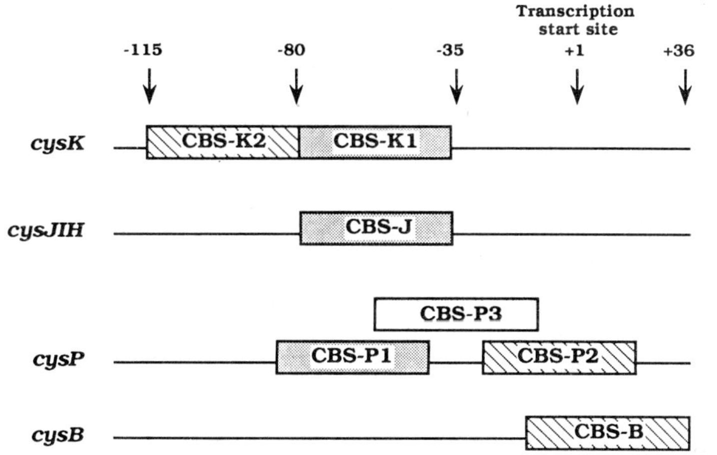

Fig. 4. Topology of binding sites for CysB protein in the *cysK*, *cysJIH*, *cysP* and *cysB* promoters. Each site extends for approximately 45 bp. The sites just upstream of the -35 region in the *cysK*, *cysJIH* and *cysP* promoters are required for positive regulation. Other sites in the *cysK* and *cysP* promoters are of unknown function may may act to sequester CysB protein at the promoter even when sulfur is replete. The *cysB* binding site is situated right over the transcription start site at + 1, where it allows CysB protein to negatively autoregulate its own expression (Ostrowski & Kredich 1991).

Molecular mechanisms of gene regulation

Considerable information now exists regarding the molecular mechanism of gene regulation in the cysteine regulon. The transcriptional activator CysB protein has been purified to homogeneity and found to be a tetramer of identical 36-kDa subunits (Miller & Kredich 1987). The deduced amino acid sequence of *cysB* (Ostrowski *et al.* 1987) identifies this protein as a member of the LysR family of activator proteins (Henikoff *et al.* 1988). In addition, the promoter regions of several *cys* genes have been isolated and sequenced, and their interactions with purified CysB protein have been studied *in vitro* by gel mobility shift binding assays, DNase I footprinting and transcription run-off experiments (Ostrowski & Kredich 1989, 1990, 1991; Monroe *et al.* 1990; Hryniewicz & Kredich 1991).

Specific binding of CysB protein to *cys* promoter DNA occurs over an expanse of approximately 45 bp and does not require inducer. A total of seven different CysB protein binding sites have been characterized in four different *cys* promoters from *S. typhimurium*. The positively regulated *cysJIH*, *cysK* and *cysP* promoters each contain a site located immediately upstream of the -35 region, and these are designated CBS-J, CBS-K1 and CBS-P1, respectively (Fig. 4). As reported for many other positively regulated promoters, these -35 regions bear little resemblance to the consensus sequence TTGACA, and require upstream binding of an activator protein (in this case CysB protein) to form a transcription initiation complex (Raibaud & Schwartz 1984; Hoopes & McClure 1987). *In vitro* studies have demonstrated that transcription initiation from these three promoters requires CysB protein and inducer, and that sulfide and thiosulfate inhibit this process in a manner that is com-

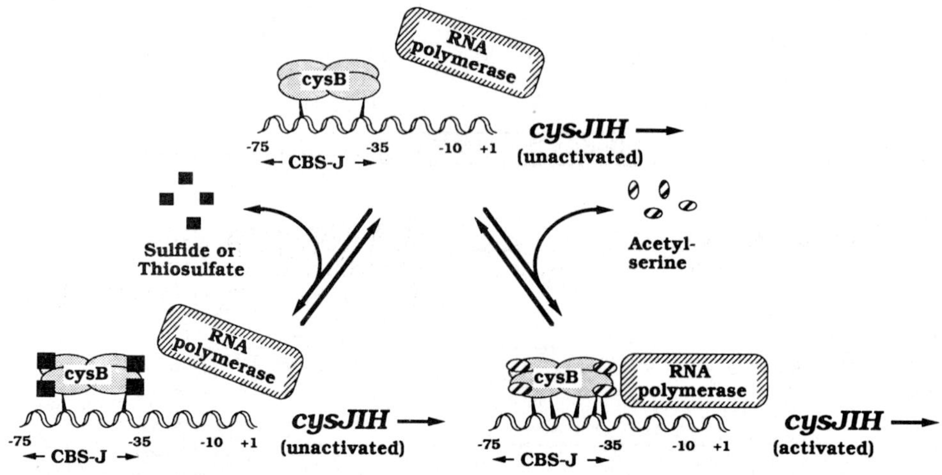

Fig. 5. A model for regulation of *cys* promoters (in this case *cysJIH*) by CysB protein, inducer (*O*-acetylserine or *N*-acetylserine) and the anti-inducers sulfide and thiosulfate. CysB protein binds to the promoter region even in the absence of inducer. In the presence of inducer binding is tighter and qualitatively different in a way that allows RNA polymerase to bind and form a transcription initiation complex.

petitive with inducer. Such experiments have provided the strongest evidence for concluding that *N*-acetylserine is an inducer and that sulfide and thiosulfate are anti-inducers.

Interactions between a *cys* promoter, CysB protein, inducer, anti-inducers and RNA polymerase are shown in Fig. 5. In this model, CysB protein binds to the promoter in the absence of inducer, but cannot activate formation of a transcription initiation complex. Inducer, either *O*-acetylserine or *N*-acetylserine, stimulates binding to the promoter, but more significantly, it causes a qualitative change in binding that is required for transcription initiation to occur. This may involve an interaction between CysB protein and RNA polymerase or a change in DNA conformation or both. Anti-inducers interfere with the effects of inducer in a competitive manner, perhaps by binding to the same site on CysB protein.

Three additional binding sites have been identified in the *cysK* and *cysP* promoters, which are of unknown function. CBS-K2 is located just upstream of CBS-K1 in the *cysK* promoter, and CBS-P2 and CBS-P3 are situated downstream of CBS-P1 in the *cysP* promoter (Fig. 4). CBS-P3 is considered to be a hybrid binding site consisting of the downstream portion of CBS-P1 and the upstream portion of CBS-P2. These "secondary" binding sites have the interesting effect of allowing a single CysB protein molecule to bend promoter DNA by binding to two sites simultaneously. DNA bending only occurs in the absence of inducer (Monroe *et al.* 1990; Hryniewicz & Kredich 1991), giving rise to complexes with anomalously slow electrophoretic mobility (Fig. 6). Addition of inducer prevents DNA bending by causing CysB protein to bind to the "primary" site alone, *i.e.* CBS-K1 or CBS-P1. It seems unlikely that DNA bending lays a direct role in promoter activation, since it occurs only in the absence of inducer. Furthermore, deletion of the CBS-K2 site abolishes DNA bending and has no measurable effect on *cysK* promoter activity *in vivo* or *in vitro*. We believe

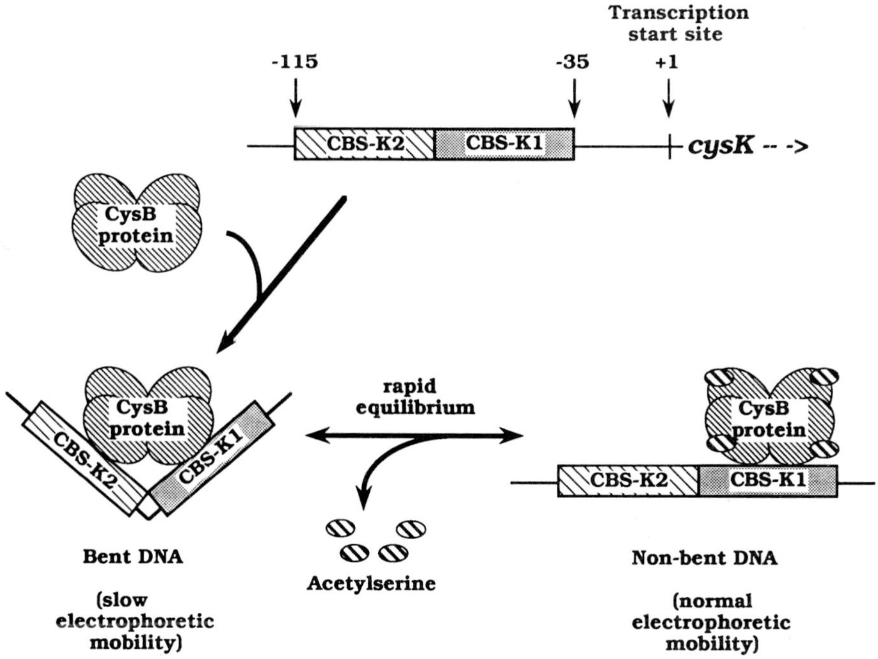

Fig. 6. DNA bending by CysB protein at the *cysK* promoter. In the absence of inducer a single CysB protein molecule binds to both CBS-K1 and CBS-K2, a distance of approximately 80 bp. Binding over this long a distance requires bending of the promoter DNA, which occurs at a point between the two binding sites. Inducer stimulates binding to CBS-K1 and inhibits binding g to CBS-K2 giving a complex in which CysB protein is bound only to CBS-K1, and the DNA is no longer bent (Monroe *et al.* 1990).

that the function of these secondary sites is to sequester CysB protein at *cys* promoters when sulfur is replete, thereby ensuring rapid gene activation in the event of a sudden decrease in sulfur availability.

A seventh binding site, designated CBS-B, has been identified in the *cysB* promoter, where it is centered over the transcription start site at approximately the position occupied by RNA polymerase (Fig. 4). *In vivo* studies have shown that *cysB* is negatively autoregulated by its own product (Bielinska & Hulanicka 1986; Jagura-Burdzy & Hulanicka 1981), and *in vitro* experiments have provided a molecular basis for this phenomenon by showing that binding of CysB protein to CBS-B inhibits transcription initiation (Ostrowski & Kredich 1991).

DNA sequence comparisons among six CysB protein binding sites in *S. typhimurium* and their counterparts in *E. coli* have shown a moderate degree of homology among the three primary sites CBS-J, CBS-K1 and CBS-P1, and with the secondary binding site CBS P2 (Hryniewicz & Kredich 1991). It may be significant in this regard that inducer stimulates binding of CysB protein to all four sites. Inducer inhibits binding to CBS-K2 and CBS-B, and the sequences of these sites are similar to each other but different from the other sites. The specific features that determine the binding properties of a site in the presence of inducer have not been determined.

Structure-function relationships in CysB protein

Very little data exists regarding functional domains of CysB protein or of other regulatory proteins of the LysR family. A high degree of homology in the amino-terminal portion of these proteins (Henikoff *et al.* 1988) and the resemblance of this region to other helix-turn-helix motifs (Brennan & Matthews 1989) suggest that this portion of the molecule binds DNA. This conclusion is supported by the observation that a mutant protein containing an amino acid substitution in this region (Colyer & Kredich, unpublished results) no longer autoregulates *cysB*, presumably because it has lost the ability to bind to CBS-B (Baptist *et al.* 1982).

Substitutions at amino acid residue Thr-149 give a CysB protein that no longer requires inducer to activate *cys* genes and is resistant to anti-inducers (Kredich 1971; Colyer & Kredich, unpublished results), suggesting that this portion of the protein is involved in the recognition of effector molecules or is influenced by their binding. Additional studies are required to explore this possibility and to determine which, if any, region of CysB protein interacts with RNA polymerase.

Gene regulation in plants

Brunold & Suter (1982) have shown that the activity of spinach leaf serine acetyltransferase activity is inhibited by 50% by 0.1 mM cysteine. It is not known whether this feedback inhibition is coupled to gene regulation as it is in bacteria, where *O*-acetylserine and *N*-acetylserine are inducers of the reductive pathway.

Sulfur assimilation in plants is generally thought to proceed via a pathway involving bound intermediates rather than free sulfite and sulfide (Tsang & Schiff 1978; see review by Schmidt 1986). Adenosine 5'-phosphosulfate (APS) sulfotransferase is a key enzyme in this pathway and transfers a sulfonyl group from APS to glutathione or some other thiol carrier. The resultant thiosulfonate is then reduced to the level of a persulfide (R-S-SH), which then transfers a sulfide equivalent to *O*-acetylserine. APS sulfotransferase expression is decreased in certain species by sulfide or cysteine (Brunold & Schmidt 1976; von Arb & Brunold 1980; Jenni *et al.* 1980) and increased in response to cadmium, which is known to induce synthesis of a cysteine-rich metallothionine (Nussbaum *et al.* 1988). Regulatory elements involved in this phenomenon have not yet been identified. For more detailed accounts of plant regulation of sulfur assimilation see Brunold (this volume) and Stulen & De Kok (this volume).

Concluding remarks

Most if not all the *S. typhimurium* and *E. coli* genes and activities required for sulfur assimilation under aerobic, laboratory growth conditions have been identified, cloned and sequenced, and appear to be regulated as part of the cysteine regulon. Considering the variety of sulfur sources encountered by these bacteria as they adapt between free-living and enteric conditions and between aerobic and anaerobic environments, one would not be surprised to find additional genes for sulfur metabolism, which might also be regulated as part of the cysteine regulon. It is also possible that other forms of regulation will be discovered, particularly for anaerobic sulfur assimilation. We may even discover some day that such mechanisms are more closely related to the regulation of sulfur assimilation in plants than the cysteine regulon.

Acknowledgment

This work was supported by research grant DK-12828 from the National Institutes of Health.

References

Bachmann, B.J. 1990. Linkage map of *Escherichia coli* K12, edition 8. Microbiol. Rev. 54: 130-197.
Baptist, E.W. & Kredich, N.M. 1977. Regulation of L-cystine transport in *Salmonella typhimurium*. J. Bacteriol. 131: 111-118.
Baptist, E.W., Hallquist, S.G. & Kredich, N.M. 1982. Identification of the *Salmonella typhimurium cysB* gene product by two-dimensional protein electrophoresis. J. Bacteriol. 151: 495-499.
Becker, M.A. & Tomkins, G.M. 1969. Pleiotropy in a cysteine-requiring mutant of *Salmonella typhimurium* resulting from an altered protein-protein interaction. J. Biol. Chem. 244: 6023-6030.
Becker, M.A., Kredich, N.M. & Tomkins, G.M. 1969. The purification and characterization of *O*-acetylserine sulfhydrylase A from *Salmonella typhimurium*. J. Biol. Chem. 244: 2418-2427.
Bielinska, A. & Hulanicka, D. 1986. Regulation of the *cysB* gene expression in *Escherichia coli*. Acta Biochim. Polon. 33: 133-137.
Brennan, R.G. & Matthews, B.W. 1989. The helix-turn-helix DNA binding motif. J. Biol. Chem. 264: 1903-1906.
Brunold, C. & Schmidt, A. 1976. Regulation of adenosine 5'-phosphosulfate sulfotransferase activity in *Lemna minor* L. Planta 133: 85-88.
Brunold, C & Suter, M. 1982. Intracellular localization of serine transacetylase in spinach leaves. Planta 155: 321-327.
Byrne, C.R., Monroe, R.S., Ward, K.A. & Kredich, N.M. 1988. DNA sequences of the *cysK* regions of *Salmonella typhimurium* and *Escherichia coli* and linkage of the *cysK* regions to *ptsH*. J. Bacteriol. 190: 3150-3157.
Clowes, R.C. 1958a. Nutritional studies of cysteineless mutants of *Salmonella typhimurium*. J. Gen. Microbiol. 18: 140-153.
Clowes, R.C. 1958b. Investigation of the genetics of cysteineless mutants of *Salmonella typhimurium* by transduction. J. Gen. Microbiol. 18: 154-172.
Demerec, M., Blomstrand, I. & Demerec, Z.E. 1955. Evidence of complex loci in Salmonella. Proc. Natl. Acad. Sci. 41: 359-364.
Dreyfuss, J. & Monty, K.J. 1963. The biochemical characterization of cysteine-requiring mutants of *Salmonella typhimurium*. J. Biol. Chem. 238: 1019-1024.
Flavin, M. & Slaughter, C. 1965. Synthesis of the succinic ester of homoserine, a new intermediate in the bacterial synthesis of methionine. Biochemistry 4: 1370-1375.
Hellinga, H.W. & Evans P.R. 1985. Nucleotide sequence and high-level expression of the major *Escherichia coli* phosphofructokinase. Eur. J. Biochem. 149: 363-373.
Henikoff, S., Haughn, G.W., Calvo, J.M. & Wallace, J.C. 1988. A large family of bacterial activator proteins. Proc. Natl. Acad. Sci. U.S.A. 85: 6602-6606.
Hoopes, B.C. & McClure, W.R. 1987. Strategies in regulation of transcription initiation. In: Neidhardt, F.C., Ingraham, J.L., Low, K.B., Magasanik, B., Schaechter, M. & Umbarger, H.E. (eds.) *Escherichia coli* and *Salmonella typhimurium*: Cellular and Molecular Biology. American Society for Microbiology, Washington, D.C, pp. 1231-1240.
Hryniewicz, M.M. & Kredich, N.M. 1991. The *cysP* promoter of *Salmonella typhimurium*: characteriza-

tion of two binding sites for CysB protein, studies of *in vivo* transcription initiation, and demonstration of the anti-inducer effects of thiosulfate. J. Bacteriol. 173: 5876-5886.

Hryniewicz, M., Sirko, A., Palucha, A, Böck, A. & Hulanicka, D. 1990. Sulfate and thiosulfate transport in *Escherichia coli* K-12: identification of a gene encoding a novel protein involved in thiosulfate binding. J. Bacteriol. 172: 3358-3366.

Jagura-Burdzy, G. & Hulanicka, D. 1981. Use of gene fusions to study expression of *cysB*, the regulatory gene of the cysteine regulon. J. Bacteriol. 147: 744-751.

Jenni, B.E., Brunold, C., Zryd, J.-P. & Lavanchy, P. 1980. Properties and regulation of adenosine 5'-phosphosulfate sulfotransferase from suspension cultures of *Nicotiana sylvestris*. Planta 150: 140-143.

Jones-Mortimer, M.C. 1968a. Positive control of sulfate reduction in *Escherichia coli*: isolation, characterization and mapping of cysteineless mutants of *E. coli* K12. Biochem. J. 110: 589-595.

Jones-Mortimer, M.C. 1968b. Positive control of sulfate reduction in *Escherichia coli*: the nature of the pleiotropic cysteineless mutants of *E. coli* K12. Biochem. J. 110: 597-602.

Jones-Mortimer, M.C. 1973. Mapping of structural genes for enzymes of cysteine biosynthesis in *Escherichia coli* K12 and *Salmonella typhimurium* LT2. Heredity 31: 213-221.

Jones-Mortimer, M.C., Wheldrake, J.R. & Pasternak, C.A. 1968. The control of sulphate reduction in *Escherichia coli* by *O*-acetyl-L-serine. Biochem. J. 107: 51-53.

Kredich, N.M. 1971. Regulation of L-cysteine biosynthesis in *Salmonella typhimurium*. I. Effects of growth on varying sulfur sources and *O*-acetyl-L-serine on gene expression. J. Biol. Chem. 246: 3474-3484.

Kredich, N.M. 1987. Biosynthesis of cysteine. In: Neidhardt, F.C., Ingraham, J.L., Low, K.B., Magasanik, B., Schaechter, M. & Umbarger, H.E. (eds.) *Escherichia coli* and *Salmonella typhimurium*: Cellular and Molecular Biology. American Society for Microbiology, Washington, D.C., pp. 419-428.

Kredich, N.M. & Tomkins, G.M. 1966. The enzymatic synthesis of L-cysteine in *Escherichia coli* and *Salmonella typhimurium*. J. Biol. Chem. 241: 4955-4965.

Kredich, N.M., Becker, M.A. & Tomkins, G.M. 1969. Purification and characterization of cysteine synthetase, a bifunctional protein complex, from *Salmonella typhimurium*. J. Biol. Chem. 244: 2428-2439.

Kredich, N.M., Keenan, B.S. & Foote, L.J. 1972. The purification and subunit structure of cysteine desulfhydrase from *Salmonella typhimurium*. J. Biol. Chem. 247: 7157-7162.

Miller, B.E. & Kredich, N.M. 1987. Purification of the cysB protein from *Salmonella typhimurium*. J. Biol. Chem. 262: 6006-6009.

Mizobuchi, K., Demerec, M. & Gillespie, D.H. 1962. Cysteine mutants of *Salmonella typhimurium*. Genetics 47: 1617-1627.

Monroe, R. S., Ostrowski, J., Hryniewicz, M. M. & Kredich, N. M. 1990. *In vitro* interactions of CysB protein with the *cysK* and *cysJIH* promoter regions of *Salmonella typhimurium*. J. Bacteriol. 172: 6919-6929.

Nakamura, T., Iwahashi, H. & Eguchi, Y. 1984. Enzymatic proof for the identity of the *S*-sulfocysteine synthase and cysteine synthase B of *Salmonella typhimurium*. J. Bacteriol. 158: 1122-1127.

Nussbaum, S., Schmutz, D. & Brunold, C. 1988. Regulation of assimilatory sulfate reduction by cadmium in *Zea maya* L. Plant Physiol. 88: 1407-1410.

Ostrowski, J. & Kredich, N.M. 1989. Molecular characterization of the *cysJIH* promoters of *Salmonella typhimurium* and *Escherichia coli*: regulation by *cysB* protein and *N*-acetyl-L-serine. J. Bacteriol. 171: 130-140.

Ostrowski, J. & Kredich, N.M. 1990. *In vitro* interactions of CysB protein with the *cysJIH* promoter of *Salmonella typhimurium*: inhibitory effects of sulfide. J. Bacteriol. 172: 779-785.

Ostrowski, J. & Kredich, N.M. 1991. Negative autoregulation of *cysB* in *Salmonella typhimurium*: *in vitro* interactions of cysB protein with the *cysB* promoter. J. Bacteriol. 173: 2212-2218.

Ostrowski, J., Jagura-Burdzy, G. & Kredich, N. M. 1987. DNA sequences of the *cysB* regions of *Salmonella typhimurium* and *Escherichia coli*. J. Biol. Chem. 262: 5999-6005.

Raibaud, O. & Schwartz, M. 1984. Positive control of transcription initiation in bacteria. Annu. Rev. Genet. 18: 173-206.

Sanderson, K.E. & Roth, J.R. 1988. Linkage map of *Salmonella typhimurium*, edition VII. Microbiol. Rev. 52: 485-532.

Schmidt, A. 1986. Regulation of sulfur metabolism in plants. In: Behnke, H.-D., Esser, K., Kubitzki, K., Runge M. & Ziegler, H. (eds.). Progress in Botany, Vol 48, Springer, Berlin, pp. 133-150.

Sirko, A., Hryniewicz, M., Hulanicka, D. & Böck, A. 1990. Sulfate and thiosulfate transport in

Escherichia coli K-12: nucleotide sequence and expression of the *cysTWAM* gene cluster. J. Bacteriol. 172: 3351-3357.

Tsang, M.L.-S. & Schiff, J.A. 1978. Studies of sulfate utilization by algae 18. Identification of glutathione as a physiological carrier in assimilatory sulfate reduction by chlorella. Plant Sci Lett. 11: 177-183.

von Arb, C. & Brunold, C. 1980. Analysis of the regulation of adenosine 5'-phosphosulfate sulfotransferase in *Lemna minor* L. using ^{15}N-density labelling. Planta 149: 355-360.

SELENIUM INTERACTIONS IN SULFUR METABOLISM*

John W. Anderson
Botany Department, La Trobe University, Bundoora, Victoria 3083, Australia

Introduction

Sulfur is an essential element for life, most importantly because it occurs in the two protein amino acids cysteine and methionine. Selenium, the higher isologue of sulfur, shares many of the characteristics of sulfur (Anderson & Scarf 1983) and, at concentrations in excess of trace amounts, interacts strongly with sulfur in plant metabolism.

In organisms, selenium and sulfur interact at several levels. In mammals and bacteria, selenium is required in trace amounts for selenium-specific processes which, strictly speaking, do not exhibit selenium/sulfur interactions as they do not support analogous reactions with sulfur. These processes include the specific incorporation of selenium into certain proteins known as selenoproteins which serve an essential function. Thus, organisms with these characteristics (*e.g.* mammals and some bacteria) require selenium as an essential micronutrient for growth.

In contrast to the selenium specific mechanisms for the synthesis of selenoproteins in mammals, many if not all of the reactions of the sulfate assimilation pathway in plants leading to the synthesis of cysteine and methionine are open to selenium thereby resulting in the synthesis of the selenium isologues of the intermediates and products of this pathway (*e.g.* selenocysteine, selenocystathionine, selenomethionine, *etc.*). Since the sulfur atom in cysteine and methionine is essential for the function of many of the molecules into which these two amino acids are incorporated (*e.g.* the thiol groups derived from cysteine in proteins and coenzyme A) incorporation of the corresponding selenoamino acids into these compounds causes loss or impairment of their biological activity resulting in selenium toxicity as the selenium concentration is increased. Incorporation of selenium in place of sulfur in this way is non-specific and proteins which contain selenium by non-specific replacement are not referred to as selenoproteins. The ratio of the selenium concentration which causes toxicity (*i.e.* non-specific replacement of sulfur) relative to the selenium concentration which is regarded as essential is much smaller than for any other element thus making control of selenium availability in biological systems extremely difficult.

Some plants support the metabolism of sulfur in secondary pathways, resulting in the accumulation of sulfur containing metabolites (*e.g.* methylcysteine in *Phaseolus lunatus*), often at quite high concentrations. In at least some species, these pathways are also open to selenium resulting in the accumulation of high concentrations of selenium isologues of the secondary metabolite (*e.g.* methylselenocysteine). Indeed, some species, known as selenium accumulators which are indigenous to seleniferous soils, accumulate very high concentrations of selenium, mostly as selenium isologues of secondary sulfur compounds although the accumulation of selenocystathionine in

* Dedicated to the late Alex Shrift, 1923-1992.

Sulfur Nutrition and Assimilation in Higher Plants
edited by L.J. De Kok et al., pp. 49–60

some selenium-accumulators is an important exception. Plants with these characteristics are extremely toxic to animals which ingest them and they are reportedly responsible for many colorful events involving the horses of explorers and armies throughout history (Rosenfeld & Beath 1964). Since selenium-accumulator plants accumulate very high concentrations of selenium isologues of organic metabolites derived from the sulfate assimilation pathway, mechanisms must exist to prevent the incorporation of selenium isologues of key intermediates of the pathway, particularly selenocysteine, from being incorporated into important functional compounds (*e.g.* proteins) to avoid selenium toxicity. Only a few details of these processes are understood.

Selenium metabolism in plants has been reviewed by Brown & Shrift (1982) and Anderson & Scarf (1984) and the biochemistry of selenium in mammals and bacteria by Stadtman (1979, 1980, 1990). Sulfur metabolism in plants has been reviewed by Anderson (1990).

Selenium-specific processes

Selenoproteins in bacteria and animals

Selenium was first recognized as an essential element for growth of the bacterium *Escherichia coli* when grown under anaerobic conditions with nitrate as the terminal electron acceptor and formate as the sole source of organic carbon (Pinsent 1954). The requirement for selenium could not be replaced by any other element. Moreover, formate dehydrogenase activity involving an Fe-S complex of 12 subunits ($\alpha 4\beta 4\gamma 4$) was also not expressed in the absence of selenium. It was subsequently established that each complex consists of 4 Se, 4 Mo, 4 heme Fe, 56 non-heme Fe and 53 acid-labile S with the 4 Se atoms believed to be associated with each of the 4 α-subunits, probably in the form of selenocysteine (for reviews, see Stadtman 1979, 1980, 1990). The enzyme complex also contains various sulfur containing cysteine residues at specific sites in the peptidyl chain. These data imply the existence of mechanisms which distinguish between selenocysteine on the one hand and cysteine on the other during formation of the peptide and also the selection of sulfur for the Fe:S clusters.

The formate dehydrogenases associated with some anaerobic bacteria (*e.g. Clostridium* spp. and *Methanococcus vanneilii*) are also selenoproteins. The formate dehydrogenases associated with obligate aerobic organisms do not contain selenium but not all anaerobic bacteria and anaerobically grown facultative anaerobes contain the selenium form of the enzyme and some (*e.g. M. vanneillii*) contain both forms of the enzyme.

The glycine reductase complex of the anaerobic bacterium *Clostridium sticklandii* also contains a selenoprotein component (protein A). It contains two sulfur-containing cysteinyl residues and one selenocysteinyl residue with the respective thiol and selenol groups occurring in their reduced forms *in vivo* (Cone *et al.* 1977).

Whereas the selenium-containing forms of formate dehydrogenase and glycine reductase are restricted to a few bacteria grown under restricted conditions, the selenium-containing form of glutathione peroxidase (GSH-Px) is common to the red blood cells of birds and mammals. GSH-Px is a critically important, high affinity enzyme which catalyses the detoxification of the organic peroxides and H_2O_2 produced through the action of oxidase and superoxide dismutase activities. Peroxides cause rapid lysis of cellular membranes if they are not reduced. Selenium-containing GSH-Px has 4 subunits, each containing 1 residue of selenocysteine in which selenium occurs in the reduced (selenol) form. The selenol group is instrumental in supporting reduction of peroxide. Each subunit also contains 2 cysteine residues which, if alkylated, cause inactivation of the enzyme indicating that the reduced thiol groups are also important determinants of enzyme activity. Again, the structure of the enzyme implies specific mechanisms for the recognition of cysteine and selenocysteine during synthesis of the enzyme. Animals also contain a selenium-independent form of GSH-Px which is active towards organic peroxides but inactive with H_2O_2.

Some geographical regions have soils with very low levels of selenium. In the absence of any corrective measures, selenium deficiency is common in livestock and, to a lesser extent, humans in these areas. Selenium deficiency in livestock is compounded by the application of superphosphate which further depletes the selenium content of pastures, the principal source of selenium for grazing animals (Gardiner 1969; Anderson & Scarf 1983). In this regard, the GSH-Px of erythrocytes is a useful indicator of the selenium status of grazing animals and humans.

Other selenoproteins have also been reported (Stadtman 1990). They include selenoprotein-P from mammalian plasma and hydrogenases from several bacteria which contain selenium as selenocysteine. Nicotinic acid hydroxylase and xanthine dehydrogenase are also selenoproteins but the selenium is not associated with selenocysteinyl or selenomethionyl residues within the peptidyl chains of these proteins.

Mechanism for the specific incorporation of selenocysteine into selenoproteins

The occurrence of selenocysteine at specific positions within a selenoprotein involves a specific selenocysteine recognition and processing system (Stadtman 1990; Böck *et al.* 1991a, 1991b). This is determined by genes containing a TGA codon which, although specifying termination in most genes, in selenoprotein genes directs the cotranslational incorporation of selenocysteine. The UGA codon in the corresponding mRNA is recognized by the anticodon of a unique $tRNA^{ser}$. The $tRNA^{ser}$ is aminocylated with serine in the presence of a corresponding seryl-tRNA synthetase and the seryl residue is phosphorylated. The resulting phosphoseryl residue reacts with an unidentified selenol to form selenocysteinyl-$tRNA^{ser}$ (Fig. 1). Thus, the acylation of the $tRNA^{ser}$ by selenocysteine is achieved by modification of serine after attachment to the appropriate tRNA and not from preformed selenocysteine. It is evident that the selenium-specific step is not the recognition of selenocysteine by an appropriate tRNA but the modification of phosphoseryl-$tRNA^{ser}$ by an appropriate selenium-donor to form selenocysteinyl-$tRNA^{ser}$.

Evidence for selenoproteins in higher plants

As a first step towards establishing whether plants contain selenoproteins, leaf extracts have been examined for various enzymes which are known to be selenoproteins in animals and bacteria. Particular emphasis has been directed at GSH-Px in pho-

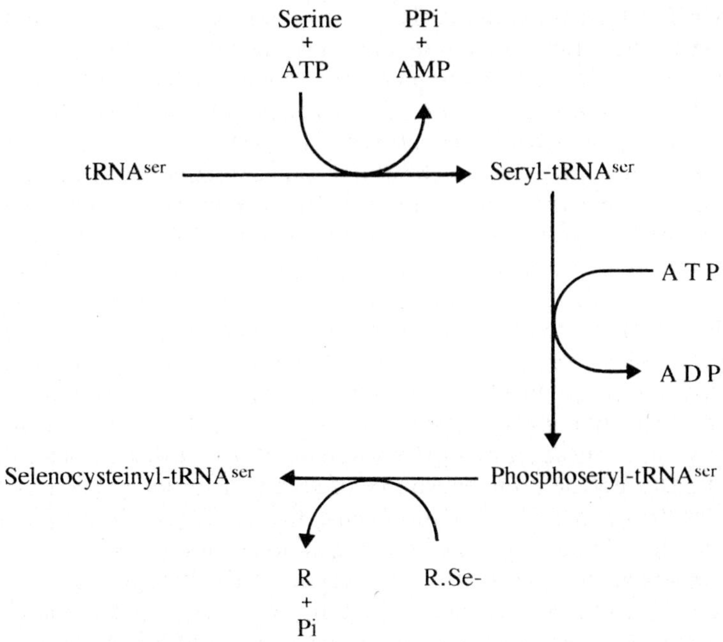

Fig. 1. Scheme for the aminoacylation of a unique tRNA^ser by selenocysteine following preliminary acylation by serine.

tosynthetic cells since on theoretical grounds this enzyme, if present, and given the very active NADPH-dependent glutathione reductase of chloroplasts, would afford a logical mechanism for detoxifying H_2O_2 formed from superoxide through superoxide dismutase activity (Wolosiuk & Buchanan 1977) according to the following sequence of reactions.

$$2O_2^- + 2H^+ \rightarrow H_2O_2 + O_2 \quad (1)$$
$$2GSH + H_2O_2 \rightarrow GSSG + 2H_2O \quad (2)$$
$$GSSG + NADPH + H^+ \rightarrow 2GSH + NADP^+ \quad (3)$$

Although there have been some reports of GSH-dependent reduction of H_2O_2 in crude extracts of higher plants which have been attributed to GSH-Px activity, Jablonski & Anderson (1984) found that an endogenous flavonoid was essential to support this activity in crude pea extracts, indicating that GSH-Px was not involved. Further, in a comprehensive survey of GSH-Px in many organisms, Smith & Shrift (1979) found no evidence of GSH-Px activity in plants. However, as discussed below, there is now strong evidence that GSH-Px occurs in various algae.

Other approaches have been used to investigate whether selenoproteins are essential for plant function. One has involved growing plants in water cultures containing highly purified salts, with and without added selenium. These experiments showed that added selenium did not enhance growth. However, since the plants gained trace amounts of selenium during the course of the experiment the possibility that selenium is essential cannot be ruled out (Broyer *et al.* 1966). In our laboratory, we have grown

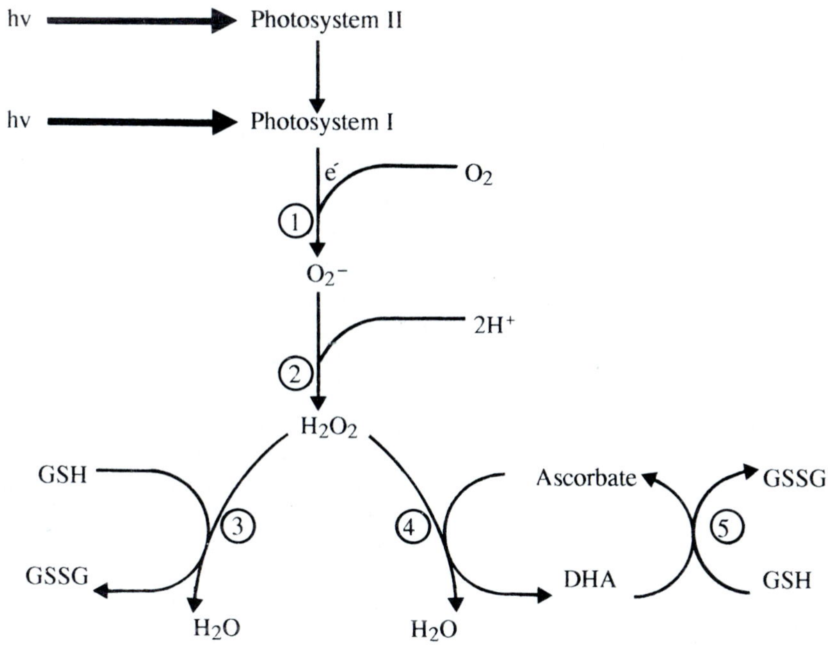

Fig. 2. Formation of H_2O_2 in reactions involving light-dependent reduction of O_2 (1) and superoxide dismutase (2) activity. Schemes for the detoxification of H_2O_2 involving GSH-Px (3) (operative in algae at low CO_2 concentrations) and ascorbate peroxidase (4) linked to glutathione dehydrogenase (5) (all green plants, including algae) are also shown. The GSSG produced by both schemes can, in theory, be reduced by NADPH in a reaction catalyzed by glutathione reductase. Abbreviation: DHA, dehydroascorbate.

plants in water cultures containing [^{75}Se]selenite or [^{35}S]sulfate in the nutrient solution and analyzed the leaf proteins for the presence of ^{75}Se and ^{35}S. We have found no evidence for a protein with a significantly higher ^{75}Se/^{35}S ratio relative to other proteins. However, the sensitivity of these experiments was confounded by the unavailability of [^{75}Se]selenite with a very high specific radioactivity. Thus, at this point in time there is no compelling evidence that selenium is essential for plant growth or plant function or that plants synthesize selenoproteins.

Evidence for selenoproteins in unicellular algae

Some progress has been made in the search for selenoproteins in unicellular algae, largely because the conditions for growing them, unlike those for higher plants, can be readily manipulated. At low concentrations of CO_2, algae divert electron flow from photosystem I into O_2 rather than CO_2 assimilation products, resulting in the production of superoxide (Fig. 2). H_2O_2 is produced from superoxide through the action of superoxide dismutase at rates of 100-150 μmol mg chlorophyll^{-1} h^{-1} (Sueltemeyer *et al.* 1986, Yokota & Kitaoka 1987). GSH-Px has been demonstrated in various unicellular algae (Overbaugh & Fall 1985; Price & Harrison 1988; Yokota *et al.* 1988) where presumably it is involved in the detoxification of H_2O_2 in association with the enzyme glutathione reductase (reactions (2) and (3), Fig. 2). In support of this proposal, Yokota *et al.* (1988) reported that the GSH-Px activity of *Chlamydomonas* grown at 5% CO_2 in the presence of 17.3 μM selenite supported H_2O_2 detoxification at *ca.* 36 μmol mg chlorophyll^{-1} h^{-1} whereas the activity of cells grown with selenite at 0.03% CO_2 was approximately 10-fold higher though only 2-fold higher in cells grown at low light intensity. GSH-Px activity was not detected in cells grown in the absence of selenium although ascorbate peroxidase, which, when linked to glutathione reductase activity provides an alternative mechanism for GSH-reduction of H_2O_2 (reactions (4) and (5), Fig. 2), was present in selenium-free cultures at an activity of 15-20 μmol mg chlorophyll^{-1} h^{-1}. These results demonstrate that conditions which favour light-dependent production of H_2O_2 at rapid rates also support the expression of GSH-Px activity, provided that selenium is available. The data are consistent with the possibility that GSH-Px is a selenoprotein. More direct evidence for this possibility is provided by studies with the marine diatom *Thalassiosira pseudonana* which has a specific requirement for selenium for growth (Price *et al.* 1987). *Thalassiosira* produces two GSH-Pxs, one of which is active with both H_2O_2 and tertiary-butyl hydroperoxide ($BuHO_2$) but the other is active with $BuHO_2$ only (Price & Harrison 1988). When the diatom was grown in medium containing 10 nM [^{75}Se]selenite, it produced two ^{75}Se-labeled proteins, one of which co-migrated with the H_2O_2-active GSH-Px on non-denaturing polyacrylamide gels. Information on the kinetic properties of this enzyme and the form of the selenium associated with it are not available. However, the *Thalassiosira* enzyme appears to be distinct from the H_2O_2-active enzyme in *Euglena gracilis* which has a high affinity for H_2O_2 (K_m = 30 μM) and is selenium independent (Overbaugh & Fall 1985). Unlike the selenium dependent enzyme from *Thalassiosira*, the activity of the selenium independent enzyme in *Euglena* was not significantly different in light- and dark-grown cells, suggesting that it has a different function.

Interactions of selenium with essential sulfur metabolism

Plants readily metabolize selenium non-specifically to form the selenium isologues of the products and intermediates of the reactions involved in the assimilation of inorganic sulfur into cysteine and methionine, processes which are essential for plant growth and function. Thus, for a particular reaction, the selenium- and sulfur substrates compete for the active site on the relevant enzyme so that the production of the selenium isologue diminishes as the concentration of the sulfur substrate increases and *vice versa*. Quantitatively, the relative yields of the selenium and sulfur products for a particular enzyme/process depend on the kinetics (K_m and V_{max}) towards the selenium- and sulfur substrates.

Several examples of nonspecific selenium/sulfur interactions have been described for the reactions of cysteine and methionine synthesis in plants (Table 1). The best known are the sulfate/selenate interaction by the sulfate uptake mechanism and the sulfate/selenate interaction by the enzyme ATP sulfurylase. The physiological significance of the selenate/sulfate interactions is unclear since there is some debate whether selenate or selenite is the form of selenium most readily available to plants grown in soil. If selenate is the most available form, then it can be taken up by the sulfate uptake mechanism and undergo activation to adenosine phosphoselenate

Table 1. Enzymes of essential sulfur metabolism from selenium-accumulator and non-accumulator plants which catalyze a corresponding reaction with the selenium isologue of the sulfur substrate.

Enzyme	References	
	Non-accumulators	Se-accumulators
Sulfate uptake	Ferrari & Renosto (1972); Smith (1976)	Ziebur & Shrift (1971)
ATP sulfurylase	Dilworth & Bandurski (1977); Shaw & Anderson (1972)	Shaw & Anderson (1974)
Cysteine synthase	Ng & Anderson (1978)	Ng & Anderson (1978)
Cystathionine γ-synthase	McCluskey *et al.* (1986)	McCluskey *et al.* (1986)
β-Cystathionase	Dawson & Anderson (1988)	Dawson & Anderson (1989)

(APSe) catalyzed by ATP sulfurylase (Shaw & Anderson 1972; Dilworth & Bandurski 1977). APSe can react non-enzymically with endogenous thiols (RSH) to produce thioselenites ($RSSeO_3^-$) which are progressively reduced to the corresponding selenols (Dilworth & Bandurski 1977). If however, selenite is the most available source of selenium, then, rather than interacting with sulfite for the enzyme sulfite reductase, it reacts non-enzymically with reduced glutathione (GSH) to form selenodiglutathione (GSSeSG) which is then reduced in two stages, either enzymically using NADPH as reductant in a reaction catalyzed by glutathione reductase or nonenzymically by further reactions with GSH (for a review, see Anderson & Scarf 1983). The reduction of selenite in this way can be linked to the very active glutathione reductase of chloroplasts which in turn is coupled to light-generated formation of NADPH. Thus, in the presence of catalytic amounts of NADP(H) and glutathione, disrupted chloroplasts in the light readily support selenite reduction with the concomitant evolution of O_2 and, in the presence of NADPH(H), GSSeSG-dependent O_2 evolution (Jablonski & Anderson 1982). The selenide produced by these processes competes with sulfide as a substrate for the enzyme cysteine synthase resulting in the production of selenocysteine/cysteine. The rate of selenocysteine synthesis by the pea enzyme is about 25% of cysteine synthesis but the affinity of the enzyme for selenide relative to sulfide is unknown. Similarly, the possibility that bound selenide (*e.g.* $GSSe^-$) competes with bound sulfide (GSS-), the presumed physiological substrate of cysteine synthase (Tsang & Schiff 1978), has not been investigated.

Some reactions of methionine synthesis have also been studied. The enzyme cystathionine γ-synthase from spinach supports the synthesis of selenocystathionine from selenocysteine and the formation of cystathionine from cysteine using phosphohomoserine as the aminobutyryl donor (Dawson & Anderson 1988). The spinach en zyme exhibited higher affinity for selenocysteine (K_m *ca.* 70 μM) than cysteine (K_m *ca.* 240 μM) and the V_{max} values for the two substrates were similar, indicating that selenocysteine was a very active substrate. The cystathionine γ-synthases from various selenium-accumulator plants showed similar properties (Dawson & Anderson 1989) indicating that any differences in selenium metabolism between selenium-accumulator and non-accumulator plants cannot be attributed to differences in substrate specificity of this enzyme. In peas and spinach, the associated enzyme, β-cystathionase, supports the α,β-elimination of both selenocystathionine and cystathionine (McCluskey *et al.* 1986). The rate of the reaction and the affinity of the enzyme for both substrates was similar. If the other enzyme of methionine synthesis, homocysteine methyltransferase, should also prove to be active towards the selenium substrate (selenohomocysteine) then the entire pathway from cysteine to methionine would be open to selenium and, on a molar basis, sulfur and selenium would compete on approximately equal terms. In contrast, the incorporation of inorganic selenium into selenocysteine competes with inorganic sulfur relatively weakly (Anderson & Scarf 1983).

Non-accumulator plants readily incorporate selenium non-specifically into protein. In mature wheat for example, selenium is incorporated into the storage proteins of developing grains where it occurs principally as selenomethionine (Olson *et al.* 1970). This not only demonstrates that the methionine biosynthetic pathway is open to selenium *in vivo* but implies that the processes involved in the incorporation of one of the amino acids (presumably methionine) permit the non-specific incorporation of selenomethionine into protein. In wheat, methionyl-tRNA synthetase supports the synthesis of both selenomethionyl-$tRNA^{met}$ and methionyl-$tRNA^{met}$ and both selenomethionyl and methionyl residues are incorporated into internal positions within peptides during peptide chain elongation (Eustice *et al.* 1980, 1981a). Peptide bond initiation, however, is more specific to the methionyl isologue (Eustice *et al.* 1981a). There has been only one report describing the non-specific occurrence of selenocysteine in proteins in plants (Brown & Shrift 1980). In this regard, the cysteinyl-tRNA synthetase from mung beans supports both selenocysteine- and cysteine-dependent PP_i-ATP exchange at approximately equal rates. However, aminoacylation of $tRNA^{cys}$ by selenocysteine and the incorporation of selenocysteine from selenocysteinyl-$tRNA^{cys}$ has not been examined in plants.

The incorporation of selenium into protein in selenium accumulators is much less than in non-accumulators (for reviews, see Anderson & Scarf 1983; Brown & Shrift 1982). This may have more to do with the diversion of selenium into secondary pathways in selenium-accumulators than some selective process which discriminates against selenium isologues since cell-free systems from both selenium-accumulators and non-accumulators incorporate selenomethionine into protein (see above).

Interactions of selenium with non-essential sulfur metabolism

Various plants produce a range of non-essential compounds which contain sulfur (*e.g.* various *S*-alkylcysteine derivatives in many members of the *Brassicaceae* and *Alliaceae*). The sulfur in these compounds is commonly derived from an essential sulfur compound (*e.g.* cysteine) thereby diverting sulfur from essential (or primary) metabolism. Plants which synthesize non-essential (or secondary) sulfur-containing compounds commonly form the selenium isologues when supplied with inorganic selenium (see Shrift 1973 for examples) and can attain quite high concentrations in some species.

The formation of selenium isologues of secondary sulfur compounds is of some interest. Most selenium-accumulator plants accumulate selenium in this form. However, all of the reactions of essential sulfur metabolism which are open to selenium in non-accumulators (see above) are also open to selenium in selenium-accumulators (Table 1). The incorporation of methionine into non-terminating positions in protein is also open to selenium (Eustice *et al.* 1981b). Thus, in the absence of mechanisms *in vivo* to either prevent selenium entering primary sulfur metabolism or to divert selenium out of the primary sulfur pathway, then selenium would be expected to be toxic to both types of plant. Selenium-accumulators, however, are extremely tolerant of selenium, both externally and internally. Most selenium-accumulators produce selenium isologues of secondary sulfur metabolites, most commonly related to intermediates of primary sulfur metabolism (*e.g.* *se*-methylselenocysteine). Thus, it has been postulated that the enzymes involved in catalyzing the synthesis of selenium isologues of secondary sulfur compounds in selenium-accumulator species could be especially active towards the selenium isologues, especially those enzymes involved in diverting intermediates of essential sulfur metabolism into secondary sulfur metabolism, thereby draining selenium isologues out of primary sulfur metabolism. Some earlier labeling experiments on the synthesis of *se*-methylselenocysteine, reviewed by Anderson & Scarf (1983), are consistent with this proposal. Also, Nigam & McConnell (1973) observed that, in the absence of selenium, developing seeds of the non-accumulator *Phaseolus lunatus* accumulate *s*-methylcysteine, but if selenium is supplied (as selenate), it is readily incorporated into *se*-methylselenocysteine but not into storage proteins. Conversely, leaves of the same species, which do not accumulate *s*-methylcysteine, incorporate selenium into selenomethionine.

Mechanisms in plants for avoiding selenium toxicity and for minimizing the selenium content of shoots

The best documented examples of plants which can survive and grow in selenium-rich environments are the selenium-accumulator plants. In water cultures they can survive and grow at Se/S ratios which are lethal to non-accumulator species. Since these plants, by definition, accumulate selenium, then differences in the rates and specificity of S/Se uptake are unlikely to explain the tolerance of selenium-accumulators to selenium. As noted above, selenium can enter the reactions of primary sulfur metabolism in both accumulators and non-accumulators, but, in selenium-accumulators selenium is thought to be diverted preferentially into secondary sulfur pathways in place of sulfur thus preventing selenium toxicity. If this hypothesis is correct, then species which accumulate high concentrations of secondary sulfur compounds derived from essential sulfur metabolism could, in theory, be active selenium-accumulators. A corollary would be that they would also be very toxic to animals. Other mechanisms could also be important in avoiding selenium toxicity in selenium accumulators. There is some evidence that the cysteinyl-tRNA synthetase from *Astragalus bisulcatus* (but not in the other accumulators studied) does not activate selenocysteine (Burnell & Shrift 1979) and that selenium accumulators are less active in incorporating selenium into protein (Brown & Shrift 1981). Also there has been speculation that the selenium isologues of secondary sulfur-compounds are accumulated within the cells of selenium accumulators in metabolic compartments which do not support essential sulfur metabolism.

Various factors influence selenium uptake by roots and its accumulation in the shoots. The most important of these are (i) the selenium and sulfur status of the soil since selenium uptake is profoundly influenced by the availability of sulfur, (ii) the form of the selenium present (Smith & Watkinson 1984), and (iii) whether the plant is a selenium-accumulator or non-accumulator. However, other factors also contribute and are of considerable practical importance in agriculturally important non-accumulator species supplied with high sulfate/saline irrigation water (see references in Shennan *et al.* 1990; Wu & Huang 1991). The shoots of plants grown under these conditions often contain quite high concentrations of selenium and pose a threat to human health. In a study with 6 taxa of *Lycopersicon* and 8 cultivars of *L. esculentum*, Shennan *et al.* (1990) found considerable variation in the rate of selenium uptake and its sensitivity to inhibition by sulfur, particularly between different species but also between cultivars. Considerable variation also occurred in the short term selenium distribution rates from root to shoot and in the root weight relative to total weight of the plant (specific root ratio). This also influenced the selenium distributed to the shoots; the higher the ratio the smaller the proportion of selenium taken up that was distributed to the shoot. Selenium accumulation in the shoot was also decreased by high chloride/salinity (Shennan *et al.* 1990; Wu & Huang 1991). Collectively, these data provide some possibilities for action to minimize the selenium content of crop plants. It would also seem pertinent to establish whether plants with an active secondary sulfur metabolism tend to accumulate selenium. This might provide important information on which plants to avoid growing for human consumption in these areas. Conversely, they could provide a way of depleting the selenium content of the soil for subsequent food crops.

Concluding remarks

The issue that continues to be of greatest interest is whether higher plants have an absolute and specific requirement for selenium. Technical difficulties have frustrated attempts to use nutritional approaches and the search for selenoproteins in plants has been confounded by non-specific replacement of sulfur by selenium. However, the discovery of a non-terminating codon in the genome of microorganisms which specifically codes for selenocysteine provides a new and novel opportunity to address the question in plants. If the genomes of higher plants are found to contain non-terminating TGA codons then this will provide strong (though indirect) evidence for the existence of selenocysteine-containing selenoproteins. Much interest will then focus on identifying the putative selenoproteins and identifying their function. Nonetheless, since selenoproteins have been described which do not contain selenocysteine (Stadtman 1990), failure to detect non-terminating TGA codons in higher plants does not necessarily rule out the possibility that selenium is essential for their growth and function.

References

Anderson, J.W. 1990. Sulfur metabolism in plants. In: Miflin, B.J. & Lea, P.J. (eds). The Biochemistry of Plants, Vol. 16. Academic Press, San Diego, pp. 327-381.

Anderson, J.W. & Scarf, A.R. 1983. Selenium and plant metabolism. In: Robb, D.A. & Pierpoint, W.S. (eds). Metals and Micronutrients: Uptake and Utilization by Plants. Academic Press, London, pp. 241-275.

Böck, A., Forchhammer, K., Heider, J. & Baron, C. 1991a. Selenoprotein synthesis: an expansion of the genetic code. TIBS 16: 463-467.

Böck, A., Forchhammer, K., Heider, J., Leinfelder W., Sawers, G., Veprek, B. & Zinoni, F. 1991b. Selenocysteine: the 21st amino acid. Molec. Microbiol. 5: 155-520.

Brown, T.A. & Shrift, A. 1980. Identification of selenocysteine in the proteins of selenate-grown *Vigna radiata*. Plant Physiol. 66: 758-761.

Brown, T.A. & Shrift, A. 1981. Exclusion of selenium from proteins of selenium-tolerant *Astragalus* species. Plant Physiol. 67: 1051-1053.

Brown, M.J. & Shrift, A. 1982. Selenium: toxicity and tolerance in higher plants. Biol. Rev. 57: 59-84.

Broyer, T.C., Lee, D.C. & Asher, C.J. 1966. Selenium nutrition of green plants. Effect of selenite supply on growth and selenium content of alfalfa and subterranean clover. Plant Physiol. 41: 1425-1428.

Burnell, J.N. & Shrift, A. 1979. Cysteinyl-tRNA synthetase from *Astragalus* species. Plant Physiol. 63: 1095-1097.

Cone, J.E., Martin del Rio, R. & Stadtman, T.C. 1977. Clostridial glycine reductase complex. Purification and characterization of the selenoprotein component. J. Biol. Chem. 252: 5337-5344.

Dawson, J.C. & Anderson, J.W. 1988. Incorporation of cysteine and selenocysteine into cystathionine and selenocystathionine by crude extracts of spinach. Phytochemistry 27: 3453-3460.

Dawson, J.C. & Anderson, J.W. 1989. Comparative enzymology of cystathionine and selenocystathionine synthesis of selenium-accumulator and non-accumulator plants. Phytochemistry 28: 51-55.

Dilworth, G.L. & Bandurski, R.S. 1977. Activation of selenate by adenosine 5'-triphosphate sulphurylase from *Saccharomyces cerevisiae*. Biochem. J. 163: 521-529.

Eustice, D.C., Foster, I., Kull, F.J. & Shrift, A. 1980. *In vitro* incorporation of selenomethionine into protein by *Vigna radiata* polysomes. Plant Physiol. 66: 182-186.

Eustice, D.C., Kull, F.J. & Shrift, A. 1981a. Selenium toxicity: aminoacylation and peptide bond formation with selenomethionine. Plant Physiol. 67: 1054-1058.

Eustice, D.C., Kull, F.J. & Shrift, A. 1981b. *In vitro* incorporation of selenomethionine into protein by *Astragalus* polysomes. Plant Physiol. 67: 1059-1060.

Ferrari, G. & Renosto, F. 1972. Regulation of sulfate uptake by excised barley roots in the presence of selenate. Plant Physiol. 49: 114-116.

Gardiner, M.R. 1969. Selenium in animal nutrition. *Outlook Agric.* 6: 19–28.

Jablonski, P.P. & Anderson, J.W. 1982. Light-dependent reduction of selenite by sonicated pea chloroplasts. Phytochemistry 21: 2179-2184.

Jablonski, P.P. & Anderson, J.W. 1984. Role of a flavonoid in the peroxide-dependent oxidation of glutathione catalyzed by pea extracts. Phytochemistry 23: 1865-1869.

McCluskey, T.J., Scarf, A.R. & Anderson, J.W. 1986. Enzyme catalyzed α,β-elimination of selenocystathionine and selenocystine and their sulphur isologues by plant extracts. Phytochemistry 25: 2063-2068.

Ng, B.H. & Anderson, J.W. 1978. Synthesis of selenocysteine by cysteine synthases from selenium-accumulator and non-accumulator plants. Phytochemistry 17: 2069-2074.

Nigam, S.N. & McConnell, W.B. 1973. Biosynthesis of Se-methylselenocysteine in lima beans. Phytochemistry 12: 359-362.

Olson, O.E., Novacek, E.J., Whitehead, E.I. & Palmer, I.S. 1970. Investigations on selenium in wheat. Phytochemistry 9: 1181-1188.

Overbaugh, J.M. & Fall, R. 1985. Characterization of a selenium-independent glutathione peroxidase from *Euglena gracilis*. Plant Physiol. 77: 437-442.

Pinsent, J. 1954. The need for selenite and molybdate in the formation of formic dehydrogenase by membranes of the *Coli-Aerogenes* group of bacteria. Biochem. J. 57: 10-16.

Price, N.M. & Harrison, P.J. 1988. Specific selenium-containing macromolecules in the marine diatom *Thalassiosira pseudonana*. Plant Physiol. 86: 192-199.

Price, N.M., Thomson, J.A. & Harrison, P.J. 1987. Selenium: an essential element for growth of the coastal marine diatom *Thalassiosira pseudonana* (*Baccillariophyceae*). J. Phycol. 23: 1-9.

Rosenfeld, I. & Beath, O.A. 1964. Selenium: Geobotany, Biochemistry, Toxicity and Nutrition. Academic Press, New York and London.

Shaw, W.H. & Anderson, J.W. 1972. Purification, properties and substrate specificity of adenosine triphosphate sulphurylase from spinach leaf tissue. Biochem. J. 127: 237-247.
Shaw, W.H. & Anderson, J.W. 1974. Comparative enzymology of the adenosine triphosphate sulphurylases from leaf tissue of selenium-accumulator and non-accumulator plants. Biochem. J. 139: 27-35.
Shennan, C., Schachtman, D.P. & Cramer, G.R. 1990. Variation in [^{75}Se]selenate uptake and partitioning among tomato cultivars and wild species. New Phytol. 115: 523-530.
Shrift, A. 1973. In: Klayman, D.L. & Gunther, W.H.H. (eds). Organic Selenium Compounds: Their Chemistry and Biology. Wiley, New York, pp. 763-814.
Smith, I.K. 1976. Characterization of sulfate transport in cultured tobacco cells. Plant Physiol. 58: 358-362.
Smith, J.A. & Shrift, A. 1979. Phylogenetic distribution of glutathione peroxidase. Comp. Biochem. Physiol. 63B: 39-44.
Smith, G.S. & Watkinson, J.H. 1984. Selenium toxicity in perennial ryegrass and white clover. New Phytol. 97: 557-564.
Stadtman, T.C. 1979. Some selenium-dependent biochemical processes. Adv. Enzymol. 48: 1-28.
Stadtman, T.C. 1980. Selenium-dependent enzymes. Annu. Rev. Biochem. 49: 93-110.
Stadtman, T.C. 1990. Selenium biochemistry. Annu. Rev. Biochem. 59: 111-127.
Sueltemeyer, D.F., Klug, K. & Fock, H.P. 1986. Effect of photon fluence rate on oxygen evolution and uptake by *Chlamydomonas reinhardtii* suspensions grown in ambient and CO_2-enriched air. Plant Physiol. 81: 372-375.
Tsang, M.L-S. & Schiff, J.A. 1978. Studies of sulfate utilization by algae 18. Identification of glutathione as a physiological carrier in assimilatory sulfate reduction by *Chlorella*. Plant Sci. Lett. 11: 177-183.
Wolosiuk, R.A. & Buchanan, B.B. 1977. Thioredoxin and glutathione regulate photosynthesis in chloroplasts. Nature 266: 565-567.
Wu, L. & Huang, Z-Z. 1991. Chloride and sulfate salinity effects on selenium accumulation by tall fescue (*Festuca arundinacea* Scureb). Crop Sci. 31: 114-118.
Yokota, A. & Kitaoka, S. 1987. Rates of glycolate synthesis and metabolism during photosynthesis of *Euglena* and microalgae grown on low CO_2. Planta 170: 181-189.
Yokota, A., Shigeoka, S., Onishi, T. & Kitaoka, S. 1988. Selenium as inducer of glutathione peroxidase in low-CO_2-grown *Chlamydomonas reinhardtii*. Plant Physiol. 86: 649-651.
Ziebur, N.K. & Shrift, A. 1971. Response to selenium by callus cultures derived from *Astragalus* species. Plant Physiol. 47: 545-550.

REGULATORY INTERACTIONS BETWEEN SULFATE AND NITRATE ASSIMILATION

Christian Brunold
Pflanzenphysiologisches Institut der Universität Bern, Altenbergrain 21, 3013 Bern, Switzerland

Introduction

Sulfate and nitrate, which contain sulfur and nitrogen in their most highly oxidized form, are the dominant species available to many plants for covering their needs for these elements (Schiff 1983; Cram 1990; Oaks 1992). Assimilatory sulfate and nitrate reduction are therefore necessary for the synthesis of amino acids including sulfur-containing amino acids like cysteine and methionine, in which both sulfur and nitrogen are present in reduced form. The dominant portion of the amino acids is used for protein synthesis. Therefore, the S/N ratio in plants is usually about 1/20 (Dijkshoorn & van Wijk 1967) reflecting the proportion of these elements in proteins. Only in species where sulfur is accumulated in the form of sulfate or of secondary plant products is the ratio significantly higher (Cram 1990; Ernst 1990). Plants appear to possess mechanisms to coordinate assimilatory sulfate and nitrate reduction so that the appropriate proportions of both sulfur containing and other amino acids are available for protein synthesis. This review focuses on these reciprocal regulatory mechanisms at the level of assimilation, but the regulation of the uptake of NO_3^- and SO_4^{2-} may be at least as important for coordinating both assimilatory pathways (Saccomani & Ferrari 1989; Cram 1990; Clarkson *et al.*, this volume).

Assimilatory nitrate (Solomonson & Barber 1990; Oaks 1992) and sulfate (Brunold 1990; Giovanelli 1990; Schmidt 1992) reduction have been reviewed very recently. Therefore, only aspects of both pathways are discussed here which form the basis for reviewing regulatory interactions between them.

Assimilatory sulfate and nitrate reduction

The first step of nitrate reduction from nitrate to nitrite is catalyzed by nitrate reductase (Fig. 1). NADH is the most common physiological electron donor for this reduction in higher plants (NADH nitrate reductase; EC 1.6.6.1), some of which also contain NAD(P)H nitrate reductase (EC 1.6.6.2.). The second step is a six-electron reduction of nitrite to ammonium catalyzed by nitrite reductase (EC 1.7.7.1). This step is coupled to the photosynthetic electron transport via reduced ferredoxin, which serves as physiological electron donor for the enzyme (Beevers & Hageman 1980; Campbell & Kinghorn 1990; Solomonson & Barber 1990). The ammonium is then incorporated into the amide nitrogen of glutamine via the action of glutamine synthetase (Miflin & Lea 1980).

The rate-limiting step of nitrate assimilation appears to be the initial reaction, catalyzed by nitrate reductase (Beevers 1981) which is considered to be a limiting factor

Sulfur Nutrition and Assimilation in Higher Plants
edited by L.J. De Kok et al., pp. 61–75

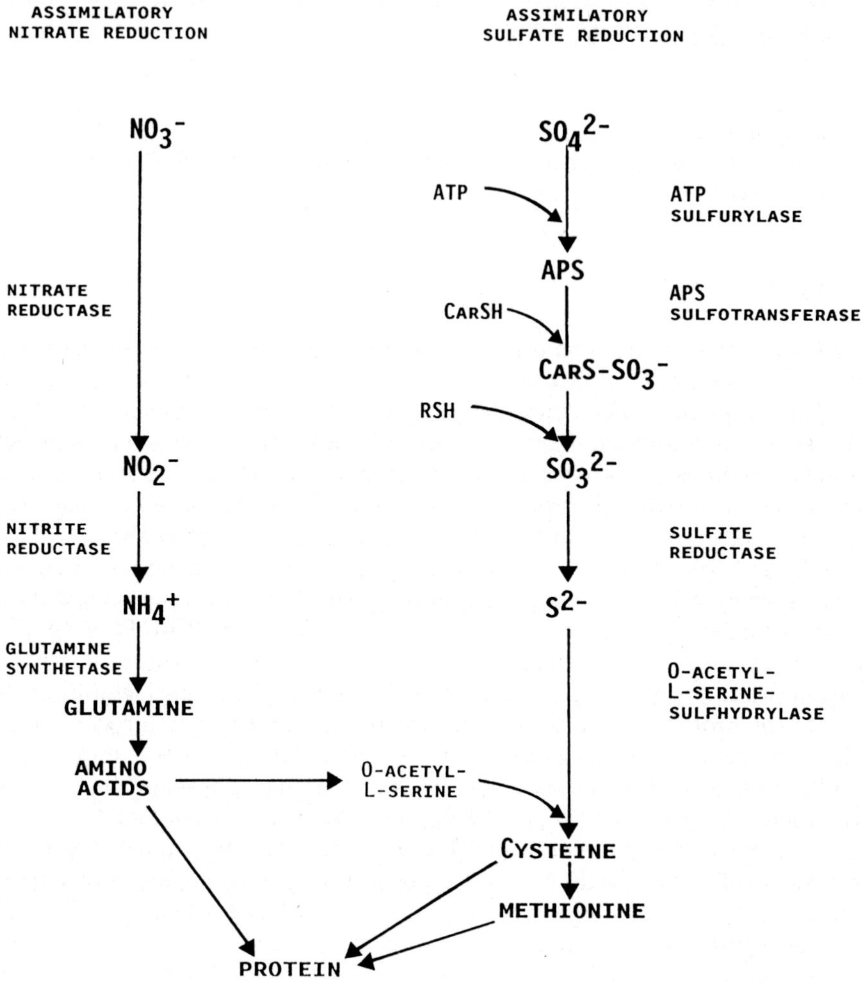

Fig. 1. Scheme of assimilatory sulfate and nitrate reduction in plants. To stress the analogy between both pathways, assimilatory sulfate reduction via adenosine 5'-phosphosulfate (APS), carrier-bound sulfite ($CARS-SO_3^-$), free sulfite (SO_3^{2-}) and free sulfide (S_2^-) is presented. The carrier thiol (CARSH) and the thiol liberating SO_3^{2-} from $CARS-SO_3^-$ (RSH) may both be glutathione (GSH). Glutamic acid functions as as acceptor for ammonium (NH_4^+), forming glutamine via glutamine synthetase. *O*-acetyl-L-serine accepts sulfide, forming cysteine via *O*-acetyl-L-serine sulfhydrylase. *O*-acetyl-L-serine is formed from serine and acetyl-CoA, catalyzed by L-serine acetyltrans-ferase (EC 2.3.1.30). For further explanations see text.

for growth, development, and protein production in plants and has therefore been extensively studied with respect to its regulation (Solomonson & Barber 1990).

The first step of assimilatory sulfate reduction (Fig. 1) is an activation catalyzed by ATP sulfurylase (EC 2.7.7.4). The adenosine 5'-phosphosulfate (APS) formed in this reaction is the substrate for adenosine 5'-phosphosulfate kinase (APS-kinase; EC 2.7.1.2.5) which forms adenosine 3'-phosphate 5'-phosphosulfate (PAPS) in a second activation step. The subsequent reaction of the pathway is catalyzed by a sulfotransferase. Two types of sulfotransferases have been described: APS sulfotransferase transfers the sulfate activated in APS to an as yet unidentified carrier molecule (CarSH) thereby forming "bound sulfite" (CarS-SO_3^-) and AMP (Schmidt 1972; Schmidt 1976 a; Tsang & Schiff 1976; Schiff 1983; Li & Schiff 1991; Schmidt 1992). PAPS sulfotransferase reacts with PAPS and reduced thioredoxin to form free sulfite, oxidized thioredoxin and adenosine 3'-phosphate 5'-phosphate (PAP) (Dreyfuss & Monty 1963; Ostrowski & Kredich 1989; Schwenn 1989; Thomas *et al.* 1990; Kredich, this volume). Free sulfite can also be formed, when CarS-SO_3^- reacts with a suitable thiol (RSH) (Tsang & Schiff 1975). Correspondingly, free $^{35}SO_3^{2-}$ was detected during cysteine formation from ^{35}S-APS in a reconstituted spinach chloroplast system containing the naturally occurring glutathione concentration (Schürmann & Brunold 1980). APS sulfotransferase activity has been detected in many higher plants (Schmidt 1975) and algae (Tsang & Schiff 1975) and has recently been purified to homogeneity from *Euglena* mitochondria (Li & Schiff 1991). Using this preparation, Li and Schiff, 1992, labeled the enzyme with radioactive ^{35}S-APS and demonstrated enzyme-bound ^{35}S-sulfite in the form of S-sulfocysteine residues. They suggest a subsequent reaction with a carrier (CarSH) to yield CarS-$^{35}SO_3^-$. PAPS sulfotransferase seems to be the only type of sulfotransferase of bacteria (Dreyfuss & Monty 1963; Ostrowski & Kredich 1989; Kredich, this volume) and yeast (Schwenn *et al.* 1988; Thomas *et al.* 1990), whereas APS (Schmidt 1976a; Tsang & Schiff 1976; Fankhauser & Brunold 1978b) as well as PAPS (Schwenn 1989) sulfotransferases were detected in spinach leaves. The reduction of sulfite to the level of sulfide may involve sulfite reductase (EC 1.8.7.1) which acts on free sulfite to form free sulfide (Krüeger & Siegel 1982), or organic thiosulfate reductase which uses carrier-bound sulfite (CarS-SO_3^-) as a substrate and forms carrier-bound sulfide (CarS-S^-) (Schmidt 1973; Schmidt 1992). For both types of reductases in chloroplasts, reduced ferredoxin has been shown to be the electron donor (Schmidt 1973; Krüeger & Siegel 1982). Free sulfide is incorporated into *O*-acetyl-L-serine via *O*-acetyl-L-serine sulfhydrylase (EC 4.2.99.8), thus forming cysteine (Giovanelli & Mudd 1968; Kredich 1971). The mechanism for formation of cysteine from CarS-S^- is not clear. It is obvious, however, that a two electron step is involved (Li & Schiff 1992; Schmidt 1992).

According to our present knowledge, the reaction steps of assimilatory sulfate reduction in higher plants most susceptible to regulatory signals appear to be those catalyzed by ATP sulfurylase and APS sulfotransferase. Therefore, these enzymes have been most intensively studied in order to understand their catalytic efficiency and regulation (Brunold 1990; Giovanelli 1990).

Both nitrate and light are required for the synthesis of nitrate reductase and nitrite reductase proteins (Beevers & Hageman 1980; Somers *et al.* 1983; Duke & Duke 1984; Gupta & Beevers 1984; Remmler & Campbell 1986). Recently with the cloning of nitrate reductase and nitrite reductase genes it has become possible to demonstrate that the induction by nitrate occurs at the level of transcription (Cheng *et al.* 1986; Crawford *et al.* 1986; Calza *et al.* 1987; Back *et al.* 1988; Lahners *et al.* 1988). A rapid and reversible decrease of nitrate reductase activity was detected in spinach leaves during darkening (Riens & Heldt 1992). Apparently in these leaves nitrate reductase is very rapidly inactivated at sudden darkness avoiding an accumulation of the toxic nitrite in the cells.

In contrast to assimilatory nitrate reduction, where the substrate nitrate functions as an inducer, assimilatory sulfate reduction seems to be repressed at normal levels of the substrate sulfate and derepressed at low concentrations (Reuveny & Filner 1977; Reuveny *et al.* 1980; Brunold *et al.* 1987). In cultured cells of tobacco (Reuveny & Filner 1977) and rose (Haller *et al.* 1986) ATP sulfurylase activity was derepressed in the absence of a sulfur source, reaching levels which were 100 and 500% higher than controls with sulfate for rose and tobacco, respectively. APS sulfotransferase activity of *Lemna minor* (Brunold *et al.* 1987) was increased by 50 to 100% after transfer to 0 or 0.0088 mM sulfate, transfer back to 0.88 mM sulfate rapidly decreased the enzyme activity to the initial level. Cultivation with 17.6 mM instead of 0.88 mM SO_4^{2-} reduced extractable APS sulfotransferase by 50%, indicating that extremely high sulfate concentrations enhance repression.

Also most evidence points to a cytoplasmic location of nitrate reductase (Solomonson & Barber 1990), recent experiments raise the interesting possibility that there is a plasma membrane bound form of this enzyme which may function as a nitrate transporter (Ward *et al.* 1988; Tischner *et al.* 1989). Nitrite reductase is clearly located in the chloroplast (Solomonson & Barber 1990).

The extractable activity of ATP sulfurylase, APS sulfotransferase and sulfite reductase increased when etiolated spinach, bean or pea seedlings were transferred from dark to light (Fankhauser & Brunold 1978b; Wyss & Brunold 1979; von Arb & Brunold 1986). These observations indicate that full development of these enzyme activities is light-dependent. This is consistent with their predominant or even exclusive location in chloroplasts as demonstrated with organelle preparations (Mayer 1967; Schwenn & Hennies 1974; Sawhney & Nicholas 1975; Fankhauser & Brunold 1978b; Gerwick *et al.* 1980; Lunn *et al.* 1990).

It is commonly considered that regardless of the culture conditions, most nitrate assimilation is carried out predominantly in the root of some genera, *e.g. Lupinus* and *Vicia* or in the shoot, in particular in leaves, of others *e.g. Xanthium* and *Stellaria* (Beevers 1981; Crafts-Brander & Harper 1982; Smirnoff & Steward 1985; Oaks 1992).

The contribution of roots to the needs of plants for reduced sulfur is not clear and may be greatly influenced by the developmental stage and the environmental conditions. Reduced sulfur compounds are normally absent from xylem sap or present only in low concentrations (Pate 1965; Rüegsegger & Brunold 1992). This is consistent with the finding that the level of APS sulfotransferase of sunflower roots was at 5% of the level in shoots (Schmidt 1976b), and that ATP sulfurylase activity of soya bean seedling is at 5 to 10 times higher in the leaves than in the roots (Adams & Rinne 1969). This situation can change dramatically, when roots are exposed to increased levels of heavy metals or to herbicide safeners and start synthesizing large amounts of glutathione for phytochelatin production (Grill *et al.* 1985; Nussbaum *et al.* 1988; Grill *et al.* 1989; Rüegsegger *et al.* 1990) or for herbicide conjugation (Farago & Brunold 1990). It may also be different in young seedlings because substantial amounts of the activity of all enzymes of assimilatory sulfate reduction were detected in the roots of 5-day-old pea seedlings (Brunold & Suter 1989).

On sucrose density gradients ATP sulfurylase, APS sulfotransferase and sulfite reductase activity from homogenates of roots of these seedlings were distributed in a manner very similar to nitrite reductase activity (Brunold & Suter 1989), an enzyme localized in the proplastids (Emes & Fowler 1979; Suzuki *et al.* 1981). This indicates that the three enzymes of sulfate assimilation have the same intracellular localization as nitrite reductase. Since proplastids from pea roots also contain *O*-acetyl-L-serine sulfhydrylase (Brunold & Suter 1989) these organelles have the capacity to synthesize cysteine from SO_4^{2-}.

Sulfur nutrition and assimilatory nitrate reduction

Sulfur deficiency induces rather consistent biochemical changes in plants, including especially a greatly expanded free amino acid pool with a composition different from that of non-deficient plants (Coleman 1957; Thompson *et al.* 1960; Smith 1980; Macnicol 1983; Macnicol & Randall 1987). The altered composition is mainly caused by the accumulation of particular amino acids like arginine and asparagine, and also *O*-acetyl-L-serine, and a depressed level of sulfur amino acids. The accumulated amino acids may be one reason for the decrease of nitrate reductase under conditions of sulfur deficiency, since this enzyme activity is low in plants cultivated with amino acids (Beevers & Hageman 1960; Suter *et al.* 1986; Neuenschwander *et al.* 1991).

The effect of the level of sulfur fertilization on nitrate reductase activity was studied in whole plants and cultured plant cells (Pal *et al.* 1976; Friedrich & Schrader 1978; Reuveny *et al.* 1980; Saccomani *et al.* 1984; Haller *et al.* 1986). In general, decreasing activities of the enzyme were detected with decreasing levels of sulfur availability (Pal *et al.* 1976; Friedrich & Schrader 1978; Reuveny *et al.* 1980; Haller *et al.* 1986) unless the experimental period was too short for establishing a clear situation of sulfur deficiency (Saccomani *et al.* 1984). The activity of leaf nitrate reductase from maize was significantly diminished by sulfur deprivation before soluble protein declined and before visual symptoms of sulfur deficiency became evident (Friedrich & Schrader 1978). By day 12 nitrate reductase activity in sulfur deprived plants was decreased to 50% of normal plants. The activity of leaf glutamine synthetase expressed on a fresh weight bases was also diminished by sulfur deprivation, but the percent reduction was less than for nitrate reductase (Friedrich & Schrader 1978).

In cultured tobacco cells sulfate and nitrate reduction were studied as influenced by nutrient solution availability of nitrate and sulfate (Reuveny *et al.* 1980). Nitrate reductase activity was induced when nitrate was available in the nutrient solution provided that there was an accumulation of reduced sulfur in the tissue. To determine the effect of the sulfur supply on the regulation of nitrate reductase stationary tobacco cells were transferred to fresh medium containing nitrate at 2.5 mM and sulfate at 0.1 (control), 0.033, 0.005 mM or "no sulfate" (Reuveny *et al.* 1980). After 24 hours there was a linear relationship between the sulfate added to the culture medium and the nitrate reductase activity induced. The decreased induction of nitrate reductase detected with the decreasing initial sulfate concentrations cannot be attributed to a decline in the ability of the cells to synthesize protein. Two pieces of evidence showed that the protein synthesizing machinery of sulfur starved cells was functional: (1) by the large increase in soluble protein during the first 24 hours for all initial sulfate concentrations and (2) the derepression of ATP sulfurylase that occurred in the same cells (Reuveny *et al.* 1980). This indicates that the regulatory interaction between assimilatory sulfate and nitrate reduction is not at the level of protein synthesis.

Nitrogen nutrition and assimilatory sulfate reduction

Derepression by sulfur limitation of ATP sulfurylase does not occur in cells starved for nitrogen. Upon addition of a nitrogen source to such cells the increase in enzyme activity begins within 12 hours (Reuveny *et al.* 1980). Thus the ATP sulfurylase of tobacco cells appears to be regulated by both negative feedback mechanisms in which sulfate or an endproduct of the sulfate assimilation pathway is the effector and by a positive mechanism which serves to couple the sulfate assimilation pathway to the cells potential for nitrogen assimilation. The rate of derepression of ATP sulfurylase is proportional to the initial nitrate concentration. Taken together with the effect of limiting sulfate concentrations on nitrate reductase these results demonstrate a reciprocal regulatory coupling between the nitrate and sulfate assimilation pathways.

The results obtained with cultured tobacco cells were corroborated using cell cultures of *Rosa* (Haller *et al.* 1986) and *Ipomoea* (Zink 1984) and tobacco and maize plants (Barney & Bush 1984; Saccomani *et al.* 1984). *Ipomoea* grown in cell suspension cultures form lower levels of ATP sulfurylase on slowly assimilated nitrogen sources such as proline, histidine or nitrate than cells grown on readily assimilated nitrogen sources such as ammonium, ammonium plus nitrate or casein hydrolysate. In tobacco plants the activity of ATP sulfurylase was derepressed when SO_4^{2-} was limited in the nutrient solution provided that there was a net accumulation of reduced nitrogen in the tissue (Barney & Bush 1985). In maize plants ATP sulfurylase activity was depressed by N deficiency (Saccomani *et al.* 1984). Even though all these results are consistent with a regulatory role of ATP sulfurylase in assimilatory sulfate reduction, it should be mentioned here that Bergmann *et al.* (1980), realized that this enzyme activity did not run parallel with the *in vivo* rates of sulfate assimilation. This was taken as evidence that ATP sulfurylase did not catalyze the main regulatory step of this reduction pathway (Bergmann *et al.* 1980). It had already been discussed by Ellis, 1969, that in higher plants APS, the product of the ATP sulfurylase reaction, is not only the substrate for sulfate reduction, leading to the formation of cysteine, but functioned also as an intermediate of sulfolipid biosynthesis. He concluded, therefore, that in higher plants regulatory mechanisms involving ATP sulfurylase had to integrate this bifunctional role of APS. Consistent with this conclusion cysteine failed to decrease ATP sulfurylase activity in *Lemna minor* (Ellis 1969), because such a repression would lead to an inhibition of sulfolipid synthesis which does not seem to proceed via cysteine (Kleppinger-Sparace & Mudd 1990) in higher plants.

Results from various systems indicate that the second enzyme of assimilatory sulfate reduction, APS sulfotransferase, is more susceptible to regulatory signals than ATP sulfurylase (Brunold 1990). Using cell suspension cultures of *Rosa* (Paul's Scarlet rose) Haller *et al.*, 1986, showed that the omission of a nitrogen source did not affect ATP sulfurylase during the experimental period of 24 hours, whereas a prominent decrease in APS sulfotransferase activity was detected after this time period. Nitrate reductase activity was already decreased after 6 hours.

Addition of NH_4Cl instead of NO_3^- to *Rosa* cell cultures induced an increase in APS sulfotransferase whereas no effect on ATP sulfurylase activity was detected under these conditions (Haller *et al.* 1986). This increase in APS sulfotransferase activity can be explained by the fact that *Rosa* cells produced more proteins (Mohanty & Fletcher 1980) and that the flow through assimilatory sulfate reduction was therefore increased to provide the necessary sulfur amino acids for the increased protein synthesis. Corresponding results were obtained in experiments using *Lemna minor* L. (Brunold & Suter 1984). After 24 hours on a medium without a nitrogen source the specific activity of APS sulfotransferase and nitrate reductase was less than 30% of the original level, whereas ATP sulfurylase was still at about 80%. Addition of 10 mM NH_4Cl to the culture medium induced a rapid increase in APS sulfotransferase activity followed by a slow decrease. Nitrate reductase activity was lost under these conditions with a half life of 14 hours, whereas the protein content increased. There was no detectable change in ATP sulfurylase activity during the experimental period of 72 hours. Labeling experiments using $^{35}SO_4^{2-}$ showed that more radioactive sulfur was incorporated into the proteins in the presence of ammonium instead of NO_3^- as nitrogen source, indicating an increased flux of sulfur through the sulfate assimilation pathway based on an increased level of APS sulfotransferase activity.

O-acetyl-L-serine

O-acetyl-L-serine is synthesized from L-serine and acetyl-CoA catalyzed by serine acetyltransferase (EC 2.3.1.30). It is well established that in microorganisms *O*-acetyl-L-serine is the substrate for the formation of cysteine according to:

$$O\text{-acetyl-L-serine} + S^{2-} \rightarrow \text{cysteine} + \text{acetate.}$$

This reaction is catalyzed by *O*-acetyl-L-serine sulfhydrylase (EC 4.2.99.8).

The central role of *O*-acetyl-L-serine and its derivative *N*-acetyl-L-serine in coordinating assimilatory sulfate and nitrate reduction has been studied in detail with *Salmonella typhimurium* (Ostrowski & Kredich 1989; Kredich, this volume). Several lines of evidence indicate that in higher plants *O*-acetyl-L-serine is also the substrate for cysteine formation and that the strategic position of this compound seems also to be exploited for coordinating assimilatory sulfate and nitrate reduction.

Both L-serine acetyltransferase and *O*-acetyl-L-serine sulfhydrylase have been detected in plants (Giovanelli & Mudd 1968; Smith & Thompson 1971; Smith 1972; Ascaño & Nicholas 1977; Fankhauser & Brunold 1978a; Brunold & Suter 1982; Nakamura *et al.* 1988; Lunn *et al.* 1990) and *O*-acetyl-L-serine has been demonstrated in cultured tobacco cells (Smith 1980). In extracts from bacteria a 50% inhibition of serine acetyltransferase activity at a cysteine concentration of 1.1 μM was detected (Kredich 1971). The enzyme from *Phaseolus vulgaris* seems to be less sensitive to L-cysteine since at 1 mM there was only a 65% inhibition (Smith & Thompson 1971). At this concentration, there was an almost complete inhibition of serine acetyltransferase from spinach chloroplasts (Fig. 2), whereas 50 μM cysteine inhibited the enzyme by more than 40% (Brunold & Suter 1982). Since cystine had no effect *in vitro* on serine acetyltransferase activity (Brunold & Suter 1982) it seems essential that oxidation of cysteine is prevented in studying inhibition of the enzyme by this amino acid.

Serine acetyltransferase was detected in mitochondrial fractions obtained from leaves of *Phaseolus vulgaris* by differential centrifugation (Smith 1972) and in an uncharacterized particulate fraction of extracts of wheat leaves (Ascaño & Nicholas 1977). Intact chloroplasts isolated from spinach leaves by density gradient centrifu-

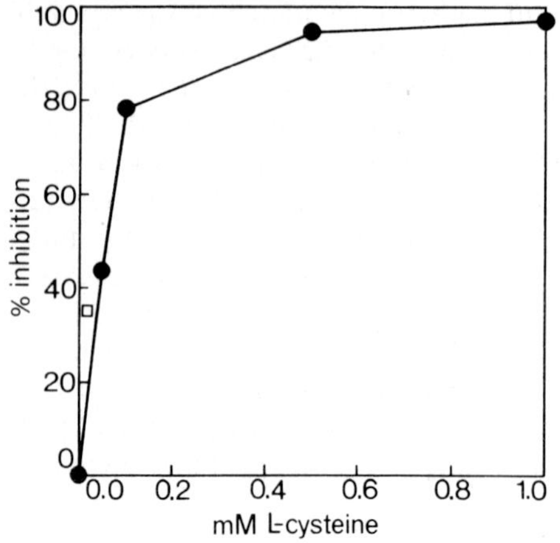

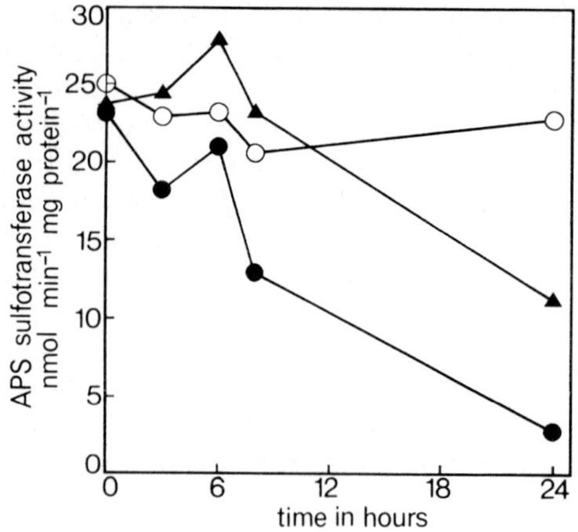

Fig. 2. Inhibition of *in vitro* L-serine acetyltransferase activity from spinach chloroplasts by L-cysteine added to the assay system at 0.05 to 1 mM. The enzyme activity obtained in an assay system without L-cysteine was 66.9 pkat mg^{-1} protein. (According to Brunold and Suter 1982).

Fig. 3. Effect of 0.5 mM *O*-acetyl-L-serine on adenosine 5'-phosphosulfate (APS) sulfotransferase activity of *Lemna minor* L. transferred to the dark (▲). Control plants were cultivated without *O*-acetyl-*L*-serine either in the dark (●) or in the light (○) (according to Neuenschwander *et al.* 1991).

gation contained about 35% of total leaf serine acetyltransferase activity (Brunold & Suter 1982) and of *O*-acetyl-L-serine sulfhydrylase (Fankhauser & Brunold 1978a). More recent experiments using spinach leaves (Lunn *et al.* 1990) make it clear that *O*-acetyl-L-serine sulfhydrylase is located primarily in the chloroplasts and the cytosol, but is also present in mitochondria. Consistent with this result, Rolland *et al.*, 1992, demonstrated this enzyme activity in the proplastids, the mitochondria and the cytosol of cauliflower inflorescence, indicating the possibility of cysteine biosynthesis in each subcellular compartment where synthesis of proteins occurs.

The function of the extrachloroplastic portion of serine acetyltransferase and *O*-acetyl-L-serine sulfhydrylase is not clear. It was suggested that plant cells may be unable to transport cysteine between the different cell compartments, so that the cysteine required for protein synthesis must be formed *in situ* (Lunn *et al.* 1990). It seems interesting in this connection that under certain conditions, plants emit H_2S (Wilson *et al.* 1978; Rennenberg 1983; Rennenberg 1984; Rennenberg 1989; Rennenberg *et al.* 1990). The H_2S emission could mean that in the chloroplasts the flux of sulfur through the pathway of assimilatory sulfate reduction to sulfide exceeds the formation of the acceptor *O*-acetyl-L-serine. Indeed, chloroplast *O*-phospho-L-serine aminotransferase (EC 2.6.1.5), which catalyses one step in the synthesis of L-serine from 3-PGA, has a rather low *in vitro* activity (Larsson & Albertsson 1979). It is tempting to speculate that the cytoplasmic serine acetyltransferase and *O*-acetyl-L-serine sulfhydrylase could be involved in incorporating at least a portion of the H_2S escaping from the chloroplast into *O*-acetyl-L-serine formed outside the chloroplast.

The possible limiting role of *O*-acetyl-L-serine in assimilatory sulfate reduction was studied in detail using pumpkin leaf discs (Rennenberg 1983), which reduce excess sulfur in the presence of 25 mM sulfate and emitted H_2S into the atmosphere.

Feeding of *O*-acetyl-L-serine or its metabolic precursor *S*-acetyl-CoA and CoA to leaf discs enhanced the incorporation of $^{35}SO_4^{2-}$ into reduced sulfur compounds, mainly into cysteine, at the cost of lowered H_2S emission. The uptake and reduction of sulfate were not affected by these treatments. β-fluoropyruvate, an inhibitor of the generation of *S*-acetyl-CoA via pyruvate dehydrogenase, stimulated H_2S emission in response to sulfate. This stimulation was overcompensated by addition of *O*-acetyl-L-serine, *S*-acetyl-Co A or CoA. Since the feeding of *S*-acetyl-CoA enhanced cysteine synthesis and reduced H_2S emission it was not the activity of serine acetyltransferase but rather the availability of *S*-acetyl-CoA that was limiting *O*-acetyl-L-serine synthesis and thus incorporation of S^{2-} into cysteine in this artificial system, in which the control function of the roots in sulfate uptake had been eliminated. A limiting role of *O*-acetyl-L-serine in assimilatory sulfate reduction can also be deduced, however, from physiological measurements in which the emission of volatile sulfur compounds from spruce trees was detected (Rennenberg *et al.* 1990). H_2S was the predominant reduced sulfur compound continuously emitted from the branches of spruce with high rates during the day and low rates in the night, indicating that also under natural conditions the formation of the acceptor *O*-acetyl-L-serine and of sulfur reduced to the thiol level may not be perfectly coordinated.

The restricted availability of *O*-acetyl-L-serine for cysteine formation is also evident from experiments in which detached spinach leaves were fumigated with H_2S (Buwalda *et al.* 1992). Fumigation with 0.75 $\mu l\ l^{-1}$ H_2S and parallel feeding with 10 mM *O*-acetyl-L-serine in tap water resulted in a fourfold increase of the cysteine content of the leaves compared to controls on tap water alone.

The dependence of sulfate assimilation on light is well established (Trebst & Schmidt 1969; Schürmann & Brunold 1980; Passera *et al.* 1989; Brunold 1990). With cucumber leaf discs, however, it has been shown that under dark conditions in the presence of 25 mM $^{35}SO_4^{2-}$ ^{35}S is incorporated into organic compounds at 60% of the rate in the light (Sekiya *et al.* 1982). When cultivated on 50 mM sulfate, leaf discs of spinach produce glutathione in the dark at a rate which is 59% of that in the light (De Kok *et al.* 1985). These findings taken together seem in contrast to the idea of a strict control of sulfate assimilation by the light reaction of photosynthesis (Schwenn 1989) and show that sulfate assimilation in green tissues can also proceed in the dark, albeit at reduced rate and at very high sulfate concentrations. This was studied in detail using *Lemna minor* at normal sulfate concentrations (Neuenschwander *et al.* 1991). When these plants were transferred to the dark there was a parallel decrease in APS sulfotransferase and nitrate reductase activity (Neuenschwander *et al.* 1991). The relatively slow decrease of APS sulfotransferase (Fig. 3) suggests that regulation by the thioredoxin system of the chloroplasts was not involved (Schürmann & Kobayashi 1984). The dark induced decrease in APS sulfotransferase activity of *Lemna minor* could be partly prevented by the addition of *O*-acetyl-L-serine to the culture medium (Fig. 3). Transfer to the dark did not affect the activity of ATP sulfurylase during the first 24 hours in darkness indicating again that this enzyme is less susceptible to regulatory signals than APS sulfotransferase. Serine and a combination of L-asparagine and L-glutamine also slowed down the dark induced decrease of APS sulfotransferase activity, a significant increase in the thiol content, however, was only detected in plants treated with *O*-acetyl-L-serine. Addition of *N*-acetyl-L-serine had no effect. In view of the importance of this compound in the regulation of the *cys JIH* promoters of *Salmonella typhimurium* and *Escherichia coli* (Ostrowski & Kredich 1989; Kredich, this volume), this finding is astonishing. It may point to a different mechanism of regulation, but also, of course, to problems in the uptake of *N*-acetyl-L-serine by *Lemna*. In plants transferred to the dark and cultivated in the presence of $^{35}SO_4^{2-}$ *O*-acetyl-L-serine caused a massive incorporation of radioactivity into cysteine and glutathione, demonstrating assimilatory sulfate reduction at high rates in the dark, which was dependent on *O*-acetyl-L-serine. When *O*-acetyl-L-serine was added to the nutrient solution of *Lemna minor* that had been precultivated in the dark for 24 hours and which contained low APS sulfotransferase activity, there was an increase of the extractable activity of the enzyme within 9 hours. A constant level of enzyme activity was reached and maintained up to 24 hours that was 50% of the controls kept in continuous light. A comparable level of APS sulfotransferase activity was also induced by L-serine and a combination of L-asparagine and L-glutamine, but the content of acid soluble thiols in the plants was only increased by *O*-acetyl-L-serine. These results lead to the conclusion that not only amino acids like serine, asparagine and glutamine, but also *O*-acetyl-L-serine are involved in regulating the extractable activity of APS sulfotransferase. This is substantiated by the fact that this enzyme activity was also increased when *O*-acetyl-L-serine was added to the medium of *Lemna minor* cultivated in light (Neuenschwander *et al.* 1991).

Concluding remarks

The general features of assimilatory nitrate reduction are well understood, whereas several open questions remain to be answered for assimilatory sulfate reduction. The most intriguing open questions are the physiological significance of the PAPS sulfotransferase pathway and of the reduction of free versus carrier-bound sulfite. The methods of molecular biology and the study of regulatory phenomena, including the coordination between sulfate and nitrate assimilation, will certainly lead to further progress, especially when the lines established during the work on assimilatory nitrate reduction of plants and on assimilatory sulfate reduction of bacteria and yeast are taken into account.

An area which may also turn out to be very rewarding in future research is the regulatory interaction between roots and shoots in coordinating sulfate and nitrate assimilation on a whole plant basis.

In view of the strategic role of *O*-acetyl-L-serine in the mutual regulation of assimilatory sulfate and nitrate reduction, the physiological knowledge about this compound needs further work, especially as far as its synthesis and concentration in the various cellular compartments and its effects in regulating enzyme levels are concerned.

Acknowledgements

I thank B. Brunold, R. Hintermann, M. Suter, and M. Zeder for their help in preparing this review, A. Schmidt for his comments and for kindly making available a preprint, and J.A. Schiff for his comments. The work cited from my own laboratory was supported by grants from the Swiss National Science Foundation, from the Körber Foundation, Hamburg, Germany, and from Ciby-Geigy AG, Basel.

References

Adams, C.A. & Rinne, R.W. 1969. Influence of age and sulfur metabolism on ATP sulfurylase activity in the soybean and survey of selected species. Plant Physiol. 44: 1241-1245.

Ascaño, A. & Nicholas, D.J.D. 1977. Purification and properties of *O*-acetyl-L-serine sulphydrylase from wheat leaves. Phytochemistry 16: 889-893.

Back, E., Burkhart, W., Moyer, M., Privalle, L. & Rothstein, S. 1988. Isolation of cDNA clones coding for spinach nitrite reductase: complete sequence and nitrate induction. Mol. Gen. Genet. 212: 20-26.

Barney, P.E. & Bush, L.P. 1985. Interaction of nitrate and sulfate reduction in tobacco. J. Plant Nutrition 8: 505-515.

Beevers, L. 1981. Nitrogen metabolism in plants. In: Bothe, H. & Trebst, A. (eds.), Biology of Inorganic Nitrogen and Sulfur. Springer, Berlin, pp. 15-29.

Beevers, L. & Hageman, R.H. 1980. Nitrate and nitrite reduction. In: Miflin B.J., (ed.), The Biochemistry of Plants. Academic Press, New York, Vol. 5, pp. 115-168.

Bergmann, L., Schwenn, J.D. & Urlaub, H. 1980. Adenosine triphosphate sulfurylase and *O*-acetylserine sulfhydrylase in photoheterotrophically and heterotrophically cultured tobacco cells. Z. Naturforsch. 35c: 952-957.

Brunold, C. 1990. Reduction of sulfate to sulfide. In: Rennenberg, H., Brunold, C., De Kok, L.J. & Stulen, I. (eds.), Sulfur nutrition and sulfur assimilation in higher plants. Academic Publishing bv, The Hague, The Netherlands, pp. 13-31.

Brunold, C. & Suter, M. 1982. Intracellular localization of serine acetyltransferase in spinach leaves. Planta 155. 321-217.

Brunold, C. & Suter, M. 1984. Regulation of sulfate assimilation by nitrogen nutrition in the duckweed *Lemna minor* L. Plant Physiol. 76: 579-583.

Brunold, C. & Suter, M. 1989. Localization of enzymes of assimilatory sulfate reduction in pea roots. Planta 179: 228-234.

Brunold, C., Suter, M. & Lavanchy, P. 1987. Effect of high and low sulfate concentrations on adenosine 5'-phosphosulfate sulfotransferase activity from *Lemna minor* L. Physiol. Plant. 70: 168-174.

Buwalda, F., De Kok, L.J. & Stulen, I. 1992. Flux of atmospheric H_2S to spinach leaves can be affected by the supply of *O*-acetylserine. Phyton 32: 13-16.

Calza, R., Huttner, E., Vincenth, M., Rouze, P., Galangau, F., Vaucheret, H., Cherel. I., Meyer, C., Kronenberger, J. & Caboche, M. 1987. Cloning of DNA fragments complementary to tobacco nitrate reductase mRNA and encoding epitopes common to the nitrate reductase from higher plants. Mol. Gen. Genet. 209: 552-562.

Campbell, W.H. & Kinghorn, J.R. 1990. Functional domains of assimilatory nitrate reductases and nitrite reductases. TIBS 15: 315-319.

Cheng, C.L., Dewdney, J., Kleinhofs, A. & Goodman, H.M. 1986. Cloning and nitrate induction of nitrate reductase mRNA. Proc. Natl. Acad. Sci. USA 83: 6825-6828.

Coleman, R.G. 1957. The effect of sulphur deficiency on the free amino acids of some plants. Aust. J. Biol. Sci. 10: 50-56.

Crafts-Brander, S.J. & Harper, J.E. 1982. Nitrate reduction by roots of soja bean (*Glycine max*. L. merr.) seedlings. Plant Physiol. 69: 1298-1303.

Cram, W.J. 1990. Uptake and transport of sulfate. In: Rennenberg, H., Brunold, C., De Kok, L.J. & Stulen, I. (eds.), Sulfur Nutrition and Sulfur assimilation in Higher Plants. Academic Publishing bv, The Hague, The Netherlands, pp. 3-11.

Crawford, N.M., Campbell, W.H. & Davis, R.W. 1986. Nitrate reductase from squash: cDNA cloning and nitrate regulation. Proc. Natl. Acad. Sci. USA 83: 8073-8076.

De Kok, L.J., Bosma, W., Maas, F.M. & Kuiper P.J.C. 1985. The effect of short-term H_2S fumigation on water-soluble sulphydryl and glutathione levels in spinach. Plant, Cell Environ. 8: 189-194.

Dijkshoorn, W. & van Wijk, A.L. 1967. The sulfur requirements of plants as evidenced by the sulfur-nitrogen ratio in the organic matter. A review of published data. Plant Soil 26: 129-157.

Dreyfuss, J. & Monty, K.J. 1963. The biochemical characterization of cysteine-requiring mutants of *Salmonella typhimurium*. J. Biol. Chem. 238: 1019-1024.

Duke, S.H. & Duke, S.O. 1984. Light control of extractable nitrate reductase activity in higher plants. Physiol. Plant. 62: 485-493.

Ellis, R.J. 1969. Sulphate activation in higher plants. Planta 88: 34-42.

Emes, M.J. & Fowler, M.W. 1979. The intracellular location of the enzymes of nitrate assimilation in the apices of seedlings pea roots. Planta 144: 249-253.

Ernst, W.H.O. 1990. Ecological aspects of sulfur metabolism. In: Rennenberg, H., Brunold, C., De Kok, L.J. & Stulen, I. (eds.), Sulfur nutrition and sulfur assimilation in higher plants, Academic Publishing bv, The Hague, The Netherlands, pp. 131-144.

Fankhauser, H. & Brunold, C. 1978a. Localization of *O*-acetyl-L-serine sulfhydrylase in *Spinacia oleracea*. Plant Sci. Lett. 44: 185-192.

Fankhauser, H. & Brunold, C. 1978b. Localization of adenosine 5'-phosphosulfate sulfotransferase in spinach leaves. Planta 143: 285-289.

Farago, S. & Brunold, C. 1990. Regulation of assimilatory sulfate reduction by herbicide safeners in *Zea mays* L. Plant Physiol. 94: 1808-1812.

Friedrich, J.W. & Schrader, L.E. 1978. Sulfur deprivation and nitrogen metabolism in maize seedlings. Plant Physiol. 61: 900-903.

Gerwick, B.C., Ku, S.B. & Black, C.C. 1980. Initiation of sulfate activation: a variation in C_4 photosynthesis plants. Science 209: 513-515.

Giovanelli, J. 1990. Regulatory aspects of cysteine and methionine synthesis. In: Rennenberg, H., Brunold, C., De Kok, L.J. & Stulen, I. (eds.), Sulfur Nutrition and Sulfur Assimilation in Higher Plants. Academic Publishing bv, The Hague, The Netherlands, pp. 33-48.

Giovanelli, J. & Mudd, S.H. 1968. Sulfurylation of *O*-acetyl-homoserine and *O*-acetylserine by two enzyme fractions from spinach. Biochem. Biophys. Res. Commun. 31: 275-280.

Grill, E., Winnacker, E.L. & Zenk, M.H. 1985. Phytochelatins: The principal heavy-metal complexing peptides of higher plants. Science 230: 674-676.

Grill, E. Löffler, S., Winnacker, E.L. & Zenk, M.H. 1989. Phytochelatins, the heavy-metal-binding peptides of plants, are synthesized from glutathione by a specific γ-glutamyl-cysteine dipeptidyl transpeptidase. Proc. Natl. Acad. Sci. USA 86: 6838-6842.

Gupta, S.C. & Beevers, L. 1984. Synthesis and degradation of nitrite reductase in pea leaves. Plant Physiol. 75: 251-252.

Haller, E., Suter, M. & Brunold, C. 1986. Regulation of ATP-sulfurylase and adenosine 5'-phosphosulfate sulfotransferase by the sulfur and the nitrogen source in heterotrophic cell suspension cultures of Paul's Scarlet rose. J. Plant Physiol. 125: 275-283.

Kleppinger-Sparace, K.F. & Mudd, J.B. 1990. Biosynthesis of sulfoquinovosyldiacylglycerol in higher plants – Use of adenosine 5'-phosphosulfate and adenosine 3'-phosphate 5'-phosphosulfate as precursors. Plant Physiol. 93: 256-263.

Kredich, N.M. 1971. Regulation of L-cysteine biosynthesis in *Salmonella typhimurium*. I. Effects of growth on varying sulfur sources and *O*-acetyl-L-serine on gene expression. J. Biol. Chem. 246: 3474-3484.

Krüeger, R.J. & Siegel, L.M. 1982. Evidence for siroheme-Fe_4S_4 interaction in spinach ferredoxin-sulfite reductase. Biochemistry 21: 2905-2909.

Lahners, K. Kramer, V., Back, E., Privalle, L. & Rothstein, S. 1988. Molecular cloning of complementary DNA encoding maize nitrite reductase. Molecular analysis and nitrate induction. Plant Physiol. 88: 741-746.

Larsson, C. & Albertsson, E. 1979. Enzymes related to serine synthesis in spinach chloroplasts. Physiol. Plant. 45: 7-10.

Li, J. & Schiff A. 1991. Purification and properties of adenosine 5'-phosphosulphate sulphotransferase from *Euglena*. Biochem. J. 274: 355-360.

Li, J. & Schiff, J.A. 1992. Adenosine 5'-phosphosulfate sulfotransferase from *Euglena*: Enzyme-bound intermediates. Plant Cell Physiol. 33: 63-72.

Lunn, J.E., Droux, M., Martin, J. & Douce, R. 1990. Localization of ATP sulfurylase and *O*-acetyl(thiol)lyase in spinach leaves. Plant Physiol. 94: 1345-1352.

Macnicol, P.K. 1983. Developmental changes in free amino acid pools and total protein amino acids of pea cotyledons (*Pisum sativum* L.). Plant Physiol. 72: 492-497.

Macnicol, P.K. & Randall, P.J. 1987. Changes in the levels of major sulfur metabolites and free amino

acids in pea cotyledons recovering from sulfur deficiency. Plant Physiol. 83: 354-359.
Mayer, A.M. 1967. Subcellular location of sulphite reductase in plant tissues. Plant Physiol. 42: 324-326.
Miflin, B.J. & Lea, P.J. 1980. Ammonia assimilation. In: Miflin, B.J. (ed.), The Biochemistry of Plants. Academic Press, New York, Vol. 5, pp. 169-202.
Mohanty, B. & Fletcher, J.S. 1980. Ammonium influence on nitrogen assimilating enzymes and protein accumulation in suspension cultures of Paul's Scarlet rose. Physiol. Plant. 48: 453-459.
Nakamura, K., Hayama, A., Masada, M., Fukushima, K. & Tamura, G. 1988. Purification and some properties of plant serine acetyltransferase. Plant Cell Physiol. 29: 689-693.
Neuenschwander, U., Suter, M. & Brunold, Ch. 1991. Regulation of sulfate assimilation by light and *O*-acetyl-L-serine in *Lemna minor* L. Plant Physiol. 97: 253-258.
Nussbaum, S., Schmutz, D. & Brunold, C. 1988. Regulation of assimilatory sulfate reduction by cadmium in *Zea mays* L. Plant Physiol. 88: 1407-1410.
Oaks, A. 1992. A re-evaluation of nitrogen assimilation in roots. BioScience 42: 103-111.
Ostrowski, J. & Kredich, N.M. 1989. Molecular characterization of the *cysJIH* promoters of *Salmonella typhimurium* and *Escherichia coli*: regulation by *cysB* protein and *N*-acetyl-L-serine. J. Bacteriol. 171: 130-140.
Pal, U.R., Gossett, D.R., Sims, J.L. & Leggett, J.E. 1976. Molybdenum and sulfur nutrition effects on nitrate reduction in Burley tobacco. Can.J. Bot. 54: 2014-2022.
Passera, C., Ghisi, R. & Ferretti, M. 1989. Light-activation of ATP-sulphurylase in leaves and chloroplasts of *Zea mays*. Photosynthetica 23: 166-172.
Pate, J.S. 1965. Roots as organs of assimilation of sulfate. Science 149: 547-548.
Remmler, J.L. & Campbell, W.H. 1986. Regulation of corn leaf nitrate reductase. II. Synthesis and turnover of the enzyme's activity and protein. Plant Physiol. 80: 442-447.
Rennenberg, H. 1983. Role of *O*-acetylserine in hydrogen sulfide emission from pumpkin leaves in response to sulfate. Plant Physiol. 73: 560-565.
Rennenberg, H. 1984. The fate of excess sulfur in higher plants. Annu. Rev. Plant Physiol. 35: 121-153.
Rennenberg, H. 1989. Synthesis and emission of hydrogen sulfide by higher plants. In: Saltzman, E.S. & Cooper, W.J. (eds.), Biogenic sulfur in the environment. Amer. Chem. Soc., Washington DC, ACS Sym. Ser. 393, pp. 44-57.
Rennenberg, H., Huber, B., Schröder, P., Stahl, K., Haunold, W., Georgii, H.-W., Slovik, S. & Pfanz, H. 1990. Emission of volatile sulfur compounds from spruce trees. Plant Physiol. 92: 560-564.
Reuveny, Z. & Filner, P. 1977. Regulation of ATP sulfurylase in cultured tobacco cells. J. Biol. Chem. 252: 1858-1864.
Reuveny, Z., Dougall, D.K. & Trinity, P.M. 1980. Regulatory coupling of nitrate and sulfate assimilation pathways in cultured tobacco cells. Proc. Natl. Acad. Sci. USA 77: 6670-6672.
Riens, B. & Heldt, W. 1992. Decrease of nitrate reductase activity in spinach leaves during a light-dark transition. Plant Physiol. 98: 573-577.
Rolland, N., Droux M. & Douce, R. 1992. Subcellular distribution of *O*-acetylserine(thiol)lyase in cauliflower (*Brassica oleracea* L.) inflorescence. Plant Physiol. 98: 927-935.
Rüegsegger, A. & Brunold, C. 1992. Effect of cadmium on γ-glutamylcysteine synthesis in maize seedlings. Plant Physiol. 99: 428-433.
Rüegsegger, A., Schmutz, D. & Brunold, C. 1990. Regulation of glutathione synthesis by cadmium in *Pisum sativum* L. Plant Physiol. 93: 1579-1589.
Saccomani, M. & Ferrari, G. 1989. Sulfate influx and efflux in seedlings of maize genotypes of different productivity. Maydica 34: 171-177.
Saccomani, M., Cacco, G. & Ferrari, G. 1984. Effect of nitrogen and/or sulfur deprivation on the uptake and assimilation steps of nitrate and sulfate in maize seedlings. J. Plant Nutrition 7: 1043-1057.
Sawhney, S.K. & Nicholas, D.J.D. 1975. Nitrite, hydroxylamine and sulphite reductases in wheat leaves. Phytochem. 14: 1499-1503.
Schiff, J.A. 1983. Reduction and other metabolic reactions of sulfate. In: Läuchli, A. & Bieleski, R. L. (eds.), Encyclopedia Plant Physiol., New Series 15A. Inorganic plant nutrition. Springer, Berlin, pp. 402-421.
Schmidt, A. 1972. On the mechanism of photosynthetic sulfate reduction. An APS-sulfotransferase from *Chlorella*. Arch. Microbiol. 84: 77-86.
Schmidt, A. 1973. Sulfate reduction in a cellfree system of *Chlorella*. The ferredoxin dependent reduction of a protein-bound intermediate by a thiosulfonate reductase. Arch. Microbiol. 93: 29-52.

Schmidt, A. 1975. Distribution of APS-sulfotransferase activity among higher plants. Plant Sci. Lett. 5: 407-415.

Schmidt, A. 1976a. The adenosin-5'-phosphosulfate sulfotransferase from spinach (*Spinacea oleracea* L.). Stabilization, partial purification and properties. Planta 130: 257-263.

Schmidt, A. 1976b. Development of adenosine-5'-phosphosulfate-sulfotransferase in sunflower (*Helianthus annuus* L.). Z. Pflanzenphysiol. 78: 164-168.

Schmidt, A. 1992. Open questions about sulfur metabolism in plants. Annu. Rev. Plant Physiol. Plant Mol. Biol. 43: 325-349.

Schürmann, P. & Brunold, C. 1980. Formation of cysteine from adenosine 5'-phosphosulfate (APS) in extracts from spinach chloroplasts. Z. Pflanzenphysiol. 100: 257-268.

Schürmann, P. & Kobayashi, Y. 1984. Regulation of chloroplast 1,6-bisphosphatase activity by the ferredoxin/thioredoxin system. In: Nijhoff, M., Junk, W. & Sybesma, C. (ed.), Advances in Photosynthesis Research, Publishers, The Hague, Vol. 3. pp. 629-632,

Schwenn, J.D. 1989. Sulphate assimilation in higher plants: A thioredoxin-dependent PAPS-reductase from spinach leaves. Z. Naturforsch. 44c: 504-508.

Schwenn, J.D. & Hennies, H.H. 1974. Enzymes and bound intermediates involved in photosynthetic sulfate reduction in spinach chloroplasts and *Chlorella*. In: Proc. 3rd inter. congress on photosynthesis, Rehovot, pp. 629-635.

Schwenn, J.D., Krone, F. & Husmann, K. 1988. Yeast PAPS-reductase: properties and requirements of the purified enzyme. Arch. Microbiol. 150: 313-319.

Sekiya, J., Schmidt, A., Rennenberg, H., Wilson, L.G. & Filner, P. 1982. Hydrogen sulfide emission by cucumber leaves in response to sulfate in light and dark. Phytochemistry 21: 2173-2178.

Smirnoff, N. & Steward, G.R. 1985. Nitrate assimilation and translocation by higher plants: Comparative physiology and ecological consequences. Physiol. Plant 64: 133-140.

Smith, I.K. 1972. Studies of L-cysteine biosynthetic enzymes in *Phaseolus vulgaris* L. Plant Physiol. 50: 477-479.

Smith, I.K. 1980. Regulation of sulfate assimilation in tobacco cells. Plant Physiol. 66: 877-883.

Smith, I.K. & Thompson, J.F. 1971. Purification and characterization of L-serine-transacetylase and O-acetyl-L-serine sulfhydrylase from kidney bean seedlings (*Phaseolus vulgaris*). Biochim. Biophys. Acta 227: 288-295.

Solomonson, L.P. & Barber, M.J. 1990. Assimilatory nitrate reductase: Functional properties and regulation. Annu. Rev. Plant Physiol. Plant Mol. Biol. 41: 225-253.

Somers, D.A., Kuo, T.M., Kleinhofs, A., Warner, R.L. & Oaks, A. 1983. Synthesis and degradation of barley nitrate reductase. Plant Physiol. 72: 949-952.

Suter, M., Lavanchy, P., von Arb, C. & Brunold, C. 1986. Regulation of sulfate assimilation by amino acids in *Lemna minor* L. Plant Sci. 44: 125-132.

Suzuki, A., Gadal, P. & Oaks A. 1981. Intracellular distribution of enzymes associated with nitrogen assimilation in roots. Planta 151: 457-461.

Thomas, D., Barbey, R. & Surdin-Kerjan, Y. 1990. Gene-enzyme relationship in the sulfate assimilation pathway of *Saccharomyces cerevisiae*. J. Biol. Chem. 265: 15518-15524.

Thompson, J.F., Morris, C.J. & Gering, R.K. 1960.The effect of mineral supply on the amino acid composition of plants. Qual. Plant Mater. Veg. 6: 261-275.

Tischner, R., Ward, M.R. & Huffaker, R.C. 1989. Evidence of a plasma membrane bound nitrate reductase involved in nitrate uptake of *Chlorella sorokiniana*. Planta 178: 19-24.

Trebst, A. & Schmidt, A. 1969. Photosynthetic sulfate and sulfite reduction by chloroplasts. In: Metzner, H. (ed.), Progr. Photosyn. Res. III, Tübingen, pp. 1510-1516.

Tsang, M.L.-S. & Schiff, J.A. 1975. Studies of sulfate utilization by algae. 14. Distribution of adenosine 5'-phosphosulfate (APS) and adenosine 3'-phosphate 5'-phosphosulfate (PAPS) sulfotransferase in assimilatory sulfate reducers. Plant Sci. Lett. 4: 301-307.

Tsang, M.L.-S. & Schiff, J.A. 1976. Studies of sulfate utilization by algae. 17. Reactions of adenosine 5'-phosphosulfate (APS) sulfotransferase from *Chlorella* and studies of model reactions which explain the diversity of side products with thiols. Plant Cell Physiol. 17: 1209-1220.

Von Arb, C. & Brunold, C. 1986. Enzymes of assimilatory sulfate reduction in leaves of *Pisum sativum*: activity changes during ontogeny and *in vivo* regulation by H_2S and cysteine. Physiol. Plant. 67: 81-86.

Ward, M.R., Tischner, R. & Huffaker, R.C. 1988. Inhibition of nitrate transport by antinitrate reductase IgG fragments and the identification of plasma membrane associated with nitrate reductase in roots of barley seedlings. Plant Physiol. 88: 1141-1145.

Wilson, L.G., Bressan, R.A. & Filner, P. 1978. Light-dependent emission of hydrogen sulfide from plants. Plant Physiol. 61: 184-189.
Wyss, H.-R. & Brunold, C. 1979. Regulation of adenosine 5'-phosphosulfate sulfotransferase activity by H_2S and cyst(e)ine in primary leaves of *Phaseolus vulgaris* L. Planta 147: 37-42.
Zink, M.W. 1984. Regulation of ATP-sulfurylase by various nitrogen and sulfur sources in cultured *Ipomoea* sp. Can. J. Bot. 62: 2107-2113.

WHOLE PLANT REGULATION OF SULFUR METABOLISM A THEORETICAL APPROACH AND COMPARISON WITH CURRENT IDEAS ON REGULATION OF NITROGEN METABOLISM

Ineke Stulen and Luit J. De Kok
Department of Plant Biology, University of Groningen, P.O. Box 14, 9750 AA Haren, The Netherlands

Introduction

Research on assimilatory sulfate reduction and metabolism has been focused on elucidating the pathways and characterizing the enzymes involved in the reduction of sulfate and incorporation of the reduced sulfur into organic sulfur compounds. Conclusions regarding the regulation of the pathways are for the greater part based on changes in extractable enzyme levels and modulation of enzyme activity *in vitro*. The observed changes in enzyme levels were a result of changes in environmental factors or occurred during development of the leaf. Modulation of enzyme activity *in vitro* was found upon addition of various compounds, as products or substrates, to the assay mixtures (Brunold 1990).

In this paper a summary is given of the current ideas on the regulation of sulfate uptake, assimilatory sulfate reduction and further metabolism. We also attempt to quantify the fluxes of various sulfur compounds in the plant, necessary to meet the demand for growth, using a model, and to calculate fluxes into reduced sulfur compounds which are then compared with data on the level and affinity of the enzymes involved. Where possible, a comparison is made with the current ideas on the regulatory control of nitrate uptake, reduction and further metabolism.

Throughout the paper the term **enzyme level** is used consistently when dealing with the amount of extractable enzyme, as measured with an *in vitro* assay. The term **enzyme activity** is used to describe: a) changes in the rate of product formation, brought about by addition of compounds during the *in vitro* assay ("modulation"); and b) to describe the **functioning** of the enzyme *in situ*.

Current ideas on the regulation of sulfate uptake, reduction and further metabolism

Uptake and transport of sulfate

Many experiments on the uptake of sulfate, and the influence of various sulfur compounds on the uptake process, have been performed with excised roots. Since uptake rates in excised roots can be less than those of intact roots (Smith & Cheema 1985), some caution is needed when extrapolating these findings to the level of the whole plant (Cram 1990).

It has been shown that both uptake and transport of sulfate in the intact plant are

Sulfur Nutrition and Assimilation in Higher Plants
edited by L.J. De Kok et al., pp. 77–91

affected by the nutritional sulfur status; roots of plants grown at a high sulfur level generally had a less efficient sulfate uptake than those grown at a low sulfur level (Holobradà 1977; Clarkson *et al.* 1983). Barley plants showed a ten-fold increase in net sulfate uptake (measured with ^{35}S-sulfate) after sulfur deprivation for one to five days, compared to plants continuously supplied with sulfur. At present it is thought that sulfate uptake may be regulated by negative feedback from sulfate itself (Cram 1990) and/or by repression of the uptake system by reduced sulfur compounds such as glutathione (Rennenberg *et al.* 1988; Herschbach & Rennenberg 1991). However, this repression appears to be rather unspecific, since for instance various amino acids at micromolar concentrations appear to have the same effect as various reduced sulfur compounds (Gunz *et al.* 1992). It should be noted that these experiments were performed by supplementing the various compounds to isolated cells or excised roots.

Sulfate reduction and assimilation

The reactions involved in assimilatory sulfate reduction of higher plants have been dealt with extensively by Brunold (1990), and the assimilation of the reduced sulfur into amino acids has been reviewed by Giovanelli (1990). The reactions of the main assimilatory sulfate reduction pathway and the synthesis of cysteine can be summarized as follows: 1) an activation step of sulfate, catalyzed by ATP sulfurylase with formation of adenosine 5'-phosphosulfate (APS); 2) APS sulfotransferase, leading to the formation of sulfite; 3) sulfite reductase, leading to the formation of sulfide; 4) incorporation of sulfide into cysteine, catalyzed by cysteine synthase.

In the next sections the current ideas on the regulatory control of these enzymes have been summarized. Attention was paid to the occurrence of the following regulatory mechanisms: 1) changes in enzyme level, 2) modulation of *in vitro* activity, and 3) substrate availability at the site of the enzyme.

ATP sulfurylase. The level of ATP sulfurylase increased with chloroplast development and development of the leaf (Schmutz & Brunold 1982; Von Arb & Brunold 1986). From the fact that the level of ATP sulfurylase was not decreased significantly by SO_2, in contrast to APS sulfotransferase (Brunold *et al.* 1983; Tschanz *et al.* 1986), the conclusion was drawn that the level of ATP sulfurylase is less subjected to regulatory control than the level of APS sulfotransferase (Brunold 1990). However, data on changes in the level of ATP sulfurylase should be interpreted with care, especially when drawing conclusions on the regulation of the enzyme *in situ*, since the *in vitro* activity of this enzyme is often measured in the reverse direction. From data on the effect of the addition of the product and other compounds (APS, 5'-AMP and 5'-ADP) (Tweedie & Segel 1971; Shaw & Anderson 1972; Schwenn & Depka 1977) to the *in vitro* assay medium, the conclusion was drawn that feed-back inhibition by APS of ATP sulfurylase might be a point of control. That the availability of the substrate, *viz.* the sulfate concentration at the ATP sulfurylase site may play a significant role in the *in situ* rate of sulfur assimilation was concluded by De Kok & Kuiper (1986) and De Kok (1989, 1990).

APS sulfotransferase. The level of APS sulfotransferase changed with age and environmental conditions; it increased during development of the leaf (Schmutz & Brunold 1982; Von Arb & Brunold 1985) and under conditions of sulfur deprivation (Barney & Bush 1985), while it decreased upon exposure to atmospheric SO_2 and H_2S (Brunold *et al.* 1983; Tschanz *et al.* 1985). These findings, together with the fact that the level of APS sulfotransferase is the lowest of the enzymes of assimilatory sulfate reduction, led to the conclusion that APS sulfotransferase is a principal site at which assimilatory sulfate reduction is regulated (Brunold 1990; Farago & Brunold 1990).

Sulfite reductase. The level of sulfite reductase also increased during development of the leaf (Schmutz & Brunold 1982; Von Arb & Brunold 1986). By addition of 18 μM sulfide to the *in vitro* assay medium, sulfite reductase activity was reduced by 50% (Von Arb & Brunold 1985). Brunold (1990) speculated that this mechanism might be of importance in situations where the acceptor for sulfide, *O*-acetyl-L-serine, is not available, thus preventing the accumulation of sulfide.

Cysteine synthase. Giovanelli (1990) reported that the changes in the level of cysteine synthase found as a result of changes in sulfur nutrition were conflicting. The level of cysteine synthase was not affected by sulfur nutrition in the experiments of Smith (1980), while Bergmann *et al.* (1980) found that sulfur starvation resulted in a ten-fold increase in the level of this enzyme. Based on the present knowledge it seems unlikely that regulation of cysteine biosynthesis might take place by changes in the level of cysteine synthase or inhibition of the activity of the enzyme *in situ*. The enzyme cysteine synthase was inhibited *in vitro* by cysteine and a number of other amino acids of the sulfur assimilation pathway, but the high concentrations required suggest that these inhibitions are not of physiological significance (Giovanelli *et al.* 1980). Giovanelli (1990) speculated about the possibility that cysteine biosynthesis might be regulated by the availability of *O*-acetyl-L-serine, in the same way as in bacteria. Brunold (this volume) postulated that *O*-acetyl-L-serine might even play a strategic role in the regulation of assimilatory sulfate reduction. However, its significance as a regulating factor is still unclear. For instance, when excess sulfur is given to foliar tissue as sulfate, SO_2 or H_2S, the plant is able to accumulate a considerable amount of cysteine and its metabolites, even without supplemental *O*-acetyl-L-serine (De Kok 1989, 1990; Buwalda *et al.* 1992; De Kok & Stulen, this volume). These findings indicate that even under conditions of an excessive sulfur supply, apparently sufficient *O*-acetyl-L-serine is available for additional cysteine synthesis.

There is circumstantial evidence that cysteine desulfhydrase in its reverse reaction might be involved in the synthesis of cysteine, at least in case of the assimilation of atmospheric H_2S (Schütz *et al.* 1991), but the significance of this reaction in the intact plant needs further investigation.

Conclusions. The current ideas on the site of regulatory control of the assimilatory sulfate reduction pathway can be summarized as follows. The pathway may be regulated in at least four ways: 1) by modulation of the *activity* of ATP sulfurylase, 2) by the *availability* of sulfate *in situ*, at the site of ATP sulfurylase, 3) by changes in the *level* of APS sulfotransferase and 4) by the availability of *O*-acetyl-L-serine for cysteine synthase.

Table 1. Estimated contents (in μmol g fresh weight^{-1}) and calculated fluxes (in nmol g fresh weight^{-1} h^{-1}) of sulfur and nitrogen compounds in a model spinach plant with a RGR of 0.2 g g^{-1} day^{-1} and a shoot/root ratio of 3.4. Concentration values are from the following references: [a]De Kok *et al.* (1985); [b]Kleppinger-Sparace *et al.* (1990); [c]Maas *et al.* (1985); [d]Steingröver (1986); [e]Steingröver *et al.* (1986a); [f]Steingröver *et al.* (1986b); [g]Van Dijk *et al.* (1986) and unpublished data. Protein N was calculated from data on total reduced N concentration, minus ammonium and amino N fraction. The value for protein S was derived by assuming an N/S ratio of 30 (Dijkshoorn & Van Wijk 1967).

Compound	Concentration		Flux		Reference
	Shoot	Root	Shoot	Plant	
sulfate	4	2.5	31	33	a,c
sulfolipid S	1	0.3	7	8	b
amino S	0.3	0.2	2	3	g
protein S	9	4.5	67	76	d,g
nitrate	50	30		380	e,f
ammonium N	0.8	0.5		6	g
amino N	6	4		46	g
protein N	270	135		1990	d,g

Partitioning of sulfur and nitrogen compounds within the plant – calculation of fluxes

Sulfur and nitrogen fluxes in relation to growth

In the absence of a complete description of the distribution of sulfur compounds in a given plant, we made a model plant to explore the relationship between enzyme levels and the sulfur fluxes needed to maintain steady-state conditions during growth. Experiments from our laboratory with spinach plants provided much of the input data (Table 1) but information on sulfolipids was taken from Kleppinger-Sparace *et al.* (1990). Using the measured plant contents sulfur fluxes (in nmol g fresh weight^{-1} h^{-1}) were calculated for a relative growth rate of 0.2 g g^{-1} day^{-1}. The partitioning of the various sulfur compounds over the plant were calculated with a shoot/root (S/R) ratio of 3.4 and are given in Fig. 1. For the sulfur compounds the fluxes were calculated on a shoot as well as a plant basis, since the major part of sulfur reduction and metabolism is still considered to take place in the leaf (Evans 1974; Brunold 1990). Similar calculations were carried out for the various nitrogen compounds, with the exception that the fluxes were calculated only on a plant basis (Table 1; Fig. 1): the extensive cycling of nitrogen compounds between shoot and root makes it hard to establish the part that shoot and root each play in the reduction and assimilation of nitrate (Simpson *et al.* 1982; Simpson 1986; Cooper & Clarkson 1989). Table 1 shows that the calculated fluxes vary considerably between the various sulfur and nitrogen compounds. For both sulfate and nitrate, fluxes into proteins were highest.

Significance of sulfur emission

Even though plants may form a sink for atmospheric sulfur gases, which may be assimilated into cysteine and its metabolites (De Kok 1989, 1990; De Kok & Stulen, this

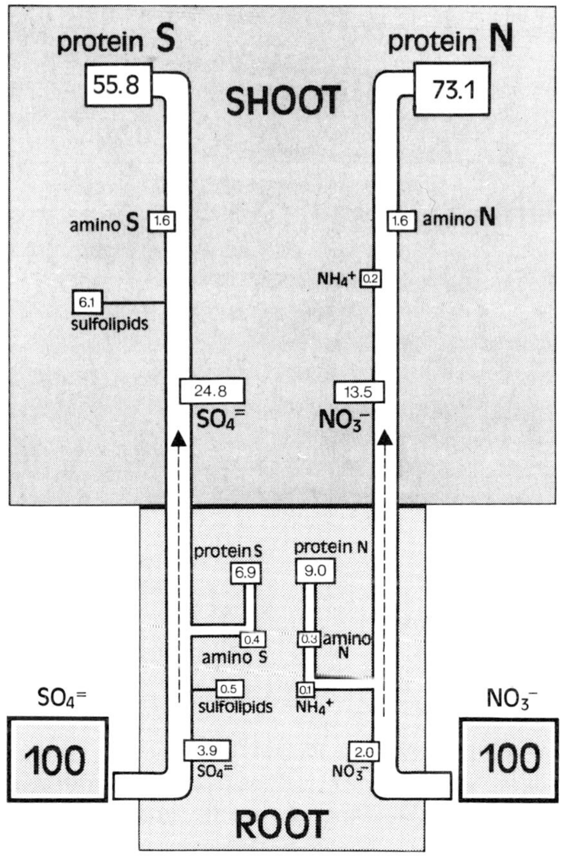

Fig. 1. Distribution of various sulfur and nitrogen compounds in a model spinach plant as % of total sulfur and nitrogen content. The values were calculated from the contents given in Table 1 and a S/R ratio of 3.4.

volume), they also may emit gaseous sulfur, mainly as H_2S (Schröder, this volume). This emission may only be substantial upon exposure of plants to non-realistic high levels of sulfur *e.g.* high levels of SO_2 (Ernst 1990). At normal sulfur supply the H_2S emission rate is rather low. For instance, the ratio between carbon and sulfur assimilation in plants is in the order of 6×10^{-3} (Cram 1990), whereas *e.g.* for spruce the ratio of photosynthetic CO_2 fixation/H_2S emission was about 3×10^{-7} (Schröder, this volume). Therefore it can be concluded that H_2S emission under a normal sulfur supply is very likely a negligible fraction of the total sulfur flux in plants.

Speculations on the regulation of sulfate reduction and further metabolism

Level of enzymes in relation to sulfur fluxes

In order to be able to link the calculated fluxes of the various sulfur compounds (Table 1), with the potential activity of the various enzymes in the plant, data on

Table 2. Level of extractable enzyme (in μmol g fresh weight^{-1} h^{-1}) of various enzymes, involved in sulfur and nitrogen metabolism, as measured with *in vitro* assays and *in vivo* assay, if necessary recalculated on a fresh weight basis with available data in the following references: [a]Barney & Bush (1985ab); [b]Bosma *et al.* (1991); [c]Brunold (1983); [d]Brunold *et al.* (1987); [e]Brunold & Schmidt (1978); [f]Brunold & Suter (1984); [g]De Kok *et al.* (1986b); [h]Fankhauser & Brunold (1978); [i]Hewitt *et al.* (1979); [j]Steingröver *et al.* (1986a); [k]Schmutz & Brunold (1982); [l]Suter *et al.* (1986); [m]Tschanz *et al.* (1986).

Enzyme	Level	Species	Reference
ATP sulfurylase	36 – 66	*Lemna minor*	c, d, e, l
	29 – 240	herbaceous species	a, k
	13 – 36	trees	b, c
APS sulfotransferase	4 – 6	*Lemna minor*	e, f, l
	1 – 2	*Spinacia oleracea*	h
	1 – 5	trees	a, d, m
sulfite reductase	5	trees	b
cysteine synthase	30	*Lemna minor*	e
	4800	herbaceous species	h
	1560	trees	b
nitrate reductase			
in vitro	3 – 4	*Lemna minor*	f, l
	16 – 29	*Spinacia oleracea*	g, j
	1 – 65	herbaceous species	a, g, i
in vivo	1 – 4	*Spinacia oleracea*	g, j
	4 – 29	herbaceous species	i

enzyme levels, expressed on a fresh weight basis, were summarized (Table 2). It should be noted that *in vitro* enzyme activities from the literature usually are assayed at a temperature of 30°C, while plants are usually grown at a much lower temperature.

By comparing the calculated sulfur fluxes in the intact plant (Table 1), with the data on enzyme levels (Table 2), the following conclusions can be drawn. To maintain the sulfur content in the protein fraction a reduced sulfur flux of 76 nmol g fresh weight^{-1} h^{-1} is required. The level of ATP sulfurylase (which *in vitro* is often measured in the reversed way) is much higher, *viz.* 13,000-29,000 nmol g fresh weight^{-1} h^{-1} at the lowest for trees and herbaceous species, respectively. The level of APS sulfotransferase is lower, but still much higher, *viz.*, minimally 700-1300 nmol g fresh weight^{-1} h^{-1} for trees and spinach, respectively, than the calculated required flux. Therefore, the level of APS sulfotransferase can be reduced substantially for trees and spinach, respectively, before the level of this enzyme becomes rate-limiting for the assimilatory sulfur flux needed for growth. It should be noted that the calculation of fluxes was based on a plant growing with a RGR of 20%. In slower growing species the flux can be expected to be lower, and the difference between the flux needed and the potential enzyme activity will be relatively higher.

Affinity and intracellular localization of enzymes in relation to intracellular concentration of sulfur compounds

Affinity and intracellular localization of enzymes. The affinity of ATP sulfurylase for sulfate seems to be relatively low. K_m values in the range of 0.5-3.1 mM (the latter value for spinach) were found (Shaw & Anderson 1972; Anderson 1980). The affinity of APS sulfotransferase from spinach for APS apparently is much greater *e.g.* Schmidt (1976) found a K_m of 10 μM. Sulfite reductase from spinach leaves also has a high affinity for sulfite. A K_m value of 6 μM was found by Aketagawa & Tamura (1980). These enzymes are predominantly or even exclusively localized in the chloroplast (Brunold 1990; Brunold this volume).

Intracellular concentrations of sulfate and cysteine. The concentration of sulfate in the chloroplast, the predominant site of sulfate reduction in the plant (Anderson 1980) appears to be strongly regulated, both at excess and at limited sulfur supply to the roots (Schröppel-Meier & Kaiser 1988; Kaiser *et al.* 1989). When barley and spinach plants were grown hydroponically on 250 and 200 mM sulfate, respectively, this resulted in an up to ten-fold increase in sulfate concentration of the leaf cells, from 5-10 to 60-70 mM, while the sulfate concentration of barley chloroplasts was only doubled, from 4-5 to 10-12 mM, (Kaiser *et al.* 1989) and that of spinach chloroplasts was unaffected (Schröppel-Meier & Kaiser 1988). Kaiser *et al.* (1989) observed that excess sulfate in the leaf cells was actively transported into the vacuole, the major storage site of sulfate in the cell up to a concentration of 100 mM (Smith 1981; Thoiron *et al.* 1981; Kaiser *et al.* 1989). De Kok (1989) also concluded that the rate of sulfate transport into the vacuole may be significant in the regulation of the sulfate concentration at the site of its activation, the chloroplast, under normal conditions of sulfur nutrition. From the observed accumulation of sulfhydryl compounds in spinach leaf discs upon continuous exposure of leaf cells to high levels of sulfate, it was concluded that the sulfate concentration in the chloroplast may increase to such an extent that sulfate reduction exceeds the metabolic needs for reduced sulfur (De Kok 1989).

Giovanelli *et al.* (1980) concluded that in Lemna the concentration of cysteine is maintained at a very low concentration (10 μM). Some authors postulated that the cysteine pool in the plant is carefully regulated (Rennenberg 1982, 1984; Filner *et al.* 1984). Rennenberg (1984) proposed that glutathione functions as a storage and transport form of sulfur in plants, helping to maintain a constant concentration of cysteine. Experiments in which excess sulfur was supplied to the shoot as H_2S, however, showed that even after two weeks of exposure cysteine concentration was still elevated (De Kok *et al.* 1986a). Further experiments showed that the cysteine fraction of the water-soluble SH pool was strongly enriched when the synthesis of glutathione was inhibited by buthionine sulfoximine (De Kok & Kuiper 1986). Based on these experiments the conclusion was drawn that the intracellular cysteine pool is poorly regulated in the presence of H_2S (De Kok 1990).

Conclusions. With respect to these conflicting results and interpretations, at present it is not possible to draw a definite conclusion on the regulation of the sulfur reduction pathway, based on data of enzyme affinities and pool sizes of substrates. In view of the fact that the sulfate pool in the chloroplast might be well regulated, however, we conclude that ATP sulfurylase, which is located in the chloroplast (Fig. 2), seems to be the most likely candidate for regulatory control.

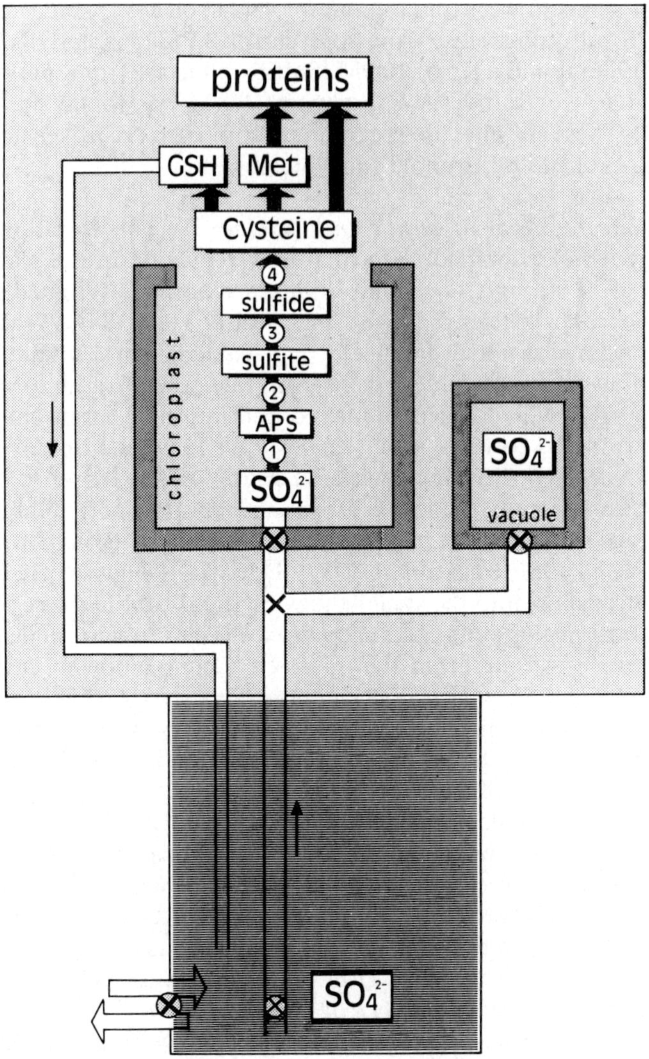

Fig. 2. Intracellular localization of enzymes of assimilatory sulfur reduction in relation to compartmentation of sulfur compounds. [1]ATP sulfurylase; [2]APS sulfotransferase; [3]sulfite reductase; [4]cysteine synthase.

Comparison with current ideas on the regulation of nitrate reduction and further metabolism

Nitrate reductase has been considered the rate-limiting step in the reduction pathway of nitrate to ammonium and the incorporation of ammonium into amino acids. This assumption is based on a) the fact that nitrate reductase is an adaptive enzyme and the first in the reduction pathway, b) the fact that the level of this enzyme is low compared to the abundant nitrite reductase level, and C) the rapid turnover rate of the enzyme (Beevers & Hageman 1969; Hewitt 1975; Hageman 1990). By comparing the level of nitrate reductase with the calculated flux of reduced nitrogen into protein in the model spinach plant (Table 1; Table 2) it can be seen that the level of the enzyme is excessive. The same conclusion can be drawn from experiments in which the incorporation of ^{15}N nitrate into the reduced nitrogen fraction was compared with the level of nitrate reductase measured *in vitro* (Stulen *et al.* 1973; Steingröver *et al.* 1986a). In some cases the *in vivo* assay for nitrate reductase provided a closer approximation of the accumulation of reduced nitrogen in the plant (Brunetti & Hageman 1976; Stulen *et al.* 1981). This conclusion is also valid for the model spinach plant (Table 1, 2). NADH availability *in vivo* might determine the rate of nitrate reduction. This conclusion was also reached from experiments in which the influence of atmospheric H_2S on nitrate reduction at low light intensity was investigated (De Kok *et al.* (1986b).

The group of Hageman carried out many experiments with wheat and maize, in order to investigate whether a correlation between nitrate reductase activity *in vitro* or *in vivo*, and protein yield could be established (summarized in Hageman 1990). For a given genotype, *in vitro* nitrate reductase activity of the whole plant, integrated over time, was closely associated with the actual amount of protein accumulated by the plant. However, the view that such measurements of *in vitro* nitrate reductase activity would provide an estimate of the input of reduced nitrogen to the plant was not valid among diverse genotypes. Likewise, no correlation was found between measurements of nitrate reductase activity *in vivo* and eventual crop protein-nitrogen yield. Based on all these experiments the conclusion was drawn that the *in situ* accumulation of reduced nitrogen could also be affected by the amount of nitrate at the induction and assimilation site of nitrate reductase, and the availability of reductant and metabolites for generation of reductant (Hageman 1990). Other authors also reached the conclusion that the amount of nitrate reduced in the plant is mainly regulated by the supply of nitrate to the plant and the light conditions, which affect the supply of reductant (Sawhney & Naik 1990; Sopory & Sharma 1990).

It has long been established that the flux of nitrate from the root to the leaves determines the level of nitrate reductase in the leaf (Shaner & Boyer 1976). The dependence of the induction of the nitrate reductase protein and the *in vitro* nitrate reductase activity was investigated by Oaks *et al.* (1988, 1990). From experiments with maize seedlings the conclusion was drawn that there is an inactive form of the nitrate reductase protein that is synthesized in the presence of low nitrate concentrations and that this inactive protein is activated at higher nitrate concentrations. They concluded that the appearance of the nitrate reductase protein and the nitrate reductase activity *in vitro* represent two phases in the induction process, and that these two phases do not occur parallel in this system. There is evidence that in barley both phases appear in parallel (Melzer *et al.* 1989; Oaks *et al.* 1990). The latter authors finally concluded that nitrate reductase protein and activity *in vitro* are indeed induced by nitrate but that there are important differences in the timing and the responses of these events to environmental cues. They stated that studies at the genetic and molecular levels will certainly help in understanding the details of these interactions. But these studies may be less successful in developing plants that use nitrate more efficiently, if the emphasis of research continues to be on the enzyme nitrate reductase and not on other aspects of nitrate assimilation (Oaks *et al.* 1990).

Summarizing we postulate that, although the level of nitrate reductase is controlled by the cytoplasmic nitrate pool, the level of the enzyme itself is not the main point of control of the reduction pathway. The availability of reductant for the functioning of the enzyme *in situ* may be important as well.

Affinity of nitrate reductase in relation to intracellular nitrate concentration. Nitrate reductase is generally considered a cytosolic enzyme, although it may be loosely associated with the chloroplast or microbody-like particles (Grant *et al.* 1970; Lips & Avissar 1972; Hewitt 1975; Sopory & Sharma 1990). For a partly purified enzyme from spinach leaves a K_m of 110 μM for nitrate was found (Hewitt *et al.* 1979). After using a more purified extract, a K_m of 40 μM for nitrate was found in squash (Campbell & Smarelli 1978; Campbell 1990). The newly developed specific nitrate electrode has made it possible to measure the cytoplasmic nitrate concentration (Zhen *et al.* 1991). For barley root cells a concentration of 5.4 and 3.2 mM was found, for epidermal and cortical cells, respectively (T. Miller, personal communication; Zhen *et al.* 1992). However, it should be noted that these experiments were done with excised roots, and that the authors themselves question the physiological significance of measurements made on excised roots. By assuming that the cytoplasmic nitrate concentration in the leaf is in the same range, we can conclude that the K_m of nitrate reductase for nitrate is well below the cytoplasmic nitrate concentration.

The catalytic efficiency of an enzyme can be expressed as the ratio of V_{max} to K_m, referred to as V/K. The V/K ratio for the squash nitrate reductase preparation was calculated to be 0.5×10^7 s^{-1} M^{-1} (Campbell 1990). The highest catalytic efficiency an enzyme can have is about 10^8 s^{-1} M^{-1}. Nitrate reductase, therefore, is among the most efficient catalysts known (Campbell 1990).

Regulation of overall nitrate reduction in the plant. Plants apparently maintain their protein concentration within a certain range. A value of 20% crude protein is representative for many plants (Hewitt *et al.* 1979). How is the overall reduction and incorporation of nitrate-nitrogen controlled, so that under non-limiting conditions of nutrient supply and light intensity nitrate reduction apparently reaches a maximum and no more reduced nitrogen is accumulated? Rodgers & Barneix (1988) showed with wheat cultivars, differing in nitrate uptake rate, that the major differences in uptake rate were related to the growth rate of the plants, which indicates that these differences were a consequence of larger plant demand, as postulated by Clarkson (1986) and Barneix (1990). But the regulation of protein synthesis in relation to nitrate availability and uptake seems to be rather complex. Nitrate accumulation in the vacuole can be considerable (Hewitt *et al.* 1979), which shows that the nitrate taken up and transported to the shoot can be channeled into the vacuole rather than being reduced. How is the partitioning of nitrate between reduction in the cytoplasm and storage in the vacuole regulated? Is regulation by the concentration of nitrate at the site of nitrate reductase involved or is the affinity of the transport mechanism of nitrate over the tonoplast the decisive factor?

Conclusions. Summarizing we can conclude that the knowledge on the regulatory control of nitrogen metabolism is more advanced than the knowledge on the regulatory control sulfur metabolism. The main similarity between the regulatory control of both pathways may lay in the fact that the nitrate concentration at the site of the enzyme is the most important factor for the functioning of nitrate reductase *in situ*, and the chloroplastic sulfate concentration for the functioning of ATP sulfurylase *in situ*. For both pathways the mechanism of the transport of nitrate and sulfate across the tonoplast and into the plastids clearly needs attention (Clarkson *et al.*, this volume).

Speculations on the regulation of sulfate uptake in relation to demand for growth

Plants apparently maintain their reduced sulfur protein concentration within a certain range, between 0.1 and 1.5% of dry weight (Duke & Reisenauer 1986). As for reduced nitrogen (see previous section) the following questions can be raised: 1) how is the overall reduction and incorporation of sulfate controlled, so that under non-limiting conditions of nutrient supply and light intensity sulfate reduction apparently reaches a maximum and no more reduced sulfur is accumulated, 2) how does the cell control the fluxes of sulfate into chloroplast and vacuole (marked with × in Fig. 2), and 3) how does the demand for growth regulate the uptake by the root, and what signals are involved – a reduced compound as glutathione or sulfate itself?

Concluding remarks

The general conclusion was drawn that data on enzyme levels and affinities alone are not sufficient to unravel the importance of various regulatory steps. Knowledge of the intracellular concentrations of metabolites, of the transport processes between cytoplasm and organelles and between the various plant organs is also required.

We concluded that the availability of sulfate in the chloroplast, at the site of the first enzyme, ATP sulfurylase, rather than the level of this enzyme, might be the most important point of control in the assimilatory sulfur reduction pathway. Likewise, the availability of nitrate for nitrate reductase, might control the assimilatory nitrogen pathway. In order to confirm these hypotheses more research, aimed at elucidating the exact intracellular compartmentation of the processes, in combination with quantitative measurements of fluxes and concentrations of the metabolites at the site of the enzymes and in the organelles, is needed. The main question then remains: how is the intracellular concentration of sulfate and nitrate regulated? In order to answer this question, more research on the mechanism of the transport of nitrate and sulfate into the vacuole and the chloroplast, and of the affinity of these processes mutually and in relation to the affinity of cytoplasmic processes as nitrate reduction, is needed. Newly developed techniques have made it possible to measure nitrate concentrations in individual cells (Zhen *et al.* 1991), but have not been developed yet for the measurement of sulfate concentrations. These techniques may also make it possible to investigate if, and to what extent a division of labour between individual cells takes place.

Based on the current knowledge of nitrogen demand in relation to nitrate uptake, the conclusion was drawn that the sulfur demand for growth probably regulates the sulfate uptake by the root, rather than *vice versa*. For regulation of both sulfate and nitrate uptake, sulfate or nitrate, or reduced compounds might act as regulators. In the regulation of sulfate uptake, glutathione seems to be a likely candidate, and knowledge on the transport and cycling of this compound needs to be expanded.

Acknowledgement

We wish to thank Mr. E. Leeuwinga for drawing the figures.

References

Aketagawa, J. & Tamura, G. 1980. Ferredoxin-sulfite reductase from spinach. Agric. Biol. Chem. 44: 2371-2378.

Anderson, J.W. 1980. Assimilation of inorganic sulphate into cysteine. In: Miflin, B.J. (ed.), The Biochemistry of Plants. Academic Press, New York, pp. 203-223.

Barneix, A. 1990. Yield variation in wheat: nitrogen accumulation, light interception and harvest index. In: Lambers, H., Cambridge, M., Konings, H. & Pons, T.L. (eds.), Causes and Consequences of Variations in Growth Rate in Higher Plants. SPB Academic Publishing, The Hague, pp. 87-101.

Barney, P.E. & Bush, L.P. 1985a. Interaction of nitrate and sulfate reduction in tobacco. 1. Influence of availability of nitrate and sulfate. Journal of Plant Nutrition 8: 505-515.

Barney, P.E. & Bush, L.P. 1985b. Interaction of nitrate and sulfate reduction in tobacco. 2. Influence of apical meristem removal. J. Plant Nutrition 8: 517- 526.

Beevers, L. & Hageman, R.H. 1969. Nitrate reduction in higher plants. Annu. Rev. Plant Physiol. 20: 485-552.

Bergmann, L., Schwenn, J.D. & Urlaub, H. 1980. Adenosine triphosphate sulfurylase and *O*-acetylserine sulfhydrylase in photoheterotrophically and heterotrophically cultured tobacco cells. Z. Naturforschung 35c: 952-957.

Bosma, W., Schupp, R., De Kok, L.J. & Rennenberg, H. 1991. Effect of selenate on assimilatory sulfate reduction and thiol content of spruce needles. Plant Physiol. Biochem. 29: 131-138.

Brunetti, N. & Hageman, R.H. 1976. Comparison of *in vivo* and *in vitro* assays of nitrate reductase in wheat (*Triticum aestivum* L. seedlings. Plant Physiol. 58: 583-587.

Brunold, C. 1983. Changes in ATP sulfurylase and adenosine 5'-phosphosulfate sulfotransferase activity during autumnal senescence of beech leaves. Physiol. Plant. 59: 319-323.

Brunold, C. 1990. Reduction of sulfate to sulfide. In: Rennenberg, H., Brunold, C., De Kok, L.J. & Stulen, I. (eds.), Sulfur Nutrition and Sulfur Assimilation in Higher Plants. SPB Academic Publishing, The Hague, pp. 13-33.

Brunold, B. & Schmidt, A. 1978. Regulation of sulfate assimilation in plants. 7. Cysteine inactivation of adenosine 5'-phosphosulfate sulfotransferase in *Lemna minor* L. Plant Physiol. 61: 342-347.

Brunold, B. & Suter, M. 1984. Regulation of sulfate assimilation by nitrogen nutrition in the duckweed *Lemna minor* L. Plant Physiol. 76: 579-583.

Brunold, C., Landolt W. & Lavanchy, P. 1983. SO_2 and assimilatory sulfate reduction in beech leaves. Physiol. Plant. 59: 313-318.

Brunold, C., Suter, M. & Lavanchy, P. 1987. Effect of high and low sulfate concentrations on adenosine 5'-phosphosulfate sulfotransferase activity from *Lemna minor*. Physiol. Plant. 70: 168-174.

Buwalda, F., De Kok, L.J. & Stulen, I. 1992. Flux of atmospheric H_2S to spinach leaves can be affected by the supply of *O*-acetylserine. Phyton 32: 13-16.

Campbell, W.H. 1990. Purification, characterization and immunochemistry of higher plant nitrate reductase. In: Abrol, Y.P. (ed.), Nitrogen in Higher Plants. John Wiley & Sons Inc., New York, pp. 65-92.

Campbell, W.H. & Smarelli, J. 1978. Purification and kinetics of higher plant NADH:nitrate reductase. Plant Physiol. 61: 61-616.

Clarkson, D.T. 1986. Regulation of the absorption and release of nitrate by plant cells: A review of current ideas and methodology. In: Lambers, H., Neeteson, J.J. & Stulen, I. (eds.), Fundamental, Ecological and Agricultural Aspects of Nitrogen Metabolism in Higher Plants. Martinus Nijhoff Publishers, The Hague, pp. 3-29.

Clarkson, D.T., Smith, F.W. & Van den Berg, P.J. 1983. Regulation of sulphate transport in a tropical legume *Macroptillum atropurpureum*, cv. Sirato. J. Exp. Bot. 34: 1463-1483.

Cooper, H.D. & Clarkson, D.T. 1989. Cycling of amino-nitrogen and other nutrients between shoots and roots in cereals, – a possible mechanism integrating shoot and root in the regulation of nutrient uptake. J. Exp. Bot. 40: 753-762.

Cram, W.J. 1990. Uptake and transport of sulfate. In: Rennenberg, H., Brunold, C., De Kok, L.J. & Stulen, I. (eds.), Sulfur Nutrition and Sulfur Assimilation in Higher Plants. SPB Academic Publishing, The Hague, pp. 3-13.

De Kok, L.J. 1989. Responses of sulfur metabolism in plants to atmospheric hydrogen sulfide. Phyton 29: 189-201.

De Kok, L.J. 1990. Sulfur metabolism in plants exposed to atmospheric sulfur. In: Rennenberg, H, Brunold, C., De Kok, L.J. & Stulen, I. (eds.), Sulfur Nutrition and Sulfur Assimilation in Higher Plants; Fundamental, Environmental and Agricultural Aspects. The Hague: SPB Academic Publishing, The Hague, pp. 111-130.

De Kok, L.J. & Kuiper, P.J.C. 1986. Effect of short-term dark incubation with sulfate, chloride and selenate on the glutathione content of spinach leaf discs. Physiol. Plant. 68: 477-482.

De Kok, L.J., Bosma, W., Maas, F.M. & Kuiper, P.J.C. 1985. The effect of short-term H_2S fumigation on water-soluble sulphydryl and glutathione levels in spinach. Plant, Cell Environ. 8: 189-194.

De Kok, L.J., Maas, F.M., Godeke, J., Haaksma, A.B. & Kuiper, P.J.C. 1986a. Glutathione, a tripeptide which may function as a temporary storage of excessive reduced sulphur in H_2S fumigated spinach plants. Plant Soil 91: 349-522.

De Kok, L.J., Stulen, I., Bosma, W. & Hibma, J. 1986b. The effect of short-term H_2S fumigation on nitrate reductase activity in spinach leaves. Plant Cell Physiol. 27: 1249-1254.

Dijkshoorn, W. & Van Wijk, A.L. 1967. The sulphur requirements of plants as evidenced by the sulphur-nitrogen ratio in the organic matter. A review of published data. Plant Soil 26: 129-157.

Duke, S.H., Reisenauer, H.M. 1986. Roles & requirements of sulfur in plant nutrition. In: Tabatabai, M.A. (ed.), Sulfur in Agriculture Am. Soc. Agron., Madison, pp. 123-168.

Evans, L.T. 1975. Sulphur in agriculture. In: McLachlan, K.D. (ed.), Sulphur in Australasian Agriculture. Sydney U.P., Sydney, pp. 3-9.

Fankhauser, H. & Brunold, C. 1978. Localization of adenosine 5'-phosphosulfate sulfotransferase in spinach leaves. Planta 148: 285-289.

Farago, S. & Brunold, B. 1990. Regulation of assimilatory sulfate reduction by herbicide safeners in *Zea mays* L. Plant Physiol. 94: 1808-1812.

Filner, P., Rennenberg, H., Sekya, J., Bressan, R.A., Wilson, L.G., Le Cureux, L. & Shimei, T. 1984. Biosynthesis and emission of hydrogen sulfide by higher plants. In: Koziol, M.J. & Whatley, F.R. (eds.), Gaseous air pollutants and Plant Metabolism. Butterworths, London, pp. 291-312.

Giovanelli, J. 1990. Regulatory aspects of cysteine and methionine synthesis. In: Rennenberg, H., Brunold, C., De Kok, L.J. & Stulen, I. (eds.), Sulfur Nutrition and Sulfur Assimilation in Higher Plants. SPB Academic Publishing, The Hague, pp. 33-49.

Giovanelli, J., Mudd, S.H. & Datko, A.H. 1980. Sulfur amino acids in plants. In: Miflin, B.J. (ed.), The Biochemistry of Plants. Academic Press, New York, pp. 453-505.

Grant, B.R., Atkins, C.A. & Canvin, D.T. 1970. Intracellular location of nitrate reductase and nitrite reductase in spinach and sunflower leaves. Planta 96: 60-72.

Gunz, G., Herschbach, C. & Rennenberg, H. 1992. Specificity of glutathione (GSH) mediated inhibition of sulfate uptake into excised tobacco roots. Phyton 32: 49-52.

Hageman, R.H. 1990. Historical perspectives of the enzymes of nitrate assimilation by crop plants and potential for biotechnological application. In: Ullrich, W.R., Rigano, C., Fuggi, A. & Aparicio, P.J. (eds.), Inorganic Nitrogen in Plants and Microorganisms. Springer Verlag, Berlin, pp. 3-12.

Herschbach, C. & Rennenberg, H. 1991. Influence of glutathione (GSH) on sulphate influx, xylem loading and exudation in excised tobacco roots. J. Exp. Bot. 42: 1021-1029.

Hewitt, E. 1975. Assimilatory nitrate-nitrite reduction. Annu. Rev. Plant. Physiol. 26: 73-100.

Hewitt, E., Hucklesby, D.P., Mann, A.F., Notton, B. A. & Rucklidge, G.F. 1979. Regulation of nitrate assimilation in plants. In: Hewitt, E. & Cutting, C.V. (eds.), Nitrogen Assimilation of Plants. Academic Press, London, pp. 255-288.

Holobradà, M. 1977. The influence of S-deficient pretreatment on sulfur uptake and assimilation by intact maize plants. II. Some aspects of metabolism. Agrochimica 21: 478-486.

Kaiser, G., Martinoia, E., Schröppel-Meier, G. & Heber, U. 1989. Active transport of sulfate into the

vacuole of plant cells provides halotolerance and can detoxify SO_2. J. Plant Physiol. 133: 756-763.
Kleppinger-Sparace, K.F., Sparace, S.A. & Mudd, J.B. 1990. Plant sulfolipids. In: Rennenberg, H., Brunold, C., De Kok, L.J. & Stulen, I. (eds.), Sulfur Nutrition and Sulfur Assimilation in Higher Plants. SPB Academic Publishing, The Hague, pp. 77-89.
Lips, H. & Avissar, Y. 1972. Plant-leaf microbodies: an intracellular site of nitrate reductase and nitrite reductase. Eur. J. Biochem. 29: 20-24.
Maas, F.M., De Kok, L.J. & Kuiper, P.J.C. 1095. The effect of H_2S fumigation on various spinach (*Spinacia oleracea* L.) cultivars. Relation between growth inhibition and accumulation of sulphur compounds in the plant. Journal of Plant Physiol. 119: 219-226.
Melzer, J.M., Kleinhofs, A. & Warner, R.L. 1989. Nitrate reductase regulation: Effects of nitrate and light on nitrate reductase mRNA accumulation. Mol. Gen. Genet. 217: 341-346.
Oaks, A., Poulle, M., Goodfellow, V.J., Cass, L.A. & Deising, H. 1988. The role of nitrate and ammonium ions and light on induction of nitrate reductase in maize leaves. Plant Physiol. 88: 1067-1072.
Oaks, A., Long, D.M., Zoumadakis, M., Li, X.-Z. & Herig, C. 1990. The role of NO_3^- and NH_4^+ ions in the regulation of nitrate reductase in higher plants. In: Ullrich, W.R., Rigano, C., Fuggi, A. & Aparicio, P.J. (eds.). Inorganic Nitrogen in Plants and Microorganisms. Springer Verlag, Berlin, pp. 165-171.
Rennenberg, H. 1982. Glutathione metabolism and possible biological roles in higher plants. Phytochemistry 21: 2271-2781.
Rennenberg, H. 1984. The fate of excess sulfur in higher plants. Annu. Rev. Plant Physiol. 35: 121-153.
Rennenberg, H., Polle, A., Martini, N. & Thoene, B. 1988. Interaction of sulfate and glutathione transport ion cultured tobacco cells. Planta 176: 68-74.
Rodgers, C.O. & Barneix, A. J. 1986. Cultivar differences in the rate of nitrate uptake by intact wheat plants as related to growth rate. Physiol. Plant. 72: 121-126.
Sawhney, S.K. & Naik, M.S. 1990. Role of light in nitrate assimilation in higher plants. In: Abrol, Y.P. (ed.), Nitrogen in Higher Plants. John Wiley & Sons Inc., New York. pp. 93-129.
Schmidt, A. 1976. The adenosine-5'-phosphosulfate-sulfotransferase in sunflower (*Helianthus annuus* L.). Z. Pflanzenphysiol. 78: 164-168.
Schmutz, D. & Brunold, C. 1982. Regulation of sulfate assimilation in plants. XIII. Assimilatory sulfate reduction during ontogenesis of primary leaves of *Phaseolus vulgaris* L. Plant Physiol. 70: 524-527.
Schröppel-Meier, G. & Kaiser, W.M. 1988. Ion homeostasis in chloroplasts under salinity and mineral deficiency. II. Solute distribution between chloroplasts and extrachloroplastic space under excess or deficiency of sulfate, phosphate or magnesium. Plant Physiol. 87: 828-832.
Schütz, B., De Kok, L.J. & Rennenberg, H. 1991. Thiol accumulation and cysteine desulfhydrase activity in H_2S-fumigated leaves and leaf homogenates of cucurbit plants. Plant Cell Physiol. 32: 733-736.
Schwenn, J.D. & Depka, B. 1977. Assimilatory sulfate reduction by chloroplasts: The regulatory influence of adenosine-mono- and adenosine-diphosphates. Z. Naturforschung 32c: 792-797.
Shaner, D.J. & Boyer, J.S. 1976. Nitrate reductase activity in maize leaves. I. Regulation by nitrate flux. Plant Physiol. 58: 499-504.
Shaw, W,H. & Anderson, J.W. 1972. Purification, properties and substrate specificity of adenosine triphosphate sulfurylase from spinach leaf tissue. Biochem J. 127: 237-247.
Simpson, R.J. 1986. Translocation and metabolism of nitrogen: whole plant aspects. In: Lambers, H., Neeteson, J.J. & Stulen, I. (eds.), Fundamental, Ecological and Agricultural Aspects of Nitrogen Metabolism in Higher Plants. Martinus Nijhoff Publishers, Dordrecht, pp. 71-97.
Simpson, R.J., Lambers, H. & Dalling, M.J. 1982. Kinetin application to roots and its effect on uptake, translocation and distribution of nitrogen in wheat (*Triticum aestivum*) grown with a split root system. Physiol. Plant. 56: 11-17.
Smith, I.K. 1980. Regulation of sulfate assimilation in tobacco cells. Effect of nitrogen and sulfur nutrition on sulfate permease and *O*-acetylserine sulfhydrylase. Plant Physiol. 66: 877-883.
Smith, I.K. 1981. Compartmentation of sulfur metabolites in tobacco cells. Use of efflux analysis. Plant Physiol. 68: 937-940.
Smith, I.K. & Cheema, H.K. 1985. Sulphate transport into plants and excised roots of soybean (*Glycine max* L.). Ann. Bot. 56: 219-224.
Sopory, S.K. & Sharma, A.K. 1990. Spectral quality of light, hormones and nitrate assimilation. In: Abrol, Y.P. (ed.), Nitrogen in Higher Plants. John Wiley & Sons Inc., New York, pp. 129-159.
Steingröver, E. 1986. Nitrate accumulation in spinach. Thesis Groningen.

Steingröver, E., Ratering, P. & Siesling, J. 1986a. Daily changes in uptake, reduction and storage of nitrate in spinach grown at low light intensity. Physiol. Plant. 66: 550-562.

Steingröver, E., Siesling, J. & Ratering, P. 1986b. Effect of one night with "low light" on uptake, reduction and storage of nitrate in spinach. Physiol. Plant. 66: 557-562.

Stulen, I., Zantinge, J. & Koster, A. 1973. Functioning of nitrate reductase in the light and in the dark in seedlings of Raphanus sativus. Acta Bot. Neerl. 22: 589-596.

Stulen, I., Lanting, L., Lambers, H., Posthumus, F., Van de Dijk, S.J. & Hofstra, R. 1981. Nitrogen metabolism of *Plantago major* ssp. *major* as dependent on the supply of mineral nutrients. Physiol. Plant. 52: 108-114.

Suter, M., Lavanchy, P., Von Arb, C. & Brunold, C. 1986. Regulation of sulfate assimilation by amino acids in *Lemna minor*. Plant Science 44: 125-132.

Thoiron, A., Thoiron, B., Demarty, M. & Thellier, M. 1981. Compartmental analysis of sulphate transport in *Lemna minor* L., taking plant growth and sulphate metabolism into consideration. Biochim. Biophys. Acta. 644: 24-35.

Tschanz, A., Landolt, W., Bleuler, P. & Brunold, C. 1986. Effect of SO_2 on the activity of adenosine 5'-phosphosulfate sulfotransferase from spruce trees (*Picea abies*) in fumigation chambers and under field conditions. Physiol. Plant. 67: 235-241.

Tweedie, J.W. & Segel, I.H. 1971. Adenosine triphosphate sulfurylase from *Penicillium chrysogenum*. II. Physical, kinetic and regulatory properties. J. Biol. Chem. 246: 2438-2446.

Van Dijk, P.J., Stulen, I. & De Kok, L.J. 1986. The effect of sulfide in the ambient air on amino acid metabolism of spinach leaves. In: Lambers, H., Neeteson, J.J. & Stulen, I. (eds.), Fundamental, Ecological and Agricultural Aspects of Nitrogen Metabolism in Higher Plants. Martinus Nijhoff Publishers, Dordrecht, pp. 207-210.

Von Arb, C. & Brunold, C. 1985. Ferredoxin-sulfite reductase and ferredoxin-nitrite reductase activities in leaves of *Pisum sativum*: changes during ontogeny and *in vitro* regulation by sulfide. Physiol. Plant. 64: 290-294.

Von Arb, C. & Brunold, C. 1986. Enzymes of assimilatory sulfate reduction in leaves of *Pisum sativum*: activity changes during ontogeny and *in vivo* regulation by H_2S and cysteine. Physiol. Plant. 67: 81-86.

Zhen, R.-G., Koyro, H.-W., Tomos, R.A. & Miller, A.J. 1991. Compartmental nitrate concentrations in barley root cells measured with nitrate-selective microelectrodes and by single-cell sap sampling. Planta 1985: 356-361.

Section 2.
Metabolism and functions of organic sulfur compounds

Contents

CATABOLISM OF SULFUR-CONTAINING AMINO ACIDS

Mendel Mazelis
Department of Food Science and Technology, University of California, Davis, CA 95616, U.S.A.

Introduction

The catabolism of methionine and cyst(e)ine in higher plants has not been the major focus of most reviews on sulfur metabolism. The emphasis has usually been on the process of incorporation of the sulfur atom and its primary precursor in the pathway, *e.g.* sulfate, sulfite, and sulfide and the role of methionine as a methylating agent (Anderson 1980; Giovanelli *et al.* 1980). More recently the involvement of *S*-adenosylmethionine in the production of polyamines has been an active field of study (Tiburcio *et al.* 1990). The catabolism of the sulfur amino acids and their derivatives was briefly discussed by Mazelis (1980). Recently Anderson (1990) has produced an excellent review on sulfur metabolism in plants in which he discusses the most recent findings on the catabolism of methionine and cysteine and attempts to incorporate and reconcile results from peripheral pertinent studies.

In this short review I will not attempt to reprise the entire literature in this area but will focus on those aspects of recent current interest and those of which some uncertainty exists as to their significance.

Methionine

Catabolism of the methyl group

The methionine molecule, $CH_3SCH_2CH_2CH(NH_2)COOH$, has three distinct moieties. These are the methyl group, the sulfur, and the 2-aminobutyryl portion. An early study by Splittstoesser & Mazelis (1967) examined the catabolic fate of the methyl group in a number of plant families with seedlings of pea, kohlrabi, pumpkin, and *Sesbania*. Using $^{14}CH_3$ methionine they found up to 4% of the total label added produced as CO_2 within 24 hours. Methionine, methyl methionine sulfonium (SMM), and serine were the major labeled compounds found in the amino acid fraction in each species. Mudd & Datko (1986) did a very thorough study of the fate of the methyl group in *Lemna*. They grew the plants in the presence of $^{14}CH_3$-labeled methionine and examined the distribution of label in the various classes of compounds extracted from the homogenate of samples removed at different time periods. They concluded that there was little, if any, oxidation of the methyl group. The greatest amounts of label were in protein methionine, methylated ethanolamine derivatives, pectin methyl esters, and chlorophyll methyl esters. They found no labeled CO_2 or serine. The possibility of differences between the monocot *Lemna* and the dicot species used in the earlier study was suggested. Although it may be surprising that serine should be labeled, thus indicating the potential for the existence

Sulfur Nutrition and Assimilation in Higher Plants
edited by L.J. De Kok et al., pp. 95–108

of a C_1 path from methionine to glycine, there is a precedent in that it has been reported that rats fed L-methionine-$^{14}CH_3$ produced serine labeled almost exclusively in the beta carbon (Kisliuk *et al.* 1956).

The S-methylmethionine (SMM) cycle

The conversion of methionine to SMM in higher plants is well-established (Sato *et al.* 1958; Greene & Davis 1960; Splittstoesser & Mazelis 1967; Baur & Yang 1972; Hanson & Kende 1976; Mudd & Datko 1986). The studies by Mudd & Datko (1986) with *Lemna* showed that the labeled methionine was converted readily to *S*-methylmethionine (SMM). The rate of incorporation of the labeled methyl group was very close to that of the incorporation into the methylated ethanolamine derivatives. However, the total amount of label in the SMM pool decreased over time. This suggested that it was acting as a precursor or intermediate in some other reactions. The following sequence of reactions was proposed to account for their findings:

(1) ATP + methionine → *S*-adenosylmethionine(SAM) + PP_i + P_i
(2) SAM + methionine → *S*-methylmethionine + *S*-adenosylhomocysteine
(3) *S*-adenosylhomocysteine → homocysteine + adenosine
(4) Homocysteine + *S*-methylmethionine → 2 methionine

SUM: ATP → adenosine + PP_i + P_i

The ability of SMM to act as a methyl donor to homocysteine and thus able to regenerate 2 molecules of methionine has been well established. The enzyme *S*-methyl-L-methionine: homocysteine *S*-methyltransferase (EC 2.1.1.10) has been demonstrated in extracts of seeds of many plant species as well as in 1-day old pea seedling cotyledons (Giovanelli *et al.* 1980). The sequence of reactions above appears to be a 'futile' cycle since it results in only a net hydrolysis of all the phosphate groups of ATP. This anomaly was addressed in the review by Giovanelli *et al.* (1980) who suggested that SMM could act as a reservoir of methyl groups for periods when methyl group synthesis *de novo* might be limiting. Since SAM is an effector of threonine synthase which competes for the precursor of homocysteine, the authors suggested that storage of methyl groups in SMM, which does not stimulate threonine biosynthesis, allows the continued accumulation of the methionine precursors, methylmethionine and homocysteine without slowing methionine biosynthesis.

In a very recent paper Mudd & Datko (1990) examined the cycle above in intact plants of *Lemna paucicostata* and cell suspension cultures of carrot (*Daucus carota*) and soybean (*Glycine max*). *Lemna* was incubated with *S*-[^3H_3C]-methyl-L-methionine sulfonium in one experiment and with labeled methylmethionine + L-[$^{14}CH_3$]-methionine in another. After a 1.5 minute incubation the radioactive medium was removed, the plants washed and placed in fresh unlabeled growth medium. At various times samples were removed and assayed for the label in various fractions. These chase experiments showed that the label from the SMM entered the same pools as that from the methionine albeit there was some time lag in entering these pools compared to the label from the methionine itself. This suggested strongly that the methyl groups from SMM were transferred to the methionine pool prior to other methylation reactions. The presence of SAM:methionine *S*-methyltransferase (EC 2.2.1.12) and SMM:homocysteine-*S*-transferase (EC 2.2.1.10), the enzymes required for the formation of SMM and its return to methionine, was demonstrated in *Lemna* homogenates.

The results with carrot suspension cultures showed that the label in the SMM pool decreased rapidly during a chase period but studies with soybean did not give this result. The authors conclude that this cycle, because of its prevalence (my interpretation), must have some physiological utility. Their suggestion is that the flux through SAM is so high and the pool of methionine so small that methionine could be trapped in SAM leaving insufficient methionine available for protein synthesis. The cycle above would regenerate methionine at the cost of ATP hydrolysis.

SMM catabolism

The enzyme *S*-methyl-L-methionine hydrolase (EC 3.3.1.2) capable of cleaving SMM by the following reaction has been found in bacteria (Tanaka & Nakamura 1964; Mazelis *et al.* 1965) and partially purified:

$$(CH_3)_2S^+CH_2CH_2CH_2(^+NH_3)COO^- + H_2O \rightarrow$$
$$(CH_3)_2S + OH\text{-}CH_2CH_2CH(^+NH_3)COO^- + H^+$$

This enzyme could also utilize SAM as a substrate. A similar activity has been reported in cabbage homogenates by Lewis *et al.* (1971) and in onion by Hattula & Granroth (1974). The results of Lewis *et al.* (1971) were obtained in studies of selenium metabolism in cabbage. This plant incorporates inorganic selenium into the selenomium analog of *S*-methylmethionine. Lewis *et al.* (1971) found dimethyl selenide in volatile compounds produced by fresh leaves of cabbage grown on nutrient media containing selenite or selenate. A homogenate of cabbage leaves was treated with $(NH_4)_2SO_4$ to 50% saturation and the precipitate obtained used as the source of enzyme. This enzyme fraction utilized both *Se*-methylselenomethionine and SMM to produce dimethylselenide or dimethylsulfide respectively, and homoserine.

Methionine catabolism

In plants and tissues which produce ethylene as a ripening or wound hormone the 2-aminobutyryl moiety of SAM is utilized to produce ethylene, HCN, and CO_2. An intermediate in this conversion is the cyclic amino acid 1-aminocyclopropane carboxylate. This sequence has been reviewed recently by Miyazaki and Yang (1987).

Several other catabolic reactions involving methionine have been reported from time to time but have not been studied to the same extent as those discussed above. Sato *et al.* (1958) found that oat seedling sections oxidized methionine to its sulfoxide *in vivo* and *in vitro*. The methionine sulfoxide was able to donate its methyl group to pectin and protopectin. It was also able to donate its methyl group to methionine to yield SMM. Doney & Thompson (1966) found L-methionine sulfoxide and *S*-methyl-L-cysteine sulfoxide were reduced to methionine and *S*-methyl cysteine in turnip (*Brassica rapa*) and bean (*Phaseolus vulgaris*) leaves. Splittstoesser & Mazelis (1967) showed that methionine sulfoxide was a major product after feeding $[^{14}CH_3]$-

methionine to various seedlings. In addition they found that these seedlings could also convert [$^{14}CH_3$]-methionine sulfoxide to methionine to some extent. The enzymatic reduction of methionine sulfoxide to methionine has been found to occur in yeast, bacteria, plants, and animals (Brot & Weissbach 1991). This reaction was analyzed in yeast by Black *et al.* (1960). They described a complex system requiring three proteins and NADPH which reduced one of the methionine sulfoxide isomers, *viz.* L-methionine sulfoxide to L-methionine. In the presence of only two of these proteins cystine as well as other disulfides could also be reduced to the sulfhydryl constituents.

The oxidation of methionine residues in proteins to methionine sulfoxide takes place both *in vivo* and *in vitro*. A separate enzyme has been found in unicellular eukaryotes, plants and animals which can reduce methionine sulfoxide in peptide linkage. The question as to whether the methylating ability of the sulfoxide is related to its prior reduction to methionine is still unanswered.

The production of methanethiol by leaf discs of pumpkin (*Cucurbita pepo*) or by homogenates of leaf discs floated on 10 mM L-methionine has been reported by Schmidt *et al.* (1985). The production of methanethiol by leaf discs began after a 6 hour lag period. The rate of production was greatly accelerated by light. The authors conjectured that these results conformed to the induction of a methionine-γ-lyase (EC 4.4.1.11).

In polyamine biosynthesis SAM decarboxylase (EC 4.1.1.50) produces the propylamine derivative of SAM which is utilized in spermine synthesis. This releases the methionine carboxyl group as CO_2. Another reaction which can decarboxylate methionine was reported by Mazelis (1959). A soluble preparation from washed cytoplasmic particles from cabbage leaf was able to decarboxylate methionine. The reaction required Mn^{2+} and pyridoxal phosphate. Since catalase was shown to inhibit the reaction (Mazelis 1960), the involvement of peroxidase was examined. Purified horseradish peroxidase was found to carry out this decarboxylation under the same conditions (Mazelis 1962). The product of the reaction was unexpected. Rather than 3-methylthiopropylamine the product was 3-methylthiopropionamide (Mazelis & Ingraham 1962). Other amino acids were also decarboxylated to the amide by this system.

A methionine decarboxylase has been solubilized from the lyophilized fronds of many fern species (Hartmann *et al.* 1984). The enzyme decarboxylates a number of alkyl amino acids, but is most active with methionine. Stevenson *et al.* (1990) has purified the enzyme from acetone powders of *Dryopteris filix-mas* and characterized its properties. The product of the methionine decarboxylation was 3-methylthiopropylamine.

Cysteine and cystine

Metabolic interconversion

Even though these two amino acids are readily interconverted chemically, it is now apparent that they can be catabolized by different routes. Cysteine can be oxidized to cystine aerobically with cytochrome c and cytochrome oxidase (Keilin 1930).

Kramer & Schmidt (1984) found that that membrane fractions of *Chlorella fusca* catalyzed the oxygen dependent oxidation of cysteine to cystine. One mole O_2 converted 4 cysteines to 2 cystines.

The reduction of cystine to cysteine has been reported in pea seed (Romano & Nickerson 1954). An extract from pea seed acetone powders was able to reduce cystine to cysteine utilizing NADH. This enzyme (EC 1.6.4.1) has not been purified to date. Mapson (1953) claimed to have verified this finding but presented no data. Very recently some preliminary efforts to repeat these results in my laboratory were unsuccessful. The status of this reaction is uncertain. Perhaps in view of the work by Black *et al.* (1960) cited above the system may be a complicated one. The reduction of cystine to cysteine has been reported recently using a cell-free extract of *Clostridium thermoaceticum* (Koesnandar *et al.* 1991). These investigators found that their preparation required molecular hydrogen and methylviologen. They suggested that the enzyme hydrogenase in the extract reduced the methylviologen which then reduced the L-cystine to L-cysteine and regenerated oxidized methylviologen. The addition of exogenous hydrogenase accelerated the rate of reduction.

Catabolism of cystine

The production of pyruvate from cystine has been well-documented. Mazelis *et al.* (1967) demonstrated the following reaction catalysed by a soluble enzyme from rutabaga (*Brassica napobrassica*):

$$^{-}OOC\text{-}CH(^{+}NH_3)\text{-}CH_2\text{-}S\text{-}S\text{-}CH_2\text{-}CH(^{+}NH_3)\text{-}COO^{-} + H_2O \rightarrow$$
$$^{-}OOC\text{-}CH(^{+}NH_3)\text{-}CH_2\text{-}S\text{-}SH + CH_3\text{-}CO\text{-}COO^{-} + {}^{+}NH_4$$

The rutabaga enzyme was highly specific for L-cystine and *S*-methyl-L-cysteine sulfoxide as substrates. L-cysteine was not a substrate but was an effective inhibitor of the reaction with L-cystine. The K_m for L-cystine was 1 mM and the K_i for L-cysteine was 0.15 mM. Glutathione was equally effective as an inhibitor. Hall & Smith (1983) obtained two isozymes of this enzyme activity from cabbage (*Brassica oleracea* var capitata). These enzymes were able to utilize not only L-cystine as a substrate but were also quite active towards *O*-acetylserine, L-cysteine, and *S*-methylcysteine sulfoxide. Hamamoto & Mazelis (1986) also found two isozymes of cystine lyase in broccoli buds (*B. oleracea* var pompejana) and purified one of them to homogeneity. The homogeneous enzyme was active only with L-cystine and alkyl cysteine sulfoxides. The enzyme was found to be a glycoprotein containing 5.8% carbohydrate on a weight basis. By gel filtration the molecular weight was estimated at 152,000 and SDS-PAGE analysis showed one subunit of 49,000 daltons. This strongly suggested the native enzyme exists as a trimer. We have recently purified cystine lyase from turnip (*Brassica rapa*), another species in this genus, for comparative purposes. Two isozymes of this enzyme are present in this species and we have purified both of them to apparent homogeneity. One isozyme is much more active than the other based on specific activity (Wongpaibool & Mazelis, unpublished results).

Catabolism of cysteine

The production of pyruvate from L-cysteine by an L-cysteine desulfhydrase (EC 4.4.1.1) in higher plants has been reported in several studies (Harrington & Smith 1980; Rennenberg *et al.* 1987; Schütz *et al.* 1991). Harrington & Smith (1980) utilized cultured tobacco cells. Acetone powders of these cells were prepared 24 hours after exposure to 0.5 mM L-cysteine. The acetone powders were extracted with pH 8.0 Bicine buffer plus dithiothreitol. After centrifugation the supernatant solution was used as the enzyme source. They found the sulfide and pyruvate were produced in equimolar amounts. Attempts at further purification of the enzyme were unsuccessful. Hall & Smith (1983) had considered that cysteine desulfhydrase activity resides in the same protein as the cystine lyase since their highly purified enzyme carried out both activities.

The other studies referenced above utilized cultured tobacco cells and cucurbit leaf discs. Homogenates from leaf discs and tobacco cells were assayed for cysteine desulfhydrase activity by incubation with L-cysteine or D-cysteine. Enzyme activity was monitored by sulfide emission. No identification of pyruvate was attempted. Sulfide was produced with either substrate. The *in vitro* studies found the D-cysteine desulfhydrase activity was significantly greater than that of the L-form. This conformed with previous studies (Schmidt 1982; Schmidt & Erdle 1983) which demonstrated the presence of D-cysteine desulfhydrase in spinach leaves and *Chlorella fusca*. In the latter plant the L- and D- specific enzymes were separated *in vitro*. The D-cysteine enzyme was much more active and was purified over 100-fold. Balance studies of the products showed much less pyruvate and NH_4^+ were produced than H_2S. Fractionation of crude homogenates (Rennenberg *et al.* 1987) found the D-cysteine enzyme was localized in the cytoplasm whereas the L-cysteine desulfhydrase was present in the chloroplasts and mitochondria.

Although a significant body of evidence has been accumulated that L- and D-cysteine desulfhydrases exist in higher plants, the properties of these enzymes have only been cursorily described as yet. The enzyme in animals and bacteria which acts as a cysteine desulfhydrase is known as cystathione-γ-lyase (EC 4.4.1.1). It is a multifunctional pyridoxal phosphate protein which cleaves L-cystathionine to yield L-cysteine, NH_3, and α-ketobutyrate. It also can utilize L-homoserine, L-cystine, and L-cysteine as a substrate. The plant enzymes so far described do not appear to have this wide specificity. Giovanelli & Mudd (1971) were unable to show this γ-cleavage in plant extracts in which cystathionine β-lyase was present, however Halaseh *et al.* (1977) did find that feeding ^{35}S-amino acids to *Astragalus pectinatus* gave products which suggested a γ-cleavage of cystathionine did occur in this species.

In a very recent study Schütz *et al.* (1991) found that fumigating cucurbit plants with H_2S increased the thiol content of the cotyledons and primary leaves but decreased the activity of the L- and D-cysteine desulfhydrases in the cotyledons very markedly. When leaf homogenates were exposed to H_2S the thiol content increased by 16-fold. The addition of pyruvate + NH_4^+ or various inhibitors of cysteine desulfhydrase like NH_2OH to the homogenate during the fumigation decreased the amount of thiols formed significantly. Since these inhibitors did not affect the *in vitro* activity of cysteine synthase, these authors suggested that the increase in thiol content may be due to the reversal of the cysteine desulfhydrase reaction. This is an interesting speculation and should be readily proven by use of ^{14}C-pyruvate and $^{15}NH_3$ even with crude preparations.

S-Substituted cysteines

A distinctive feature of several plant families is the presence of a number of *S*-alkyl cysteines as secondary non-protein amino acids. These may be present as the thioether or the sulfoxide and also, in many instances, as the γ-glutamyl dipeptide. They are often of significant commercial interest since these are the precursors of the major flavor constituents in several important economic crops. The monocot families *Amaryllidaceae* and *Liliaceae*, and the dicot families *Cruciferae* and *Leguminosae*, have many species which give off a characteristic odor upon maceration or bruising. Garlic (*Allium* sativum), onion (*Allium cepa*), cabbage, cauliflower, broccoli (varieties of *Brassica oleracea*), turnip (*Brassica rapa*), rutabaga (*Brassica napobrassica*), and kidney bean (*Phaseolus vulgaris*) are well-known examples of common vegetables which have a significant amount of these derivatives as secondary amino acids.

Alliin lyase

Stoll & Seebeck (1951) described the isolation and characterization of the sulfur amino acid of garlic responsible for the characteristic odor and flavor and the enzyme which degraded it. The amino acid was given the trivial name of alliin and was determined to be (+)-*S*-allyl-L-cysteine sulfoxide. The enzyme was named alliinase or alliin lyase and the overall reaction is shown below:

$$2\ CH_2{=}CH\text{-}CH_2\text{-}SO\text{-}CH_2\text{-}CH(^{+}NH_3)\text{-}COO^- + H_2O \rightarrow$$
$$CH_2{=}CH\text{-}CH_2\text{-}SO\text{-}S\text{-}CH_2\text{-}CH{=}CH_2 + 2\ CH_3\text{-}CO\text{-}COO^- + 2\ ^{+}NH_4$$

The alliin lyase (EC 4.4.1.4) of garlic has been purified to apparent homogeneity by several groups (Kazaryan & Goryachenkova 1975; Nock & Mazelis 1986, 1987; Jansen *et al.* 1989a). Each of these preparations had different physical characteristics and properties. The molecular weights reported were 130,000, 85,000 and 108,000 in the order of references above. The isoelectric points in order were 6.2, 7.0, and 4.9 respectively. All agreed that the enzyme had a pyridoxal phosphate prosthetic group and consisted of two equal subunits. A recent brief abstract by Rabinkov *et al.* (1991), however, describes a homogeneous garlic alliinase as a tetramer of 45,000 dalton subunits. Nock & Mazelis (1986, 1987) found the enzyme to be a glycoprotein however the homogeneous enzyme of Jansen *et al.* (1989a) was not. Confirmation of the glycoprotein nature of the garlic enzyme and a subunit molecular weight of 40,000 was recently obtained in a personal communication from Dr. P. Simon of the Horticulture Department at the University of Wisconsin, Madison.

There were some unusual aspects of the homogeneous protein examined by Jansen *et al.* (1989a). The electrophoretically homogeneous fraction used for characterization had a specific activity only 1/3 that of the previous purification step. It is conceivable that their results might be explained by the isolation and characteriazation

Table 1. Classification of Alliin Lyases of *Allium* species by pH optimum.

	Species	Reference
I. pH 5.6-6.5	*A. sativum* (garlic)	Stoll & Seebeck (1951)
	A. chinense (rakkyo)	Tsuno (1958)
	A. odorum	Tsuno (1958)
II. pH 8.0-8.5	*A. cepa* (onion)	Schwimmer & Mazelis (1963)
	A. porrum (leek)	Won & Mazelis (1989)
	A. fistulosum (Welsh onion)	Fujita *et al.* (1990)

of a small fraction of modified or deglycosylated enzyme. They also reported that a flavin coenzyme was present in their preparation (Jansen *et al.* 1989b) and that rotenone inhibited very strongly at nanomolar concentrations. Efforts to confirm this in my laboratory were unsuccessful.

Onion, a closely related *Allium* species, unlike garlic produces a very irritating compound upon cutting or macerating onion tissue. This product is a transient lachyramator. The major sulfur amino acid of the onion has been found to be an isomer of the garlic alliin, *viz.* trans-(+)-*S*-(1-propenyl)-L-cysteine sulfoxide (Whitaker & Mazelis 1991). Although it was assumed that onion must possess an alliin lyase, it was 11 years after the finding by Stoll & Seebeck (1951) before a cell-free preparation from onions capable of alliin lyase activity was demonstrated. Almost simultaneously two laboratories reported these results (Schwimmer *et al.* 1960; Kupiecki & Virtanen 1960).

The onion alliinase has been purified to homogeneity and characterized (Nock & Mazelis 1987). It was similar to the garlic enzyme in that it was a glycoprotein. The native enzyme was, however, a tetramer consisting of subunits of 50,000 daltons. Pyridoxal phosphate equivalent to one mole/mole of subunit was bound to both enzymes. A distinctive difference between these two enzymes was in the pH optimum for the assay. The garlic enzyme had an optimum at pH 6.5 whereas that of the onion alliinase was at pH 8.0.

Alliin lyase from two other *Allium* species, leek (*A. porrum*) and welsh onion (*A. fistulosum*), have also been purified to homogeneity (Won & Mazelis 1989; Fujita *et al.* 1990). Both were glycoproteins and were similar in their properties to the onion enzyme. The alliin lyases from various species appear to divide into two distinct classes in regard to their pH optima for activity (Jacobsen *et al.* 1968). A brief table of the alliin lyases of various *Allium* species is shown in Table 1.

S-methyl-L-cysteine sulfoxide is a common secondary amino acid in the *Cruciferae*. The enzymatic degradation of this substrate by cell-free extracts of *Brassica* species was first reported by Mazelis (1963). After cystine lyase had been discovered, further purification of this enzyme could not separate the two activities in the *Brassica* species described previously. The homogeneous cystine lyase of broccoli utilized *S*-methyl-L-cysteine sulfoxide very effectively as a substrate (Hamamoto & Mazelis 1986). The *Allium* enzymes were unable to utilize cystine as a substrate.

Other C-S lyases

The seeds of many species of *Leguminosae* contain a variety of *S*-alkyl cysteine derivatives as free non-protein amino acids. On germination they often give rise to volatile compounds of strong odor. Schwimmer & Kjaer (1960) obtained an enzyme from the ground endosperm of *Albizzia lophanta* which had a broad specificity in regard to substrate. It utilized both *S*-alkylcysteines and their sulfoxides very effectively. L-Djenkolic acid, which is a natural constituent in the seed, was the most active substrate.

A similar enzyme was purified to homogeneity from seedlings of *Acacia farnesiana* (Mazelis & Creveling 1975). Its substrate specificity was very similar to that of *A. lophanta*. An unusual feature is that L-djenkolate is the preferred substrate, but L-cystathionine, which might be considered almost a methylene analogue of djenkolate, is an extremely poor substrate. The non-specific character of the enzyme was demonstrated by its ability to cleave *O*-methyl-L-serine, the oxygen analogue of *S*-methyl-L-cysteine, very readily (Mazelis 1975). It also had a significant activity with the nitrogen analogue, β-methylamino-α-aminopropionate. Like all the other C-S lyases described previously pyridoxal phosphate was a cofactor.

Homology by immunology

It is apparent that plant C-S lyases that utilize the *S*-substituted cysteines fall into three categories. They are the *Allium* type which are limited only to the sulfoxide; the *Brassica* type which not only utilizes the sulfoxide but also cystine even more effectively; and the *Acacia* class which has a very broad specificity using the thioether as readily as the sulfoxide, and can even utilize the *O*- and *N*-analogues.

The question arises as to the degree of homology present in the enzyme proteins from such different families. In fact even more curious, at least to me, was the fact that two closely related species such as onion and garlic had enzymes of such different physical and kinetic properties. The enzymes differ greatly in molecular weight, number of subunits, and pH optimum (Nock & Mazelis 1987), yet their substrate specificity is similar. Polyclonal antibodies were obtained from rabbits challenged with either homogeneous garlic alliinase or onion alliinase. Using immunodiffusion the garlic and onion antisera were tested against homogeneous onion, garlic, and leek enzyme. The only reaction occurred with the lyase and the antiserum derived from it. The leek enzyme did not interact with either the garlic or onion antibodies (Nock & Mazelis 1989).

The antigens used were the native enzymes which are glycoproteins. The carbohydrate moieties are very potent antigenic determinants and it is conceivable that they could be the dominant feature in the interactions with the antibodies. Purified alliin lyase from both garlic and onion was deglycosylated chemically and the denatured protein used to produce antibodies. Using a more sensitive ELISA test for antibody-antigen reaction, crude extracts of garlic, onion, and leek were tested for their response to glycosylated garlic antibodies (Ab_g) and deglycosylated enzyme antibodies (Ab_d) after SDS-PAGE. Ab_g reacted strongly with garlic extract with one major band. There was also some response with onion and leek in the same location. When Ab_d was similarly tested, there was only a reaction with crude garlic extract, purified garlic enzyme, and with one of the subunits of the leek enzyme, but not with the onion lyase (Ho & Mazelis 1993). This strongly suggests that the different *Allium* lyases are quite different in protein structure and makes them interesting subjects for studies in molecular evolution.

Cystathionine

This is a non-protein *S*-substituted cysteine which is ubiquitous in the autotrophic plant world. As an intermediate in methionine biosynthesis it is cleaved by a β-elimination to form homocysteine and pyruvate. Cystathionine-β-lyase (EC 4.4.1.8) has been demonstrated in crude preparations from several plants and partially purified from spinach leaves by Giovanelli & Mudd (1971). Because of its relationship to the C-S lyases discussed above, Staton & Mazelis (1991) purified the enzyme from spinach to homogeneity for comparison to alliin lyase and cystine lyase. The enzyme had a very limited specificity in that it could only use cystathionine, djenkolic acid, and cystine as substrates. The first two were much more active as substrates than cystine. This does mean, however, that cystine cleavage is potentially a universal plant phenomenon.

The molecular weight was estimated at 210,000 consisting of a tetramer of subunits of 53,000 daltons. Polyclonal antibodies were made to this antigen in rabbits and crude extracts of garlic, onion, leek, and spinach were examined by immunoblotting techniques for interaction with the antiserum. All of them showed a single band at the same position. When purified alliin lyase from garlic and onion was tested there was no reaction (Ho & Mazelis 1993). This suggests that there is a homologous protein present in the crude extract which is not related to the other C-S lyase, alliin lyase.

Concluding remarks

The catabolism of methionine in higher plants differs significantly from that found in animals. Methionine is an essential amino acid for mammals and, depending on the diet, may be ingested in excess of requirements for growth and maintenance. In higher plants methionine is synthesized as needed by the organism and its production is under very strict regulatory control, primarily by feedback inhibition at certain critical steps in the biosynthetic pathway. One of the major differences between plant and animal mechanisms in methionine metabolism is at the level of cystathionine utilization. Animals can only utilize this important intermediate via a γ-cleavage to produce cysteine and α-ketobutyrate. Plants on the other hand invoke a β-elimination to form homocysteine and pyruvate. Although there has been a report of the γ-cleavage of cystathionine in *Astragalus pectinatus* (Halaseh *et al.* 1977) this appears to be a minor aspect of cystathionine degradation in plants. An additional function of methionine in plants, lacking in animals, is the production of the ripening hormone ethylene.

Because of the strict regulatory controls of methionine biosynthesis, plants appear to have established what might be called the Law of the Conservation of Methionine.

The presence of the SMM cycle and the regeneration of methionine from 5-methylthioadenosine, a product of ethylene production from methionine, would support this contention.

Although cysteine and cystine are readily interconvertible chemically, they are catabolized by independent routes in higher plants. The purification and characterization of cystine lyase has been thoroughly documented. L- and D-cysteine desulfhydrases have been found and characterized to some extent but the level of purification required for critical specificity studies and physical characterization has not been obtained to date. A real question is what is the role of such active D-cysteine desulfhydrases biologically?

The substituted *S*-alkyl cysteine lyases vary from the very limited specificity of the enzyme in the *Allium* species to the very broad specificity exhibited by the enzymes in the *Leguminosae* species studied. In every case L-cystathionine is a poor substrate for these enzymes and there is a specific C-S lyase, cystathionine-β-lyase. It is of interest that this latter enzyme can utilize L-cystine, thus making a cystine lyase activity ubiquitous in all autotrophs.

The biological role of the *S*-alkyl cysteine lyases is uncertain. It is known that volatile sulfur compounds are toxic to many fungi. The amino acid substrate and the enzyme are separated spatially within the cell in onion (Lancaster & Collin 1981) and the enzyme action does not take place until the tissue is ruptured. An invasive pathogen could cause such a reaction, in effect, the plant creates its own antiseptic.

Some final comments should be made concerning the potential utility of the products of the alliin lyase reaction in onion and garlic. In recent years a spate of articles has been published in regard to the use of garlic and onion as a source of medical and pharmaceutical compounds (Block 1985; Jain & Apitz-Castro 1987). The activity of ajoene, a compound derived from allicin, as an antithrombotic agent has been well documented. It has been suggested that it could be a very effective substitute for heparin during surgery (Jain & Apitz-Castro 1987). Ajoene has also been found to have a cytotoxic effect on various animal cell lines, with the tumorigenic line being the most sensitive (Scharfenburg *et al.* 1990). Allicin itself was found to be a specific inhibitor of the acetyl-CoA synthetases from plants, yeast, and mammals (Focke *et al.* 1990).

The "old wives' tales" concerning the health and dietetic benefits of garlic may be another example of the intuitive wisdom of our forbears.

References

Anderson, J.W. 1980. Incorporation of inorganic sulfate into cysteine. In: Miflin, B.J. (ed.), The Biochemistry of Plants, Vol. 5. Academic Press, New York, pp. 203-223.

Anderson, J.W. 1990. Sulfur metabolism in plants. In: Miflin, B.J. & Lea, P.J. (eds.), The Biochemistry of Plants, Vol 16. Academic Press, New York, pp. 327-381.

Baur, A.H. & Yang, S.F. 1972. Methionine metabolism in apple tissue in relation to ethylene biosynthesis. Phytochemistry 11: 3207-3214.

Black, S., Harte, E.M., Hudson, B. & Wartofsky, L. 1960. A specific reduction of L(-)methionine sulfoxide and a related nonspecific reduction of disulfides. J. Biol. Chem. 235: 2910-2916.

Block, E. 1985. The chemistry of garlic and onion. Scientific American 252: 114-119.

Brot, N. & Weissbach, H. 1991. Biochemistry of methionine sulfoxide residues in proteins. Biofactors 3: 91-96.

Doney, R.C. & Thompson, J.F. 1966. The reduction of *S*-methyl-L-cysteine sulfoxide and L-methionine sulfoxide in turnip and bean leaves. Biochim. Biophys. Acta 124: 39-49.

Focke, M., Feld, A. & Lichtenthaler, H.K. 1990. Allicin, a naturally occurring antibiotic from garlic, specifically inhibits acetyl-CoA synthetase. FEBS Lett. 261: 106-108.

Fujita, M., Endo, M. & Sano, M. 1990. Purification and characterization of alliin lyase from Welsh onion, *Allium fistulosum* L. Agric. Biol. Chem. 54: 1077-1079.

Giovanelli, J.G. & Mudd, S.H. 1971. Transsulfuration in higher plants. Partial purification and properties of β-cystathionase of spinach. Biochim. Biophys. Acta 227: 654-670.

Giovanelli, J., Mudd, S.H., & Datko, A.H. 1980. Sulfur amino acids in plants. In: Miflin, B.J. (ed.), The Biochemistry of Plants, Vol. 5. Academic Press, New York, pp. 453-505.

Greene, R.C. & Davis, N.B. 1960. Biosynthesis of *S*-methylmethionine in the jack bean. Biochim. Biophys. Acta 43: 360-362.

Halaseh, A., Nigam, S.N. & McConnell, W.B. 1977. Biosynthesis and metabolism of cystathionine in *Astragalus pectinatus*. Biochim. Biophys. Acta 496: 272-277.

Hall, D.I. & Smith, I.K. 1983. Partial purification and characterization of cystine lyase from cabbage (*Brassica oleracea* var *capitata*). Plant Physiol. 72: 654-658.

Hamamoto, A. & Mazelis, M. 1986. The C-S lyases of higher plants. Isolation and properties of homogeneous cystine lyase from broccoli (*Brassica oleracea* var. *botrytis*) buds. Plant Physiol. 80: 702-706.

Hanson, A.D. & Kende, H. 1976. Methionine metabolism and ethylene biosynthesis in senescent flower tissue of morning-glory. Plant Physiol. 57: 528-537.

Harrington, H.M. & Smith, I.K.1980. Cysteine metabolism in cultured tobacco cells. Plant Physiol. 65: 151-155.

Hartmann, T., Bax, K., & Scholz, R., 1984. Aliphatic L-amino acid decarboxylase from ferns (*Filicopsida*). Z. Naturforsch. 39C: 24-30.

Hattula, T. & Granroth, B. 1974. Formation of dimethyl sulphide from *S*-methylmethionine in onion seedlings (*Allium cepa*). J. Sci. Food Agric. 25: 1517-1521.

Ho, M.F. & Mazelis, M. 1993. The C-S lyases of higher plants. Determination of homology by immunological procedures. Phytochemistry, *in press*.

Jain, M.K & Apitz-Castro, R. 1987. Garlic: molecular basis of the putative 'vampire-repellant' action and other matters related to heart and blood. Trends in Biochem. Sci. 12: 252-254.

Jacobsen, J.V., Yamaguchi, Y, Mann, L.K. & Howard, F.D. 1968. An alkyl-cysteine lyase in *Tulbaghia violacea* and its relation to other alliinase-like enzymes. Phytochemistry 7: 1099-1108.

Jansen, H., Muller, B. & Knobloch, K. 1989a. Characterization of an alliin lyase preparation from garlic (*Allium sativum*). Planta Medica 55: 434-439.

Jansen, H., Muller, B. & Knobloch, K. 1989b. Alliin lyase from garlic, *Allium sativum*: Investigations on enzyme/substrate, enzyme/inhibitor interactions, and on a new coenzyme. Planta Medica 55: 440-445.

Kazaryan, R.A. & Goryachenkova, E.V. 1978. Alliinase: Purification and characterization. Biokhimya (Eng.) 43: 1905-1913.

Keilin, D. 1930. Cytochrome and intracellular oxidase. Proc. Roy. (London) B106: 418-444.

Kisliuk, R.L., Sakami, W.& Patwardhan, M.J. 1956. The metabolism of the methionine carbon chain in the intact rat. J. Biol. Chem. 221: 885-893.

Koesnandar, Nishio, N., Yamamoto, A. & Nagai, S. 1991. Enzymatic reduction of cystine into cysteine by a cell-free extract of *Clostridium thermoaceticum*. J. Fermentation & Bioengineering 72: 11-14.

Kramer, E. & Schmidt, A. 1984. Oxidation of cysteine to cystine by membrane fractions of *Chlorella fusca*. Planta 160: 235-241.

Kupiecki, F.P. & Virtanen, A.I. 1960. Cleavage of alkyl cysteine sulfoxides by an enzyme in onion. Acta Chem. Scand. 14: 1913-1918.

Lancaster, J.E. & Collin, H.A. 1981. Presence of alliinase in isolated vacuoles and of alkyl cysteine sulphoxides in the cytoplasm of onion (*Allium cepa*). Plant Sci. Lett. 22, 169-176.

Lewis, B.G., Johnsom, C.M. & Broyer, T.C. 1971. Cleavage of *Se*-methylselenomethionine salt by a cabbage leaf enzyme fraction. Biochim. Biophys. Acta 237: 603-605.

Mapson, L.W. 1953. The estimation of oxidized glutathione. Biochem. J. 44: 714-717.

Mazelis, M. 1959. Decarboxylation of methionine by an enzyme system from cabbage leaf. Biochem. Biophys. Res. Com. 1: 59-62.

Mazelis, M. 1960. Enzymic formation of CO_2 from the carboxyl group of methionine. Federation

Proceedings 19: 4.
Mazelis, M. 1962. The pyridoxal phosphate-dependent oxidative decarboxylation of methionine by peroxidase. I. Characteristics and properties of the reaction. J. Biol. Chem. 237: 104-108.
Mazelis, M. 1963. Demonstration and characterization of cysteine sulfoxide lyase in the Cruciferae. Phytochemistry 2: 15-22.
Mazelis, M. 1975. The cleavage of the ether linkage of *O*-methyl serine by the alkyl cysteine lyase of *Acacia farnesiana*. FEBS Lett. 54: 193-195.
Mazelis, M. 1980. Amino acid catabolism. In: Miflin, B.J. (ed.), The Biochemistry of Plants, Vol. 5. Academic Press, New York, pp. 541-567.
Mazelis, M. & Ingraham, L.L. 1962. The pyridoxal phosphate- dependent oxidative decarboxylation of methionine by peroxidase. II. Identification of 3-methylthiopropionamide as a product of the reaction. J. Biol. Chem. 237: 109-112.
Mazelis, M. & Creveling, R.K. 1975. Purification and properties of *S*-alkyl-L-cysteine lyase from seedlings of *Acacia farnesiana* Willd. Biochem. J. 147: 485-491.
Mazelis, M., Levin, B. & Mallinson, N. 1965. Decomposition of methyl methionine sulfonium salts by a bacterial enzyme. Biochim. Biophys. Acta 195: 106-114.
Mazelis, M., Beimer, N., & Creveling, R.K. 1967. Cleavage of L-cystine by soluble enzyme preparations from *Brassica* species. Arch. Biochem. Biophys. 120: 371-378.
Miyazaki, J.H. & Yang, S.F. 1987. The methionine salvage pathway in relation to ethylene and polyamine biosynthesis. Physiol. Plant. 69: 366-370.
Mudd, S.H. & Datko, A.H. 1986. Methionine methyl group metabolism in *Lemna*. Plant Physiol. 81: 103-114.
Mudd, S.H. & Datko, A.H. 1990. The *S*-methylmethionine cycle in *Lemna paucicostata*. Plant Physiol. 93: 623-630.
Nock, L.P. & Mazelis, M. 1986. The C-S lyases of higher plants: Preparation and properties of homogeneous alliin lyase from garlic (*Allium sativum*). Arch. Biochem. Biophys. 249: 27-33.
Nock, L.P. & Mazelis, M. 1987. The C-S lyases of higher plants: Direct comparison of the physical properties of homogeneous alliin lyase of garlic (*Allium sativum*) and onion (*Allium cepa*). Plant Physiol. 85: 1079-1083.
Nock, L.P. & Mazelis, M. 1989. Lack of homology between the alliin lyases of garlic and onion. Phytochemistry 28: 729-731.
Rabinkov, A., Wilchek, M. & Mirelman, D. 1991. Alliinase from *Allium sativum* (Garlic): Isolation and properties. FASEB Journal 5(5): A1149.
Rennenberg, H., Arabatzis, N. & Grundel, I. 1987. Cysteine desulphydrase in higher plants: Evidence for the action of L- and D-cysteine specific enzymes. Phytochemistry 26: 1583-1589.
Romano, A.H. & Nickerson, W.J. 1954. Cystine reductase of pea seeds and yeasts. J. Biol. Chem. 208: 409-416.
Sato, C.S., Byerrum, R.U., Albersheim, P. & Bonner, J. 1958. Metabolism of methionine and pectin esterification in a plant tissue. J. Biol. Chem. 233: 128-131.
Scharfenberg, K., Wagner, R. & Wagner, K.G. 1990. The cytotoxic effect of ajoene, a natural product from garlic, investigated with different cell lines. Cancer Lett. 53: 103-108.
Schmidt, A. 1982. A cysteine desulfhydrase from spinach leaves specific for D-cysteine. Z. Pflanzenphysiol. 107: 301-312.
Schmidt, A. & Erdle, I. 1983. A cysteine desulfhydrase specific for D-cysteine from the green alga *Chlorella fusca*. Z. Naturforsch. 38C: 428-435.
Schmidt, A., Rennenberg, H., Wilson, L.G., & Filner, P. 1985. Formation of methanethiol from methionine by leaf tissue. Phytochemistry 24: 1181-1185.
Schütz, B., De Kok, L.J., & Rennenberg, H. 1991. Thiol accumulation and cysteine desulfhydrase activity in H_2S-fumigated leaves and leaf homogenates of cucurbit plants. Plant Cell Physiol. 32: 733-736.
Schwimmer, S. & Kjaer, A. 1960. Purification and specificty of the C-S lyase of *Albizzia lophanta*. Biochim. Biophys. Acta 42: 316-324.
Schwimmer, S. & Mazelis, M. 1963. Characterization of alliinase of *Allium cepa*. Arch. Biochem. Biophys. 100: 66-73.
Schwimmer, S., Carson, J.F., Makower, R.U., Mazelis, M., & Wong, F.F. 1960. Demonstration of alliinase in a protein preparation from onion. Experientia 16: 449-450.
Splittstoesser, W.E. & Mazelis, M. 1967. The metabolism of methionine in higher plants. Catabolism of the methyl group by seedlings of various species. Phytochemistry 6: 39-47.

Staton, A.L. & Mazelis, M. 1991. The C-S lyases of higher plants: Homogeneous β-cystathionase of spinach leaves. Arch. Biochem. Biophys. 290: 46-50.
Stevenson, D.E., Akhtar, M. & Gani, D. 1990. L-methionine decarboxylase from *Dryopteris filix-mas*: Purification, characterization, substrate specificity, abortive transamination of the coenzyme, stereochemical courses of substrate decarboxylation, coenzyme transamination. Biochemistry 29: 7631-7647.
Tsuno, S. 1958. The nutritional value of *Allium* plants. XVI Alliinase in *Allium* plants. Bitamin 14: 659-664.
Whitaker, J.R. & Mazelis M. 1991. Enzymes and flavor development in the *Alliums*. In: Fox, P.F. (ed.), Food Enzymology, Vol 1. Elsevier, Essex, England, pp. 479-497.
Won, T. & Mazelis, M. 1989. The C-S lyases of higher plants. Purification and characterization of homogeneous alliin lyase of leek (*Allium* porrum). Physiol. Plant. 77: 87-92.

GLUTATHIONE METABOLISM IN PLANTS

Ludwig Bergmann[1] & Heinz Rennenberg[2]
[1]*Botanisches Institut der Universität zu Köln, Gyrhofstrasse 15, D-5000 Köln 41, F.R.G.;* [2]*Institut für Forstbotanik und Baumphysiologie, Professur für Baumphysiologie, Am Flughafen 17, D-7800 Freiburg i. Br., F.R.G.*

Introduction

Reviews on glutathione (γ-L-glutamyl-L-cysteinyl-glycine) usually start with the statement that glutathione is the major low molecular weight thiol in most organisms. For plants, this statement is not fully correct. There are at least two groups of higher plants in which another γ-glutamyl-cysteinyl-tripeptide is found either in addition to, or instead of glutathione. As first reported by Price (1957) homoglutathione, a glutathione-homologue tripeptide in which the C-terminal amino acid is replaced by β-alanine, is found in several members of the order *Fabales*. In gramineous plants that belong to the family *Poaceae* a second homologue of glutathione with the sequence γ-L-glutamyl-L-cysteinyl-serine has recently been detected and identified by Klapheck *et al.* (1992); for this tripeptide the name "hydroxymethylglutathione" has been proposed. It cannot be excluded that further γ-glutamyl-cysteinyl tripeptides will be found in extended surveys of the Plant Kingdom.

The functions of both glutathione-homologues seem to be similar to those of glutathione (Table 1). Leaves of *Phaseolus vulgaris* and *Trifolium pratense* exposed to H_2S show increased levels of homoglutathione, indicating its function in the storage of reduced sulfur (Buwalda *et al.* 1993). Homoglutathione is used in the long-distance transport of reduced sulfur in *Vigna radiata* (Macnicol & Bergmann 1984). Like glutathione disulfide, oxidized homoglutathione is reduced by NADPH-dependent glutathione reductase (Klapheck 1988). In the two species tested (*Phaseolus coccineus* and *Glycine max*) homoglutathione is present in the chloroplasts. However, it is still doubtful whether homoglutathione participates in the H_2O_2-scavenging pathway in this organelle (Zopes 1990; Zopes *et al.* 1993). Like glutathione, homoglutathione serves as the substrate for the formation of homo-phytochelatins (Grill *et al.* 1986) and can be used for the detoxification of xenobiotics by glutathione-*S*-transferases (Lamoureux & Russness 1989b).

Leaves of *Triticum* supplied with high concentrations of sulfate contained elevated levels of hydroxymethylglutathione and glutathione indicating a function of both tripeptides as storage forms of reduced sulfur (Klapheck *et al.* 1992). In its oxidized form hydroxymethylglutathione is reduced in a NADPH-dependent reaction by glutathione reductase from yeast (Klapheck *et al.* 1992). It is present in isolated chloroplasts of *Triticum aestivum* and probably can be used in the H_2O_2-scavenging pathway together with glutathione. The presence of serine in hydrolysates of phytochelatins from *Agrostis gigantea* (Rauser *et al.* 1986) implies that hydroxymethylglutathione is used as a precursor for the synthesis of phytochelatins with the structure (γ-Glu-Cys)nSer.

Sulfur Nutrition and Assimilation in Higher Plants
edited by L.J. De Kok et al., pp. 109–123

Table 1. Functions of γ-glutamyl-cysteinyl-tripeptides in higher plants.

Function	γ-Glu-Cys-Gly	γ-Glu-Cys-β-Ala	γ-Gly-Cys-Ser
Storage and transport of reduced sulfur	Smith (1975) Rennenberg *et al.* (1979) Bonas *et al.* (1982) Buwalda *et al.* (1990) Rauser *et al.* (1991)	Buwalda (pers. com.) Macnicol & Bergmann (1984)	Klapheck *et al.* (1992)
Scavenging of H_2O_2 by the ascorbate-GSH reduction pathway	Foyer & Halliwell (1976) Jablonski & Anderson (1978) Gilham & Dodge (1986) Hossain & Asada (1987)	Klapheck *et al.* (1988) Zopes *et al.* 1993	Klapheck *et al.* (1992)
Detoxification of xenobiotics	Breaux *et al.* (1987) Lamoureux & Rusness (1989a)	Breaux *et al.* (1987) Lamoureux & Rusness (1989b)	
Phytochelatin synthesis	Grill *et al.* (1990) Rauser (1990) Steffens (1990)	Grill *et al.* (1986)	Klapheck *et al.* (1992) Rauser *et al.* (1986)

Apparently, the homologues of glutathione play an important role in the thiol metabolism of numerous plants. Therefore, the present article will deal not only with glutathione, but will include our knowledge on synthesis and degradation of homoglutathione and hydroxymethylglutathione.

Synthesis of glutathione and glutathione homologues

The results of experiments with several species show that plant cells contain glutathione synthetase- and γ-glutamyl-cysteine synthetase activity. Apparently, the synthesis of glutathione proceeds in the same two-step reaction sequence described for animal cells (Meister & Anderson 1983):

$$\text{Glu} + \text{Cys} + \text{ATP} \rightarrow \gamma\text{-Glu-Cys} + \text{ADP} + \text{Pi} \quad (1)$$
$$\gamma\text{-Glu-Cys} + \text{Gly} + \text{ATP} \rightarrow \gamma\text{-Glu-Cys-Gly} + \text{ADP} + \text{Pi} \quad (2)$$

γ-Glu-Cys synthetase

The first reaction involves the ATP-dependent formation of γ-glutamyl-cysteine catalyzed by γ-glutamyl-cysteine synthetase (EC 6.3.2.2). The plant enzyme has first been analyzed in Sephacryl eluents of extracts from suspension cultures of *N. tabacum* by determination of γ-glutamyl-cysteine as its monobromobiman derivative (Hell & Bergmann 1990). The enzyme shows maximum activity at pH 8.0 and has an absolute requirement for Mg^{2+}. Gel filtration yielded a relative molecular mass (M_r) of 60 kDa. Treatment with 5 mM dithioerythritol led to a heavy loss of activity and dissociation into subunits (M_r 34 kDa). Incubation with low concentrations of mercaptoethanol also resulted in a heavy loss of enzyme activity. As both, mercaptoethanol and dithioerythritol are frequently used to protect thiol groups in plant

	K_m (mM)			K_i (mM)	
	Glu	Cys	α-Amino butyrate	GSH	
Nicotiana tabacum	10.4	0.19	6.36	0.42	Hell & Bergmann (1990)
Petroselinum crispum	4.0	0.074		0.27	Schneider (1992)

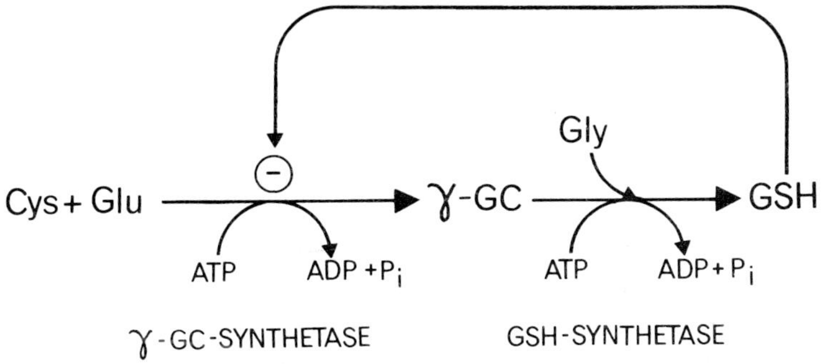

Fig. 1. γ-Glutamyl-cysteine synthetase: kinetic properties and regulation.

extracts, the loss of γ-glutamyl-cysteine synthetase activity in the presence of these compounds is probably a reason why earlier attempts to detect this enzyme in plants did not succeed.

The enzymes from tobacco and parsley show high affinities for cysteine and low affinities for glutamate. The enzyme activity is inhibited by physiological GSH concentrations. These data indicate that the rate of GSH synthesis may be regulated *in vivo* by feedback inhibition of γ-glutamyl-cysteine synthetase by GSH (Fig. 1).

Like γ-glutamyl-cysteine synthetase from mammalian cells the plant enzyme is inhibited by buthionine sulfoximine (BSO) and L-methionine sulfoximine (Hell & Bergmann 1990). As a consequence, GSH synthesis can be blocked in plant cells by both inhibitors. However, inhibition of glutathione synthesis by BSO can be a transient effect in plant cells and, therefore, has to be verified in each individual experiment.

Glutathione synthetase

In the second step of GSH synthesis glycine is added to the C-terminal site of γ-glutamyl-cysteine in an ATP-dependent reaction. This reaction is catalyzed by glutathione synthetase (EC 6.3.2.3). The enzyme has been demonstrated in several plant species (Table 2). Its characterization showed similar affinities for γ-glutamyl-cysteine and glycine. The apparent K_m values for γ-glutamyl-cysteine measured *in vitro* are close to the concentrations of this dipeptide found in plant cells. The enzyme showed maximum activity at pH 8.0 to 9.0, an absolute requirement for Mg^{2+}, and was slightly stimulated by K^+.

The molecular masses (M_r) of the enzyme as estimated by gelfiltration were 113 ± 3 kDa for GSH synthetase from photoheterotrophic and heterotrophic sus-

Table 2. Apparent K_m values (in mM) for the various substrates and molecular mass (M_r; kDa) of glutathione synthetase. (n.a. = not accepted as substrate). References: I, Hell & Bergmann (1988); II, Schneider (1990); III, Zimmer, personal communication; IV, Schlunz (1991).

	K_m					
	γ-GC	γ-GA	Gly	β-Ala	M_r	Ref.
Nicotiana tabacum suspension cultures	0.022	0.030	0.30	n.a.	113 ± 3	I
Petroselinum crispum suspension cultures	0.018	–	0.093	n.a.		II
Spinacia oleracea	0.032	–	0.72	n.a.		III
leaves	–	0.091	1.00	–		
Pisum sativum leaves	0.071	–	0.23	n.a.	117 ± 5	IV

pension cultures of *N. tabacum* and 117 ± 5 kDa for the enzyme from leaves and roots of *Pisum sativum*. In former experiments a molecular mass of 68 kDa has been determined for GSH synthetase from tobacco suspension cultures (Hell & Bergmann 1988), which is nearly half the molecular mass found in recent experiments. In the experiments of Hell & Bergmann (1988) the enzyme preparation has been dialyzed for several hours against EDTA-containing buffer, which probably resulted in a dissociation of the enzyme into two subunits. GSH synthetase from rat kidney showed a molecular mass of 118 kDa and also dissociated into two, apparently identical subunits at SDS gel electrophoresis (Meister 1985).

With one exception all GSH synthetases so far tested were specific for glycine and did not accept β-alanine as a substrate. Therefore, the enzyme does not catalyze the formation of homoglutathione.

Homoglutathione synthetase:

There has been a controversial discussion as to whether the formation of homoglutathione in legumes is catalyzed by a specific homoglutathione synthetase or by a glutathione synthetase with a broad substrate specificity using glycine and β-alanine as substrates. Macnicol (1987) described a synthetase isolated from Pisum sativum that showed a high affinity for glycine and a 100 times lower affinity for β-alanine (Table 3). He also isolated a synthetase from *Vigna radiata* exhibiting a high affinity for β-alanine and a low affinity for glycine. Analyzing the enzyme responsible for the synthesis of hGSH in primary leaves of *Phaseolus coccineus* Klapheck *et al.* (1988) reported a homoglutathione synthetase with a high specificity for β-alanine as compared to glycine. This enzyme synthesized hGSH and GSH *in vitro* in a ratio of 100 : 1 when supplied with equal concentrations of β-alanine and glycine.

For *Vigna radiata* and *Phaseolus* coccineus the enzyme characteristics are in agreement with the thiol composition found in both species. As the careful analysis of Klapheck (1988) revealed, leaves, seeds, and roots of these species contain high levels of hGSH and only traces of GSH. The ratio of GSH: hGSH found was in the order of 1 : 100.

Table 3. Apparent K_m values (in mM) for the various substrates of homoglutathione synthetase. References: I, Klapheck *et al.* (1988); II, Macnicol (1987).

	β-Ala	Gly	γ-GC	Ref.
Phaseolus coccineus	1.34	98.0	0.073	I
Vigna radiata	0.33	7.5	0.75	II
Pisum sativum	14.0	0.19	0.35	II

Table 4. Concentrations (in nmol g fresh weight^{-1}) of glutathione and homoglutathione in leaves and roots of *Pisum sativum* L.

Varieties	Leaves		Roots	
	GSH	hGSH	GSH	hGSH
rheinländerin	445 ± 83	0-3	117 ± 48	79 ± 16
onward	418 ± 31	0-3	90 ± 46	81 ± 13
siegerin	392 ± 27	0-3	74 ± 29	83 ± 5
allerfrüheste	441 ± 70	0-3	140 ± 32	88 ± 8
übereich	414 ± 10	0-3	88 ± 49	56 ± 3

The pH optimum, the Mg^{2+} requirement, and the stimulation by K^+ of the homoglutathione synthetases showed close similarities to the glutathione synthetases described above. The homoglutathione synthetase from *P. coccineus* exhibited a high affinity for γ-glutamyl-cysteine which is in agreement with the data reported for GSH synthetase (Table 2). In contrast the apparent K_m reported for the enzymes of *V. radiata* and *P. sativum* are relatively high as compared to the γ-glutamyl-cysteine concentrations of 5-10 μM in plant cells.

In *Pisum sativum* the thiol composition and content differ considerably between leaves and roots. In leaves of several varieties 400-450 nmol GSH g fresh weight^{-1} and only traces of hGSH have been found; in contrast, roots of 14 days old plants contain 70-140 nmol GSH and 60-90 nmol hGSH g fresh weight^{-1} (Table 4). As the β-alanine content in the leaves is 5 times higher than in the roots at almost equal amounts of glycine, it is difficult to understand how this organspecific thiol composition can be obtained by an individual glutathione synthetase with a broad substrate specificity.

Therefore, the enzymes responsible for the synthesis of glutathione and homoglutathione were analyzed in leaves and roots of *P. sativum* by product determination of monobromobimane derivatives (Schlunz 1991). In leaves and shoots a glycine-specific GSH synthetase was found that did not accept β-alanine as a substrate. In roots GSH synthetase activity as well as hGSH synthetase activity were detected. The extractable hGSH synthetase activity in young roots increased by almost 400% within 12 days, whereas the GSH synthetase activity remained constant (Fig. 2). This finding suggests that at least two enzymes are involved in the synthesis of GSH and hGSH in roots. This assumption is supported by investigations on the effect of cadmium on the GSH and hGSH synthetase activities in pea roots. Incubation of hydroponically grown plants with Cd resulted in an increase of GSH synthetase activity by a factor of 3.5 as also observed by Rüegsegger *et al.* (1990). In contrast, hGSH synthetase activity did not change as compared to the controls (Fig. 3). This effect of Cd

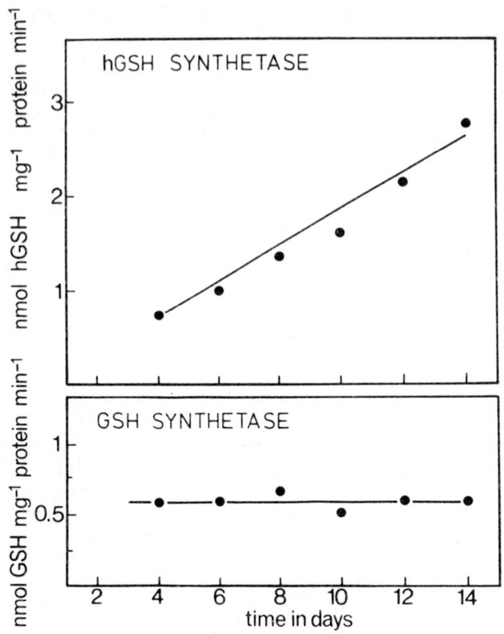

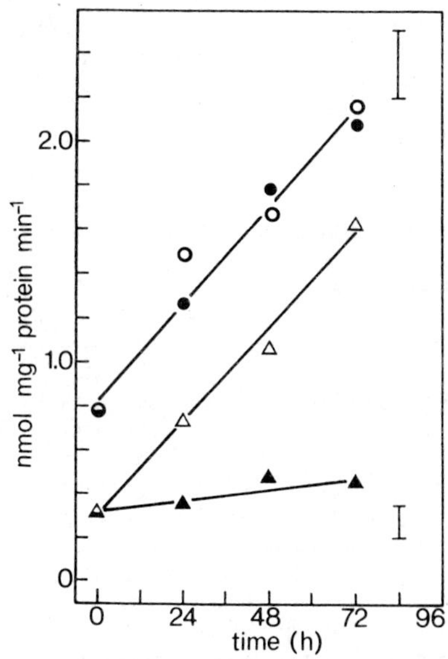

Fig. 2. Activities of glutathionesynthetase and homoglutathionesynthetase in pea seedlings (from Schlunz 1991).

Fig. 3. Glutathione- and homoglutathionesynthetase activity in Cd-exposed roots of *Pisum sativum* (from Schlunz 1991). GSH synthetase, △ + Cd, ▲ − Cd; hGSH synthetase, ○ + Cd, ● + Cd.

Table 5. Apparent K_m values (in mM) for the various substrates and molecular mass (M_r; kDa) of glutathione-/homoglutathione synthetase in *Pisum sativum* L. (in roots a = glutathione synthetase, b = homoglutathione synthetase). References: I, Macnicol (1987); II, Schlunz (1991).

	K_m				
Varieties	Gly	β-Ala	γ-GC	M_r	Ref.
greenfast	0.19	14.0	0.35	85	I
rheinländerin					
leaves	0.23	–	0.071	117 ± 5	II
roots a	0.21	–	0.069	120	II
roots b	4.5-6.0	0.83	0.113	120	II

cannot be explained by the assumption that an individual enzyme is responsible for GSH and hGSH synthesis.

In addition, measurements of enzyme kinetics have demonstrated the presence of two enzymes with different substrate specificities in root extracts: a GSH synthetase with substrate affinities corresponding with those of GSH synthetase found in the leaves; and a hGSH synthetase with a high affinity for β-alanine and a low affinity for glycine corresponding well with the substrate affinities of hGSH synthetase of *Vigna radiata* (Table 5).

These results indicate that legumes possess a specific hGSH synthetase that is responsible for the synthesis of hGSH and the small amounts of GSH found in plants of the tribe *Phaseoleae*. In *Pisum sativum*, a typical representative of the *Vicieae*, two enzymes are expressed, a GSH synthetase in the shoots and the roots, and a hGSH synthetase that is only formed in the roots (Schlunz 1991). The organ-specific distribution of the enzyme activities as well as the changes in enzyme activities during development and at exposure to Cd suggest that GSH- and hGSH synthetase are encoded on different sites and are subject to activation by different processes. It will certainly by a challenge to unravel the molecular mechanisms behind this phenomenon.

Synthesis of hydroxymethylglutathione

The synthesis of hydroxymethylglutathione has so far not been elucidated. Inhibition of γ-glutamyl-cysteine synthetase in leaves of *T. aestivum* with BSO results in a drastic drop of both, GSH and hydroxymethyl-GSH levels. This finding suggests that γ-glutamylcysteine is a precursor of hydroxymethyl-GSH synthesis (Klapheck *et al.* 1992). However, all experiments to add serine to the C-terminal site of γ-glutamyl-cysteine similar to GSH or hGSH synthesis have failed so far. As hydroxymethyl-GSH is always found together with glutathione in the species analyzed, it may be produced by processing the GSH molecule, *e.g.* by hydroxymethylation of the terminal glycine moiety.

Localization of the enzymes of glutathione and homoglutathione synthesis

The localization of the enzymes of glutathione and homoglutathione synthesis has only been studied in few plant species. In all investigations the enzymes were found in isolated chloroplasts and in the cytosol. Chloroplasts of *P. sativum* contained about 72% of the γ-glutamyl-cysteine synthetase activity and about 48% of the GSH-synthetase activity found in leaf homogenates. In chloroplasts of *Spinacia oleracea* about 61% of the total γ-glutamyl-cysteine synthetase activity and 58% of the total GSH-synthetase activity of the cells were detected (Hell & Bergmann 1990).

γ-Glutamyl-cysteine synthetase has also been demonstrated by Zopes (1990) in isolated chloroplasts of *P. coccineus*, but an exact quantification of the subcellular distribution has not been achieved due to the high instability of the enzyme. About 17% of the total hGSH-synthetase activity of leaf cells were found in isolated chloroplasts of *P. coccineus*.

These data are consistent with the intracellular distribution of the γ-glutamyl-cysteinyl-tripeptides (Table 6) and indicate that glutathione and homoglutathione are synthesized at least in two cellular compartments. So far information on the existence of isozymes of glutathione and homoglutathione synthesis is lacking. Despite the localization of glutathione reductase in plant mitochondria (Klapheck *et al.* 1990) data on the synthesis of glutathione in this organelle have so far not been reported.

Is homoglutathione involved in H_2O_2 scavenging in chloroplasts?

Glutathione and glutathione reductase are thought to participate in the H_2O_2-scavenging pathway in chloroplasts as well as in the cytoplasm (Anderson 1990; Polle &

Table 6. Intracellular distribution of glutathione and homoglutathione in leaves. Data for whole-leaf extracts and isolated chloroplasts. Concentrations were calculated on the assumption of average volume/chlorophyll ratios as indicated. GSH (hGSH) contents of the cytosol were calculated by subtracting the GSH (hGSH) content of the isolated chloroplasts from that of the whole-leaf extracts. References: (1), Bielawski & Joy (1986); (2) Zopes (1990); (3) Foyer & Halliwell (1976); (4) Zopes *et al.* 1993.

		GSH content (nmol mg Chl.$^{-1}$)		GSH concentration (mM)		
Species		Chloroplasts	Leaves	Stroma (21 μl mg Chl.$^{-1}$)	Cytosol (10 μl mg Chl.$^{-1}$)	Cytosol (25 μl mg Chl.$^{-1}$)
Pisum sativum	(1)	26	264	1.24	23.7	9.5
	(2)	27	221	1.28	19.4	7.8
Spinacia oleracea	(2)	49	313	2.33	26.3	10.5
	(3)	74.3		3.54		
Phaseolus coccineus	(2)	8.5	147	0.4	13.6	5.5
Glycine max	(4)	16.2	186	0.77	17.0	6.8

Table 7. Enzyme activities of the ascorbate-glutathione-reduction pathway and thiol concentration in isolated chloroplasts of glutathione- and homoglutathione-containing plants (Zopes *et al.* 1993).

	Spinacia oleracea	*Pisum sativum*	*Phaseolus coccineus*	*Glycine max*
thiol concentration (nmol mg chl^{-1})	49 (GSH)	27 (GSH)	9 (hGSH)	16 (hGSH)
(mM)	2.33	1.3	0.43	0.65
ascorbate peroxidase (μmol ASA mg chl^{-1})	213.6	207.6	70.8	187.6
monodehydro-ascorbate reductase (μmol NADH mg chl^{-1})	60.6	44.4	60.6	33.6
GSH dehydrogenase (μmol ASA mg chl^{-1})	19.2	15.6	n.d.	n.d.
GSSG reductase (μmol mg chl^{-1})	84.0	54.0	12.6	n.d.
intactness (%)	90	96	75	75

Rennenberg 1993). In this context the question has to be addressed as to whether homoglutathione and hydroxymethylglutathione also function in this pathway. Table 6 shows that the concentrations of hGSH in chloroplasts of *P. coccineus* and *Glycine max* are much lower than the concentrations found in chloroplasts of *S. oleracea* and *P. sativum*, where the ascorbate-glutathione-reduction pathway has been established. The activities of GSSG reductase and GSH dehydrogenase were found to be very low in leaf extracts of *P. coccineus* and *Glycine max* and could not be detected in isolated chloroplasts from these species (Table 7). In contrast the activities of ascorbate peroxidase and monodehydroascorbate reductase in the chloroplasts of

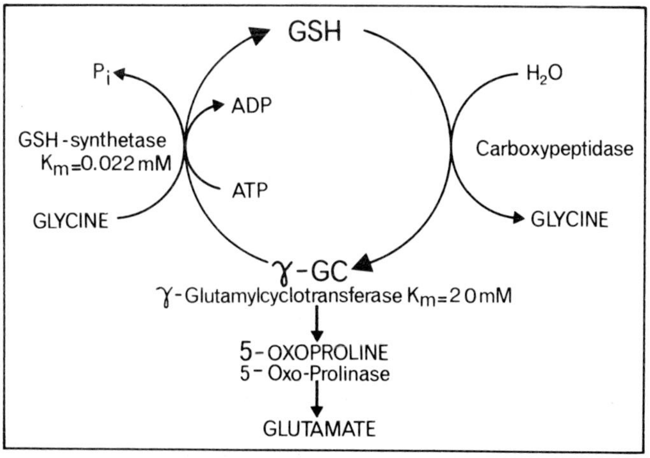

Fig. 4. "Futile cycle" of glutathione degradation and synthesis.

P. coccineus and *Glycine max* were comparable to those found in *S. oleracea* and *P. sativum* chloroplasts. These findings strongly suggest that regeneration of ascorbate in chloroplasts of *P. coccineus* and *Glycine max* mainly proceeds by monodehydroascorbate reductase activity, whereas hGSH plays a secondary role in H_2O_2-scavenging.

Degradation of glutathione

Based on a series of experiments with suspension cultures of *Nicotiana tabacum* it has been suggested that GSH degradation in plants is initiated by the cleavage of the C-terminal peptide bound of GSH yielding γ-glutamyl-cysteine and glycine (Rennenberg & Lamoureux 1990). The γ-glutamyl-cysteine produced in this reaction is further degraded to cysteine and glutamate by the successive action of γ-glutamylcyclotransferase (γ-GCT) and 5-oxo-prolinase which are both present in tobacco cells (Rennenberg & Lamoureux 1990). The first step of this pathway is consistent with the metabolism of GSH-conjugates of pesticides in plants, where γ-glutamyl-cysteine conjugates have consistently been observed as first degradation products (Lamoureux & Russness 1989a). Therefore, this pathway has generally been accepted, although carboxypeptidase activities specific for GSH and/or GSH-conjugates have not been demonstrated. However, as γ-glutamyl-cysteine is also a substrate of GSH synthetase, the assumption of γ-glutamyl-cysteine as an intermediate of glutathione degradation generates some theoretical difficulties. Both processes, glutathione degradation and glutathione synthesis, are supposed to take place in the cytoplasm. Therefore, it is not easy to conceive a mechanism by which a futile cycle of glutathione degradation and glutathione synthesis is avoided (Fig. 4). Further, the affinity of γ-GCT for γ-glutamyl-cysteine seems to be much too low to compete successfully with GSH synthetase that showed a 100 times higher substrate affinity in tobacco cells (γ-GCT: Steinkamp *et al.* 1987; GSH synthetase: this paper, Table 2).

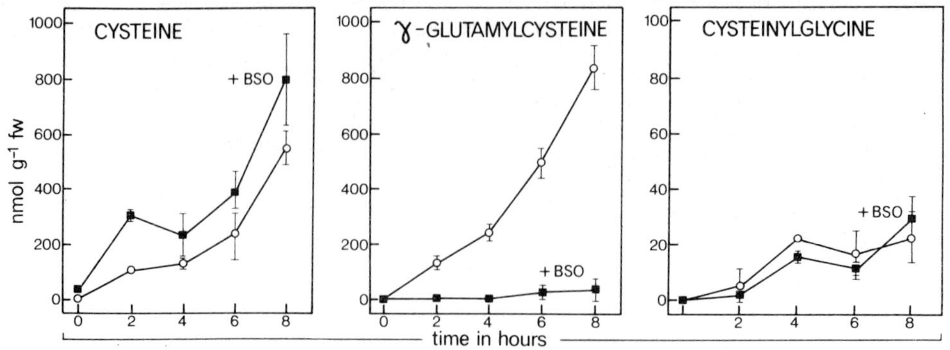

Fig. 5. Products of glutathione degradation in leaves of *Glycine max* L. (from Sommer 1991).

Cysteinyl-glycine, a new intermediate in glutathione metabolism of plants

Recent experiments with higher plants indicate another pathway of GSH degradation. In this pathway degradation of GSH is initiated by the removal of the γ-glutamyl-moiety catalyzed by γ-glutamyltranspeptidase (γ-GT; EC 2.3.2.2). The resulting dipeptide is then hydrolyzed by a dipeptidase, yielding cysteine and glycine. Evidence for this pathway that bypasses the formation of γ-glutamyl-cysteine and, thus, the interference with GSH synthesis comes from the finding of cysteinyl-glycine as an intermediate of GSH degradation and from studies of the enzymes involved (Sommer 1991, Schneider & Rennenberg 1992).

Detached primary leaves of *Glycine max* fed GSH with the transpiration stream degrade GSH at a rate of about 225 nmol g fresh weight^{-1} h^{-1}. The degradation of GSH results in considerably increased levels of cysteine and γ-glutamyl-cysteine and the appearance of small amounts of cysteinyl-glycine (Fig. 5). Inhibition of γ-glutamyl-cysteine synthetase by BSO completely prevents the accumulation of γ-glutamyl-cysteine. At the same time the cysteine content increases, whereas the cysteinyl-glycine level does not change significantly. These results indicate that the accumulation of γ-glutamyl-cysteine is not caused by GSH degradation but is the result of stimulated γ-glutamyl-cysteine synthesis caused by an increase in the cysteine and glutamate levels of the cells. Similar results were obtained with suspension cultures of *Hordeum vulgare* incubated with GSH. Feeding of cysteinyl-glycine to these cultures resulted in an accumulation of cysteine and glutathione. Apparently, cysteinyl-glycine is hydrolyzed at a high rate and the reaction products are used for GSH synthesis (Table 8). These results suggest that GSH is degraded in *Glycine max* and *H. vulgare* via the γ-glutamyltranspeptidase-dipeptidase pathway that has previously been described by Jaspers *et al.* (1985).

The enzymes needed for these reactions are present in leaves of *Glycine max* and suspension cultures of *H. vulgare*. In soybean leaves 63% of the γ-GT activity are soluble and have been found in the cytosolic fraction; 37% of the activity were associated with the 100,000 g fraction and could be solubilized with Triton X 100. The cytosolic γ-GT exhibited a higher affinity to GSH (K_m = 0.21 ± 0.03 mM) than to the artificial substrate γ-glutamyl-p-nitroanilide (K_m = 0.54 ± 0.09 mM). Also the rate of transfer of the γ-glutamyl group to an amino acid acceptor was more than

Table 8. Degradation of glutathione and cysteinyl-glycine in suspension cultures of *Hordeum vulgare* L. Suspension cultures in GAMBORG 2B5 medium were incubated with and without 5 mM glutathione or 1 mM cysteinyl-glycine. After 12 h of incubation the cells were harvested, washed and extracted with 0.1 N HCl. Thiols were determined as monobromobimane derivatives by reversed phase HPLC as described by Klapheck (1988).

	Control	5 mM GSH	1 mM Cys-gly
GSH	461 ± 45	1309 ± 65	789 ± 16
Cys-gly	13	98 ± 18	51 ± 14
Cysteine	99 ± 13	508 ± 38	412 ± 14
γ-Glu-cys	15 ± 3	52 ± 7	8 ± 3

three times faster with GSH as a donor than with γ-glutamyl-p-nitroanilide. The γ-GT of *Glycine max* shows a broad substrate specificity: in addition to the transfer of the γ-glutamyl moiety to amino acid acceptors, it catalyzes the transfer of the γ-glutamyl group to GSH itself, and the hydrolytic cleavage of the γ-glutamyl moiety of GSH. The relative proportion of transpeptidation and hydrolysis is dependent on the concentration of the acceptor amino acids and the pH. High rates of transpeptidation are observed at high concentrations of acceptor amino acids and alkaline pH, high rates of hydrolysis are found at low concentrations of acceptor amino acids and slightly acidic pH. These results indicate that GSH hydrolysis and not transpeptidation is the major reaction catalyzed by γ-GT *in vivo*, as also concluded for the γ-GT from rat kidney by McIntyre & Curthoys (1979).

Dipeptidase activity catalyzing the hydrolytic cleavage of cys-gly *in vitro* was detected in leaf- and cell-extracts of *Glycine max* and *H. vulgare*. Incubation of suspension cultures of H. vulgare with cysgly resulted in a marked increase in cellular cysteine and glycine, indicating the hydrolytic cleavage of cys-gly *in vivo* (Table 8).

Pathways of glutathione degradation in plant cells

The present results indicate that glutathione can be degraded in plant cells via different pathways. In *Glycine max* and *H. vulgare* GSH degradation is initiated by γ-GT catalyzing the hydrolysis of GSH into glutamate and cys-gly. Subsequently, cysgly is hydrolyzed by a dipeptidase activity to yield cysteine and glycine (Fig. 6, II). The transfer of the γ-glutamyl moiety of GSH to an amino acids acceptor, thought to initiate GSH degradation in animal cells (Fig. 6, I; Meister 1981), appears not to play a significant role in GSH degradation in plants. As cysgly-conjugates of pesticides have recently been found in plant cells (Lamoureux, personal communication), it appears possible that the degradation of GSH-conjugates can also proceed via the γ-GT-dipeptidase-pathway. In *N. tabacum* GSH degradation seems to be initiated by a carboxypeptidase catalyzing the hydrolysis of GSH into γ-glutamyl-cysteine and glycine (Fig. 6, III). The γ-glutamyl-cysteine produced in this reaction can be further degraded to cysteine and glutamate by the successive action of γ-GCT and 5-oxoprolinase. In several plant species GSH-conjugates of pesticides also seem to be degraded via this pathway. To verify the pathways proposed for the degradation of GSH in plants, labeling experiments with radioactive GSH with different plant species are required. A first attempt with *N. tabacum* using ^{35}S-GSH resulted in inten-

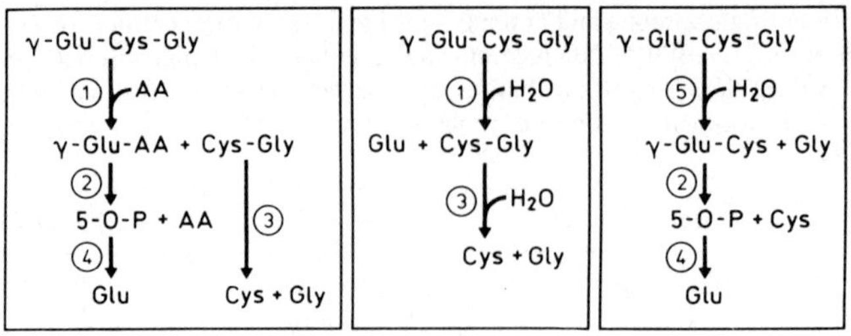

Fig. 6. Pathways of glutathione degradation. 1, γ-glutamyl transpeptidase; 2, γ-glutamyl cyclotransferase; 3, dipeptidase; 4, 5-oxo-prolinase; 5, carboxypeptidase.

sive labeling of the cysteine pool of the cells, but not of the γ-glutamyl-cysteine or the cysteinyl-glycine pools (Schneider 1992). The question as to whether degradation of GSH and GSH-conjugates proceeds via the same or via different pathways in the same species remains to be elucidated. This question is of particular interest for the regulation of GSH metabolism.

Regulation of glutathione content

The glutathione content of plant cells is determined by (1) GSH-synthesis, (2) GSH-degradation, (3) its use in biosynthetic processes, *e.g.* the formation of GSH-conjugates and phytochelatins, and (4) its transport out of, and into the cell. Neither the regulation of these individual processes, nor their regulatory interaction is presently understood. Most of the information available deals with the regulation of glutathione synthesis. As described above the first step of glutathione synthesis is subject to feedback inhibition by physiological GSH concentrations *in vitro*. Experiments in which the GSH level is reduced by Cd-induced phytochelatin synthesis show that this feedback inhibition also operates *in vivo*. Exposure of plant cells to Cd results in a decline in the GSH content and an extensive synthesis of phytochelatins (Rauser 1990; Steffens 1990). In cultured cells of *Petroselinum* and *N. tabacum*, where long-distance transport processes can be neglected, the rate of GSH synthesis can be calculated from the increase in the phytochelatin content. Fig. 7 shows that during a time, the GSH content itself is maintained at a low level, the rate of GSH synthesis is increased 5- to 8-fold. Simultaneously, the γ-glutamyl-cysteine content increases by a factor of 10 as compared to the controls and the cysteine content is doubled (Fig. 7). A similar accumulation of γ-glutamyl-cysteine has been observed in roots of maize seedlings exposed to Cd (Rauser *et al.* 1991). Apparently, a new steady state is reached upon Cd exposure, in which low GSH concentrations are maintained at enhanced GSH synthesis. This may be obtained by a partial release of the feedback inhibition of γ-glutamyl-cysteine synthetase by GSH. This assumption is consistent with the finding of an increase in the γ-glutamyl-cysteine content of the cells. In addition, this increase shows that GSH synthetase is partially inhibited by exposure to Cd *in vivo*, as previously observed *in vitro* (Hell *et al.* 1990).

The release of the feedback inhibition of γ-glutamyl-cysteine synthetase by GSH may be a means of plants to rapidly react to enhanced requirements of GSH caused

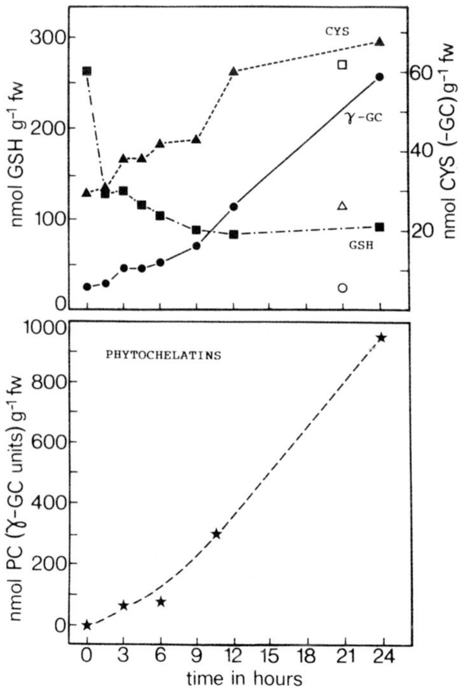

Fig. 7. Thiol- and phytochelatin contents of Cd-exposed cells of *Petroselinum crispum* (from Schneider 1992).

by environmental stresses. In the long term range enhanced GSH synthesis upon Cd-exposure may also be obtained by an increase in the amount of GSH synthetase (Rüegsegger *et al.* 1990; Schlunz 1991). Such an increase has also been observed in crop plants treated with safeners, *i.e.* compounds that decrease herbicide injury to crop species without reducing herbicide injury to weeds (Hatzios & Hoagland 1989; Lamoureux & Rusness 1989a). This increase in GSH synthetase reflects an enhanced potential of crops for the conjugation of herbicides with GSH, a major path of herbicide detoxification. The finding of elevated GSH levels in safener treated crops supports the idea that GSH levels are controlled at the molecular level not only by feedback regulation but also by other processes. However, all these regulatory systems can only operate in an appropriate way, if sufficient sulfate reduction and cysteine synthesis takes place. Experiments of Nussbaum *et al.* (1988) provide evidence that sulfate reduction is enhanced upon Cd exposure. In corn, sulfate reduction was also found to be stimulated by exposure to a safener (Lamoureux & Rusness 1989a). Apparently, assimilatory sulfate reduction and glutathione synthesis can be regulated in a coordinated way. The mechanisms involved in this coordination have so far not been identified.

References

Anderson, J.W. 1990. Sulfur metabolism in plants. In: Miflin, B.J. & Lea, P.J. (eds.), The Biochemistry of plants, Vol. 16. Academic Press, New York, pp. 327-381.

Bielawski, W. & Joy, K.W. 1986. Reduced and oxidized glutathione and glutathione reductase activity in tissues of *Pisum sativum*. Planta 169: 267-272.

Bonas, U., Schmitz, K., Rennenberg, H. & Bergmann, L. 1982. Phloem transport of sulfur in *Ricinus*. Planta 155: 82-88.

Breaux, E.J., Patanella, J.E. & Sanders, E.F. 1987. Chloroacetanilide herbicide selectivity: analysis of glutathione and homoglutathione in tolerant, susceptible, and safened seedlings. J. Agric. Food Chem. 35: 474-478.

Buwalda, F., Stulen, I., De Kok, L.J. & Kuiper, P.J.C. 1990. Cysteine, γ-glutamyl-cysteine and glutathione contents of spinach leaves as affected by darkness and application of excess sulfur. II. Glutathione accumulation in detached leaves exposed to H_2S in the absence of light is stimulated by the supply of glycine to the petiole. Physiol. Plant. 80: 196-204.

Buwalda, F., De Kok, L.J. & Stulen, I. 1993. Effects of atmospheric H_2S on thiol composition of crop plants. J. Plant Physiol. 142: *in press*

Foyer, C.H. & Halliwell, B. 1976. The presence of glutathione and glutathione reductase in chloroplasts: a proposed role in ascorbic acid metabolism. Planta 133: 21-25.

Grill, E., Winnacker, E.-L. & Zenk, M.H. 1990. Phytochelatins, the heavy metal binding peptides of the plant kingdom. In: Rennenberg, H., Brunold, C., De Kok, L.J. & Stulen, I. (eds.), Sulfur Nutrition and Sulfur Assimilation in Higher Plants. SPB Acad. Publ., The Hague, pp. 89-95.

Hatzios, K.K. & Hoagland, R.E. 1989. Herbicide Safeners: Progress and Prospects. Academic Press, New York.

Hell, R. & Bergmann, L. 1988. Glutathione synthetase in tobacco suspension cultures: catalytic properties and localization. Physiol. Plant. 72: 70-76.

Hell, R. & Bergmann, L. 1990. γ-Glutamylcysteine synthetase in higher plants: catalytic properties and subcellular localization. Planta 180: 603-612.

Hell, R., Schneider, S. & Bergmann, L. 1990. Glutathione synthesis and cadmium induced phytochelatin formation in suspension cultures. Physiol. Plant. 79: A121.

Hossain, M.A. & Asada, K. 1987. Ascorbate regenerating enzymes in chloroplasts. Ind. J. Biochem. Biophys. 24: 52-56.

Jablonski, P.P. & Anderson, J.W. 1978. Light dependent reduction of oxidized glutathione by ruptured chloroplasts. Plant Physiol. 61: 221-225.

Jaspers, C.H., Gigot, D. & Penninckx, M.J. 1985. Pathways of glutathione degradation in the yeast *Saccharomyces cerevisiae*. Phytochemistry 24: 703-707.

Klapheck, S. 1988. Homoglutathione: isolation, quantification and occurrence in legumes. Physiol. Plant. 74: 727-732.

Klapheck, S., Zopes, H., Levels, H.-G. & Bergmann, L. 1988. Properties and localization of the homoglutathione synthetase from *Phaseolus coccineus* leaves. Physiol. Plant. 74: 733-739.

Klapheck, S., Zimmer, I. & Cosse, H. 1990. Scavenging of hydrogen peroxide in the endosperm of *Ricinus communis* by ascorbate peroxidase. Plant Cell Physiol. 31: 1005-1013.

Klapheck, S., Chrost, B., Starke, J. & Zimmermann, H. 1992. γ-Glutamylcysteinylserine – a new homologue of glutathione in plants of the family *Poaceae*. Bot. Acta 105: 174-179.

Lamoureux, G.L. & Rusness, D.G. 1989a. The role of glutathione and glutathione-*S*-transferases in pesticide metabolism, selectivity, and mode of action in plants and insects. In: Dolphin, D., Poulson, R. & Awamovic, O., (eds.), Glutathione: Chemical, Biochemical and Medical Aspects, Vol. IIIB. J. Wiley & Sons Publ., New York, pp. 153-196.

Lamoureux, G.L. & Rusness, D.G. 1989b. Propachlor metabolism in soybean plants, excised soybean tissues and soil. Pestic. Biochem. Physiol. 34: 187-204.

Macnicol, P.K. 1987. Homoglutathione and glutathione synthetase of legume seedlings: partial purification and substrate specificity. Plant Sci. 53: 229-235.

Macnicol, P.K. & Bergmann, L. 1984. A role for homoglutathione in organic sulfur transport to the developing mung bean seed. Plant Sci. 36: 219-223.

McIntyre, T.M. & Curthoys, N.P. 1979. Comparison of the hydrolytic and transfer activities of rat renal γ-glutamyltranspeptidase. J. Biol. Chem. 254: 6499-6504.

Meister, A. 1981. On the cycles of glutathione metabolism and transport. In: Estabrook, R.W. & Srere, P. (eds.), Biological Cycles. Current Topics in Cellular Regulation 18. Academic Press, New York, pp. 21-58.
Meister, A. 1985. Glutathione synthetase from rat kidney. Meth. Enzymol. 113: 393-399.
Meister, A. & Anderson, M.E. 1983. Glutathione. Annu. Rev. Biochem. 52: 711-760.
Nussbaum, S., Schmutz, D. & Brunold, C. 1988. Regulation of assimilatory sulfate reduction by cadmium in *Zea mays* L. Plant Physiol. 88: 1407-1410.
Polle, A. & Rennenberg, H. 1993. Photo-oxidative stress in trees. In: Foyer, C. & Mullineaux, P. (eds.), Photooxidative Stresses on Plants: Causes and Amelioration. CRC Press, London, *in press*.
Price, C.A. 1957. A new thiol in legumes. Nature 180: 148-149.
Rauser, W.E. 1990. Phytochelatins. Annu. Rev. Biochem. 59: 61-86.
Rauser, W.E., Hunziker, P.E. & Kägi, I.H.R. 1986. Reverse phase high-performance liquid chromatography of Cd-binding proteins from the grass *Agrostis gigantea*. Plant Sci. 45: 105-109.
Rauser, W.E., Schupp, R. & Rennenberg, H. 1991. Cysteine, γ-glutamylcysteine and glutathione levels in maize seedlings. Plant Physiol. 97: 128-138.
Rennenberg, H. & Lamoureux, G.L. 1990. Physiological processes that modulate the concentration of glutathione in plant cells. In: Rennenberg, H., Brunold, C., De Kok, L.J. & Stulen, I. (eds.), Sulfur Nutrition and Sulfur Assimilation in Higher Plants. SPB Acad. Publ., The Hague, pp. 53-65.
Rennenberg, H., Schmitz, K. & Bergmann, L. 1979. Long-distance transport of sulfur in *Nicotiana tabacum*. Planta 147: 57-62.
Rüegsegger, A., Schmutz, D. & Brunold, C. 1990. Regulation of glutathione synthesis by cadmium in *Pisum sativum*. Plant Physiol. 93: 1579-1584.
Schlunz, S. 1991. Glutathion- und Homoglutathion-Synthese in *Pisum sativum L.*, Thesis, University of Cologne.
Schneider, A. 1992. Untersuchungen zum Glutathion-Transport und -Stoffwechsel in Suspensionskulturen von Nicotiana tabacum L. var. "Samsun". Thesis, Technical University, Munich.
Schneider, A. & Rennenberg, H. 1992. Degradation of glutathione (GSH) in heterotrophic tobacco cells. Phyton 32: 111-115.
Schneider, S. 1992. Die Regulation der Glutathion-Synthese in Zellkulturen von *Nicotiana tabacum* und *Petroselinum crispum*. Thesis, University of Cologne.
Smith, I.K. 1975. Sulfate transport in cultured tobacco cells. Plant Physiol. 55: 303-307.
Sommer, M. 1991. Eigenschaften und Funktion der L-γ-Glutamyltranspeptidase aus *Glycine max*. Thesis, University of Cologne.
Steffens, I.C. 1990. The heavy metal-binding peptides of plants. Annu. Rev. Plant Physiol. 41:535-575.
Steinkamp, R., Schweihofen, B. & Rennenberg, H. 1987. γ-Glutamylcyclotransferase in tobacco suspension cultures: catalytic properties and subcellular localization. Physiol. Plant. 69: 499-503.
Zopes, H. 1990. Die Synthese und Funktion des Homoglutathions in Chloroplasten von *Phaseolus coccineus*. Thesis, University of Cologne.
Zopes, H., Klapheck, S. & Bergmann, L. 1993. The function of homoglutathione and hydroxymethylglutathione for the scavenging of hydrogen peroxide. Plant Cell Physiol. 34: *in press*

ROLE OF GLUTATHIONE IN PLANTS UNDER OXIDATIVE STRESS

Luit J. De Kok & Ineke Stulen
Department of Plant Biology, University of Groningen, P.O. Box 14, 9750 AA Haren, The Netherlands

Introduction

A small proportion of the organic reduced sulfur in the plant (about 2 %) is present in the water-soluble non-protein thiol fraction (Stulen & De Kok, this volume). Glutathione (γ-glu-cys-gly) is generally the major thiol compound of this fraction, and under normal conditions its content accounts for more than 90 % of the water-soluble non-protein thiols (Rennenberg 1982, 1987; Alscher 1989; Smith *et al.* 1989, 1990; Rennenberg & Lamoureux 1990). Some plant families contain homologues of glutathione: homoglutathione (γ-glu-cys-β-ala) in *Leguminosae* and hydroxymethyl-glutathione (γ-glu-cys-ser) in *Poaceae*, instead of glutathione (Bergmann & Rennenberg, this volume).

Glutathione (and its homologues) can be found intracellularly in all parts of the plant. High levels of glutathione can be found in both chloroplasts and cytosol (Bergmann & Rennenberg, this volume). In general, glutathione is predominantly present in the reduced form with GSH/GSSG ratios higher than 10 (Smith *et al.* 1989). Glutathione reductase, an NADPH dependent enzyme which is abundant throughout the plant, is involved in keeping glutathione (and its homologues) in the reduced state (Smith *et al.* 1989; Bergmann & Rennenberg, this volume).

Possible physiological functions of glutathione have been discussed in various reviews in the last decade (Rennenberg 1982, 1984, 1987, Alscher 1989, Smith *et al.* 1989, 1990; Rennenberg & Lamoreux 1990). Summarizing, glutathione may function in:

* Sulfur metabolism:
  - as regulator of sulfate uptake and transport
  - as carrier and reductant in sulfate reduction
  - as transport and storage form of reduced sulfur
* Selenium metabolism:
  - as carrier and reductant in selenite reduction
* Modulation of enzyme activity and gene expression
* Protection against oxidative and environmental stress:
  - in the enzymatic detoxification of H_2O_2
  - as protectant against oxidative stress
  - as precursor of heavy-metal binding peptides
  - in the detoxification and metabolism of xenobiotics

Homoglutathione and hydroxymethylglutathione are likely to fulfill similar roles in plants as glutathione (Bergmann & Rennenberg, this volume). Several of the above

Sulfur Nutrition and Assimilation in Higher Plants
edited by L.J. De Kok et al., pp. 125–138

mentioned functions of glutathione in plants are also discussed in detail in other sections of this volume. For instance, synthesis, degradation and intracellular distribution of glutathione (and its homologues) are discussed by Bergmann & Rennenberg. Its involvement in selenium metabolism is discussed by Anderson and its role in thiol/disulfide exchange reactions and the modulation of enzyme and gene expression by Kunert & Foyer. Also, the significance of glutathione metabolism in the functioning of plants under environmental stress is reviewed in part by others in this volume. For instance, the role of glutathione in synthesis of the heavy metal binding phytochelatins is discussed by Rauser, and its role in metabolism and detoxification of xenobiotics by Lamoureux & Rusness.

For the presumed significance of glutathione in the various processes, a precise regulation of the intracellular glutathione level would be a prerequisite. However, it is obvious, that *in situ*, the glutathione pool in the plant tissue is very dynamic and the resultant of the rate of its synthesis, degradation, translocation and utilization for biosynthetic processes (Rennenberg & Lamoureux 1990; Rauser *et al.* 1991). The glutathione levels in plants can be strongly affected by physiological and environmental factors, *e.g.* sulfur supply, temperature, pollutants, heavy metals and xenobiotics (Rennenberg & Lamoureux 1990; De Kok 1990; Bergmann & Rennenberg, this volume; Lamoureux & Rusness, this volume; Rauser, this volume).

In the present paper the presumed significance of glutathione in the reaction/adaptation of plants to oxidative stress is evaluated.

Active oxygen species in plants

Formation and significance

Singlet oxygen, superoxide, H_2O_2, but also radicals can be formed at various sites in the plant cell, and even may have a function in plant metabolism, as in cell wall synthesis (Winston 1990). The chloroplast is a major site where active oxygen species may be formed, especially under conditions where the Calvin cycle is not in tune with photosynthetic electron transport. Singlet oxygen may be photosynthesized by chlorophyll, oxygen may be photo-reduced at the site of photosystem 1 yielding superoxide (Elstner 1982; Asada 1984; Siefermann-Harms 1987; Robinson 1988; Van Hasselt 1990). There is no consensus on the *in vivo* rate of synthesis of active oxygen species. Robinson (1988) estimated that at high light intensity, the photosynthetic oxygen reduction may range from 10 to 40 μmol O_2 mg chlorophyll^{-1} h^{-1} under optimal conditions for photosynthetic CO_2 fixation (100-300 μmol mg chlorophyll^{-1} h^{-1}). In case they are not scavenged, activated oxygen species may react with various cell constituents, which may result in the formation of toxic free radicals.

Detoxification

Plants contain various compounds, for instance, amino acids, ascorbic acid, carotenoids, α-tocopherol and enzymes, as superoxide dismutases, catalase, peroxidases, which have been proved, mainly *in vitro*, to quench or inhibit formation of active oxygen species and free radicals (Elstner 1982; Asada 1984; Salin 1987; Siefermann-

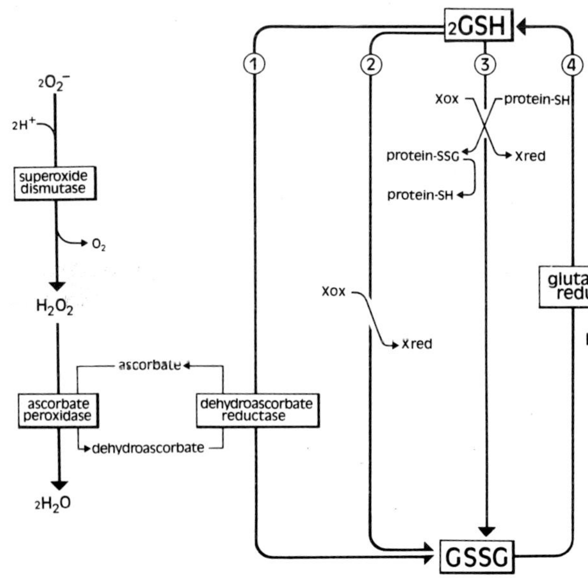

Fig. 1. Possible roles of glutathione in the protection of plants against the harmful effects of active oxygen species and free radicals. As reductant in the enzymatic detoxification of H_2O_2 (1), by stabilization and protection of *e.g.* protein thiol groups through direct reaction with active oxygen species (*X*ox) (2) or by the formation of mixed disulfides (3), and reduction of the oxidized glutathione by glutathione reductase (4).

Harms 1987; Larson 1988; Robinson 1988). It has been proposed that glutathione may also be of significance in the protection of plants against the harmful effects of active oxygen species and free radicals (Fig. 1). Thiol compounds as glutathione may directly react with singlet oxygen, superoxide, H_2O_2 and hydroxyl radicals (Allen 1977; Larson 1978, 1988, Wefers & Sies 1983). Besides, glutathione may stabilize and protect protein thiol groups by acting as a thiol buffer (Kunert & Foyer, this volume). Glutathione may also form mixed disulfides with proteins, which have been detected in dry plant embryos (Fahey *et al.* 1980) and matured seeds (Butt & Ohlrogge 1991). Whether this conjugation also occurs in other tissue, and is of significance in protection of proteins against oxidative stress, is still unclear and needs further investigation. *In situ*, glutathione reductase plays a crucial role in reduction of the formed oxidized glutathione (or mixed disulfides; Smith *et al.* 1989; Fig. 1). Glutathione is also assumed to play a role in the enzymatic detoxification of H_2O_2 in the chloroplast. Here, the superoxide formed after the photosynthetic reduction of oxygen is enzymatically dismutated by superoxide dismutase, yielding H_2O_2 (Halliwell & Foyer 1978; Foyer *et al.* 1989; Monk *et al.* 1989; Kunert & Foyer, this volume). From *in vitro* measurements there is evidence that in the chloroplast H_2O_2 is subsequently degraded by the ascorbate peroxidase-dehydroascorbate reductase-glutathione reductase cycle (Foyer & Halliwell 1976; Alscher & Amthor 1988; Halliwell & Foyer 1978; Alscher 1989; Foyer *et al.* 1989; Smith *et al.* 1989, 1990, Fig. 2). This cycle appears to have physiological significance in the chloroplast, because of the lack of catalase. The latter is responsible for the degradation of H_2O_2 formed during the glycolate oxidation in the peroxisomes. From the average *in vitro* activity of superoxide dismutase and the H_2O_2 degrading enzymes of the ascorbate/glutathione cycle, their substrate affinities and substrate concentrations, it has been concluded that the chloroplast is able to remove superoxide and H_2O_2 very efficiently (Robinson 1988; Foyer *et al.* 1989, 1991; Kunert and Foyer, this volume). Even the addition of methylviologen to leaves, resulting in a boost of superoxide formation, hardly affected the metabolite pool size of the ascorbate-glutathione cycle (Foyer *et al.* 1991). This illustrated that even under severe oxidative stress the available enzymatic scavenging mechanism may be sufficient to avoid negative effects of active oxygen species.

Recently, it has been demonstrated that the ascorbate-glutathione cycle also may have significance outside the chloroplast (Bielawski & Joy 1986), *e.g.* in the apoplast (Castillo & Greppin 1988), endosperm (Klapheck *et al.* 1990) and root nodules (Dalton *et al.* 1991). Even though there are some reports that a glutathione peroxidase may be active, at least in some plant species, its real significance needs to be further established (Smith *et al.* 1990).

Oxidative stress; role of glutathione

During their life cycle plants may have to deal with periods of low temperature, frost, high radiation, heat, drought or pollutant stress. It has been postulated that under such stress conditions plants have to deal with an increased strain of active oxygen species and oxidative reactions, which may be responsible for the onset of injury or induced premature senescence of the plant (Thompson *et al.* 1987; Dhindsa 1987, 1991; Alscher & Amthor 1988; Alscher 1989; Monk *et al.* 1989; Smith *et al.* 1989, 1990; Tanaka *et al.* 1990; Kuroda *et al.* 1991). In this view, an increment of the levels of antioxidants, *e.g.* glutathione and/or enhanced activities of the protecting enzymes, *e.g.* glutathione reductase, would be of great importance in the adaption/protection of the plant against oxidative stress. However, to what extent the various environmental factors definitely result in enhanced levels of active oxygen species is not always clear, since it is sometimes questionable whether the formation of active oxygen species is the cause or the consequence of cellular injury. For instance, during foliar senescence the presence of active oxygen species and free radicals are an inherent feature during degradation of the cellular tissue (Thompson *et al.* 1987). However, there is no direct evidence that active oxygen species are causally involved in the onset of senescence and their formation is more likely the consequence of cellular degradation (Kar & Feierabend 1984). The loss of thiols/glutathione, which is generally observed during foliar senescence (Pauls & Thompson 1984; De Kok & Graham 1989; Gullner & Tyihák 1991) has been used as proof for the occurrence of active oxygen species. However, also the loss of thiols/glutathione is likely not a primary event, since this only occurred when the majority of the pigments was already degraded (De Kok & Graham 1989). Also, the actual supporting experimental data on the assumed role of glutathione and glutathione reductase in the protection of plants against environmental stress are often only circumstantial or relative and sometimes even conflicting.

Low temperature

Chilling. Chilling-sensitive plants may already be injured at temperatures of 10 °C and lower, especially upon illumination. Under these conditions some chilling sensitive plants *e.g.* cucumber may suffer from photooxidative injury most likely initiated by the formation of active oxygen species and the subsequent radical formation (Van Hasselt 1990). Low temperature-induced photooxidation was characterized by a loss of pigments, breakdown of unsaturated lipids and accompanied with a loss of the endogenous antioxidants (Van Hasselt 1990), including glutathione, which was both oxidized and degraded (Wise & Naylor 1987). However, the level of antioxidants/glutathione itself is not directly the determining factor in the protection of plants against high radiation stress at low temperature. For instance, there were no substantial differences in the content of the antioxidants between chilling-sensitive cucumber and chilling-resistant pea leaves; the glutathione content in pea leaves was even two times lower than that in cucumber leaves (Wise & Naylor 1987). In pea leaves no photooxidation of the pigments occurred at low temperature and light and the content of the antioxidants was not substantially affected. It is more likely that differences in energy dissipation at low temperature, for instance in low temperature-inducible structural changes of the chloroplasts membrane proteins and in operation of the xanthophyll cycle, have greater significance in the protection of plants against irradiation injury (Van Hasselt 1990; Öquist & Huner 1991).

Freezing. The physiological background of freezing injury in plant tissue and the factors determining freezing tolerance of plant tissue are still subject of discussion. Frost hardening of plants is accompanied with numerous changes in cellular constituents, enzymes and structure and it remains unclear which are essential in the development of freezing tolerance (Heber & Santarius 1973; Steponkus 1984; Guy 1990). Protein denaturation due to oxidation of protein thiol groups during freezing-induced cellular dehydration, resulting in the subsequent formation of intra- and inter-disulfide bonds, has been proposed to be a significant factor in the onset of freezing injury (Levitt 1980). In this view, antioxidants as glutathione would function as a cryoprotectant against this type of protein denaturation and play an important role in the freezing tolerance of plants (Levitt 1980). However, it is still doubtful whether oxidative reactions are essential for the onset of freezing injury, since anaerobic conditions during freezing did not prevent injury (Heber & Santarius 1973). Indeed, the development of foliar freezing injury is accompanied with a loss of soluble proteins and protein thiol, but it is unclear whether the loss in protein-thiol is the cause or consequence of protein denaturation, since they decreased at a similar rate (Stuiver *et al.* 1988). The development of freezing injury was not preceded or accompanied with a substantial loss of glutathione; its content only strongly decreased after the subsequent thawing of the tissue (Stuiver *et al.* 1988). It has been proposed that enhanced levels of glutathione have adaptive value in the development of freezing tolerance of plants, however, this is only supported by correlative data. Certainly, high levels of glutathione are characteristic for plants during fall and during winter (Esterbauer & Grill 1978; Guy & Carter 1984; Guy *et al.* 1984; Grill *et al.* 1988, 1990; Schupp & Rennenberg 1988, 1989; Madamanchi *et al.* 1991) and for frost-hardened plants (De Kok & Oosterhuis 1983; Guy *et al.* 1984; Stuiver *et al.* 1992a). However, a direct relation between high levels of glutathione and freezing-tolerance of the tissue is very doubtful. Enhancement of glutathione levels upon low temperature exposure occurred very rapidly, reaching a maximum after 2 to 3 days (Stuiver *et al.* 1992a,b), whereas substantial increase in freezing tolerance upon low temperature exposure only occurred after two weeks and longer (Stuiver *et al.* 1992a). Besides, a low temperature-induced increase of the glutathione level was not restricted to freezing tolerant plants but it also occurred in the chilling-sensitive soybean, with the difference that here also substantial amounts of oxidized glutathione accumulated, probably due to an inadequate supply of NADPH to glutathione reductase (Smith *et al.* 1989; Vierheller & Smith 1990). Enhancements of glutathione level itself, by sulfate incubation (De Kok *et al.* 1981), by herbicide action (Guy *et al.* 1984) or by H_2S exposure (Stuiver *et al.* 1992a) also did not improve the freezing tolerance of the foliar tissue. It is more likely that the enhanced level of glutathione upon low temperature exposure is a reflection of an altered equilibrium between the sulfur assimilation and protein synthesis and/or a reduced translocation of glutathione from the shoots to the roots at low temperature (Stuiver *et al.* 1992a,b).

Enhanced levels of glutathione reductase have also been observed in plants during winter and in frost-hardened plants (Esterbauer & Grill 1978; De Kok & Oosterhuis 1983; Guy & Carter 1984; Guy *et al.* 1984). Besides, during low temperature acclimation the affinity of the enzyme for oxidized glutathione increased, likely by the formation of new isoenzymes (Guy an Carter 1984). The thermal dependency of the apparent K_m of glutathione reductase for NADPH may strongly vary between species (Kidambi *et al.* 1990; Mahan *et al.* 1990). Still, the adaptive value and its significance in low temperature tolerance of plants is unclear, since the actual activity of glutathione reductase and the supply of sufficient reductant NADPH at freezing temperatures has still to be demonstrated (Smith *et al.* 1990).

It has been observed that in spruce the levels of glutathione and glutathione reductase may vary with the altitude. High levels were characteristic for plants growing at high altitudes, and it has been proposed that it may have possible adaptive value in the protection against injurious effects of high irradiation and low temperature (Grill *et al.* 1988, 1990; Bermadinger *et al.* 1989; Madamanchi *et al.* 1991; Guttenberger *et al.* 1992). In view of the above discussion, more research is needed to establish the physiological implications of the observed variation in glutathione levels with the altitude.

Air pollution

Photochemical air pollutants. Glutathione and glutathione reductase are also presumed to play a role in the protection of plants against air pollutant stress. The toxic effects of the photochemical air pollutants ozone and peroxyacetyl nitrates (PAN) can directly be related to their oxidative potency and reactivity with cellular constituents. In addition, upon absorption/reaction they may yield in the formation of singlet oxygen, superoxide and free radicals (Heath 1975, 1980; Murphy *et al.* 1977; Mudd 1982; Mudd *et al.* 1984; Bennett *et al.* 1984; Rennenberg & Polle 1989). This means that mechanisms that are involved in the detoxification of active oxygen species also may be involved in the detoxification of photochemical air pollutants.

However, because of their reactivity it is largely unclear to what extent ozone and PAN at realistic concentrations really enter the cell, since they already may react with various compounds including antioxidants or extracellular peroxidases in the apoplast and be detoxified in the apoplast (Rennenberg & Polle 1989).

Thiol groups of proteins and glutathione appeared to be very susceptible for direct oxidation by ozone and PAN (Heath 1975; Mudd 1975, 1982; Murphy *et al.* 1977; Heck *et al.* 1977; Freeman *et al.* 1979; Mudd *et al.* 1984; Castillo & Greppin 1988; Chevrier *et al.* 1988). It has been observed that upon reaction with ozone, thiol groups are not only oxidized yielding disulfides, but even sulfenic and sulfonic acid groups were formed, the latter oxidation is irreversible (Heath 1975; Mudd 1982; Castillo & Greppin 1988). In plants, ozone exposure sometimes resulted in a decrease in glutathione content of the leaves (Guri 1983; Tanaka *et al.* 1985; Mehlhorn *et al.* 1986; Castillo & Greppin 1988; Price *et al.* 1990), on the other hand, increases in glutathione level have also been observed (Smidt 1984; Hausladen *et al.* 1990; Sen Gupta *et al.* 1990, 1991). In addition, the level of oxidized glutathione is increased upon ozone exposure (Tanaka *et al.* 1985; Hausladen *et al.* 1990; Sen Gupta *et al.* 1990, 1991). There are some reports on the enhancement of glutathione reductase levels upon ozone (Tanaka *et al.* 1988; Price *et al.* 1990). Guri (1983) observed that tolerant bean cultivars contained higher glutathione reductase levels than sensitive cultivars. In addition, the glutathione levels in tolerant cultivars were much less decreased by ozone exposure than in sensitive cultivars. However, the glutathione reductase level was not affected in all cultivars after ozone fumigation (Guri 1983). In addition to enhanced levels of glutathione reductase, Price *et al.* (1990), observed even more enhanced levels of glutathione-*S*-transferase in ozone-exposed barley leaves. The latter enzyme catalyzes the conjugation of glutathione with various xenobiotics (Rennenberg & Lamoureux 1990; Lamoureux & Rusness, this volume), but is also shown to possess glutathione peroxidase activity (Prohaska 1980). Price *et al.* (1990) proposed that this enzyme may function in the protection against the toxic effects of lipid ozonolysis.

As illustrated, glutathione levels may be affected by photochemical air pollutants. However, at present there are still too little sound experimental data that assure the significance of glutathione, glutathione reductase and glutathione-*S*-transferase as determining factors in the protection against ozone and PAN and their significance in differences in tolerance between species.

Sulfur gases. In general, exposure of plants to SO_2 results in addition to strongly increased sulfate levels, in a rapid accumulation of water-soluble non-protein thiol compounds, including glutathione, in the shoots (De Kok 1990). The observed effects of SO_2 on glutathione reductase levels are rather inconsistent (Tanaka *et al.* 1982; Grill *et al.* 1982; Madamanchi & Alscher 1991; Soldatini *et al.* 1992). The cause and consequence of enhanced glutathione in plants upon SO_2 exposure are interpreted in different ways. Chronically enhanced thiol levels have been suggested as one of the causes of the phytotoxicity of SO_2 (Grill *et al.* 1979). In this view it would result in a deregulation of cellular metabolism. Contrary, it has also been proposed that the increase in levels of glutathione and glutathione reductase have adaptive value in the protection of plants against the toxic effects of sulfite (Chiment *et al.* 1986; Mehlhorn *et al.* 1986; Alscher *et al.* 1987; Alscher 1989; Madamanchi & Alscher 1991; Soldatini

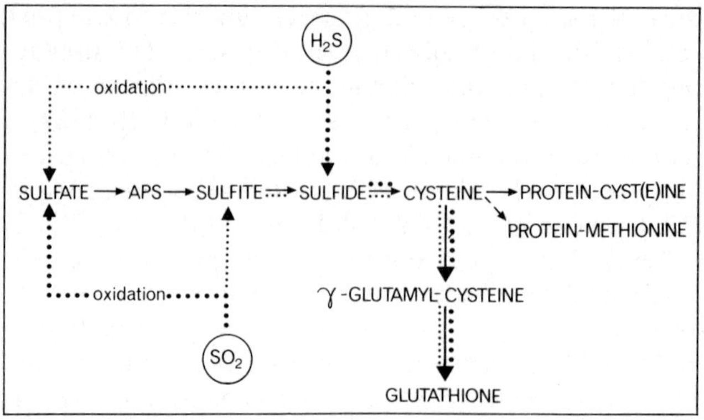

Fig. 2. Metabolism of SO_2 and H_2S in plants.

et al. 1992). In their view, after absorption and hydration of SO_2, part of the formed sulfite would undergo free radical chain oxidation in the chloroplast. This free radical chain oxidation would be mediated by superoxide, formed in the chloroplast upon illumination and would result in an amplified production of active oxygen species and radicals, which has been suggested to be the possible basis for the injurious effects of SO_2 in plants. Enhanced levels of glutathione, glutathione reductase and superoxide dismutase would avoid SO_2-induced oxidative injury (Madamanchi & Alscher 1991). Indeed, *in vitro*, superoxide-triggered oxidation of reduced sulfur compounds, sulfite but also sulfide, has been observed in illuminated isolated spinach chloroplasts, but only at relatively high concentrations (0.05 mM and higher) (Asada & Kiso 1973; De Kok *et al.* 1983; Ghisi *et al.* 1990; Dittrich *et al.* 1992). Light-induced oxidation of sulfite by chloroplasts was only substantial in broken chloroplasts (Ghisi *et al.* 1990) and in absence of an electron acceptor of photosystem I and it was negligible if ferredoxin and NADP were added to the isolated chloroplasts (Asada & Kiso 1973). The light-triggered sulfite (and sulfide) oxidation could be suppressed by the addition of superoxide dismutase or other scavengers of active oxygen species, glutathione included (Asada & Kiso 1973; De Kok *et al.* 1983; Ghishi *et al.* 1990; Dittrich *et al.* 1992). Besides, the significance of sulfite oxidation in the *in situ* chloroplast and its involvement in the development of oxidative injury at realistic outside SO_2 levels still needs to be established. First, it has been demonstrated that light is not an essential factor for the development of plant injury by SO_2 (Olszyk & Tingey 1984). Second, it is questionable whether high sulfite concentrations really occur in chloroplasts under realistic SO_2 pollutant levels. Significant intracellular sulfite concentrations could only be detected at relatively high SO_2 levels ($> 0.5\ \mu l\ l^{-1}$) (De Kok 1990). Before entering the cell the greater part of sulfite may already have been oxidized in the apoplast *e.g.* by extracellular peroxidases (Pfanz *et al.* 1990; 1992). Even in case significant amounts of sulfite would enter the chloroplast, sulfite-induced radical formation in the light would be very unlikely. For instance, even *in vitro*, isolated intact spinach chloroplasts contained sufficient active oxygen scavenging capacity to prevent light-triggered sulfite oxidation (Ghisi *et al.* 1990; Dittrich *et al.* 1992). Furthermore, they observed that the sulfite was rapidly reduced by the chloroplasts and subsequently metabolized.

The enhanced glutathione levels in plant shoots upon SO_2 exposure are most likely due to a direct assimilation of part of the absorbed sulfur (Fig. 2). In case regulation of sulfate uptake by the roots was bypassed and sulfur, either as sulfate, SO_2 or H_2S, was directly supplied to leaves or twigs these treatments resulted in a rapid accumulation of water-soluble non-protein thiol compounds both in light and in darkness (Grill *et al.* 1979; De Kok *et al.* 1981, 1988; De Kok & Kuiper 1986; Buwalda *et al.* 1988, 1990, 1992, 1993; De Kok 1989, 1990; Fig. 2). Accumulation of thiol compounds varied with the concentration of the sulfur supplied and in general it reached a maximum after one or two days of the exposure (De Kok *et al.* 1985; Maas *et al.* 1987, Bosma *et al.* 1990, Stuiver *et al.* 1992). Upon exposure to excessive sulfur, glutathione (or homoglutathione in legumes; Buwalda *et al.* 1993) was not the sole thiol compound accumulating in leaf tissue. Also substantial amounts of cysteine and in darkness also of γ-glutamyl-cysteine accumulated (Buwalda *et al.* 1988, 1990; De Kok *et al.* 1988; De Kok 1989, 1990; Stuiver *et al.* 1992a; Soldatini *et al.* 1992; Fig. 2). However, the effect of excess sulfur on thiol composition strongly differed between species (Buwalda *et al.* 1988, 1990, 1993; De Kok *et al.* 1988; Stuiver *et al.* 1992).

Apparently, if the regulation of sulfate uptake by the root is by-passed, there is no strict regulation of the size and composition of the thiol pool. From the sulfate-induced accumulation of thiol compounds demonstrated in foliar tissue it is obvious, that the sulfate concentration at its activation site, yielding APS, is of great importance in the regulation of the rate of sulfate assimilation (De Kok & Kuiper 1986; Stulen & De Kok, this volume). The SO_2-induced thiol (and glutathione) accumulation may be explained by a direct reduction and assimilation of the absorbed SO_2 or reduction and assimilation of sulfate which is formed after oxidation of the absorbed SO_2 in the leaves (De Kok 1990). There is strong evidence that atmospheric H_2S is directly assimilated into cysteine and subsequently into its metabolites in the leaves; the H_2S uptake by foliar tissue even directly depends on its metabolism in the plant (De Kok 1989, 1991; Buwalda *et al.* 1992).

Concluding remarks

In order to ascertain the significance of the suggested functions of glutathione in plants, more information is needed on its subcellular distribution and the dynamics in level and metabolism as affected by physiological and environmental conditions. Glutathione is a potent antioxidant, which in combination with glutathione reductase, is likely to have significance in the enzymatic detoxification of active oxygen species, especially in the chloroplast. However, the significance of enhanced levels of glutathione and glutathione reductase as factors in the protection against oxidative stress need still to be established. At present the supporting evidence for such a role is largely based on correlative or indirect data, and is often conflicting. Even though it will be a very difficult task, future research on elucidating the role of glutathione needs to be focussed on the actual flux of active oxygen species and the *in situ* activity of the enzymes involved in their detoxification, including glutathione reductase, under normal conditions and under various environmental stress conditions.

Acknowledgement

We wish to thank Mr. E. Leeuwinga for drawing the figures.

References

Allen, J.F. 1977. Superoxide and photosynthetic reduction of oxygen. In: Michelson, A.M, McCord, J.M. & Fridovich, I. (eds.), Superoxide and Superoxide Dismutases. Academic Press, London, pp. 417-436.

Alscher, R.G. 1989. Biosynthesis and antioxidant function of glutathione in plants. Physiol. Plant. 77: 457-464.

Alscher, R.G. & Amthor, J.S. 1988. The physiology of free-radical scavenging: maintenance and repair processes. In: Schulte-Hostede, S., Darrall, N.M., Blank, L.W. & Wellburn, A.R. (eds.), Air Pollution and Plant Metabolism. Elsevier Applied Science, London, pp. 94-115.

Alscher, R.G., Bower, J.L. & Zipler, W. 1987. The basis for different sensitivities of photosynthesis to SO_2 in two cultivars of pea. J. Exp. Bot. 38: 99-108.

Asada, K. 1984. Chloroplasts: formation of active oxygen species and its scavenging. In: Methods in Enzymology. Vol 105. Academic Press, New York, pp. 422-429.

Asada, K. & Kiso, K. 1973. Initiation of aerobic oxidation of sulfite by illuminated spinach chloroplasts. Eur. J. Biochem. 33: 253-257.

Bennett, J.H., Lee, E.H. & Heggestad, H.E. 1984. Biochemical aspects of plant tolerance to ozone and oxyradicals: superoxide dismutase. In: Koziol, M.J. & Whatley, F.R. (eds.), Gaseous Air Pollutants and Plant Metabolism. Butterworths, London, pp. 413-424.

Bermadinger, E., Grill, D. & Guttenberger, H. 1989. Thiole, Ascorbinsäure, Pigmente und Epikutikularwachse in Fichtennadeln aus dem Höhenprofil "Zillertal". Phyton 29: 163-185.

Bielawski, C. & Joy, K.W. 1986. Reduced and oxidised glutathione and glutathione reductase activity in tissue of *Pisum sativum*. Planta 169: 267-272.

Bosma, W., Kamminga, G. & De Kok, L.J. 1990. H_2S-induced accumulation of sulfhydryl-compounds in the leaves of plants under field and laboratory exposure. In: Rennenberg, H., Brunold, C., De Kok, L.J. & Stulen, I. (eds.), Sulfur Nutrition and Sulfur Assimilation in Higher Plants; Fundamental, Environmental and Agricultural Aspects. SPB Academic Publishing, The Hague, pp. 173-175.

Butt, A.D. & Ohlrogge, J.B. 1991. Acyl carrier protein is conjugated to glutathione in spinach seed. Plant Physiol. 96: 937-942.

Buwalda, F., De Kok, L.J., Stulen, I. & Kuiper, P.J.C. 1988. Cysteine, γ-glutamyl-cysteine and glutathione contents of spinach leaves as affected by darkness and application of excess sulfur. Physiol. Plant. 74: 663-668.

Buwalda, F., Stulen, I., De Kok, L.J. & Kuiper, P.J.C. 1990. Cysteine, γ-glutamyl-cysteine and glutathione contents of spinach leaves as affected by darkness and application of excess sulfur. II. Glutathione accumulation in detached leaves exposed to H_2S in the absence of light is stimulated by the supply of glycine to the petiole. Physiol. Plant. 80: 196-204.

Buwalda, F., De Kok, L.J. & Stulen, I. 1992. The flux of atmospheric H_2S to spinach leaves can be affected by the supply of *O*-acetylserine. Phyton 32: 12-16.

Buwalda, F., De Kok, L.J. & Stulen, I. 1993. Effects of atmospheric H_2S on thiol composition of crop plants. J. Plant Physiol. 142: *in press*

Castillo, F.J. & Greppin, H. 1988. Extracellular ascorbic acid and enzyme activities related to ascorbic acid metabolism in *Sedum album* L. leaves after ozone exposure. Environ. Exp. Bot. 28: 231-238.

Chevrier, N., Sarhan, F. & Chung, Y.S. 1988. Oxidative damages and repair in *Euglena gracilis* exposed to ozone. I. SH groups and lipids. Plant Cell Physiol. 29: 321-327.

Chiment, J.J., Alscher, R. & Hughes, P.R. 1986. Glutathione as an indicator of SO_2-induced stress in soybean. Environ. Exp. Bot. 26: 147-152.

Dalton, D.A., Post, C.J. & Langeberg, L. 1991. Effects of ambient oxygen and of fixed nitrogen on concentrations of glutathione, ascorbate, and associated enzymes in soybean root nodules. Plant Physiol. 96: 812-818.

De Kok, L.J. 1989. Responses of sulfur metabolism in plants to atmospheric hydrogen sulfide. Phyton 29: 189-201.

De Kok, L.J. 1990. Sulfur metabolism in plants exposed to atmospheric sulfur. In: Rennenberg, H., Brunold, C., De Kok, L.J. & Stulen, I. (eds.), Sulfur Nutrition and Sulfur Assimilation in Higher

Plants; Fundamental, Environmental and Agricultural Aspects. SPB Academic Publishing, The Hague, pp. 111-130.
De Kok, L.J. & Oosterhuis, F.A. 1983. Effects of frost-hardening and salinity on glutathione and sulfhydryl levels and on glutathione reductase activity in spinach leaves. Physiol. Plant. 58: 47-51.
De Kok, L.J. & Kuiper, P.J.C. 1986. Effect of short-term dark incubation with sulfate, chloride and selenate on the glutathione content of spinach leaf discs. Physiol. Plant. 68: 477-482.
De Kok, L.J. & Graham, M. 1989. Levels of pigments, soluble proteins, amino acids and sulfhydryl compounds in foliar tissue of *Arabidopsis thaliana* during dark-induced and natural senescence. Plant. Physiol. Biochem. 27: 203-209.
De Kok, L.J., De Kan, P.J.L., Tánczos, O.G. & Kuiper, P.J.C. 1981. Sulphate-induced accumulation of glutathione and frost-tolerance of spinach leaf tissue. Physiol. Plant. 53: 435-438.
De Kok, L.J., Thompson, C.R. & Kuiper, P.J.C. 1983. Sulfide-induced oxygen uptake by isolated spinach chloroplasts catalyzed by photosynthetic electron transport. Physiol. Plant. 59: 19-22.
De Kok, L.J., Bosma, W., Maas, F.M. & Kuiper, P.J.C. 1985. The effect of short-term H_2S fumigation on water-soluble sulphydryl and glutathione levels in spinach. Plant Cell Environ. 8: 189-194.
De Kok, L.J., Buwalda, F. & Bosma, W. 1988. Determination of cysteine and its accumulation in spinach leaf tissue upon exposure to excess sulfur. J. Plant Physiol. 133: 502-505.
De Kok, L.J., Stahl, K. & Rennenberg, H. 1989. Fluxes of atmospheric hydrogen sulfide to plant shoots. New Phytol. 112: 533-542.
De Kok, L.J., Rennenberg, H. & Kuiper, P.J.C. 1991. The internal resistance in spinach shoots to atmospheric H_2S deposition is determined by metabolism processes. Plant Physiol. Biochem. 29: 463-470.
Dhindsa, R.S. 1987. Glutathione status and protein synthesis during drought and subsequent rehydration in *Tortula ruralis*. Plant Physiol. 83: 816-819.
Dhindsa, R.S. 1991. Drought stress, enzymes of glutathione metabolism, oxidation injury, and protein synthesis in *Tortula ruralis*. Plant Physiol. 95: 648-651.
Dittrich, A.P.M., Pfanz, H. & Heber, U. 1992. Oxidation and reduction of sulfite by chloroplasts and formation of sulfite addition compounds. Plant Physiol. 98: 738-744.
Elstner, E.F. 1982. Oxygen activation and oxygen toxicity. Annu. Rev. Plant Physiol. 33: 73-96.
Esterbauer, H. & Grill, D. 1978. Seasonal variation of glutathione and glutathione reductase in needles of *Picea abies*. Plant Physiol. 61: 119-121.
Fahey, R.C., Di Stefano, D.L., Meier, G.P. & Bryan, R.N. 1980. Role of hydration state and thiol-disulfide status in the control of thermal stability and protein synthesis in wheat embryo. Plant Physiol. 65: 1062-1066.
Foyer, C.H. & Halliwell, B. 1976. The presence of glutathione and glutathione reductase in chloroplasts. A proposed role in ascorbic acid metabolism. Planta 133: 21-25.
Foyer, C.H., Dujardyn, M. & Lemoine, Y. 1989. Responses of photosynthesis and the xanthophyll and ascorbate-glutathione cycles to changes in irradiance, photoinhibition and recovery. Plant Physiol. Biochem. 27: 751-760.
Foyer, C.H., Lelandais, M., Galap, C. & Kunert, K.J. 1991. Effects of elevated cytosolic glutathione reductase activity on the cellular glutathione pool and photosynthesis in leaves under normal and stress conditions. Plant Physiol. 97: 863-872.
Freeman, B.A., Miller, B.E. & Mudd, J.B. 1979. Reactions of ozone with human erythrocytes. In: Lee, S.D. & Mudd, J.B. (eds.), Assessing Toxic Effects of Environmental Pollutants. Ann Arbor Science, Michigan, pp. 151-171.
Ghisi, R., Dittrich, A.P.M. & Heber, U. 1990. Oxidation *versus* reductive detoxification of SO_2 by chloroplasts. Plant Physiol. 92: 846-849.
Grill, D., Esterbauer, H. & Klösch, U. 1979. Effect of sulfur dioxide on glutathione in leaves of plants. Environ. Pollut. 17, 187-194.
Grill, D., Esterbauer, H. & Hellig, K. 1982. Further studies on the effect of SO_2-pollution on the sulfhydryl-system of plants. Phytopath. Z. 104: 264-271.
Grill, D. Pfeifhofer, H., Tschulik, A., Hellig, K. & Holzer, K. 1988. Thiol content of spruce needles at forest limits. Oecologia 76: 294-297.
Grill, D., Guttenberger, H. & Bermadinger, E. 1990. Thiol content of plants depending on the altitude. In: Rennenberg, H., Brunold, C., De Kok, L.J. & Stulen, I. (eds.), Sulfur Nutrition and Sulfur Assimilation in Higher Plants; Fundamental, Environmental and Agricultural Aspects. SPB Academic Publishing, The Hague, pp. 191-193.

Gullner, G. & Tyikák, E. 1991. Measurement of formaldehyde, hydrogen peroxide and non-protein thiols in tobacco leaves during ageing. Biochem. Physiol. Pflanzen 187: 131-138.

Guri, A. 1983. Variation in glutathione and ascorbic acid content among selected cultivars of *Phaseolus vulgaris* prior to and after exposure to ozone. Can. J. Plant Sci. 63: 733-737.

Guttenberger, H., Bermadinger, E., Grill, D. & Holzer, K. 1992. Water-soluble thiols in spruce seedlings from provenances of different sea levels. Phyton 31: 217-226.

Guy, C.L. 1990. Cold acclimation and freezing stress tolerance: role of protein metabolism. Annu. Rev. Plant Physiol. Plant Mol. Biol. 41: 187-223.

Guy, C.L. & Carter, J.V. 1984. Characterization of partially purified glutathione reductase from cold-hardened and non-hardened spinach leaf tissue. Cryobiol. 21: 454-464.

Guy, C.L., Carter, J.V., Yelenosky, G. & Guy, C.T. 1984. Changes in glutathione content during cold acclimation in *Cornus sericea* and *Citrus sinensis*. Cryobiol. 21: 443-453.

Halliwell, B. & Foyer, C.H. 1978. Properties and physiological function of a glutathione reductase purified from spinach leaves by affinity chromatography. Planta 139: 9-17.

Hausladen, A., Madamanchi, N.R., Fellows, S., Alscher, R.G. & Amundson, R.G. 1990. Seasonal changes in antioxidants in red spruce as affected by ozone. New. Phytol. 115: 447-458.

Heath, R.L. 1975. Ozone. In: Mudd, J.B. & Kozlowski, T.T. (eds.), Responses of Plants to Air Pollutants. Academic Press, New York, pp. 23-55.

Heath, R.L. 1980. Initial events in injury to plants by air pollutants. Annu. Rev. Plant Physiol. 3. 31: 395-431.

Heber, U. & Santarius, K.A. 1973. Cell death by cold and heat and resistance to extreme temperatures. Mechanism of hardening and dehardening. In: Precht, H., Christophersen, J., Hensel, H. & Larcher, W. (eds.), Temperature and Life. Springer-Verlag, Berlin, pp. 232-263.

Heck, W.W., Mudd, J.B. & Miller, P.R. 1977. Plants and microorganisms. In: Ozone and other Photochemical Oxidants. Natl. Acad. Sci., Washington, D.C., pp. 437-585.

Kar, M. & Feierabend, J. 1984. Metabolism of activated oxygen in detached wheat and rye leaves and its relevance to the initiation of senescence. Planta 160: 385-391.

Kidambi, S.P., Mahan, J.R. & Matches, A.G. 1990. Purification and thermal dependence of glutathione reductase from two forage legume species. Plant Physiol. 92: 363-367.

Klapheck, S., Zimmer, I. & Cosse, H. 1990. Scavenging of hydrogen peroxide in the endosperm of *Ricinus communis* by ascorbate peroxidase. Plant Cell Physiol. 31: 1005-1013.

Kuroda, H., Sagisaka, S., Adasa, M. & Chiba, K. 1991. Peroxide-scavenging systems during cold acclimation of apple callus in culture. Plant Cell Physiol. 32: 635-641.

Larson, R.A. 1978. Environmental chemistry of reactive oxygen species. CRC Critical Rev. Environ. Control 8: 197-246.

Larson, R.A. 1988. The antioxidants of higher plants. Phytochemistry 27: 969-978.

Levitt, J. 1980. Molecular basis of freezing injury and tolerance. In: Reponses of Plants to Environmental Stress. Vol. I. Academic Press, New York, pp. 248-289.

Maas, F.M., De Kok, L.J., Strik-Timmer, W. & Kuiper, P.J.C. 1987. Plant responses to H_2S and SO_2 fumigation. II. Differences in metabolism of H_2S and SO_2 in spinach. Physiol. Plant. 70: 722-728.

Madamanchi, N.R. & Alscher, R.G. 1991. Metabolic bases for differences in sensitivity of two pea cultivars to sulfur dioxide. Plant Physiol. 97: 88-93.

Madamanchi, N.R., Hausladen, A., Alscher, R.G., Amundsen, R.G. & Fellows, S. 1991. Seasonal changes in antioxidants in red spruce (*Picea rubens* Sarg.) from three field sites in the northeastern United States. New. Phytol. 118: 331-338.

Mahan, J.R., Burke, J.J. & Orzech, K.A. 1990. Thermal dependence of the apparent K_m of glutathione reductases from three plant species. Plant Physiol. 93: 822-824.

Mehlhorn, H., Seufert, G., Schmidt, A. & Kunert, K.J. 1986. Effect of SO_2 and O_3 on production of antioxidants in conifers. Plant Physiol. 82: 336-338.

Monk, L.S., Fagerstedt, K.V. & Crawford, R.M.M. 1989. Oxygen toxicity and superoxide dismutase as an antioxidant in physiological stress. Physiol. Plant. 76: 456-459.

Mudd, J.B. 1975. Peroxyacyl nitrates. In: Mudd, J.B. & Kozlowski, T.T. (eds.), Responses of Plants to Air Pollutants. Academic Press, New York, pp. 97-119.

Mudd, J.B. 1982. Effects of oxidants on metabolic function. In: Unsworth, M.H. & Ormrod, M.H. (eds.), Effects of Gaseous Air Pollutants in Agriculture and Horticulture. Butterworth Scientific, London, pp. 189-203.

Mudd, J.B., Banerjee, S.K., Dooley, M.M. & Knight, K.L. 1984. Pollutants and plant cells: effects on membranes. In: Koziol, M.J. & Whatley, F.R. (eds.), Gaseous Air Pollutants and Plant Metabolism, Butterworths, London, pp. 105-116.

Murphy, S.D., Goldstein, B.D. & Hitchcock, M. 1977. Toxicology. In: Ozone and other Photochemical Oxidants. Natl. Acad. Sci., Washington, D.C., pp. 323-387.

Olszyk, D.M. & Tingey, D.T. 1984. Phytotoxicity of air pollutants. Evidence for the photodetoxification of SO_2 but not O_3. Plant Physiol. 74: 999-1005.

Öquist, G. & Huner, N.P.A. 1991. Effects of cold acclimation on the susceptibility of photosynthesis to photoinhibition in Scots pine and in winter and spring cereals: a fluorescence analysis. Functional Ecol. 5: 91-100.

Pauls, K.P. & Thompson, J.E. 1984. Evidence for the accumulation of peroxidized lipids in membranes of scenescing cotyledons. Plant Physiol. 75: 1152-1157.

Pfanz, H., Dietz, K.-J., Weinerth, I. & Oppmann, B. 1990. Detoxification of sulfur dioxide by apoplastic peroxidases. In: Rennenberg, H., Brunold, C., De Kok, L.J. & Stulen, I. (eds.), Sulfur Nutrition and Sulfur Assimilation in Higher Plants; Fundamental, Environmental and Agricultural Aspects. SPB Academic Publishing, The Hague, pp. 229-233.

Pfanz, H., Würth, G., Opmann, B. & Schultz, G. 1992. Sulfite oxidation in, and sulfate uptake from the cell wall of leaves. *In muro* studies. Phyton 32: 87-90.

Price, A., Lucas, P.W. & Lea, P.J. 1990. Age dependent damage and glutathione metabolism in ozone fumigated barley: a leaf section approach. J. Exp. Bot. 41: 1309-1317.

Prohaska, J.R. 1980. The glutathione peroxidase activity of glutathione-*S*-transferase. Biochim. Biophys. Acta 611: 87-98.

Rauser, W.E., Schupp, R. & Rennenberg, H. 1991. Cysteine, γ-glutamylcysteine, and glutathione levels in maize seedlings. Plant Physiol. 97: 128-138.

Rennenberg, H. 1982. Glutathione metabolism and possible biological roles in higher plants. Phytochemistry 21: 2771-2781.

Rennenberg, H. 1984. The fate of excess sulfur in higher plants. Annu. Rev. Plant Physiol. 35: 121-153.

Rennenberg, H. 1987. Aspects of glutathione function and metabolism in plants. In: von Wettstein, D. & Chua, N.-H. (eds.), Plant Molecular Biology. Plenum Publishing Corporation, pp. 279-292.

Rennenberg, H. & Polle, A. 1989. Effects of photooxidants on plants. In: Georgii, H.-W. (ed.), Mechanisms and Effects of Pollutant-Transfer into Forests. Kluwer Academic Publishers, Dordrecht, pp. 252-258.

Rennenberg, H. & Lamoureux, G.L. 1990. Physiological processes that modulate the concentration of glutathione in plant cells. In: Rennenberg, H., Brunold, C., De Kok, L.J. & Stulen, I. (eds.), Sulfur Nutrition and Sulfur Assimilation in Higher Plants; Fundamental, Environmental and Agricultural Aspects. SPB Academic Publishing, The Hague, pp. 53-65.

Robinson, J.M. 1988. Does O_2 photoreduction occur within chloroplasts *in vivo*? Physiol. Plant. 72: 666-680.

Salin, M.L. 1987. Toxic species and protective systems of the chloroplast. Physiol. Plant. 72: 681- 689.

Schupp, R. & Rennenberg, H. 1988. Diurnal changes in the glutathione content of spruce needles (*Picea abies*. L.). Plant Sci. 57: 113-117.

Schupp, R. & Rennenberg, H. 1989. Light-dependent changes in the glutathione content of Norway spruce (*Picea abies* (L.) Karst.). Ann. Sci. For. 46 suppl.: 837s-841s.

Sen Gupta, A., Alscher, R.G. & McCune, D.C. 1990. Ozone exposure, glutathione levels and photosynthesis in hybrid poplar. In: Rennenberg, H., Brunold, C., De Kok, L.J. & Stulen, I. (eds.), Sulfur Nutrition and Sulfur Assimilation in Higher Plants; Fundamental, Environmental and Agricultural Aspects. SPB Academic Publishing, The Hague, pp. 195-197.

Sen Gupta, A., Alscher, R.G. & McCune, D.C. 1991. Response of photosynthesis and antioxidants to ozone in *Populus* leaves. Plant Physiol. 96: 650-655.

Siefermann Harms, D. 1987. The light-harvesting and protective functions of carotenoids in photosynthetic membranes. Physiol. Plant. 69: 561-568.

Smidt, S. 1984. Begasungsversuche mit SO_2 und Ozon an jungen Fichten. Eur. J. For. Path. 14: 241-248.

Smith, I.K., Vierheller, T.L. & Thorne, C.A. 1989. Properties and functions of glutathione reductase in plants. Physiol. Plant. 77: 449-456.

Smith, I.K., Polle, A. & Rennenberg, H. 1990. Glutathione. In: Alscher, R.G. & Cumming, J.R. (eds.), Stress Responses in Plants: Adaptation and Acclimation Mechanisms. Wiley-Liss, New York, pp. 201-215.

Soldatini, G.F., Ranieri, A., Lencioni, L. & Lorenzini, G. 1992. Effects of continuous SO_2 fumigation on SH-containing compounds in two wheat cultivars of different sensitivities. J. Exp. Bot. 251: 797-801.

Steponkus, P.L. 1984. Role of the plasma membrane in f freezing injury and cold acclimation. Ann. Rev. Plant Physiol. 35: 543 – 584.

Stuiver, C.E.E., De Kok, L.J. & Kuiper, P.J.C. 1988. Freezing injury in spinach leaf tissue: effects on water-soluble proteins, protein-sulfhydryl and water-soluble non-protein sulfhydryl compounds. Physiol. Plant. 74: 72-76.

Stuiver, C.E.E., De Kok, L.J. & Kuiper, P.J.C. 1992a. Freezing tolerance and biochemical changes in wheat shoots as affected by H_2S fumigation. Plant Physiol. Biochem. 30: 47-55.

Stuiver, C.E.E., De Kok, L.J. & Kuiper, P.J.C. 1992b. Does glutathione play a role in freezing tolerance of plants? Phyton 32: 137-140.

Tanaka, K., Kondo, N. & Sugahara, K. 1982. Accumulation of hydrogen peroxide in chloroplasts of SO_2-fumigated spinach leaves. Plant Cell Physiol. 23: 999-1007.

Tanaka, K., Suda, Y., Kondo, N. & Sugahara, K. 1985. O_3 tolerance and the ascorbate-dependent H_2O_2 decomposing system in chloroplasts. Plant Cell Physiol. 26: 1425-1431.

Tanaka, K., Saji, H. & Kondo, N. 1988. Immunological properties of spinach glutathione reductase and inductive biosynthesis of the enzyme with ozone. Plant Cell Physiol. 29: 637-642.

Tanaka, K., Masuda, R., Sugimoto, T., Omasa, K. & Sakaki, T. 1990. Water deficiency-induced changes in the contents of defensive substances against active oxygen species in spinach leaves. Agric. Biol. Chem. 54: 2629-2634.

Thompson, J.E., Legge, R.L. & Barber, R.F. 1987. The role of free radicals in senescence and wounding. New Phytol. 105: 317-344.

Vierheller, T.L. & Smith, I.K. 1990. Effect of chilling on glutathione reductase and total glutathione in soybean leaves (*Glycine max* (L) Merr). In: Rennenberg, H., Brunold, C., De Kok, L.J. & Stulen, I. (eds.), Sulfur Nutrition and Sulfur Assimilation in Higher Plants; Fundamental, Environmental and Agricultural Aspects. SPB Academic Publishing, The Hague, pp. 261-265.

Van Hasselt, P.R. 1990. Light-induced damage during chilling. In: Wang, C.Y. (ed.), Chilling Injury of Horticultural Crops. CRC Press, Boca Raton, Florida, pp. 113-129.

Wefers, H. & Sies, H. 1983. Oxidation of glutathione by the superoxide radical to the disulfide and the sulfonate yielding singlet oxygen. Eur. J. Biochem. 137: 29-36.

Winston, G.W. 1990. Physiochemical basis for free radical formation in cells: production and defenses. In: Alscher, R.G. & Cumming, J.R. (eds.), Stress Responses in Plants: Adaptation and Acclimation Mechanisms. Wiley-Liss, New York, pp. 57-86.

Wise, R.R. & Naylor, A.W. 1987. Chilling-enhanced photooxidation. Evidence for the role of singlet oxygen and superoxide in the breakdown of pigments and endogenous antioxidants. Plant Physiol. 83: 278-282.

THIOL/DISULFIDE EXCHANGE IN PLANTS

Karl Josef Kunert[1] & Christine Foyer[2]
[1]*AECI Limited, R&D Department, Private Bag X2, PO Modderfontein 1645, South Africa;* [2]*INRA Versailles, Laboratoire du Metabolisme, F-78026 Versailles Cedex, France*

Introduction

Plant cells contain considerable amounts of free low molecular weight thiols. Thiols are known to be directly involved in many important metabolic and physiological functions via thiol/disulfide exchange reactions. These include protein synthesis and activation and inactivation of enzymes as well as synthesis of DNA and reduction of cystine. Thiols are also important as antioxidants in the scavenging of cytotoxic oxygen radicals to prevent oxidative damage to cell components. Although thiol/disulfide exchange reactions are quite similar in both plant and animal systems and also, in part, in bacterial systems, the amount of data available on such exchange reactions in plants is rather small compared to data from animal or bacterial systems. For example, the lipoic acid system in plants has not been intensively studied.

The importance of a thiol/disulfide exchange for the plant cell was first considered several years ago in Levitt's thiol/disulfide hypothesis of frost injury and resistance in plants (Levitt 1962). According to this hypothesis, frost resistance would result from prevention of thiol oxidation, thiol/disulfide interchange and formation of intermolecular disulfides. Since that time several research groups have become interested in the roles of thiol compounds, such as glutathione, which is the major free low molecular weight non-protein plant thiol compound, and thioredoxins that are low molecular weight colorless disulfide proteins. Both glutathione and thioredoxins have the potential to undergo thiol/disulfide exchange reactions that are vital for many cellular processes.

In this paper we review recent developments in our understanding of the thiol/disulfide exchange in plants at the biochemical, physiological and molecular levels. We will concentrate on the two major classes of free thiol compounds in plants, glutathione and thioredoxins. We will specifically focus, in this review, on the importance of the thiol/disulfide exchange for the plant cell and on cellular mechanisms that either prevent thiol oxidation or that increase thiol levels via genetic manipulation. Finally, we will give a brief view of future prospects for strategies to manipulate thiol/disulfide reactions in plants.

General characteristics of thiols

Glutathione

It is generally found that glutathione (L-γ-glutamyl-L-cysteinyl-glycine; GSH) is the major thiol compound in plants (for an overview see: Rennenberg 1982; Hausladen

Sulfur Nutrition and Assimilation in Higher Plants
edited by L.J. De Kok et al., pp. 139–151

& Alscher 1992). In most investigations, however, glutathione has been determined only as total cellular thiol (-SH) content although measurement of glutathione has been claimed. There is considerable evidence that many thiol/disulfide exchange reactions involve participation of GSH. Glutathione, which is not translated on mRNA, is synthesized in a two-step reaction sequence (Meister & Anderson 1983). In the first step, a dipeptide is synthesized from L-glutamate and L-cysteine by the enzyme γ-glutamyl-cysteine synthetase (EC 6.3.2.2). Glycine is finally added to the C-terminal end of the dipeptide to form GSH. This reaction is catalyzed by the enzyme glutathione synthetase (EC 6.3.2.3).

Under most conditions the intracellular glutathione pool is maintained in the reduced form (GSH). In plant tissues, GSH accounts for at least 90% of the total number of GSH equivalents (Foyer *et al.* 1991). The intracellular content of glutathione disulfide (GSSG) is generally very low but it may be increased as a result of severe stress (Wise & Naylor 1987). The importance of glutathione as a thiol compound is evidenced by its high concentration in parts of the plant. The glutathione content is, however, dependent on developmental and environmental factors. Cellular glutathione concentrations of about 0.1 mM have been reported in some tissues (Smith *et al.* 1990), but green leaves can contain up to 10 mM of the thiol compound. The glutathione contents may vary by several orders of magnitude between different species and between different tissues and organs within the same species and, indeed, within different compartments of the same cell. Bergmann (1981) reported that in tobacco leaf protoplasts 76% of the glutathione was in the chloroplasts, 7% in the cytosol and 17% in the vacuole giving concentrations of 20 mM, 60 μM and 20 μM, respectively. Foyer & Halliwell (1976) measured a stromal glutathione concentration of 3.5 mM for spinach chloroplasts, while Anderson *et al.* (1983) reported that washed intact chloroplasts contained an endogenous glutathione pool of 10 mM. Gillham & Dodge (1986) reported a chloroplast glutathione concentration of about 4 mM but they found that this glutathione pool of the chloroplasts in pea leaves amounted to only 10% of the leaf glutathione.

Thioredoxins

A second very important family of thiols are thioredoxins. These are low molecular weight SH-containing proteins with an active site composed of a cys-gly-pro-cys region that undergoes reversible reduction and oxidation (Buchanan 1980; Holmgren 1985). Two thioredoxin systems are known: the NADP/thioredoxin and the ferredoxin/thioredoxin system. In these systems thioredoxins are reduced enzymically with either NADPH or reduced ferredoxin by NADP-thioredoxin reductase (NTR; EC 1.6.4.5) or ferredoxin-thioredoxin reductase (FTR), respectively.

All components of the ferredoxin/thioredoxin system, which is present in higher plants, algae, cyanobacteria and fermentative bacteria, have been purified and characterized (Crawford *et al.* 1989). The plant and cyanobacterial FTR is an iron sulfur protein which is composed of two dissimilar protein subunits and reduces the two thioredoxins present in oxygenic photosynthetic systems thioredoxins *f* and *m* (Droux *et al.* 1987). Thioredoxin *f*, which preferentially activates the fructose-1,6-biphosphatase (EC 3.1.3.11) exists in monomeric and dimeric forms. Thioredoxins *mb* and *mc* preferentially activate the chloroplast enzyme NADPH-malate dehydrogenase (EC 1.1.1.82). These thioredoxins possess a single reducible disulfide bridge per monomer.

The second thioredoxin system in plants is the NADP/thioredoxin system consisting of NADPH, NTR and a thioredoxin *h* which is characteristic of heterotrophic systems (Florencio *et al.* 1988). Spinach thioredoxin *h* exists in two forms and apparently differs from thioredoxin *m* or *f* from the same source, but thioredoxin *h* is, however, similar to the thioredoxin described from wheat kernels (Berstermann *et al.* 1983). Thioredoxin *h* is reduced by NADPH via the flavoprotein NTR. This system is localized outside the chloroplast but may reside in other types of organelles. NTR and thioredoxin *h* appear to decrease by about 50% during greening although all components are still present in fully green tissues (Florencio *et al.* 1988).

Thiol/Disulfide exchange reactions

Glutathione

Proteins. Reduced glutathione possesses two important structural features the γ-glu linkage and the SH-group of the cysteine which provide its intracellular stability and which are intimately associated with function (Rennenberg 1982, Meister 1988). There is also considerable evidence that glutathione plays a critical role as a reductant in the defense against oxidants because of its potential to undergo an oxidation/reduction cycle (E_0^s = -0.34 V). The role of glutathione has been investigated to some extent in a number of plant stresses which act partly via activated oxygen species. Stresses investigated include drought, senescence, elevated oxygen, bipyridyl and diphenylether herbicide treatment, sulfur dioxide fumigation and chilling (for an overview see: Alscher 1989). Functions, such as defense against oxidants, are performed by hydrogen atom donation from the thiol moiety. The transfer of hydrogen atoms rather than separate electron transfer and protonation reactions enables glutathione to reduce both enzymes and damaging free radicals while minimizing the possibility of the reduction of molecular oxygen to superoxide. Furthermore, reduced glutathione can also scavenge superoxide itself and the potent hydroxyl radical derived from this oxygen species (Kosower & Kosower 1978; Wefers & Sies 1983). Depletion of the glutathione pool may not be, in itself, lethal but will seriously impair the antioxidative and protective systems of plant cells which will thus be rendered susceptible to oxidative damage and metabolic failure.

Many enzymes function only in a reduced form by virtue of component thiol groups (Ziegler 1985). Oxidation of these thiol groups frequently leads to loss of activity and occasionally degradation of enzymes. Molecular oxygen and reactive oxygen species are the natural oxidant for many of these thiol groups (Davies *et al.* 1987). A major role of thiols in plants is, therefore, to modulate the reduction state in the plant and to protect the thiol groups on susceptible enzymes by a thiol/disulfide exchange reaction. The regulatory sulfhydryl groups on enzymes, however, represent only a relatively small redox pool. A model representing the approximate redox potentials of the major components involved in the thiol/disulfide exchange is illustrated in Fig. 1. Since glutathione is the abundant thiol compound in organelles, such as the chloroplast, it may be that this compound is thus more available to oxygen than

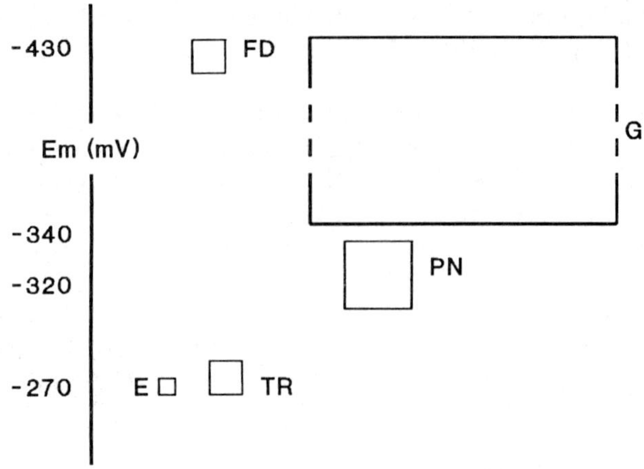

Fig. 1. A model representing the relative pool sizes and the approximate redox potentials of the major components involved in thiol/disulfide exchange reactions. (E) Thiol-regulated enzymes such as fructose-1,6-biphosphatase; (FD) ferredoxin; (G) glutathione; (TR) thioredoxin; (PN) pyridine nucleotides. The sizes of the boxes represent the relative sizes of the redox pools.

the enzyme thiol groups (Meister 1988) and may thus be preferentially oxidized. In addition, GSH can also reactivate some enzymes that have become inactive following exposure to elevated oxygen. Through the mediation of the ferredoxin and pyridine nucleotide pools the regulatory thiol groups on thioredoxins may equilibrate with the glutathione redox couple directly. However, GSH cannot replace thioredoxin in the mechanisms of activation of enzymes involved in carbon assimilation, although Vivekanandan & Edwards (1987) seem to have evidence that GSH may activate at least one light-modulated enzyme (NADP-malate dehydrogenase) that normally is activated by thioredoxin.

In contrast to GSH, glutathione disulfide (GSSG) can inactivate enzymes by the formation of mixed disulfides (Fig. 2). Mixed disulfides formation provides one good reason why plant cells, and chloroplasts in particular, keep GSH/GSSG ratios high. If the GSH/GSSG ratios of the cell were to fall as a result of stress, the capacity of metabolism could be severely impaired by the formation of mixed disulfides. However, there is also evidence that GSSG may activate an enzyme. *In vitro* protein synthesis is inhibited by GSSG possible via activation of a translational inhibitor which is normally kept in the inactive form by a thiol-reducing system (Jackson *et al.* 1983).

Gene activation. Recently, Wingate *et al.* (1988) reported that reduced glutathione may have a role in gene activation via a receptor protein which may undergo a thiol/disulfide exchange reaction. They found that GSH supplied exogenously to cell suspension cultures of bean caused a massive and selective induction of the transcription of defense genes. These genes encode enzymes involved in the phytoalexin and lignin biosynthesis such as phenylalanine ammonia-lyase (EC 4.3.1.5) and chalcone synthase (EC 2.3.1.74). Both enzymes are involved in protection against microbial attack. Transcriptional activation of these genes resulted in a massive change in the overall pattern of protein synthesis which closely resembled the response to a funcal

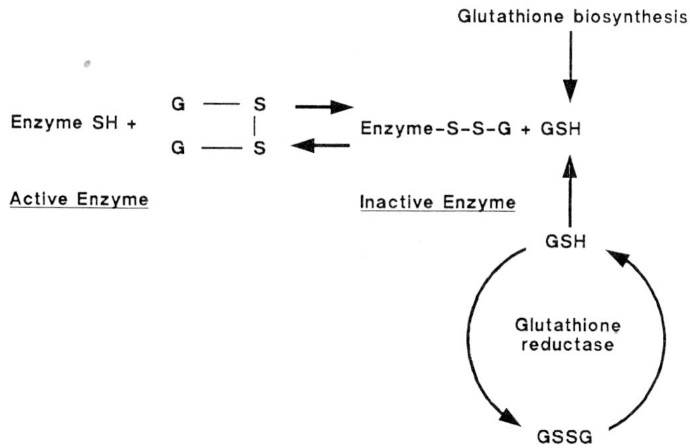

Fig. 2. Glutathione disulfide (GSSG)-dependent inactivation of enzymes by formation of mixed disulfides and activation of enzymes by reduced glutathione (GSH). GSH is generated either directly by biosynthesis or by reduction of GSSG by the action of the enzyme glutathione reductase.

elicitor. Edwards *et al.* (1991) found that both oxidized and reduced exogenous glutathione can elicite the phytoalexin response in cell suspension cultures of bean. However, defense gene activation did not occur when only the intracelluar thiol level was significantly increased after treatment of cells with the compound L-oxo-thiazolidine-4-carboxylate (OTC), an oxoproline analog (Williamson & Meister 1981), which elevates glutathione levels. Lack of gene activation after OTC treatment was also found by Hausladen & Kunert (1990) when the extractable activity of enzymes that are involved in cellular defense against oxidation, such as monodehydroascorbate reductase (EC1.6.5.4), dehydroascorbate reductase (EC 1.8.5.1) and glutathione reductase (EC 1.6.4.2), was measured.

Overall, the role of glutathione in the activation of defense genes is still obscure. One hypothesis outlined by Wingate *et al.* (1988) to explain the action of reduced glutathione has been recently developed by Malbon *et al.* (1987) for animal cells and β-adrenergic receptors. In this hypothesis, in order to explain how cells manage to respond to extracellular signals, it has been postulated that disulfide bridges and free sulfhydryl groups play an integral role in the structure of cell surface receptors that are coupled to G proteins. A receptor might possess intramolecular disulfide bridges that can be cleaved by thiol compounds activating the receptor in a manner similar to agonist binding. A similar receptor may also be present in the cell wall of plants. Oxidized glutathione may work in a similar way to provide an intracellular response, via a plant receptor, as described for certain receptor proteins on erythrocytes (Reglinski *et al.* 1988). In erythrocytes, a mechanism has been proposed which involves the formation of mixed disulfides at sulfhydryl group receptor sites by exchange. Clearly, more detailed studies are necessary to elucidate the role of glutathione in gene activation.

Thioredoxins

In general, the thioredoxin system is a disulfide reductase and catalyzes NADPH- or ferredoxin-dependent reductions of exposed S-S bridges in a variety of proteins

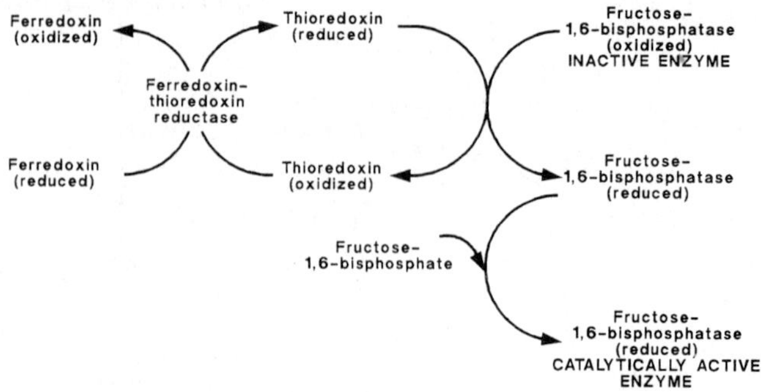

Fig. 3. The mechanism of activation of stromal fructose-1,6-biphosphatase showing that activation is regulated by the ferredoxin-thioredoxin system and that it is a reversible two-stage process.

(Crawford *et al.* 1989; Holmgren 1985). The proposed role of thioredoxins in plants is to activate biosynthetic pathways in the light and catabolic pathways in the dark. Thioredoxins appear to be absolutely essential for photosynthetic growth (Muller & Buchanan 1989) and disruption of the gene coding for thioredoxin *m* in the blue-green alga *Anacystis nidulans* is a lethal mutation.

In plants, the ferredoxin/thioredoxin system functions via a thiol/disulfide exchange reaction in the regulation of certain enzymes involved in photosynthesis: thioredoxin activates fructose-1,6-biphosphatase, phosphoribulokinase (EC 2.7.1.19), sedoheptulose-1,7-biphosphatase (EC 3.1.3.37) and NADP-dependent glyceraldehyde phosphate dehydrogenase (EC 1.2.1.9) (Cséke & Buchanan 1986; Crawford *et al.* 1989; Knaff 1989). Each reduced thioredoxin reduces specific disulfide bridges on biosynthetic target enzymes, such as on the chloroplast enzyme fructose-1,6-biphosphatase, converting them between inactive and active forms (Fig. 3). In the case of fructose-1,6-biphosphatase each subunit contains a regulatory disulfide group that is reduced to the sulfhydryl level by reduced thioredoxin thereby leading to activation of the enzyme. Indeed most of the available sulfhydryl groups in the chloroplast appear to be associated with stromal protein as opposed to the membrane fraction (Slovacek & Vaughn 1982). Increase in sulfhydryl content of approximately 50% occur upon illumination. The increase was found to be kinetically similar to the rise in the activity of fructose-1,6-biphosphatase.

Prevention of thiol oxidation

Ascorbate-glutathione cycle

Plants have developed specific systems to regulate the thiol/disulfide ratios in the cell. A high GSH/GSSG ratio seems to be necessary to sustain the role of the major cellular thiol compound, glutathione, as a reductant. This allows glutathione to undergo thiol/disulfide exchange reactions with cellular compounds thereby avoiding direct oxidation of enzymes. Continuous reduction of oxidized glutathione can be achieved by the enzyme glutathione reductase. Both glutathione and glutathione

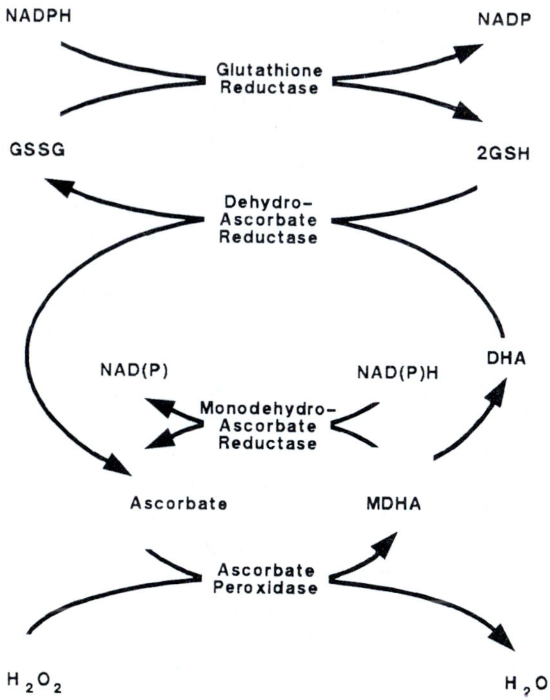

Fig. 4. The pathway of elimination of phytotoxic H_2O_2; the ascorbate-glutathione cycle. (GSH) reduced glutathione; (GSSG) oxidized glutathione; (MDHA) monodehydroascorbate; (DHA) dehydroascorbate. Monodehydroascorbate is unstable and disproportionates *in vivo* to yield ascorbate and dehydro-ascorbate.

reductase are components of a protective cellular system called the ascorbate-glutathione cycle (Fig. 4) (Foyer & Halliwell 1976). This cycle is best known for the elimination of the risk of oxidation of enzymes by H_2O_2. However, this cycle is also connected to α-tocopherol, a lipid-soluble antioxidant, which protects membrane proteins against toxic free-radical reaction in biomembranes (Finckh & Kunert 1985; Kunert *et al.* 1985). Thus glutathione is important in sustaining the defense systems in both the hydrophilic and hydrophobic environments of plant cells.

An excellent example of the protective role of the ascorbate-glutathione cycle is the protection of enzymes involved in photosynthetic carbon assimilation. Carbon assimilation is dependent on a stromal H_2O_2-scavenging system for its continued functioning in the light. Several of the enzymes of the Calvin cycle are activated in the light by thiol modulation via the thioredoxin system. They are, however, extremely sensitive to oxidation by H_2O_2, and loss of thioredoxin-mediated activation of fructose-1,6-biphosphatase and sedoheptulose-1,7-biphosphatase appears to be relatively specific (Leegood *et al.* 1985). The reduction state of these enzymes, which determines the activation state, reflects the balance between the flux of reducing equivalents through the electron transport chain (which causes their activation) and the oxidizing environment of the stroma that continuously favors inactivation (Leegood 1990). H_2O_2 and other reactive oxygen species interfere with the delicate balance of this system, since they favour oxidation and prevent thioredoxin-dependent reductive activation of these enzymes.

The redox pairs dehydroascorbate(DHA)/ascorbate and GSSG/GSH are intermediate electron carriers in the reduction of H_2O_2 by NADPH in chloroplasts. The role of the DHA/ascorbate and GSSG/GSH redox pairs in the reduction of H_2O_2 was supported by changes in the concentrations of ascorbate and GSH both in intact and ruptured chloroplasts in response to light and exogenously supplied H_2O_2 (Anderson *et al.* 1983). Illuminated chloroplasts reduce O_2 to H_2O_2 using reducing equivalents from photosystem I in the process known as pseudocyclic electron flow. H_2O_2 is then metabolised using NADPH as the electron donor via the reaction sequence of the ascorbate-glutathione cycle as illustrated in Fig. 4. It is clear that the ascorbate-glutathione cycle also operates efficiently in other cellular compartments as well as in the chloroplast, and the cycle is present in the cytosol of both green and non-green tissues (Foyer 1992). When the *Escherichia coli* (*E. coli*) glutathione reductase gene (*gor* gene) was overexpressed in the cytosol of tobacco plants, the resultant high level of glutathione reductase gave added protection to the ascorbate pool of plants suffering oxidative stress caused by treatment with methylviologen (Foyer *et al.* 1991). Protection was achieved presumably because of cycling of ascorbate between the cytosolic and chloroplast compartments. These experiments demonstrate that the cytosolic reactions of the ascorbate-glutathione cycle can help to support the chloroplast ascorbate-glutathione cycle in the stroma, possibly by the transport system across the chloroplast envelope (Foyer 1992).

Glutathione reductase

Both glutathione and glutathione reductase appear to be ubiquitous in the biosphere. The enzyme seems to play a major role in preventing thiol oxidation and it maintains high thiol levels necessary for thiol/disulfide exchange reactions. The idea that glutathione reductase is involved in a reaction to avoid the consequences of environmental stress was first developed by Levitt (1962). This hypothesis to explain frost resistance clearly requires an enzyme like glutathione reductase in order to maintain glutathione almost exclusively in the reduced form according to the equation:

$$GSSG + NADPH + H^+ \rightarrow 2GSH + NADP$$

NADPH is the preferred reductant for the plant enzyme, as it is for the enzymes from animal and microbial sources. Glutathione reductase is localized mainly in the chloroplast stroma, but is also found in the mitochondria and cytosol (Edwards *et al.* 1990). The occurrence of multiple forms of the enzyme in plant tissues is well documented (Edwards *et al.* 1990; Foyer *et al.* 1991) and this may be necessitated not only by compartmentational differences but also by the metabolic needs of the plant. Total glutathione reductase activity is often found to increase in response to several environmental stresses including low and high temperature or oxidative and water stress (Smith *et al.* 1989). A two- to three-fold increase in the level of the enzyme appears to be the maximum that can be generated in plants under environmental stress conditions (Schmidt & Kunert 1986; Foyer *et al.* 1991). It is generally assumed that this increase is a specific plant response to overcome glutathione oxidation. This

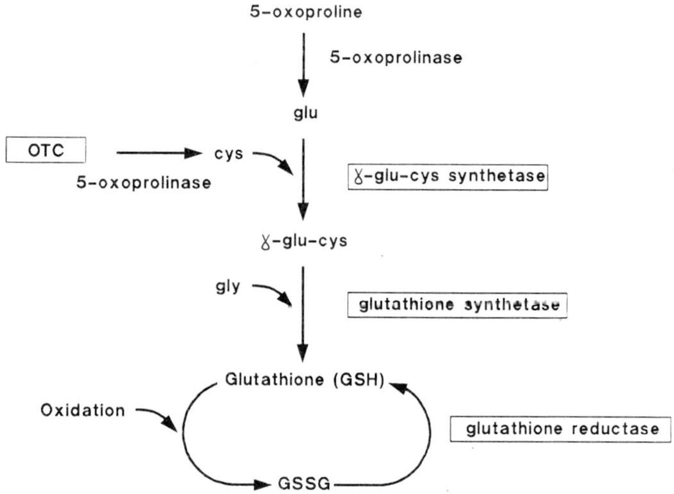

Fig. 5. Different routes for the production of reduced glutathione (GSH) in the plant cell. GSH arises either directly by biosynthesis, which can be affected by the compound OTC, or by reduction of oxidized glutathione (GSSG) by the action of glutathione reductase. Cellular consequences of overproduction of the enzymes indicated in boxes are currently being investigated.

response of the enzyme seems, however, to be plant species specific (Hausladen & Kunert 1990). In addition, pattern of isoenzyme forms varies after exposure to low temperatures as does the kinetic properties of the enzyme (Esterbauer & Grill 1978; Guy & Carter 1984; Hausladen *et al.* 1991).

Manipulation of thiol content

Optimization of thiol/disulfide exchange reactions under stress conditions may be achieved in plants by increase of the cellular content of thiols. Beside treatment of a plant with a chemical, such as OTC (Hausladen & Kunert 1990), in order to increase the glutathione level, this may also involve two different genetic strategies as outlined in Fig. 5. A first strategy may be based on a more efficient recycling of a disulfide compound, such as GSSG, to its respective active thiol form. This may be achieved by genetic manipulation of the actual level of an recycling enzyme such as glutathione reductase. A second strategy may be based on the increase of the potential of the cell to produce higher absolute levels of specific thiol compounds. Both strategies applied either alone or together may finally result in increased tolerance of cellular processes to oxidation through improved thiol/disulfide exchange reactions.

In the first approach using glutathione as a model for a thiol compound we have manipulated the cellular level of glutathione reductase via the technique of genetic engineering. We have specifically investigated whether an increased glutathione reductase level in the cytosol influences the overall glutathione level and the GSH/GSSG ratio in plants. The *gor* gene of *E. coli* coding for bacterial glutathione reductase was transferred into tobacco (*Nicotiana tabacum* L. var Samsun). Leaves from transgenic tobacco plants had an age-dependent two- to tenfold increase in cytosol glutathione reductase activity compared to non-trangenic plants (Foyer *et al.* 1991). However, increased enzymatic activity had no effect on the amount or reduction state of the GSH/GSSG pool under optimal conditions or oxidative conditions induced by the herbicide methylviologen. This clearly demonstrates that glutathione reductase activity in tobacco leaves neither limits the glutathione content nor affects the reduction state of the glutathione pool, even under severe oxidative stress. An identical result was also obtained in studies with *E. coli* where overproduction of glutathione reductase in the bacterium had no effect on the glutathione content or the GSH/GSSG ratio (Kunert *et al.* 1990). Overall, from our studies with tobacco it appeared that the plant might generally produce sufficient levels of antioxidative enzymes to maintain the balance of cellular compounds. Thus, any further increase in enzyme level is without significant effect. Furthermore, changes in the distribution pattern of individual glutathione reductase isozymes may be more significant than changes in the total pool size of glutathione reductase.

Recently, we have started a second study in order to investigate the consequences of overexpression of the *E. coli* genes coding for the two other enzymes involved in glutathione metabolism in plants, γ-glutamyl-cysteine synthetase and glutathione synthetase. This strategy to increase the capacity of glutathione production is based on recent findings by Moore *et al.* (1989) that an *E. coli* strain enriched with the genes for the two synthetases was significantly more resistant to radiation than was the corresponding wild strain. They also found that the enhanced resistance exhibited by the gene-enriched strain was associated with increased capacity to synthesize glutathione rather than solely to their increased cellular levels of glutathione.

Intracellular glutathione synthesis in plants requires the two enzymes γ-glutamyl-cysteine synthetase and glutathione synthetase and the amino acid substrates glutamate, cysteine and glycine. Both synthetases have already been characterized in plants (Klapheck *et al.* 1987; Hell & Bergmann 1988; Hell & Bergmann 1990). The enzymes are located in both the chloroplast and cytoplasm. Data obtained by Hell & Bergmann (1990) indicate that the rate of glutathione synthesis *in vivo* may be influenced substantially by the concentration of the two substrates cysteine and glutamate and may be further regulated via feedback inhibition of the enzyme γ-glutamyl-cysteine synthetase by glutathione itself. Such regulation of biosynthesis by feedback inhibition will have advantages. For example, under GSH consuming conditions the plant cell will be able to produce additional GSH immediately. In addition to feedback inhibition glutathione biosynthesis may also be regulated by the amount of biosynthetic enzymes available. This might be controlled at the transcriptional level. The importance of the actual enzyme levels present in plants was recently demonstrated for glutathione synthetase after induction of phytochelatin synthesis by cadmium in pea roots (Rüegsegger *et al.* 1990). With our present studies investigating the consequences of overproduction of both synthetases we might be able to elucidate the significance of both feedback inhibition and actual enzyme levels.

Concluding remarks

Engineering stress tolerance is undoubtedly of great benefit in terms of applications for the improvement of the potential for thiol/disulfide exchange reactions in plants.

This is an attractive idea because it presents possibilities for overcoming phytotoxic consequences such as enzyme inactivation by oxidation. Also it has been shown that the efficiency of enzyme activation by the thioredoxin system can be improved by genetic manipulation leading to an increase in the reactivity of thioredoxin with chloroplast enzymes (de Lamotte-Guery *et al.* 1991).

From data obtained it is clear that manipulation of a single component, such as glutathione reductase, seems not to be a feasable way to overcome the consequences of cellular oxidation. Recently, Teppermann & Dunsmuir (1990) came to a similar conclusion for superoxide dismutase. It is possible that in plants the absolute glutathione content and, therefore, protection is much more determined by the relative rates of synthesis and degradation of glutathione, which appear to be related to growth conditions, leaf age, the NADPH level (Burke & Hatfield 1987) and biosynthetic enzyme levels, than by simple recycling of GSSG via glutathione reductase. A more detailed answer concerning the significance of recycling of glutathione in plants by glutathione reductase will only become available when plants that lack the activity of the enzyme are produced and studied. Such plants can be produced using antisense technology. A first step in the direction of using antisense technology has been recently reported by Creissen *et al.* (1991) who have cloned and sequenced the *gor* gene from pea.

In order to genetically engineer a plant in the future and to improve its performance, we need further a much better understanding of the biochemical and physiological processes in such a complex system. Our present understanding allows us already to put forward a strategy which will be followed in order to manipulate thiol/disulfide exchange reactions for the benefit of the plant. However, our understanding might be still fragmentary and we are only just beginning to understand what effects the introduction of foreign genes into plants might have.

References

Alscher, R.G. 1989. Biosynthesis and antioxidant function of glutathione in plants. Physiol. Plant. 77: 457-464.

Anderson, J.W., Foyer, C.H. & Walker, D.A. 1983. Light dependent reduction of hydrogen peroxide by intact spinach chloroplasts. Biochim. Biophys. Acta 724: 69-74.

Bergmann L. 1981. Aspects of S- and N-metabolism in tissue cultures. In: Bothe, H. & Trebst, A. (eds.), Biology of Nitrogen and Sulfur. Springer Verlag, Berlin, pp. 341-351.

Berstermann, A., Vogt, K. & Follmann, H. 1983. Plant seeds contain several thioredoxins of regular size. Eur. J. Biochem. 131: 339-344.

Buchanan, B.B. 1980. Role of light in the regulation of chloroplast enzymes. Annu. Rev. Plant Physiol. 31: 341-374.

Burke, J.J. & Hatfield, J.L. 1987. Plant morphological and biochemical responses to field water deficits. III Effect of foliage temperature on the potential activity of glutathione reductase. Plant Physiol. 85: 100-103.

Creissen, G., Edwards, E.A., Enard, C., Wellburn, A. & Mullineaux, P. 1991. Molecular characterization of glutathione reductase cDNAs from pea (*Pisum sativum* L.). The Plant Journal 2: 129-131.

Davies, K.J.A., Delsignore, M.A & Lin, S.W. 1987. Protein damage and degradation by oxygen radicals. Modification of amino acids. J. Biol. Chem. 262: 9902-9907.

Crawford, N.A., Droux, M., Kosower, N.S. & Buchanan, B.B. 1989. Evidence for function of the ferredoxin/thioredoxin system in the reductive activation of target enzymes of isolated intact chloroplasts. Arch. Biochem. Biophys. 271: 223-239.

Cséke, C. & Buchanan, B.B. 1986. Regulation of the formation and utilization of photosynthate in leaves. Biochim. Biophys. Acta 853: 43-63.

Droux, M., Jacquot, J.P., Miginiac-Maslow, M., Gadal, P., Huet, J.C., Crawford, N.A., Yee B.C. & Buchanan, B.B. 1987. Ferredoxin-thioredoxin reductase: an iron-sulfur enzyme linking light to enzyme regulation in oxygenic photosynthesis. Purification and properties of the enzyme from C_3, C_4 and cyanobacteria species. Arch. Biochem. Biophys. 252: 426-439.

Edwards, E.A., Rawsthorne, S. & Mullineaux, P.M. 1990. Subcellular distribution of multiple forms of glutathione reductase in leaves of pea (*Pisum sativum* L.). Planta 180: 278-284.

Edwards, R., Blount, J.W. & Dixon, R.A. 1991. Glutathione and elicitation of the phytoalexin response in legume cell cultures. Planta 184: 403-409.

Esterbauer, H. & Grill, D. 1978. Seasonal variation of glutathione and glutathione reductase in needles of *Picea abies*. Plant. Physiol. 61: 119-121.

Finckh, B.F. & Kunert, K.J. 1985. Vitamin C and E: an antioxidative system against herbicide-induced lipid peroxidation in higher plants. J. Agric. Food Chem. 33: 574-577.

Florencio, F.J., Yee, B.C., Johnson, T.C. & Buchanan, B.B. 1988. An NADP/thioredoxin system in leaves: purification and characterization of NADP-thioredoxin reductase and thioredoxin *h* from spinach. Arch. Biochem. Biophys. 266: 496-507.

Foyer, C.H. 1992. Ascorbic acid. In: Alscher, R.G. & Hess, J.L. (eds.), Antioxidants in Higher Plants, CRC Press, Boca Raton, *in press*.

Foyer, C.H. & Halliwell, B. 1976. The presence of glutathione and glutathione reductase in chloroplasts: a proposed role in ascorbic acid metabolism. Planta 133: 21-25.

Foyer, C.H., Lelandais, M., Galap, C. & Kunert, K.J. 1991. Effects of elevated cytosolic glutathione reductase activity on the cellular glutathione pool and photosynthesis in leaves under normal and stress conditions. Plant Physiol. 97: 863-872.

Gillham, D.J. & Dodge, A.D. 1986. Hydrogen peroxide-scavenging systems within pea chloroplasts. A quantitative study. Planta 167: 246-251.

Guy, C.L. & Carter, J.V. 1984. Characterisation of partially purified glutathione reductase from cold-hardened and non-hardened spinach leaf tissue. Cryobiology 21: 454-464.

Hausladen, A. & Kunert, K.J. 1990. Effects of artifically enhanced levels of ascorbate and glutathione on the enzymes monodehydroascorbate reductase, dehydroascorbate reductase and glutathione reductase in spinach (*Spinacia oleracea*). Physiol. Plant. 79: 384-388.

Hausladen, A., Doulis, A., Chevone, B.I. & Alscher, R.G. 1991. Glutathione reductase and cold tolerance of red spruce. In: Pell, E. & Steffen, K. (eds.), Active Oxygen/Oxidative Stress and Plant Metabolism. American Society of Plant Physiologists, Rockville, pp. 241-243.

Hell, R. & Bergmann, L. 1988. Glutathione synthetase in tobacco suspension cultures. Catalytic properties and localization. Physiol. Plant. 72: 70-76.

Hell, R. & Bergmann, L. 1990. γ-Glutamylcysteine synthetase in higher plants: catalytic properties and subcellular localization. Planta 180: 603-612.

Holmgren, A. 1985. Thioredoxin. Annu. Rev. Biochem. 54: 237-271.

Jackson, R.J., Campbell, E.A., Herbert, P. & Hunt, T. 1983. The roles of sugar phosphates and thiol reducing systems in the control of reticulocyte protein synthesis. Eur. J. Biochem. 131: 313-324.

Klapheck, S., Latus, C. & Bergmann, L. 1987. Localization of glutathione synthetase and distribution of glutathione in leaf cells of *Pisum sativum* L. J. Plant Physiol. 131: 123-131.

Knaff, D.B. 1989. The regulatory role of thioredoxin in chloroplasts. TIBS. 14: 433-434.

Kosower, N.S. & Kosower, E.W. 1978. The glutathione status of cells. Int. Rev. Cytol. 54: 109-160.

Kunert, K.J., Homrighausen, C., Böhme, H. & Böger, P. 1985. Oxyfluorfen and lipid peroxidation: protein damage as a phytotoxic consequence. Weed Sci. 33: 766-770.

Kunert, K.J., Cresswell, C.F., Schmidt, A., Mullineaux P.M. & Foyer, C.H. 1990. Variations in the activity of glutathione reductase and the cellular glutathione content in relation to sensitivity to methylviologen in *Escherichia coli*. Arch. Biochem. Biophys. 282: 233-238.

de Lamotte-Guery, F., Miginiac-Maslow, M., Decottignies, P., Stein, M. & Jaquot, J.P. 1991. Mutation of a negatively charged amino acid in thioredoxin modifies its reactivity with chloroplastic enzymes. Eur. J. Biochem. 196: 287-294.

Leegood, R.C., Walker, D.A. & Foyer, C.H. 1985. Regulation of the Benson-Calvin cycle. In: Barber, J. & Baker, N.R. (eds.), Photosynthetic Mechanisms and the Environment. Elsevier Science Publishers BV, The Netherlands, pp. 191-238.

Leegood, R.C. 1990. Enzymes of the Calvin cycle. In: Lea, P.J. (ed.), Methods in Plant Biochemistry,

Vol. 3, Chapter 2. Academic Press, London, pp. 15-37.
Levitt, J. 1962. A sulfhydryl-disulfide hypothesis of frost injury and resistance in plants. J. Theor. Biol. 3: 355-391.
Malbon, C.C., Shaji, T.G. & Moxham, C.P. 1987. Intramolecular disulfide bridges: avenues to receptor activation? TIBS. 12: 172-175.
Muller, E.G.D. & Buchanan, B.B. 1989. Thioredoxin is essential for photosynthetic growth. The thioredoxin *m* gene of *Anacystis nidulans*. J. Biol. Chem. 264: 4008-4014.
Meister, A. 1988. Glutathione metabolism and its selective modification. J. Biol. Chem. 263: 17205-17208.
Meister, A. & Anderson, M.E. 1983. Glutathione. Annu. Rev. Biochem. 52: 711-760.
Moore, W.R., Anderson, M.E., Meister, A., Murata, K. & Kimura, A. 1989. Increased capacity for glutathione synthesis enhances resistance to radiation in *Escherichia coli*: a possible model for mammalian cell protection. Proc. Natl. Acad. Sci. USA. 86: 1461-1464.
Reglinski, J., Hoey, S., Smith, W.E. & Sturrock, R.D. 1988. Cellular response to oxidative stress at sulfhydryl group receptor sites on the erythrocyte membrane. J. Biol. Chem. 263: 12360-12366.
Rennenberg, H. 1982. Glutathione metabolism and possible biological roles in higher plants. Phytochemistry 21: 2771-2781.
Rüegsegger, A., Schmutz, D. & Brunold, C. 1990. Regulation of glutathione synthesis by cadmium in *Pisum sativum* L. Plant Physiol. 93: 1579-1584.
Schmidt, A. & Kunert, K.J. 1986. Lipid peroxidation in higher plants. The role of glutathione reductase. Plant Physiol. 82: 700-702.
Slovacek, R.E. & Vaughn, S. 1982. Chloroplast sulfhydryl groups and the light activation of fructose-1,6-biphosphatase. Plant Physiol. 70: 978-981.
Smith, I.K., Vierheller, T.L. & Thorne, C.A. 1989. Properties and functions of glutathione reductase in plants. Physiol. Plant. 77: 449-456.
Smith, I.K., Polle, A. & Rennenberg, H. 1990. Glutathione. In: Alscher, R.G. & Cumming, J.R. (eds.), Stress Responses in Plants: Adaptation and Acclimation Mechanisms. Wiley-Liss Inc, pp. 201-215
Tepperman, J.M. & Dunsmuir, P. 1990. Transformed plants with elevated levels of chloroplastic SOD are not more resistant to superoxide toxicity. Plant Mol. Biol. 14: 501-511.
Vivekanandan, M. & Edwards, G.E. 1987. Activation of NADP-malate dehydrogenase in C_3 plants by reduced glutathione. Photosyn. Res. 14: 113-124.
Wefers, H. & Sies, H. 1988. The protection by ascorbate and glutathione against microsomal lipid peroxidation is dependent on vitamin E. Eur. J. Biochem. 174: 353-357.
Williamson, J.M. & Meister, A. 1981. Stimulation of hepatic glutathione formation by administration of L-2-oxothiazolidine-4-carboxylate, a 5-oxo-L-prolinase substrate. Proc. Natl. Acad. Sci. USA. 78: 936-939.
Wingate, V.P.M., Lawton, M.A. & Lamb, C.J. 1988. Glutathione causes a massive and selective induction of plant defense genes. Plant Physiol. 31: 205-211.
Wise, R.R. & Naylor, A.W. 1987. Chilling-enhanced photooxidation. Evidence for the role of singlet oxygen and superoxide in the breakdown of pigments and endogenous antioxidants. Plant Physiol. 83: 278-282.
Ziegler, D.M. 1985. Role of reversible oxidation-reduction of enzyme thiol-disulfides in metabolic regulation. Annu. Rev. Biochem. 54: 305-329.

PLANT THIOREDOXINS

Peter Schürmann
Laboratoire de Biochimie végétale, Université de Neuchâtel, CH-2000 Neuchâtel, Switzerland

Introduction

Thioredoxins are ubiquitous, low molecular weight (approx. M_r 12,000) proteins with a well conserved, characteristic active site structure: -Trp-Cys-Gly-Pro-Cys-. The two cysteine residues form a redox-active, intramolecular disulfide bridge. Thioredoxins were shown to have many different functional roles in a large variety of organisms (Holmgren 1985; Gleason & Holmgren 1988; Holmgren 1989). Originally thioredoxin was isolated from *Escherichia coli* and shown to act as hydrogen donor for the ribonucleotide reductase (Laurent *et al.* 1964) an essential enzyme for DNA replication. *E. coli* cells contain one unique type of thioredoxin which has been extensively studied and its three-dimensional structure determined by x-ray crystallography (Holmgren *et al.* 1975; Katti *et al.* 1990). The molecule is highly structured with more than 90% of its residues involved in secondary structure. It is built of 5 β-strands forming a twisted, central β-sheet which is surrounded by 4 α-helices. The active site is at the surface of the protein with one of its cysteines (Cys-32) exposed to the solvent. One particular area close to the active site is hydrophobic and has been suggested to be the main interaction site with other proteins (Eklund *et al.* 1984). Thioredoxin is reduced in *E. coli* by a NADP-dependent thioredoxin reductase, a flavoprotein composed of two identical subunits (Thelander 1967).

Plant cells contain multiple forms of thioredoxins (Jacquot *et al.* 1978; Wolosiuk *et al.* 1979). They have been classified on the basis of functional properties and of their location within the cell. In green algae as well as higher plants three distinct types of thioredoxins have been characterized: thioredoxin *f*, thioredoxin *m* and thioredoxin *h*. Thioredoxins *f* and *m* are located in the chloroplasts where they function primarily as regulatory proteins in carbon metabolism. Thioredoxin *h* is present outside the chloroplast.

This paper summarizes knowledge on higher plant and algal thioredoxins. It will emphasize more the structural aspects that have emerged during the last years and focus for that purpose on the best known systems. The reader is referred to recently published reviews on thioredoxin and related aspects for further information (Gleason & Holmgren 1988; Holmgren 1989; Knaff 1989; Buchanan 1991; Eklund *et al.* 1991).

Thioredoxin *f*

Thioredoxin *f* has originally been described as the chloroplast thioredoxin capable of activating fructose 1,6-bisphosphatase (Wolosiuk *et al.* 1979). In addition to fructose 1,6-bisphosphatase several other chloroplast enzymes have been reported to be

Sulfur Nutrition and Assimilation in Higher Plants
edited by L.J. De Kok et al., pp. 153–162

activated by reduced thioredoxin *f*. These include sedoheptulose 1,7-bisphosphatase (Nishizawa & Buchanan 1981), phosphoribulokinase and NADP-glyceraldehyde-3-phosphate dehydrogenase (Wolosiuk *et al.* 1979), NADP-malate dehydrogenase (Schürmann *et al.* 1981; Tsugita *et al.* 1983) and CF_1-ATPase (Mills *et al.* 1981). However, the primary function of thioredoxin *f* seems to be the activation of three important regulatory enzymes of carbohydrate synthesis, catalyzing reactions of the reductive pentose phosphate cycle with large free energy differences (Krause & Bassham 1969), *i.e.* fructose 1,6-bisphosphatase, sedoheptulose 1,7-bisphosphatase and phosphoribulokinase. These three enzymes are exclusively activated by reduced thioredoxin *f* whereas the other enzymes mentioned can also be activated by thioredoxin *m*.

Thioredoxin *f* has been purified and characterized from several higher plants and green algae, including spinach (Wolosiuk *et al.* 1979; Schürmann *et al.* 1981; Soulié *et al.* 1981), corn (Crawford *et al.* 1986), *Kalanchoë daigremontiana* (Hutcheson & Buchanan, 1983), *Scenedesmus obliquus* (Langlotz *et al.* 1985), *Chlamydomonas reinhardtii* (Huppe *et al.* 1990) and *Acetabularia mediterranea* (Van Langendonckt & Vanden Driessche, 1992). Most of these thioredoxins were found to have molecular weights of 12–13 kDa, typical for thioredoxins. An exception is the thioredoxin from *Scenedesmus* which was shown, by SDS electrophoresis, to be larger (28 kDa), but exhibits the same properties as other *f*-type thioredoxins. A larger *f*-type thioredoxin has also been purified from the cyanobacterium *Anabaena* sp. 7119 (Whittacker & Gleason 1984). The *f*-type thioredoxins are reported to be rather unstable. This may, in part, be due to their low solubility in buffer solution which can be somewhat improved by the addition of dimethylsulfoxide (Aguilar *et al.* 1992).

Four *f*-type isoforms have been isolated from corn and two from *Kalanchoë*. They differ slightly in size and/or isoelectric point but are functionally indistiguishable. It is not known whether these variants are *in vivo* isoforms or whether they originate from proteolysis. However, it is interesting to note that these isoforms are found in plants with specialized carbon assimilation pathways, *i.e.* in corn, a C_4 plant, and in *Kalanchoë*, a CAM plant.

The best known thioredoxin *f* is the one from spinach. Its amino acid sequence (Fig. 1A) has been determined and a cDNA clone coding for this nuclear encoded protein has been isolated (Kamo *et al.* 1989). Its primary structure is distinct from other thioredoxins with less than 30% sequence identity with its companion, thioredoxin *m*. A recently isolated cDNA clone from pea codes for a thioredoxin f which is almost identical (84% similarity) to the spinach protein (Lepiniec *et al.* 1992). This very strongly supports the idea, that certain features found for the thioredoxin *f* from spinach are general properties of most *f*-type thioredoxins. One such feature is the presence of a third Cys, also reported for *Scenedesmus* (Langlotz *et al.* 1985). It is located in the C-terminal half of the protein flanked by polar residues. This stretch of amino acids may be important in the interaction of thioredoxin *f* with its target enzymes as has been suggested from experiments with mutated thioredoxin (Lamotte-Guéry *et al.* 1991) and has been observed after chemical modification of the third Cys (unpublished results).

The comparison of the three dimensional structures of thioredoxin *f* and *E. coli* thioredoxin might give some insights into the target enzyme specificity. For that purpose thioredoxin f has been crystallized (Génovésio-Taverne *et al.* 1991). This opens the possibility to study the three dimensional structure by X-ray analysis. These structural and further functional studies are facilitated by the availability of large amounts of a recombinant thioredoxin *f* (Aguilar *et al.* 1992) which has been shown to have the same biological properties as the native chloroplast protein.

Thioredoxin *m*

The *m*-type thioredoxin has been described as the thioredoxin showing greater activity with chloroplast NADP-malate dehydrogenase (Wolosiuk *et al.* 1979). Thioredoxin *m* can regulate other enzyme activities outside the reductive pentose phosphate cycle and has been demonstrated to be essential for photosynthetic growth of a cyanobacterium (Muller & Buchanan 1989). Its main functions probably are the light-deactivation and dark-activation of glucose 6-phosphate dehydrogenase, a key enzyme of the oxidative pentose phosphate pathway in chloroplasts and the light-activation of NADP-malate dehydrogenase. This latter enzyme is especially important in C_4 plants where C_4 dicarboxylic acids, like malate, are early carbon fixation products. In the chloroplasts of C_3 plants NADP-malate dehydrogenase is proposed to function as part of a light-dependent mechanism for the export of reducing equivalents (Scheibe 1987). Additional reported functions of thioredoxin *m* are the light-activation of chloroplast coupling factor (Mills *et al.* 1981), PAPS-reductase (Schwenn 1989) and of two enzymes of nitrogen assimilation (Schmidt 1981; Tischner & Schmidt 1982).

Thioredoxin *m* has been purified and characterized from the same higher plants and green algae as already mentioned for thioredoxin *f*. All *m*-type thioredoxins are of similar size (≈12 kDa). From spinach (Schürmann *et al.* 1981), corn (Crawford *et al.* 1986) and Acetabularia (Van Langendonckt & Vanden Driessche 1992) isoforms have been isolated that are functionally indistinguishable, differing only in apparent molecular weight and/or charge.

The primary structures of *m*-type thioredoxins from spinach (Maeda *et al.* 1986; Wedel *et al.* 1992), *Chlamydomonas* (Decottignies *et al.* 1990; Jacquot *et al.* 1992) and two cyanobacteria, *Anabaena* sp. 7119 (Gleason *et al.* 1985) and *Anacystis* (Muller & Buchanan 1989), are known (Fig. 1 C). The eukaryotic thioredoxins show extensive similarities between each other (76% similarity) and with the cyanobacterial (72%) and the *E. coli* proteins (59%). This suggests that the *m*-type thioredoxins of the eukaryotic cells are of bacterial origin (Maeda *et al.* 1986). These structural similarities are in line with functional similarities. In reactions where thioredoxin *m* is involved it can usually be replaced by the thioredoxin from *E. coli* with good efficiency (Tsugita *et al.* 1983).

In spinach chloroplasts three functionally indistinguishable isoforms of thioredoxin *m* have been found (Maeda *et al.* 1986). These isomers are due to a N-terminal heterogeneity which is no purification artefact caused by proteolytic degradation of the N-terminus, nor the result of the expression of different genes. In the eukaryotic cells thioredoxins are nuclear encoded. They are synthesized on cytoplasmic ribosomes as precursor proteins with a transit peptide directing them towards the chloroplasts. After the import the transit peptide is cut off by a processing enzyme. It has recently been shown that the spinach thioredoxin *m* isomers are the products of processing

```
A.            f-type thioredoxins

Spinach       MEQALGTQEMEAIVGKVTEVNKDTFWPIVKAAGDKPVVLDMFTQWCGPCKAMAPKYEKLAEEYLDVIFLKLDCNQENKTLAKELGIRVVPTFKILKENSVVGEVTGAKYDKLLEAIQAARSS
Pisum         VRSSLETAGPTVTVGKVTEVNKDTFWPIVNAAGDKTVVLDMFTKWCGPCKVTAPLYEELSQKYLDVVFLKLDCNQDNKSLAKELGIKVVPTFKILKDNKIVKEVTGAKFDDLVAAIDTVRSS
              .  .* *      *****************.****** *******.******  ** ** *.. ****.*********.**.********.**********.* .* *******.* *. **.. ***

B.            h-type thioredoxins

Chlam.reinh.  ----------GGSVIVIDSKAAWDAQLAKGKEEHKPIVVDFTATWCGPCKMIAPLFETLSNDYAGKVIFLKVDVDAVAAVAEAAGITAMPTFHVYKDGVKADDLVGASQDKLKALVAKHAA------
Tobacco       MAANDATSSEEGQVFGCHKVEEWNEYFKKGVETKKLVVVDFTASWCGPCRFIAPILADIAKKMP-HVIFLKVDVDELKTVSAEWSVEAMPTFVFIKDGKEVDRVVGAKKEELQQTIVKHAAPATVTA
                        * *.  .    *.  . ** *  * .******.*****. ***..   ...   .  ********** . .*.    .. *****   ***   * .*** .. *.   . ****

Spinach root  -----------------------------------LVVIDFTASYCAPCRFISPILTDLAK----------VDVDELKTVAEDFKVDAMPTFVLFKDGMEIDRVVGAK-EALVELVAK--------
                                                 .*.****. *.**. *.*..   ...         **** . .*.    . *****   ***   * .***  . *   . *

C.            m-type thioredoxins

Spinach m            KASEAVKEVQDVNDSSWKEFVLESEVPVMVDFWAPWCGPCKLIAPVIDELAKEYSGKIAVTKLNTDEAPGIATQYNIRSIPTVLFFKNGERKESIIGAVPKSTLTDSIEKYLSP
Chlamydomo m         EAGA-VN------DDTFKNVVLESSVPVLVDFWAPWCGPCRIIAPVVDEIAGEYKDKLKCVKLNTDESPNVASEYGIRSIPTIMVFKGGKKCETIIGAVPKATIVQTVEKYLN-
Anabaena m           SAAAQVT------DSTFKQEVLDSDVPVLVDFWAPWCGPCRMVAPVVDEIAQQYEGKIKVVKVNTDENPQVASQYGIRSIPTLMIFKGGQKVDMVVGAVPKTTLSQTLEKHL--
Anacystis m          SVAAAVT------DATFKQEVLESSIPVLVDFWAPWCGPCRMVAPVVDEIAQQYSDQVKVVKVNTDENPSVASQYGIRSIPTLMIFKDGQRVDTVVGAVPKTTLANTLDKHL--
                      .   *      * . *. **.* .**.************...***.**.* .* ...   *.**** * .*..* ******.. ** * .  . ..*****.*. ....* *

E coli               S-----DKIIHLTDDSFDTDVLKADGAILVDFWAEWCGPCKMIAPILDEIADEYQGKLTVAKLNIDQNPGTAPKYGIRGIPTLLLFKNGEVAATKVGALSKGQLKEFLDANLA-
                             * .    ** .  ...***** *****...**..**.* .* ...   *.* *. *   *   * **.***.. ** *      .**..* . . .. *

D.            All plant and some bacterial thioredoxins

Spinach f     MEQALGTQEMEAIVGKVTEVNKDTFWPIVKAAG---DKPVVLDMFTQWCGPCKAMAPKYEKLAEEYLD-VIFLKLDCNQENKTLAKELGIRVVPTFKILKENSVVGEVTGA-KYDKLLEAIQ-----AARSS
Pisum f       VRSSLETAGPTVTVGKVTEVNKDTFWPIVNAAG---DKTVVLDMFTKWCGPCKVTAPLYEELSQKYLD-VVFLKLDCNQDNKSLAKELGIKVVPTFKILKDNKIVKEVTGA-KFDDLVAAID-----TVRSS
Chlamydomo h  --------------GGSVIVIDSKAAWDAQLAKGKEEHKPIVVDFTATWCGPCKMIAPLFETLSNDYAGKVIFLKVDVD-AVAAVAEAAGITAMPTFHVYKDGVKADDLVGA-SQDKLKALVAKHAA------
Nicotiana h   MAAN---DATSSEEGQVFGCHKVEEWNEYFKKGVETKKLVVVDFTASWCGPCRFIAPILADIAKKMP-HVIFLKVDVD-ELKTVSAEWSVEAMPTFVFIKDGKEVDRVVGA-KKEELQQTIVKHAAPATVTA
Spinach h     --------------------------------------LVVIDFTASYCAPCRFISPILTDLAK----------VDVD-ELKTVAEDFKVDAMPTFVLFKDGMEIDRVVGA-K-EALVELVAK---------
Spinach m     --------KASEAVKEVQDVNDSSWKEFVL----ESEVPVMVDFWAPWCGPCKLIAPVIDELAKEYSGKIAVTKLNTD-EAPGIATQYNIRSIPTVLFFKNGERKESIIGAVPKSTLTDSIEKYLSP-----
Chlamydomo m  --------EAGA-VN------DDTFKNVVL----ESSVPVLVDFWAPWCGPCRIIAPVVDEIAGEYKDKLKCVKLNTD-ESPNVASEYGIRSIPTIMVFKGGKKCETIIGAVPKATIVQTVEKYLN------
Anabaena m    --------SAAAQVT------DSTFKQEVL----DSDVPVLVDFWAPWCGPCRMVAPVVDEIAQQYEGKIKVVKVNTD-ENPQVASQYGIRSIPTLMIFKGGQKVDMVVGAVPKTTLSQTLEKHL-------
Anacystis m   --------SVAAAVT------DATFKQEVL----ESSIPVLVDFWAPWCGPCRMVAPVVDEIAQQYSDQVKVVKVNTD-ENPSVASQYGIRSIPTLMIFKDGQRVDTVVGAVPKTTLANTLDKHL-------
E coli        -------------SDKIIHLTDDSFDTDVLKA----DGAILVDFWAEWCGPCKMIAPILDEIADEYQGKLTVAKLNIDQN-PGTAPKYGIRGIPTLLLFKNGEVAATKVGALSKGQLKEFLDANLA------
                                                 ...* .  *.**.  .*    ..          .. .       .    . .**   *          **    .   .
```

Fig. 1. Primary structures of plant thioredoxins. The amino acid sequences have been aligned using Clustal V (Higgins *et al.* 1992). Conserved residues are indicated by *, conservative replacements by ·. The sequence data were taken from the following sources: spinach chloroplast thioredoxin *f* (Kamo *et al.* 1989; Aguilar *et al.* 1992), pea chloroplast thioredoxin *f* (Lepiniec *et al.* 1992), *Chlamydomonas reinhardtii* thioredoxin *h* (Decottignies *et al.* 1991), *Nicotiana tabacum* thioredoxin*h* (Marty & Meyer 1991), spinach root thioredoxin*h* (Marcus *et al.* 1991), spinach chloroplast thioredoxin *m* (Wedel *et al.* 1992), *Chlamydomonas reinhardtii* chloroplast thioredoxin *m* (Decottignies *et al.* 1990), *Anabaena* sp. 7119 (Gleason *et al.* 1985), *Anacystis nidulans* (Muller & Buchanan 1989), *E. coli* (Höög *et al.* 1984).

after the import (Wedel *et al.* 1992). It is possible that the isoforms observed with other thioredoxins are also due to the processing.

Thioredoxin *h*

Thioredoxin *h* has been purified from several organisms, from two algae *Acetabularia mediterranea* (Van Langendonckt & Vanden Driessche 1992) and *Chlamydomonas reinhardtii* (Decottignies *et al.* 1991), and from higher plant tissues, cultured carrot root cells (Johnson *et al.* 1987a), spinach roots (Marcus *et al.* 1991), spinach leaves (Florencio *et al.* 1988) and wheat seeds (Johnson *et al.* 1987b). Since this protein was found in dark grown tissue it was designated thioredoxin *h*, for heterotrophic (Johnson *et al.* 1987a). Thioredoxin *h* is not present in chloroplasts, but is found in the cytosol, in the endoplasmatic reticulum and in mitochondria (Marcus *et al.* 1991; Bodenstein-Lang *et al.* 1989). Cytosolic thioredoxins isolated earlier from *Chlorella* (Tsang 1981), wheat and soybean (Suske *et al.* 1979; Berstermann *et al.* 1983; Vogt & Follmann 1986) can also be classified as *h*-type thioredoxins.

Two complete primary structures have been reported, one for *Chlamydomonas* (Decottignies *et al.* 1991), based on protein sequencing, and one for tobacco (Marty & Meyer 1991), deduced from the nucleotide sequence of a cDNA (Fig. 1B). A partial amino acid sequence has been determined for a spinach root thioredoxin *h* (Marcus *et al.* 1991). All three proteins are very similar with $\approx 60\%$ similarity (Fig. 1). There is one noteworthy difference: the spinach root thioredoxin has the modified active site sequence -Trp-Cys-Ala-Pro-Cys-, where Gly is replaced by Ala. Such a substitution has already been reported for a thioredoxin from *Corynebacterium nephridii* where it has little impact on the redox function of the protein (McFarlan *et al.* 1989), indicating that the Gly to Ala substitution is a relatively minor change that can be easily accommodated. The modified active site sequence does so far not seem to be typical for all *h*-type thioredoxins since two of the three known members of this group have the usual active site. The tobacco protein sequence leaves no doubt that it represents a cytoplasmic thioredoxin *h* since the cDNA encoding this thioredoxin does not contain any information for a transit peptide.

The biological functions of thioredoxin *h* are yet not clear since mostly *in vitro* activities have been reported sofar. There is one recent report suggesting that during germination thioredoxin *h* is reducing wheat storage proteins making them thus susceptible to proteolytic degradation (Kobrehel *et al.* 1992). *In vitro* thioredoxin *h* reduces purothionin, a small basic protein from the seed endosperm of cereals (Johnson *et al.* 1987b) and α-amylase and trypsin inhibitor proteins (Kobrehel *et al.* 1991), functions as hydrogen donor for ribonucleotide reductase (Berstermann *et al.* 1983; Langlotz *et al.* 1985; Bodenstein-Lang *et al.* 1989) and will activate NADP-dependent malate dehydrogenase although less efficiently than thioredoxin *m*. However, thioredoxin *h* is unable to activate fructose 1,6-bisphosphatase. It is quite probable that the electron transfer to the ribonucleotide reductase is the primary function of thioredoxin *h* in the cytosol of cells in vegetative tissue.

Structure comparison

A comparison of the primary structures clearly shows that there are three distinct groups of plant thioredoxins (Fig. 1D) as has been concluded from the functional properties of the various thioredoxins. In general plant thioredoxins tend to be a few amino acids longer than bacterial thioredoxins. Within one particular group of thioredoxins there are extensive sequence similarities of between 64 and 84%, the higher values being found between phylogenetically closer related species. This has already been suggested based on immunological crossreactivity of spinach thioredoxin antibodies with thioredoxins from higher plants and algae. Higher plant thioredoxins are well recognized by spinach thioredoxin antibodies whereas algal thioredoxins are weekly or not at all recognized by the same antibodies.

Between the different groups of thioredoxins there are rather large differences and the similarities drop to below 50%. The most striking example are the chloroplast thioredoxins where there are only 27% identical residues, strongly suggesting that the *f*- and *m*-type thioredoxins are of different origin.

Besides the active site area which is well conserved a small number of residues are strictly conserved throughout all plant thioredoxins and also throughout bacterial and animal thioredoxins. A comparison shows that these are the important residues for the maintenance of the overall structure and the redox function of the thioredoxins (Eklund *et al.* 1991) indicating that the plant thioredoxins have the same general three-dimensional structure as the one known for the protein from *E. coli*. The observed specificity of each group must be the consequence of specific details in the surface-structure of its members reflecting the differences in the amino acid sequences between the groups.

It will be of considerable interest to obtain three-dimensional structures of thioredoxins representative of the different groups and to try to relate their detail structures to their specific functions.

Thioredoxin reduction

Plant cells contain two different enzyme systems capable of reducing thioredoxins, one located in the chloroplasts and known as the ferredoxin/thioredoxin system and the second one in the cytoplasm known as the NADP/thioredoxin system.

The ferredoxin/thioredoxin system uses electrons produced by the photosynthetic electron transport chain to reduce thioredoxins via ferredoxin and the enzyme ferredoxin-thioredoxin reductase (FTR). The functioning of this system *in vivo* is very well documented (Crawford *et al.* 1989; Buchanan 1991). The FTR is a protein of about 30 kDa, composed of two nonidentical subunits, containing a Fe-S cluster and a redox-active disulfide bridge (Droux *et al.* 1987; Tsugita *et al.* 1991). The enzyme exhibits a certain thioredoxin specificity reducing preferentially chloroplast thioredoxins whereas thioredoxin *h* is reduced less efficiently (Huppe *et al.* 1991).

The NADP/thioredoxin system reduces thioredoxins with electrons from NADPH via the enzyme NADP-thioredoxin reductase (NTR). This system has been characterized in wheat (Suske *et al.* 1979), carrot (Johnson *et al.* 1987a), spinach (Florencio *et al.* 1988), *Chlorella* (Tsang 1981), *Chlamydomonas* (Huppe *et al.* 1991) and

Acetabularia (Van Langendonckt & Vanden Driessche 1992). The plant NTR is comparable to the well known bacterial and mammalian enzymes. It is a flavoprotein of about 70 kDa, composed of two identical subunits. NTR exhibits rather high thioredoxin specificity. It will reduce *h*-type thioredoxins best, *m*-type thioredoxins only weakly and *f*-type thioredoxins not at all.

Target enzymes

With several target enzymes their thioredoxin-activation has been studied at the molecular level. It has been demonstrated that, upon activation, specific regulatory disulfide bridges are reduced. The amino acid sequences of these regulatory sites, containing the Cys involved, have been determined for four enzymes, including fructose 1,6-bisphosphatase (Marcus *et al.* 1988; Raines *et al.* 1988), NADP-malate dehydrogenase (Decottignies *et al.* 1988), phosphoribulokinase (Porter *et al.* 1988) and chloroplast coupling factor (Miki *et al.* 1988). The presence of a regulatory disulfide bridge in the target enzyme seems to be a prerequisite for thioredoxin-activation. The demonstration of the presence of such a thioredoxin-reducible regulatory site will therefore be a necessary criterion for the characterization of thioredoxin target enzymes.

Thioredoxin and sulfur metabolism

There is ample evidence that thioredoxin is an important electron carrier in sulfur metabolism (Holmgren 1989). It has recently been demonstrated that in *E. coli* thioredoxin is essential for sulfate reduction (Russel *et al.* 1990). An involvement of thioredoxin in the sulfate assimilation in plants is less well documented. It has been shown that in *Chlorella* thioredoxin cannot be used as reductant by the adenosine 5'-phosphosulfate sulfotransferase (Tsang 1981). Only in two instances a possible involvement of plant thioredoxins has been reported. The APS-kinase isolated from *Chlamydomonas* was stimulated in a heterologous system by thioredoxin *f* isolated from spinach (Schwenn & Schriek 1984) and a PAPS-reductase isolated from spinach leaves was activated by homologous thioredoxin m or heterologous thioredoxin from *E. coli* (Schwenn 1989). Further experiments will have to confirm and extend these results.

Acknowledgement

The research done in the authors laboratory is supported by the Schweizerischer Nationalfonds (31-28811.90).

References

Aguilar, F., Brunner, B., Gardet-Salvi, L., Stutz, E. & Schürmann, P. 1992. Biosynthesis of active spinach-chloroplast thioredoxin *f* in transformed *E. coli*. Plant Mol. Biol. 20: 301-306.

Berstermann, A., Vogt, K. & Follmann, H. 1983. Plant seeds contain several thioredoxins of regular size. Eur. J. Biochem. 131: 339-344.

Bodenstein-Lang, J., Buch, A. & Follmann, H. 1989. Animal and plant mitochondria contain specific thioredoxins. FEBS (Fed. Eur. Biochem. Soc.) Lett. 258: 22-26.

Buchanan, B.B. 1991. Regulation of CO_2 assimilation in oxygenic photosynthesis: The ferredoxin/thioredoxin system: Perspective on its discovery, present status, and future development. Arch. Biochem. Biophys. 288: 1-9.

Crawford, N.A., Yee, B.C., Hutcheson, S.W., Wolosiuk, R.A. & Buchanan, B.B. 1986. Enzyme regulation in C_4 photosynthesis: purification properties and activities of thioredoxins from C_4 and C_3 plants. Arch. Biochem. Biophys. 244: 1-15.

Crawford, N.A., Droux, M., Kosower, N.S. & Buchanan, B.B. 1989. Evidence for function of the ferredoxin/thioredoxin system in the reductive activation of target enzymes of isolated intact chloroplasts. Arch. Biochem. Biophys. 271: 223-239.

Decottignies, P., Schmitter, J.M., Miginiac-Maslow, M., Le Marechal, P., Jacquot, J.P. & Gadal, P. 1988. Primary structure of the light-dependent regulatory site of corn NADP-malate dehydrogenase. J. Biol. Chem. 263: 11780-11785.

Decottignies, P., Schmitter, J.-M., Jacquot, J.-P., Dutka, S., Picaud, A. & Gadal, P. 1990. Purification, characterization, and complete amino acid sequence of a thioredoxin from a green alga, *Chlamydomonas reinhardtii*. Arch. Biochem. Biophys. 280: 112-121.

Decottignies, P., Schmitter, J.-M., Dutka, S., Jacquot, J.-P. & Miginiac-Maslow, M. 1991. Characterization and primary structure of a second thioredoxin from the green alga, *Chlamydomonas reinhardtii*. Eur. J. Biochem. 198: 505-512.

Droux, M., Jacquot, J., Miginiac-Maslow, M., Gadal, P., Huet, J.C., Crawford, N.A. & Yee, B.C.B. 1987. Ferredoxin-thioredoxin reductase an iron-sulfur enzyme linking light to enzyme regulation in oxygenic photosynthesis: purification and properties of the enzyme from C_3 C_4 and cyanobacterial species. Arch. Biochem. Biophys. 252: 426-439.

Eklund, H., Cambillau, C., Sjöberg, B.M., Holmgren, A., Jörnvall, H., Höög, J.O. & Bränden, C.I. 1984. Conformational and functional similarities between glutaredoxin and thioredoxins. EMBO (Eur. Mol. Biol. Organ.) J. 3: 1443-1449.

Eklund, H., Gleason, F.K. & Holmgren, A. 1991. Structural and functional relations among thioredoxins of different species. Proteins 11: 13-28.

Florencio, F.J., Yee, B.C., Johnson, T.C. & Buchanan, B.B. 1988. An NADP/thioredoxin system in leaves. Purification and characterization of NADP-thioredoxin reductase and thioredoxin *h* from spinach. Arch. Biochem. Biophys. 266: 496-507.

Génovésio-Taverne, J.-C., Jetzer, Y., Sauder, U., Hohenester, E., Huguet, Ch., Jansonius, J.N., Gardet-Salvi, L. & Schürmann, P. 1991. Crystallization and preliminary X-ray diffraction studies of the spinach-chloroplast thioredoxin *f*. J. Mol. Biol. 222: 459-461.

Gleason, F.K. & Holmgren, A. 1988. Thioredoxin and related proteins in procaryotes. FEMS. Microbiology Reviews. 54: 271-298.

Gleason, F.K., Whittacker, M.M., Holmgren, A. & Jörnvall, H. 1985. The primary structure of thioredoxin from the filamentous cyanobacterium *Anabaena* Sp 7119. J. Biol. Chem. 260: 9567-9573.

Higgins, D.G., Bleasby, A.J. & Fuchs, R. 1992. CLUSTAL V: improved software for multiple sequence alignment. CABIOS 8: 189-191.

Holmgren, A. 1985. Thioredoxin. Annu. Rev. Biochem. 54: 237-271.

Holmgren, A. 1989. Thioredoxin and glutaredoxin systems. J. Biol. Chem. 264: 13963-13966.

Holmgren, A., Söderberg, B.O., Eklund, H. & Bränden, C.I. 1975. Three-dimensional structure of *Escherichia coli* thioredoxin-S2 to 2.8 Å resolution. Proc. Natl. Acad. Sci. USA 72: 2305-2309.

Höög, J.O., Von Bahr-Lindström, H., Josephson, S., Wallace, B.J., Kushner, S.R., Jörnvall, H. & Holmgren, A. 1984. Nucleotide sequence of the thioredoxin gene from *Escherichia coli*. Bioscience Reports. 4: 917-923.

Huppe, H.C., Lamotte-Guéry, F.de, Jacquot, J.-P. & Buchanan, B.B. 1990. The ferredoxin-thioredoxin system of a green alga, *Chlamydomonas reinhardtii*. Identification and characterization of thioredoxins and ferredoxin-thioredoxin reductase components. Planta 180: 341-351.

Huppe, H.C., Picaud, A., Buchanan, B.B. & Miginiac-Maslow, M. 1991. Identification of an NADP/thioredoxin system in *Chlamydomonas reinhardtii*. Planta 186: 115-121.

Hutcheson, S.W. & Buchanan, B.B. 1983. Enzyme regulation in crassulacean acid metabolism photosynthesis. Studies of the ferredoxin/thioredoxin system of *Kalanchoë daigremontiana*. Plant Physiol. 72: 870-876.

Jacquot, J.-P., Vidal, J., Gadal, P. & Schürmann, P. 1978. Evidence for the existence of several enzyme-specific thioredoxins in plants. FEBS Lett. 96: 243-246.

Jacquot, J.-P., Stein, M., Hodges, M. & Miginiac-Maslow, M. 1992. PCR cloning of a nucleotidic sequence coding for the mature part of *Chlamydomonas reinhardtii* thioredoxin Ch2. Nucleic Acids Res. 20: 617-617.

Johnson, T.C., Cao, R.Q., Kung, J.E. & Buchanan, B.B. 1987a. Thioredoxin and NADP-thioredoxin reductase from cultured carrot cells. Planta 171: 321-331.

Johnson, T.C., Wada, K., Buchanan, B.B. & Holmgren, A. 1987b. Reduction of purothionin by the wheat seed thioredoxin system. Plant Physiol. 85: 446-451.

Kamo, M., Tsugita, A., Wiessner, Ch., Wedel, N., Bartling, D., Herrmann, R.G., Aguilar, F., Gardet-Salvi, L. & Schürmann, P. 1989. Primary structure of spinach-chloroplast thioredoxin *f*. Protein sequencing and analysis of complete cDNA clones for spinach-chloroplast thioredoxin *f*. Eur. J. Biochem. 182: 315-322.

Katti, S.K., LeMaster, D.M. & Eklund, H. 1990. Crystal structure of thioredoxin from *Escherichia coli* at 1.68 Å resolution. J. Mol. Biol. 212: 167-184.

Knaff, D.B. 1989. The regulatory role of thioredoxin in chloroplasts. Trends Biochem. Sci. 14: 433-434.

Kobrehel, K., Yee, B.C. & Buchanan, B.B. 1991. Role of the NADP/thioredoxin system in the reduction of α-amylase and trypsin inhibitor proteins. J. Biol. Chem. 266: 16135-16140.

Kobrehel, K., Wong, J.H., Balogh, A., Kiss, F., Yee, B.C. & Buchanan, B.B. 1992. Specific reduction of wheat storage proteins by thioredoxin *h*. Plant Physiol. 99: 919-924.

Krause, G.H. & Bassham, J.A. 1969. Free energy changes and metabolic regulation in steady-state photosynthetic carbon reduction. Biochim. Biophys. Acta Bio-Energetics 189: 207-221.

Lamotte-Guéry, F. de, Miginiac-Maslow, M., Decottignies, P., Stein, M., Minard, P. & Jacquot, J.-P. 1991. Mutation of a negatively charged amino acid in thioredoxin modifies its reactivity with chloroplastic enzymes. Eur. J. Biochem. 196: 287-294.

Langlotz, P., Wagner, W. & Follmann, H. 1985. A large chloroplast thioredoxin *f* found in green algae. Z. Naturforsch. 41c: 275-283.

Laurent, T.C., Moore, E.C. & Reichard, P. 1964. Enzymatic synthesis of deoxyribonucleotides. IV. Isolation and characterization of thioredoxin the hydrogen donor from *Escherichia coli* B. J. Biol. Chem. 239: 3436-3444.

Lepiniec, L., Hodges, M., Gadal, P. & Crétin, C. 1992. Isolation, characterization and nucleotide sequence of a full-length pea cDNA encoding thioredoxin-*f*. Plant Mol. Biol. 18: 1023-1025.

Maeda, K., Tsugita, A., Dalzoppo, D., Vilbois, F. & Schürmann, P. 1986. Further characterization and amino acid sequence of *m*-type thioredoxins from spinach chloroplasts. Eur. J. Biochem. 154: 197-203.

Marcus, F., Moberly, L. & Latshaw, S.P. 1988. Comparative amino acid sequence of fructose-1 6-bisphosphatases: Identification of a region unique to the light-regulated chloroplast enzyme.
Proc. Natl. Acad. Sci. USA 85: 5379-5383.

Marcus, F., Chamberlain, S.H., Chu, C., Masiarz, F.R., Shin, S., Yee, B.C. & Buchanan, B.B. 1991. Plant thioredoxin *h*: An animal-like thioredoxin occurring in multiple cell compartments. Arch. Biochem. Biophys. 287: 195-198.

Marty, I. & Meyer, Y. 1991. Nucleotide sequence of a cDNA encoding a tobacco thioredoxin. Plant Mol.Biol. 17: 143-147.

McFarlan, S.C., Hogenkamp, H.P.C., Eccleston, E.D., Howard, J.B. & Fuchs, J.A. 1989. Purification, characterization and revised amino acid sequence of a second thioredoxin from *Corynebacterium nephridii*. Eur. J. Biochem. 179: 389-398.

Mikl, J., Maeda, M., Mukohata, Y. & Futai, M. 1988. The y-subunit of ATP synthase from spinach chloroplasts Primary structure deduced from the cloned cDNA sequence. FEBS Lett. 232: 221-226.

Mills, J.D., Mitchell, P. & Schürmann, P. 1981. The role of the thioredoxin system in the modulation of coupling factor ATPase activity in intact chloroplasts. In: Akoyunoglou, G. (ed.), Photosynthesis II. Electron Transport and Photophosphorylation, Balaban International Science Services, Philadelphia, pp. 839-848.

Muller, E.G.D. & Buchanan, B.B. 1989. Thioredoxin is essential for photosynthetic growth. The thioredoxin *m* gene of *Anacystis nidulans*. J. Biol. Chem. 264: 4008-4014.

Nishizawa, A.N. & Buchanan, B.B. 1981. Enzyme regulation in C_4 photosynthesis. Purification and properties of thioredoxin-linked fructose bisphosphatase and sedoheptulose bisphosphatase from corn leaves. J. Biol. Chem. 256: 6119-6126.

Porter, M.A., Stringer, C.D. & Hartmann, F.C. 1988. Characterization of the regulatory thioredoxin site of phosphoribulokinase. J. Biol. Chem. 263: 123-129.

Raines, C.A., Lloyd, J.C., Longstaff, M., Bradley, D. & Dyer, T. 1988. Chloroplast fructose-1 6-bisphosphatase: The product of a mosaic gene. Nucleic Acids Res. 16: 7931-7942.

Russel, M., Model, P. & Holmgren, A. 1990. Thioredoxin or glutaredoxin in *Escherichia coli* is essential for sulfate reduction but not for deoxyribonucleotide synthesis. J. Bacteriol. 172: 1923-1929.

Scheibe, R. 1987. $NADP^+$-malate dehydrogenase in C_3-plants: Regulation and role of a light-activated enzyme. Physiol. Plant. 71: 393-400.

Schmidt, A. 1981. A thioredoxin activated glutamine synthetase in *Chlorella*. Z. Naturforsch. 36c: 396-399.

Schürmann, P., Maeda, K. & Tsugita, A. 1981. Isomers in thioredoxins of spinach chloroplasts. Eur. J. Biochem. 116: 37-45.

Schwenn, J. D. 1989. Sulphate assimilation in higher plants: a thioredoxin-dependent PAPS-reductase from spinach leaves. Z. Naturforsch. 44c: 504-508.

Schwenn, J.D. & Schriek, U. 1984. A new role for thioredoxin in assimilatory sulphate reduction. Activation of the adenylylsulphate kinase from the green alga *Chlamydomonas reinhardtii* CW 15. FEBS Lett. 170: 76-80.

Soulié, J.M., Buc, J., Meunier, J.C., Pradel, J. & Ricard, J. 1981. Molecular properties of chloroplastic thioredoxin *f* and the photoregulation of the activity of fructose 1,6-bisphosphatase. Eur. J. Biochem., 119: 497-502.

Suske, G., Wagner, W. & Follmann, H. 1979. NADP-dependent thioredoxin reductase and a new thioredoxin from wheat. Z. Naturforsch. 34c: 214-221.

Thelander, L. 1967. Thioredoxin reductase. Characterization of a homogeneous preparation from *Escherichia coli* B. J. Biol. Chem., 242: 852-859.

Tischner, R. & Schmidt, A. 1982. A thioredoxin-mediated activation of glutamine synthetase and glutamate synthase in synchronous *Chlorella sorokiniana*. Plant Physiol. 70: 113-116.

Tsang, M.L. 1981. Thioredoxin/glutaredoxin system of *Chlorella*. Plant Physiol. 68: 1098-1104.

Tsugita, A., Maeda, K. & Schürmann, P. 1983. Spinach chloroplast thioredoxins in evolutionary drift. Biochem. Biophys. Res. Comm. 115: 1-7.

Tsugita, A., Yano, K., Gardet-Salvi, L. & Schürmann, P. 1991. Characterization of spinach ferredoxin-thioredoxin reductase. Protein Seq. Data Anal. 4: 9-13.

Van Langendonckt, A. & Vanden Driessche, T. 1992. Isolation and characterization of different forms of thioredoxins from the green alga *Acetabularia mediterranea*: Identification of an NADP/thioredoxin system in the extrachloroplastic fraction. Arch. Biochem. Biophys. 292: 156-164.

Vogt, K. & Follmann, H. 1986. Characterization of three different thioredoxins in wheat. Biochim. Biophys. Acta 873: 415-418.

Wedel, N., Clausmeyer, S., Herrmann, R.G., Gardet-Salvi, L. & Schürmann, P. 1992. Nucleotide sequence of cDNAs encoding the entire precursor polypeptide for thioredoxin *m* from spinach chloroplasts. Plant Mol. Biol. 18: 527-533.

Whittacker, M.M. & Gleason, F.K. 1984. Isolation and characterization of thioredoxin *f* from the filamentous cyanobacterium Anabaena sp 7119. J. Biol. Chem. 259: 14088-14093.

Wolosiuk, R.A., Crawford, N.A., Yee, B.C. & Buchanan, B.B. 1979. Isolation of three thioredoxins from spinach leaves. J. Biol. Chem. 254: 1627-1632.

RECENT INVESTIGATIONS ON THE BIOSYNTHESIS OF THE PLANT SULFOLIPID

Ernst Heinz
Institut für Allgemeine Botanik, Universität Hamburg, Ohnhorststrasse 18, D-2000 Hamburg 52, F.R.G.

Introduction

In the preceding volume of this series, most aspects of the biochemistry and biosynthesis of plant sulfolipids and in particular of sulfoquinovosyl diacylglycerol (SQD) have been reviewed in comprehensive form (Kleppinger-Sparace *et al.* 1990). Therefore, there is no need to repeat these and other well known data (Mudd & Kleppinger-Sparace 1987), and the following article should be considered as a continuation of previous reviews concentrating on work published in the last years and supplementing some aspects covered only marginally before. In addition, an attempt is made to correlate sulfolipid quantities and formation with corresponding data from other sulfur-containing constituents of leaves.

All organisms carrying out oxygenic photosynthesis contain the "sulfolipid" (SQD) as membrane constituent of thylakoids, whereas its occurrence in anoxygenic, photosynthetically active bacteria is limited to a few genera (Wood *et al.* 1965; Radunz 1969; Russell & Harwood 1979; Imhoff *et al.* 1982). Benson and coworkers discovered and analyzed this lipid and carried out the first experiments on its biosynthesis (Benson 1963). The characteristic feature of SQD with the structure of 1,2-di-0-acyl-3-0-(6-deoxy-6-sulfo-α-D-glucopyranosyl)-*sn*-glycerol(sulfoquinovosyl diacylglycerol, Fig. 1) is the sulfonic acid residue at C6 of 6-deoxyglucose (= quinovose) which at physiological pH carries a negative charge. Apart from this ubiquitous membrane component photosynthetic organisms may produce a variety of other sulfur-containing lipophilic compounds (Haines 1973; Kleppinger-Sparace *et al.* 1990). A membrane lipid of comparable abundance as SQD is phosphatidylsulfocholine in diatoms (Bisseret *et al.* 1984), where this compound may replace the zwitterionic phosphatidylcholine in extraplastidic membranes. These algal cells, therefore, contain two sulfur-containing membrane lipids of similar abundance. But none of the other sulfur-containing lipids can compete with SQD in its quantitative importance in the biosphere.

Analysis of sulfoquinovosyl diacylglycerol

Investigations of SQD biochemistry usually necessitate separation of this compound, and often data on absolute quantities are required as well. Due to the presence of both glyco- and phospholipids in plant lipid extracts satisfactory resolution of all compounds by a single TLC run is nearly impossible, and only a few solvent systems have proven suitable for the separation of SQD (Nichols 1965; Pohl & Wagner 1972). Quantitation may be carried out by fatty acid analysis of chromatographically

Sulfur Nutrition and Assimilation in Higher Plants
edited by L.J. De Kok et al., pp. 163–178

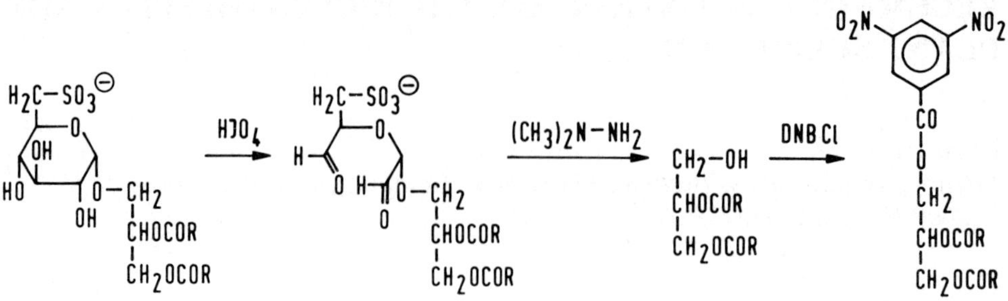

Fig. 1. Conversion of sulfoquinovosyl diacylglycerol into 1,2-diacyl-3-dinitrobenzoyl-*sn*-glycerol as required for analysis of molecular species by HPLC. This method is also applicable to other glycosyl diacylglycerols. Periodate oxidation is followed by hydrazinolysis and esterification of released diacylglycerol with dinitrobenzoylchloride (DNBCl).

resolved SQD bands or by colorimetry based on sugar derivatization. The reaction of sulfoquinovosyl residues with anthrone results in a colored product with an absorbance maximum which is about 30 nm shorter than measured for the corresponding glucose derivative (Weenink 1963; Russell 1966; Radunz 1969; Hoch *et al.* 1989). With phenol as reagent sulfoquinovose and galactose derivatives have the same extinction coefficient at 485 nm (Roughan & Batt 1968). A lower limit of these methods is reached at about 30 nmol representing 25 μg of SQD. Depending on set up and handling, quantitation by GLC of fatty acids may be more sensitive (by a factor of 10). SQD from algae with complex mixtures of fatty acids can separate into two bands differing in fatty acid composition (Araki *et al.* 1989) which may complicate analytical work. Separation of SQD has also been accomplished by HPLC (Demandre *et al.* 1985), but a limitation of this method is quantitation, since optical recording at short wavelength reflects the content of double bonds in fatty acids which may vary with the organism and several other parameters. A combination of HPLC with a mass sensitive detector compatible with complicated solvent systems and not requiring specific calibration for different lipids would be the method of choice (Stolyhwo *et al.* 1987). With ternary gradients all plant lipids including SQD were resolved on normal phase silica columns with a detection limit of about 1 μg per component (Moreau *et al.* 1990). On the other hand, acidic phospholipids and SQD could not be eluted from NH_2-modified columns (Heemskerk *et al.* 1986). Solvent systems for TLC of sulfoquinovose (Araki *et al.* 1989; Hoch *et al.* 1989), sulfoquinovosyl phosphate (Hoch *et al.* 1989) and sulfoquinovosyl glycerol (Joyard *et al.* 1986) have been developed.

Investigations of lipids from chloroplasts are confronted with the phenomenon, that diacylglycerol (DAG) backbones have pro- and eucaryotic structures depending on the site of assembly by plastidic or microsomal enzyme systems, respectively (Roughan & Slack 1984). Accordingly, there is a demand for resolution and quantitation of fatty acid pairings (molecular species) and positional analysis of these fatty acids. The assignment of fatty acid mixtures to the *sn*-1 and *sn*-2 position of SQD is carried out by enzymatic hydrolysis of the primary ester group using the lipase from *Rhizopus* (Fischer *et al.* 1973). This is the only enzyme known so far to be specific for the *sn*-1 position of SQD (and of other glyco- and phospholipids as well).

Compared to other glycolipids from chloroplasts, the *sn*-1 position of SQD from higher plants is enriched in palmitic acid (Heinz 1977). Apart from a contribution by the eucaryotic pathway, recent results (see below) have shown that this can be a consequence of the selectivities of the galactosyl- and sulfoquinovosyltransferases.

The analysis of fatty acid pairings or molecular species can be carried out by several methods. Separation by argentation TLC of free (Heinz & Harwood 1977; Siebertz *et al.* 1980) and acetylated SQD (Nishihara *et al.* 1980) or of its sulfonic acid methyl ester (SQD-Me, prepared by treatment of protonated SQD with diazomethane, Seifert & Heinz 1992) is simple but identification and quantitation require subsequent fatty acid analysis. The complex patterns revealed after *in vivo* labeling were first evidence for the participation of SQD-bound fatty acids in lipid-linked desaturation. GLC of trimethylsilyl ether derivatives of SQD-Me afforded a separation of groups according to the sum of carbon atoms irrespective of degree of unsaturation (Tulloch *et al.* 1973). This technique demonstrated the existence of dipalmitoyl species in SQD of higher plants. Individual fatty acid combinations can be identified by direct inlet-EI-mass spectroscopy of the peracetylated SQD-Me (Siebertz *et al.* 1979). The spectra contained signals for DAG fragments of high intensity and again proved the presence of dipalmitoyl combinations. In addition, fragmentation between C1 and C2 of the glycerol backbone and charge retention in the fragment having lost C1 of glycerol with its esterified acyl group [M-CH_2OCOR] provided an independent confirmation (Budzikiewicz *et al.* 1973) of the positional distribution of fatty acids as analyzed by lipase hydrolysis. Recently the application of modern techniques of mass spectroscopy confirmed the structural identity of SQD from *Rhodobacter* (Gage *et al.* 1992). Spectra obtained from underivatized lipids (by FAB-CAD-MS/MS) contained molecular ions as the most intensive signals which are suitable for a direct analysis of fatty acid combinations. At this point it may be mentioned that NMR spectra of SQD confirmed the gluco-configuration of the sugar residue (Tulloch *et al.* 1973; Johns *et al.* 1978; Araki *et al.* 1989).

Molecular species of SQD can be separated by HPLC without prior derivatization (Giroud *et al.* 1988), but as mentioned above the absorbance at low wavelength used for detection correlates with double bonds and, therefore, species with decreasing number of double bonds have decreasing extinction coefficients and fully saturated species escape detection. Release of the DAG portion from SQD (and from other glycolipids) by chemical degradation via periodate oxidation and hydrazinolysis (Heinze *et al.* 1984) followed by attachment of a dinitrobenzoyl group to the free hydroxyl group at C3 eliminate these problems (Fig. 1), since the new chromophore can be detected at 250 nm without interference from double bonds (Kesselmeier & Heinz 1987; Bishop 1987). The high resolution of reversed-phase HPLC and the availability of many reference data from various DAG (Takamura *et al.* 1986) as well as the ease of collecting and identifying peaks via fatty acid analysis makes this method most versatile.

Antibodies have been raised against SQD (Radunz & Berzborn 1970) and shown to inhibit photosynthetic electron transport reactions in thylakoid membranes at the entrance sites of both photosystem I and II and thus interfere at similar sites as do the antibodies against other thylakoid membrane lipids (Radunz et al. 1984). Recently SQD antibodies have been used in western blotting experiments to investigate a specific association of SQD with protein complexes from thylakoid membranes (Voss

Table 1. Quantitative data on sulfoquinovosyl diacylglycerol contents of leaves. To avoid excessive literature citation only such data have been selected which cover ranges found in other independent investigations (if available). To allow checking by cross calculation, independent data based on fresh weight (FW), dry weight (DW), chlorophyll (chl) and leaf area (cm^2) were compiled. Parallel values may differ by as much as several factors. The molar ratio of S:N in leaf protein varies from 0.028-0.038 (Dijkshoorn & van Wijk 1967); Kjeldahl-N is multiplied by 6.3 to yield protein mass and *vice versa*; a molecular weight of 830 was used for SQD (mixture of fatty acids) and of 900 for chlorophyll. References: **1** Davies *et al.* (1991); **2** Schröppel-Meier & Kaiser (1988a,b); **3** Koenig (1971); **4** Hudson *et al.* (1992); **5** calculated via 44 μg chl cm^{-2} from Foyer *et al.* 1991; **6** calculated via chl cm^{-2} and soluble protein: chl = 13.6 according to Meyers *et al.* (1982); **7** calculated via chl cm^{-2} and total protein: chl = 20 according to Haas *et al.* (1979); **8** calculated as 880:6.3:14; **9** measured as acetone-insoluble N by Possingham & Saurer (1969); **10** calculated as 10 x 0.033; **11** Roughan & Batt (1969); **12** Russell (1966); **13** Siebertz & Heinz (1977); **14** Rauser *et al.* (1991); **15** Leech *et al.* (1973); **16** when data based on chl are used, a ratio of 7.5-25 is obtained (see Table 3).

Parameter	Unit	Range	Reference
per cm^2 leaf area			
fresh weight	(mg)	14-49	1,2
dry weight	(mg)	1.2-2.9	3,4
chlorophyll	(μg)	25-44	1,3,4
SQD	(nmol)	7-12	1,2
glutathione	(nmol)	11-32	5
soluble protein	(mg)	0.3-0.6	4,6
total protein	(μg)	880	7
protein-N	(μmol)	10	8
insoluble N	(μmol)	7-9	9
protein-S	(nmol)	330	10
ratios			
chl/FW	(mg g^{-1})	0.7-3	1,2,6
chl/DW	(mg g^{-1})	12-19	4,3
SQD/FW	($\mu mol\ g^{-1}$)	0.3-1.7	11
SQD/DW	($\mu mol\ g^{-1}$)	1.5-5.8	12,3
SQD/chl	($\mu mol\ mg^{-1}$)	0.3-1.0	11
SQD/protein	(nmol mg^{-1})	5.7-14	13,7
glutathione/FW	($\mu mol\ g^{-1}$)	0.5	14
soluble protein/FW	(mg g^{-1})	38-47	5
DW/FW	(g g^{-1})	10-18	15
protein-S/SQD	(mol mol^{-1})	26-66	16

et al. 1992). It was found that SQD was specifically associated with a D_1D_2-dimer from photosystem II in accordance with previous investigations (Gounaris & Barber 1985).

Sulfur in sulfolipid and proteins from leaves

In Table 1, data from various references have been compiled to evaluate the quantity of leaf SQD in correlation to other relevant constituents of photosynthetic tissues. Results from different investigations vary considerably and therefore, only data from a few investigations approaching means from others have been listed in the Table. Figures for the most interesting correlation between sulfur in SQD, in free sulfate,

Table 2. Proportion of sulfoquinovosyl diacylglycerol in membrane lipids from photosynthetic tissues to demonstrate the exceptionally high values recently found in algae. Some data had to be recalculated on the basis of total lipids.

Organism	mol% SQD in total lipids	Reference
Spermatophyta		
Angiosperm leaf	4.1	Roughan & Batt (1969)
Raphidophyceae		
Chattonella antiqua	29.0	Sato *et al.* (1988)
Rhodophyceae		
Gracilaria bursa-pastoris	26.9	Araki *et al.* (1990)
Phaeophyceae		
Hizikia fusiformis	29.5	Araki *et al.* (1991)
Ectocarpus arabicus	35.3	Dembitsky *et al.* (1990)
Fucus serratus	37.5	Smith & Harwood (1983)
Cystoseira crinita	41.5	Dembitsky *et al.* (1990)
Pylaiella littoralis	45.6	Dembitsky *et al.* (1990)
Padina paventia	49.0	Dembitsky *et al.* (1990)

Table 3. Metabolic rates and pools relevant for sulfur metabolism in leaves. References: **1** Schröppel-Meier & Kaiser (1988a,b); Hudson *et al.* (1992); **2** Trebst & Schmidt (1969); **3** Soll & Roughan (1982); Masterson *et al.* (1990); **4** Joyard *et al.* (1986); Kleppinger-Sparace & Mudd (1987, 1990); **5** calculated according to Rennenberg (1984) with 44 μg chl cm^{-2} (Table 2); **6** from Table 2 with protein: chl = 20; 7 Foyer *et al.* (1991), Rauser *et al.* (1991) with 1 mg chl g fresh weight^{-1}; **8** Roughan & Batt (1969).

Reaction	Rate nmol mg chl^{-1} h^{-1}	Ref.	Pool	Size nmol mg chl^{-1}	Ref.
CO_2 reduction	$198\text{-}418 \times 10^3$	1	protein-S	7.5×10^3	6
sulfate reduction	3×10^3	2	GSH-S	$3\text{-}7 \times 10^2$	7
fatty acid synthesis	$0.2\text{-}1.2 \times 10^3$	3	SQD-S	$2\text{-}10 \times 10^2$	8
SQD synthesis	1-5	4			
sulfide release	23-46	5			

in protein and in other organic compounds do not seem to be available for the same tissue and, in particular, the ratio of SQD/protein has been determined directly in a very few cases only (Siebertz & Heinz 1977; Haas *et al.* 1979). Therefore, data from different references were used for additional and independent, though indirect calculation of this and other ratios. In leaves the molar ratio of SQD:chlorophyll:protein-N:protein-S:glutathione-S may be 1:5:10^3:30:2 corresponding to a weight ratio of SQD:chlorophyll:protein of about 1:5:90. SQD and glutathione occur in about the same molar range, whereas the assumption of equal quantities of sulfur in SQD and protein in leaves (Harwood & Nicholls 1979) may be too high by about one order of magnitude or even more (last line of Table 1 and right column of Table 3). Some of the figures of Table 1 which have to be regarded as representing steady state levels of continuously turning over constituents, will be compared with enzymatic rates in Table 3. On the other hand, equimolar quantities of sulfur in SQD and proteins may actually be present in marine algae (Table 2), where recent investigations have shown that SQD may represent as much as 40 mol % of lipids as compared to 4 mol % in angiosperm leaves. Therefore, in some of these organisms SQD is the predominant membrane lipid. The reason and structural consequence of this increase in a negatively charged membrane lipid are unknown.

Whatever the predominating ratios are, SQD represents a prominent member in the global sulfur chain to be passed and recycled every year (Harwood & Nicholls 1979). From an ecological point of view, the knowledge of both biosynthetic and degradative reactions are of equal importance (reviewed by Mudd & Kleppinger-Sparace 1987; Kleppinger-Sparace *et al.* 1990). A completely different interest in SQD was raised by the observation that this lipid could protect human cells against infection by HIV-virus (Gustafson *et al.* 1989). This observation prompted an independent chemical resynthesis of SQD which confirmed previously elucidated structural details as well as the protective effect with respect to the viral infection (Gordon & Danishevsky 1992). Furthermore, during a search for repellents against fouling organisms which interfere with the operation of power plants and aquiculture it was found that SQD was particularly effective against blue mussels (*Mytilus edulis*) (Katsuoka *et al.* 1990).

Sulfolipid and sulfur metabolism in leaves

Plant cells try to maintain a homeostatic state in the cytoplasm regarding anions such as phosphate, nitrate and sulfate, excesses of which are actively loaded into the vacuole (Martinoia *et al.* 1981; Schröppel-Meier & Kaiser 1988a,b; Kaiser *et al.* 1989; Gout *et al.* 1990). In *Neurospora* the putative plasma membrane sulfate transporter is already under investigation (Jarai & Marzluf 1991). The negative charge of sulfate anions in the vacuole is compensated for by equivalent K^+-import via K^+/H^+-exchange. In one case where it has been investigated an increased sulfate supply did not result in an elevated level of SQD in chloroplasts (Stuiver *et al.* 1984). In this experiment the highest sulfate concentration used was 100 mM, whereas marine algae such as those listed in Table 2 are continuously exposed to about 28 mM sulfate in the seawater. In this context it may be mentioned that transcription of sulfur-rich protein genes during development of seeds is regulated by sulfur availability (Spencer *et al.* 1990). Uptake of sulfate into chloroplasts for activation, reduction and incorporation into SQD occurs through the envelope via a specific sulfate carrier (Mourioux & Douce 1979) and/or the phosphate translocator (Hampp & Ziegler 1977). Subsequent experiments on a further characterization of a separate sulfate carrier have not been carried out. The K_m for sulfate uptake into chloroplasts is in the same range as the sulfate concentration in the cytoplasm (several mM, Schröppel-Meier & Kaiser 1988) and its capacity (1-25 μmoles mg chlorophyll^{-1} h^{-1}, Hampp & Ziegler 1977; Mourioux & Douce 1979) appears to be adjusted to the rate of sulfate reduction in chloroplasts (3 μmoles mg chlorophyll^{-1}, Trebst & Schmidt 1969). This is far in excess of the supply required for SQD synthesis. Maximal rates of fatty acid synthesis by isolated chloroplasts from acetate or acetyl carnitine (2-10 μmoles acetate mg chlorophyll^{-1} h^{-1}, Soll & Roughan 1982; Masterson *et al.* 1990) correspond to 0.1-0.6 μmoles mg chlorophyll^{-1} h^{-1} of phosphatidic acid synthesis which represents the upper limit of DAG formation to become available via pro- and eucaryotic routes for incorporation into glycolipids. But part of phosphatidic acid is channeled into phosphatidylglycerol and the major portion of DAG is used for synthesis of other phospholipids and galactolipids, so that only a few percent of DAG are actually incorporated into SQD (Browse *et al.* 1986). This branching of the DAG stream results in rates of SQD formation which compare well with directly measured SQD formation in isolated chloroplasts from acetate, sulfate or adenosine-5'-phosphosulfate (APS) (1-5 nmol SQD mg $chlorophyll^{-1}$ h^{-1}, Joyard *et al.* 1986; Kleppinger-Sparace & Mudd 1987, 1990). Therefore, activation and reduction of sulfate as well as DAG formation do not seem to be rate-limiting in SQD biosynthesis which rather may be controlled by formation of sulfoquinovose, its subsequent activation and the final transfer of this sugar from its activated form to DAG, if SQD synthesis occurs via these steps (see below).

In Table 3 some of these and other rates relevant for sulfate metabolism in leaves have been compiled. Despite the uncertainties mentioned above and the fact that enzymatic rates often represent V_{max} values which are not reached *in vivo* these data reveal some interesting comparisons. Sulfate reduction proceeds at about 1 $^{o}/_{0}$ of the rate of CO_2 reduction, whereas SQD formation uses only about 1 $^{o}/_{00}$ of the sulfate reduction capacity. On the other hand, the release of sulfide from leaves (Rennenberg 1984) can be about 10 times higher than SQD synthesis. The pool of sulfur in leaf proteins represents a quantity which can be supplied by sulfate reduction in a few hours.

Biosynthesis of water-soluble precursors

Despite the recent progress (see below), the intermediates between APS or carrier-bound thiosulfonate and SQD are still not known. It should be remembered that the first reduction of the sulfur oxidation state from +6 to +4 as found in SQD is initiated by the nucleophilic attack of the sulfate sulfur in APS by a thiol-S of the carrier protein involved in the sulfonyltransferase reaction (Schmidt 1986). In this reaction reducing agents such as NADPH, ferredoxin or thioredoxin are not directly required. The resulting protein-bound thiosulfonate is not considered to have a changed sulfur oxidation number, despite the fact that it readily exchanges with free sulfite and may be the actual source of the SQD-sulfonate group. Even attack of sulfur in APS by a nucleophilic carbon atom such as C3 of phosphoenolpyruvate could result in direct formation of a C-sulfonic acid group with a concomitant change in the oxidation number of sulfur and of C2 in pyruvate (after loss of phosphate), whereas similar reactions with inorganic sulfur-containing ions require sulfite. But the identity of the first compound having the C-S bond is not known. It is not completely unlikely that the synthesis of sulfoquinovose involves D-lactaldehyde 3-sulfonate (Davies *et al.* 1966), since even normal aldolase (EC 4.1.2.13) combines dihydroxyacetone phosphate with a wide variety of aldehydes including glyceraldehyde 3-sulfate, whereas dihydroxyacetone sulfonate is not accepted (Martensen & Mansour 1976; Bischofberger *et al.* 1988; Toone *et al.* 1989). Therefore, the first sulfohexose in this pathway could be 6-deoxy-6-sulfo-D-fructose 1-phosphate. On the other hand, the observation that washed membranes from *Chlamydomonas* incorporated radioactive sulfite into SQD (Hoppe & Schwenn 1981) could indicate that in addition to water-soluble sulfite acceptors with a double bond (such as phosphoenolpyruvate, 6-deoxy-hex-5,6-enopyranosyl 1-phosphate or UDP-6-deoxy-hex-5,6-enose) even a lipophilic membrane-bound 1,2-diacyl-3-(6'-deoxy-hex-5',6'-enopyranosyl)-*sn*-glycerol could be the actual sulfite acceptor (Lehmann & Benson 1964). As has been pointed out (Kleppinger-Sparace & Mudd 1990), the interpretation of sulfite labeling data may be complicated by various side reactions and in particular by exchange of sulfite with the SO_3-group of thiosulfonate. Among others this has been demonstrated to occur with S-sulfocysteine (Saidha & Schiff 1989) which may closely resemble the thiosulfonate group of the carrier protein participating in the sulfonyltransferase reaction. On the other hand, $^{35}SO_3^{2-}$ does not exchange with sulfoquinovose (Lehmann & Weckerle 1972). Efficient incorporation of sulfite into SQD is also found in *Euglena* (Saidha & Schiff 1989), where sulfate activation and reduction occur in mitochondria (Saidha *et al.* 1988). These organelles release sulfite (and cysteine) which is taken up by chloroplasts and incorporated into SQD. Whether unicellular green algae have a completely different way for SQD synthesis is unknown. The translocation of the sulfate reduction process into mitochondria may reflect the adaptability of *Euglena* and represent a precaution to retain this autotrophic trait in organelles which are not reduced under heterotrophic conditions as are chloroplasts. Also in higher plants, sulfate activation is not completely confined to plastids, since 17% of the cellular ATP-sulfurylase activity was found to be present in the cytoplasm (Lunn *et al.* 1990) where it may be required for flavonoid sulfation, for example (Varin *et al.* 1992). In view of this subcellular distribution, previous experiments on the labeling of SQD in intact chloroplasts from exogenous APS (Kleppinger-Sparace & Mudd 1990) could receive unexpected relevance. In heterotrophic tissues, for example, sulfate could be activated in the cytoplasm to be transported into plastids for SQD biosynthesis. For this sequence a carrier for activated sulfate would be required in envelopes as is present in Golgi membranes (Perez & Hirschberg 1986).

An even more complicated situation may exist in C_4 plants such as maize which contain SQD in both mesophyll and bundle sheath chloroplasts (Poincelot 1973). Since almost all ATP sulfurylase activity from leaves is located in bundle sheath chloroplasts (Gerwick *et al.* 1980; Burnell 1984), mesophyll cells and their chloroplasts may depend on APS or sulfite exported from bundle sheath cells or chloroplasts. Also in this system the participation of sulfite in SQD biosynthesis cannot be excluded. On the other hand, in view of the large difference between the capacities to reduce sulfate and to synthesize SQD, even a largely reduced activity in mesophyll chloroplasts would be sufficient to support normal SQD biosynthesis.

Despite the sulfite labeling data (Hoppe & Schwenn 1981) we followed Benson's suggestion and thought that sulfoquinovose activated in the form of a sulfosugar nucleotide would be an interesting compound for experiments on the origin of the SQD headgroup, particularly, since such a compound had been tentatively identified nearly 30 years ago (Shibuya *et al.* 1963). Therefore, we synthesized α-D-sulfoquinovopyranosyl 1-phosphate (Hoch *et al.* 1989) which is the intermediate required for sugar nucleotide synthesis (Fig. 2). The sulfonic acid residue was introduced by replacement of a triflyl group with tetrabutylammonium hydrogensulfide followed by oxidation of the thiol (or actually the disulfide) with perbenzoic acid. During the synthesis of SQD (Gigg *et al.* 1980; Gordon & Danishevsky 1992) the corresponding sequence comprised tosyl substitution by thioacetate and subsequent oxidation.

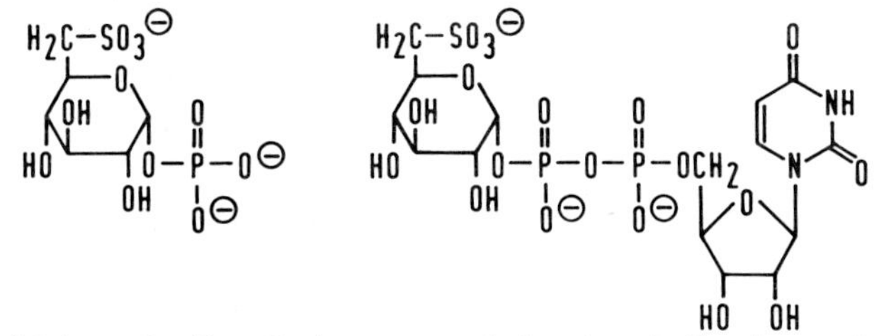

Fig. 2. Structure of 6-deoxy-6-sulfo-α-D-glucopyranosyl phosphate (sulfoquinovosyl phosphate) and of UDP-sulfoquinovose. These compounds have been synthesized for use as possible precursors in experiments on sulfolipid biosynthesis.

Regarding the hydrolysis of the phosphate group at C1, sulfoquinovose 1-phosphate is somewhat more stable than glucose 1-phosphate. pKa_1 for phosphate and sulfonate are low and very similar, whereas pKa_2 of the phosphate group is 6.7 and, therefore, slightly higher than pKa_2 of glucose 1-phosphate (6.4). Accordingly, sulfoquinovosyl 1-phosphate can be expected to be a fairly stable intermediate.

The sugar phosphate was used to synthesize a series of nucleoside diphosphosulfoquinovoses which differed in the nucleotide part (Heinz *et al.* 1989). Using the corresponding mononucleotide morpholidates ADP-, CDP-, GDP- and UDP-sulfoquinovose (UDPS) were prepared, purified by reversed-phase HPLC and used in experiments on SQD biosynthesis. For this purpose, intact spinach chloroplasts were prelabeled with acetate and glycerol 3 phosphate resulting in the accumulation of labeled DAG. Addition of sulfosugar nucleotides to such pre-labeled intact organelles did not raise the background levels of SQD labeling. On the other hand, after osmotic shock of pre-labeled chloroplasts, SQD labeling was significantly increased when UDP- or GDP-sulfoquinovose were added. The same stimulation was observed with isolated pre-labeled envelope membranes, the UDP-derivative being again more efficient than the GDP-compound. From these results it was concluded that the final step in SQD biosynthesis is catalyzed by a UDPS:diacylglycerol sulfoquinovosyl transferase. Since this envelope-bound enzyme cannot use exogenously added sulfosugar nucleotides, we assume that its active site is oriented towards the chloroplast stroma which provides UDPS. These results demonstrate the existence of sulfoquinovosyl transferase activity in chloroplasts which in turn requires the availability of UDPS in these organelles. Unresolved questions remain the biosynthesis of UDPS (see alternatives mentioned above), the entrance of sulfite into this sequence (Saidha & Schiff 1989) and, most intriguingly, the incorporation of sulfite into SQD in *Chlamydomonas* membranes (Hoppe & Schwenn 1981).

Sulfolipid-deficient mutants

A completely different approach to solve the open questions in sulfolipid and precursor biosynthesis applies the methods of biochemistry and molecular biology (Benning 1991; Benning & Somerville 1992 a,b) to *Rhodobacter spaeroides* which is one of the photosynthetic bacteria containing SQD. Following nitrosoguanidine treatment several mutants with reduced levels of SQD were isolated by TLC screening of lipid extracts from more than 1500 colonies. From these only a few could grow phototrophically and had no other phenotype than very low levels of SQD and an accordingly reduced capacity to incorporate labeled sulfate into SQD. Since sulfate-labeling of proteins was unimpaired, the reactions for sulfate activation and reduction are not altered. A closer inspection showed that two of these mutants accumulated water-soluble sulfate-labeled compounds. It is very likely, that these compounds which were resolved by electrophoresis/chromatography may represent precursors for SQD up to UDPS. Furthermore, the defect in SQD biosynthesis could be cured by complementation of these mutants with cosmids from a library of wild type genomic DNA. By subcloning and DNA sequencing several genes (*sqdA,B,C, ORF2*) were identified which obviously code for proteins and enzymes involved in SQD and precursor synthesis. So far no homologies to proteins required for sulfate activation and reduction from organisms such as *Rhizobium, E. coli* and *Saccharomyces* (Schwedock & Long 1990; Krone *et al.* 1991; Cherest & Surdin-Kerjan 1992) have been found. *sqdA* is not linked to the other three genes, which are organized in an operon. The sequence *sqdB*(46)-*ORF2*(30)-*sqdC*(26) is preceded by a single ribosome binding site in front of *sqdB*. Two of the proteins encoded (molecular weights 46 and 26 kDa, in brackets) seem to represent cytoplasmic enzymes lacking hydrophobic membrane-spanning helices. The N-terminal stretch of the *sqdB*-encoded sequence resembles the N-terminal part of UDP-glucose epimerases. The *ORF2* is separated from its neighbors by 57 and 68 bp and has sequence similarity with a gene coding for rabbit muscle glycogenin. This protein is a UDP-glucose dependent glucosyltransferase in charge of priming glycogen biosynthesis. Therefore, the 30 kDa protein encoded by *ORF2* may represent the sulfoquinovosyltransferase catalyzing the final step in SQD biosynthesis. The future results of this most powerful approach will be very interesting.

An unexpected aspect of this work is the fact that these mutants despite severely reduced levels or even absence of SQD can perform normal photoheterotrophic growth. This raises questions concerning the functions of SQD which apparently are dispensable or transferable to other membrane lipids. This effect is not confined to organisms carrying out anoxygenic photosynthesis and therefore lacking photosystem II, since similarly SQD-free mutants from *Chlamydomonas* also grow phototrophically (Satoh *et al.* 1992). But in contrast to *Rhodobacter*, the SQD mutants of *Chlamydomonas* are characterized by a high fluorescence phenotype which was used for screening and which indicates an impairment in the energy transduction chain. In this context previous results should be recalled which indicated a more or less specific and tight binding of SQD to various protein components of the thylakoid membrane such as ATPase (Pick *et al.* 1985), LHC II (Sigrist *et al.* 1988) and PS II reaction center (Gounaris & Barber 1985; Voss *et al.* 1992).

Properties of the UDP-sulfoquinovose:diacylglycerol sulfoquinovosyl transferase (SQT)

The availability of unlabeled UDPS and DAG-labeled envelope membranes allowed an investigation of some properties of SQT (Seifert & Heinz 1990, 1992) without the complications involved by working with sulfate and intact chloroplasts. Envelope membranes were labeled *in situ* with DAG and thus provide enzyme and one of the two substrates at the same time. The higher efficiency of UDPS as compared to GDPS and mentioned above can be ascribed to a difference in K_m. Furthermore, the experiments revealed that SQT is stimulated by Mg^{2+} ions. This stimulation is due to a decrease of the K_m for UDPS from 80 μM (in the absence of Mg^{2+}) to 10 μM (in the presence of Mg^{2+}). The Mg^{2+}-stimulation itself has a K_m of 0.7 mM Mg^{2+} and may be relevant in a light/dark modulation of the enzyme (Buchanan 1980). On the other hand, a light regulation via the thioredoxin system as operating for APS kinase (Schwenn & Schriek 1984) is unlikely, since the *in vitro* substituting low molecular weight thiols were without effect. Using optimized and linearized conditions it was found that SQT activity in various subchloroplast fractions behaved like a component of envelope membranes. This distribution requires export of SQD from envelopes into thylakoid membranes, where most of this lipid is localized. The inaccessibility of exogenous UDPS to SQT in intact chloroplasts suggests that SQT is localized in the inner of the two envelope membranes. Further experiments have to prove this conclusion. A comparison of SQT with the activity of UDP-galactose:diacylglycerol galactosyltransferase in isolated envelope membranes showed a ten-fold higher activity of the galactosyltransferase. This comparison was carried out at sugar nucleotide concentrations (0.4 mM UDPS and 0.6 mM UDP-galactose) which in terms of K_m favoured SQT and which particularly with respect to UDPS may significantly exceed *in vivo* concentrations. In this context it has to be remembered again, that UDPS has never been demonstrated in chloroplasts from higher plants. But the availability of reference compounds and their resolution by HPLC will be of great use in such experiments. The large difference between SQT and galactosyltransferase activities obviously contributes to the unequal splitting of the DAG usage in envelope membranes.

The possibility to measure SQT *in vitro* in envelope membranes enabled a series of experiments on the DAG-selectivity of this enzyme. As mentioned above, SQD differs from galactolipid by a higher proportion of palmitic acid (16:0). Even dipalmitoyl species (16:0/16:0) are present (Tulloch *et al.* 1973; Siebertz *et al.* 1979; Nishihara *et al.* 1980; Murata & Hoshi 1984; Bishop 1987; Giroud *et al.* 1988) which have not been detected in galactolipids. Based on the selectivities of the two acyltransferases controlling the incorporation of oleoyl (18:1) and/or 16:0 residues into the *sn* 1 and *sn*-2 position during the biosynthesis of DAG (Frentzen *et al.* 1983), different acyl-CoA mixtures were used to label envelope membranes separately with different DAG pools: 18:1/16:0, 16:0/16:0 + 18:1/16:0 and 16:0/16:0. In subsequent incubations with sugar nucleotides it was shown that the galactosyltransferase was selective and preferred 18:1/16:0, whereas 16:0/16:0 was discriminated against and only used when it was present as the only acceptor. In contrast the SQT was unselective and used both DAG species to the same extent. When envelope membranes prelabeled with an equimolar DAG-mixture of 18:1/16:0 + 16:0/16:0 were incubated with a mixture of UDPS and UDP-galactose to allow simultaneous and concurrent synthesis of galactolipid and SQD, it was found that SQD contained nearly exclusively 16:0/16:0-species, apparently as left overs from the galactosyltransferase reaction. Quantitative evaluation of a similar experiment showed that the activity ratio of the two glycosyltransferases depended on the DAG-composition of envelope membranes. The galactolipid/SQD ratio of 13.3 with 18:1/16:0 as substrate dropped to 1.4, when 16:0/16:0 was the only acceptor available. Therefore, the molecular species composition of the DAG pool may contribute to its distribution between galactolipids and SQD. Since the species composition of DAG is determined in the first acylation step by the selectivity of the acyl-ACP:*sn*-glycerol-3-phosphate acyltransferase (Frentzen *et al.* 1983), this stroma enzyme may in fact exert control on the quantity of SQD in chloroplast membranes. Whether this selectivity is required to produce SQD with a high content of 16:0 (Kenrick & Bishop 1986a,b) or to reduce the entrance of 16:0 into galactolipids or whether it has no physiological relevance at all, is an open question. Inspection of fatty acids in SQD from the algae listed in Table 2 shows that most of them have indeed a rather increased proportion of 16:0, but a clear-cut correlation is not obvious and in most cases data of molecular species composition or positional distribution of fatty acids are not available. Therefore, additional factors such as higher SQT activity or higher concentrations of UDPS may favour increased synthesis of SQD in these organisms.

Acknowledgment

Financial support by the Deutsche Forschungsgemeinschaft is gratefully acknowledged.

References

Araki, S., Eichenberger, W., Sakurai, T. & Sato, N. 1991. Distribution of diacylglycerylhydroxymethyltrimethyl-β-alanine (DGTA) and phosphatidylcholine in brown algae. Plant Cell Physiol. 32: 623-628.

Araki, S., Sakurai, T., Oohusa, T., Kayama, M. & Nisizawa, K. 1990. Content of arachidonic and eicosapentaenoic acids in polar lipids from *Gracilaria* (*Gracilariales, Rhodophyta*). Hydrobiologia 204/205: 513-519.

Araki,S., Sakurai, T., Oohusa, T., Kayama, M. & Sato, N. 1989. Characterization of sulfoquinovosyl diacylglycerol from marine red algae. Plant Cell Physiol. 30: 775-781.

Benning, C. 1991. Genetic analysis of the pathway for the biosynthesis of the plant sulfolipid in the purple bacterium *Rhodobacter sphaeroides*. PhD thesis, Michigan State University, East Lansing.

Benning, C. & Somerville, C.R. 1992. Isolation and genetic complementation of a sulfolipid-deficient mutant of *Rhodobacter sphaeroides*. J. Bacteriol. 174: 2352-2360.

Benning, C. & Somerville, C.R. 1992. Identification of an operon involved in sulfolipid biosynthesis in *Rhodobacter sphaeroides*. J. Bacteriol. 174: 6479-6487.

Benson, A.A. 1963. The plant sulfolipid. Adv. Lipid Res. 1: 387-394.

Bischofberger, N., Waldmann, H., Saito, T., Simon, E.S., Lees, W,. Bednarski, M.D. & Whitesides, G.M. 1988. Synthesis of analogues of 1,3-dihydroxyacetone phosphate and glyceraldehyde 3-phosphate for use in studies of fructose-1,6-diphosphate aldolase. J. Org. Chem. 53: 3457-3465.

Bishop, D.G. 1987. Analysis of molecular species of chloroplast lipids by HPLC. J. Liquid Chromatogr. 10: 1497-1505.

Bisseret, P., Ito, S., Tremblay, P.A., Volcani, B.E., Dessort, D. & Kates, M. 1984. Occurrence of phosphatidylsulfocholine, the sulfonium analog of phosphatidylcholine in some diatoms and algae. Biochim. Biophys. Acta 796: 320-327.

Browse, J., Warwick, N., Somerville, C.R. & Slack, C.R. 1986. Fluxes through the prokaryotic and eukaryotic pathways of lipid synthesis in the '16:3' plant *Arabidopsis thaliana*. Biochem. J. 235: 25-31.

Buchanan, B.B. 1980. Role of the light on the regulation of chloroplast enzymes. Annu. Rev. Plant Physiol. 31: 341-374.

Budzikiewicz, H., Rullkötter, J. & Heinz, E. 1973. Massenspektroskopische Untersuchungen an Glycosylglyceriden. Z. Naturforsch. 28c: 499-504.

Burnell, J.N. 1984. Sulfate assimilation in C_4 plants. Intercellular and intracellular location of ATP sul-

furylase, cysteine synthase, and cystathionine β-lyase in maize leaves. Plant Physiol. 75: 873-875.
Cherest, H. & Surdin-Kerjan, Y. 1992. Genetic analysis of a new mutation conferring cysteine auxotrophy in *Saccharomyces cerevisiae*: updating of the sulfur metabolism pathway. Genetics 130: 51-58.
Davies, W.H., Mercer, E.I. & Goodwin, T.W. 1966. Some observations on the biosynthesis of the plant sulpholipid by *Euglena gracilis*. Biochem. J. 98: 369-373.
Davies, H., Jordan, B.R. & Harwood, J.L. 1991. Changes in acyl-lipid, protein and chlorophyll content of lettuce and tomato chloroplasts during acclimation to changes in irradiance. Plant Physiol. Biochem. 29: 281-288.
Demandre, C., Trémolières, A., Justin, A.M. & Mazliak, P. 1985. Analysis of molecular species of plant polar lipids by high-performance and gas liquid chromatography. Phytochemistry 24: 481-485.
Dembitsky, V.M., Rozentsvet, O.A. & Pechenkina, E.E. 1990. Glycolipids, phospholipids and fatty acids of brown algae species. Phytochemistry 29: 3417-3421.
Dijkshoorn, W. & van Wijk, A.L. 1967. The sulphur requirements of plants as evidenced by the sulphur-nitrogen ratio in the organic matter. A review of published data. Plant and Soil 26: 129-157.
Fischer, W., Heinz, E. & Zeus, M. 1973. The suitability of lipase from *Rhizopus arrhizus delemar* for analysis of fatty acid distribution in dihexosyl diglycerides, phospholipids and plant sulfolipids. Hoppe-Seyler's Z. Physiol. Chem. 354: 1115-1123.
Foyer, C., Lelandais, M., Galap, C. & Kunert, K.J. 1991. Effects of elevated cytosolic glutathione reductase activity on the cellular glutathione pool and photosynthesis in leaves under normal and stress conditions. Plant Physiol. 97: 863-872.
Frentzen, M., Heinz, E., McKeon, T.A. & Stumpf, P.K. 1983. Specificities and selectivities of glycerol-3-phosphate acyltransferase and monoacylglycerol-3-phosphate acyltransferase from pea and spinach chloroplasts. Eur. J. Biochem. 129: 629-636.
Gage, D.A., Huang, Z.H. & Benning, C. 1992. Comparison of sulfoquinovosyl-diacylglycerol from spinach and the purple bacterium *Rhodobacter sphaeroides* by FAB-CAD-MS/MS. Lipids, 27: 632-636.
Gerwick, B.C., Ku, S.B. & Black, C.C. 1980. Initiation of sulfate activation: A variation in C_4 photosynthesis plants. Science 209: 513-515.
Gigg, R., Penglis, A.A.E. & Conant, R. 1980. Synthesis of 3-0-(6-deoxy-6-sulpho-α-D-glucopyranosyl)-1,2-di-0-hexadecanoyl-L-glycerol, 'sulphoquinovosyl diglyceride'. J. Chem. Soc. Perk. Transact. I: 2490-2493.
Giroud, C., Gerber, A. & Eichenberger, W. 1988. Lipids of *Chlamydomonas reinhardtii*. Analysis of molecular species and intracellular site(s) of biosynthesis. Plant Cell Physiol. 29: 587-595.
Gordon, D.M. & Danishevsky, S.J. 1992. Synthesis of a cyanobacterial sulfolipid. Confirmation of its structure, stereochemistry, and anti-HIV-1 activity. J. Amer. Chem. Soc. 114: 659-662.
Gounaris, K. & Barber, J. 1985. Isolation and characterization of a photosystem II reaction centre lipoprotein complex. FEBS Lett. 188: 68-72.
Gout, E., Bligny, R., Roby, C. & Douce, R. 1990. Transport of phosphocholine in higher plants: ^{31}P nuclear magnetic resonance studies. Proc. Natl. Acad. Sci. *USA* 87: 4280-4283.
Gustafson, K.R., Cardellina II, J.H., Fuller, R.W., Weislow, O.S., Kiser, R.F., Snader, K.M., Patterson, G.M.L. & Boyd M.R. 1989. AIDS antiviral sulfolipids from cyanobacteria (blue-green algae). J. Natl. Cancer Inst. 81: 1254-1258.
Haas, R., Heinz, E., Popovici, G. & Weissenböck, G. 1979. Protoplasts from oat primary leaves as tools for experiments on the compartmentation in lipid and flavonoid metabolism. Z. Naturforsch. 34c: 854-864.
Haines, T.H. 1973. Sulfolipids and halosulfolipids. In: Erwin, J.A. (ed.), Lipids and Biomembranes of Eucaryotic Microorganisms. Academic Press, New York, pp. 197-232.
Hampp, R. & Ziegler, I. 1977. Sulfate and sulfite translocation via the phosphate translocator of the inner envelope membrane of chloroplasts. Planta 137: 309-312.
Harwood., J.L. & Nicholls, R.G. 1979. The plant sulfolipid - a major component of the sulphur cycle. Biochem. Soc. Transact. 7: 440-447.
Heemskerk, J.W.M., Bögemann, G., Scheijen, M.A.M. & Wintermans, J.F.G.M. 1986. Separation of chloroplast polar lipids and measurement of galactolipid metabolism by high performance liquid chromatography. Anal. Biochem. 154: 85-91.
Heinz, E. 1977. Enzymatic reactions in galactolipid biosynthesis. In: Tevini, M. & Lichtenthaler, H.K. (eds.), Lipids and Lipid Polymers in Higher Plants. Springer-Verlag, Berlin, pp. 102-120.
Heinz, E. & Harwood, J.L. 1977. Incorporation of carbon dioxide, acetate and sulfate into glycerolipids of *Vicia faba* leaves. Hoppe-Seyler's Z. Physiol. Chem. 358: 897-908.

Heinz, E., Schmidt, H., Hoch, M., Jung, K.H., Binder, H. & Schmidt, R.R. 1989. Synthesis of different nucleoside 5'-diphospho-sulfoquinovoses and their use for studies on sulfolipid biosynthesis in chloroplasts. Eur. J. Biochem. 184: 445-453.

Heinze, F.J., Linscheid, M. & Heinz, E. 1984. Release of diacylglycerol moieties from various glycosyl diacylglycerols. Anal. Biochem. 139: 126-133.

Hoch, M., Heinz, E. & Schmidt, R.R. 1989. Synthesis of 6-deoxy-6-sulfo-α-D-glucopyranosyl phosphate. Carbohydr. Res. 191: 21-28.

Hoppe, W. & Schwenn, J.D. 1981. *In vitro* biosynthesis of the plant sulpholipid: on the origin of the sulphonate group. Z. Naturforsch. 36c: 820-826.

Hudson, G.S., Evans, J.R., von Caemmerer, S., Arvidsson, Y.B.C. & Andrews, T.J. 1992. Reduction of ribulose-1,5-bisphosphate carboxylase/oxygenase content by antisense RNA reduces photosynthesis in transgenic tobacco plants. Plant Physiol. 98: 294-302.

Imhoff, J.F., Kushner, D.J., Kushwaha, S.C. & Kates, M. 1982. Polar lipids in phototrophic bacteria of the *Rhodospirillaceae* and *Chromatiaceae* families. J. Bacteriol. 150: 1192-1201.

Jarai, G. & Marzluf, G.A. 1991. Sulfate transport in *Neurospora crassa*: Regulation, turnover, and cellular localization of the CYS-14 protein. Biochemistry 30: 4768-4773.

Johns, S.R., Leslie, D.R., Willing, R.I. & Bishop, D.G. 1978. Studies on chloroplast membranes. III. ^{13}C chemical shifts and longitudinal relaxation times of 1,2-diacyl-3-(6-sulpho-α-quinovosyl)-*sn*-glycerol. Aust. J. Chem. 31: 65-72.

Joyard, J., Blée, E. & Douce, R. 1986. Sulfolipid synthesis from $^{35}SO_4^{2-}$ and [1-^{14}C]acetate in isolated intact spinach chloroplasts. Biochim. Biophys. Acta 879: 78-87.

Kaiser, G., Martinoia, E., Schröppel-Meier, G. & Heber, U. 1989. Active transport of sulfate into the vacuole of plant cells provides halotolerance and can detoxify SO_2. J. Plant Physiol. 133: 756-763.

Katsuoka, M., Ogura, C., Etoh, H., Sakata, K. & Ina, K. 1990. Galactosyl- and sulfoquinovosyldiacylglycerols isolated from the brown algae, *Undaria pinnatifida* and *Costaria costata* as repellents of the blue mussel, *Mytilus edulis*. Agric. Biol. Chem. 54: 3043-3044.

Kenrick, J.R. & Bishop, D.G. 1986b. Phosphatidylglycerol and sulfoquinovosyldiacylglycerol in leaves and fruits of chilling-sensitive plants. Phytochemistry 25: 1293-1295.

Kenrick, J.R. & Bishop, D.G., 1986a. The fatty acid composition of phosphatidylglycerol and sulfoquinovosyldiacylglycerol of higher plants in relation to chilling sensitivity. Plant Physiol. 81: 946-949.

Kesselmeier, J. & Heinz, E. 1987. Separation of molecular species of plant glycolipids and phospholipids by high-performance liquid chromatography. Meth. Enzymol. 148: 650-661.

Kleppinger-Sparace, K.F. & Mudd, J.B. 1987. Biosynthesis of sulfoquinovosyldiacylglycerol in higher plants. The incorporation of $^{35}SO_4$ by intact chloroplasts in darkness. Plant Physiol. 84: 682-687.

Kleppinger-Sparace, K.F. & Mudd, J.B. 1990. Biosynthesis of sulfoquinovosyldiacylglycerol in higher plants. Use of adenosine-5'-phosphosulfate and adenosine-3'-phosphate 5'-phosphosulfate as precursors. Plant Physiol. 93: 256-263.

Kleppinger-Sparace, K.F., Sparace, S.A. & Mudd, J.B. 1990. Plant sulfolipids. In: Rennenberg, H., Brunold, C., De Kok, L.J. & Stulen, I. (eds.), Sulfur Nutrition and Sulfur Assimilation in Higher Plants. SPB Academic Publishers, The Hague, pp. 77-88.

Koenig, F. 1971. Konzentration einiger Lipide in den Chloroplasten von *Zea mays* und *Antirrhinum majus*. Z. Naturforsch. 26b: 1180-1187.

Krone, F.A., Westphal, G. & Schwenn, J.D. 1991. Characterization of the gene *cys* H and of its product phospho-adenylylsulfate reductase from *Escherichia coli*. Mol. Gen. Genet. 225: 314-319.

Leech, R.M., Rumsby, M.G. & Thomson, W.W. 1973. Plastid differentiation, acyl lipid, and fatty acid changes in developing green maize leaves. Plant Physiol. 52: 240-245.

Lehmann, J. & Benson, A.A. 1964. The plant sulfolipid. IX. Sulfosugar syntheses from methyl hexoseenides. J. Amer. Chem. Soc. 86: 4469-4472.

Lehmann, J. & Weckerle, W. 1972. Zuckersulfonsäuren. Teil I. Möglichkeiten der Konformationsanalyse durch ^{35}S-Bisulfitaddition an Hex-5-enopyranoside. Carbohydr. Res. 22: 23-35.

Lunn, J.E., Droux, M., Martin, J. & Douce, R. 1990. Localization of ATP sulfurylase and *O*-acetylserine(thiol)lyase in spinach leaves. Plant Physiol. 94: 1345-1352.

Martensen, T.M. & Mansour, T.E. 1976. Fructose-1-phosphate-6-sulfate as an alternative substrate for aldolase and fructose-1,6-diphosphatase. Biochem. Biophys. Res. Commun. 69: 844-851.

Martinoia, E., Heck, U. & Wiemken, A. 1981. Vacuoles as storage compartments for nitrate in barley leaves. Nature 289: 292-294.

Masterson, C., Wood, C. & Thomas, D.R. 1990. L-acetyl-carnitine, a substrate for chloroplast fatty acid synthesis. Plant Cell Environ. 13: 755-765.

Meyers, S.P., Nichols, S.L., Baer, G.R., Molin, W.T. & Schrader, L.E. 1982. Ploidy effects in isogenic populations of Alfalfa. I. Ribulose-1,5-bisphosphate carboxylase, soluble protein, chlorophyll, and DNA in leaves. Plant Physiol. 70: 1704-1709.

Moreau, R.A., Asmann, P.T. & Norman, H.A. 1990. Analysis of major classes of plant lipids by high-performance liquid chromatography with flame ionization detection. Phytochemistry 29: 2461-2466.

Mourioux, G. & Douce, R. 1979. Transport du sulfate à travers la double membrane limitante, ou enveloppe, des chloroplastes d'épinard. Biochimie 61: 1283-1292.

Mudd, J.B. & Kleppinger-Sparace, K.F. 1987. Sulfolipids. In: Stumpf, P.K. (ed.), The Biochemistry of Plants. A Comprehensive Treatise. Volume 9. Lipids: Structure and Function. Academic Press, Orlando, pp. 275-289.

Murata, N. & Hoshi, H. 1984. Sulfoquinovosyl diacylglycerols in chilling-sensitive and chilling-resistant plants. Plant Cell Physiol. 25: 1241-1245.

Nichols, B.W. 1965. Light induced changes in the lipids of *Chlorella vulgaris*. Biochim. Biophys. Acta 106: 274-279.

Nishihara, M., Yokota, K. & Kito, M. 1980. Lipid molecular species composition of thylakoid membranes. Biochim. Biophys. Acta 617: 12-19.

Perez, M. & Hirschberg, C.B. 1986. Transport of sugar nucleotides and adenosine 3'-phosphate 5'-phosphosulfate into vesicles derived from the Golgi apparatus. Biochim. Biophys. Acta 864: 213-222.

Pick, U., Gounaris, K., Weiss, M. & Barber, J. 1985. Tightly bound sulpholipids in chloroplast CF_0-CF_1. Biochim. Biophys. Acta 808: 415-420.

Pohl, P. & Wagner, H. 1972. Control of fatty acid and lipid biosynthesis in *Euglena gracilis* by ammonia, light and DCMU. Z. Naturforsch. 27b: 53-61.

Poincelot, R.P. 1973. Differences in lipid composition between undifferentiated and mature maize chloroplasts. Plant. Physiol. 51: 802-804.

Possingham, J.V. & Saurer, W. 1969. Changes in chloroplast number per cell during leaf development in spinach. Planta 86: 186-194.

Radunz, A. 1969. Über das Sulfochinovosyl-diacylglycerin aus höheren Pflanzen, Algen und Purpurbakterien. Hoppe-Seyler's Z. Physiol. Chem. 350: 411-417.

Radunz, A. & Berzborn, R. 1970. Antibodies against sulphoquinovosyl-diacylglycerol and their reactions with chloroplasts. Z. Naturforsch. 25b: 412-419.

Radunz, A., Bader, K.P. & Schmid, G.H. 1984. Influence of antisera to sulfoquinovosyl diglyceride and to β-sitosterol on the photosynthetic electron transport in chloroplasts from higher plants. In: Siegenthaler, P.A. & Eichenberger, W. (eds.), Structure, Function and Metabolism of Plant Lipids. Elsevier Sci. Publ. B.V., Amsterdam, pp. 479-484.

Rauser, W.E., Schupp, R. & Rennenberg, H. 1991. Cysteine, γ-glutamylcysteine, and glutathione levels in maize seedlings. Distribution and translocation in normal and cadmium-exposed plants. Plant Physiol. 97: 128-138.

Rennenberg, H. 1984. The fate of excess sulfur in higher plants. Annu. Rev. Plant Physiol. 35: 121-153.

Roughan, P.G. & Batt, R.D. 1968. Quantitative analysis of sulfolipid (sulfoquinovosyl diglyceride) and galactolipids (monogalactosyl and digalactosyl diglycerides) in plant tissues. Anal. Biochem. 22: 74-88.

Roughan, P.G. & Batt, R.D. 1969. The glycerolipid composition of leaves. Phytochemistry 8: 363-369.

Roughan, P.G. & Slack, C.R. 1984. Glycerolipid synthesis in leaves. TIBS 9: 383-386.

Russell, G.B. 1966. The routine measurement of lipid-bound sulfoquinovose in plants. Anal. Biochem. 14: 205-214.

Russell, N.J. & Harwood, J.L. 1979. Changes in the acyl lipid composition of photosynthetic bacteria grown under photosynthetic and non-photosynthetic conditions. Biochem. J. 181: 339-345.

Saidha, T. & Schiff, J.A. 1989. The role of mitochondria in sulfolipid biosynthesis by *Euglena* chloroplasts. Biochim. Biophys. Acta 1001: 268-273.

Saidha, T., Na, S.Q., Li, J. & Schiff, J.A. 1988. A sulphate metabolizing centre in *Euglena* mitochondria. Biochem. J. 253: 533-539.

Sato, N., Nemoto, Y. & Furuya, M. 1988. Lipids of *Chattonella antiqua* (*Raphidophyceae*). Plant Physiol. Biochem. 26: 93-98.

Satoh, N., Tsuzuki, M. & Kawaguchi, A. 1992. Isolation of mutants affected in lipid metabolism from *Chlamydomonas reinhardtii*. Photosynth. Res. 34: 170.

Schmidt, A. 1986. Regulation of sulfur metabolism in plants. Progr. Bot. 48: 133-150.

Schröppel-Meier, G. & Kaiser, W.M. 1988a. Ion homeostasis in chloroplasts under salinity and mineral deficiency. I. Solute concentrations in leaves and chloroplasts from spinach plants under NaCl or $NaNO_3$ salinity. Plant Physiol. 87: 822-827.

Schröppel-Meier, G. & Kaiser, W.M. 1988b. Ion homeostasis in chloroplasts under salinity and mineral deficiency. II. Solute distribution between chloroplasts and extrachloroplastic space under excess or deficiency of sulfate, phosphate, or magnesium. Plant Physiol. 87: 828-832.

Schwedock, J. & Long, S.R. 1990. ATP sulphurylase activity of the *nodP* and *nodQ* gene products of *Rhizobium meliloti*. Nature 348: 644-647.

Schwenn, J.D. & Schriek, U. 1984. A new role of thioredoxin in assimilatory sulphate reduction. Activation of the adenylylsulphate kinase from the green alga *Chlamydomonas reinhardii* CW 15. FEBS Lett. 170: 76-80.

Seifert, U. & Heinz, E. 1990. Experiments on sulfolipid biosynthesis in chloroplasts. In: Quinn, P.J. & Harwood, J.L. (eds.), Plant Lipid Biochemistry, Structure and Utilization. Portland Press Ltd., London, pp. 241-244.

Seifert, U. & Heinz, E. 1992. Enzymatic characteristics of UDP-sulfoquinovose:diacylglycerol sulfoquinovosyltransferase from chloroplast envelopes. Bot. Acta 105: 197-205.

Shibuya, I., Yagi, T. & Benson, A.A. 1963. Sulfonic acids in algae. In: Jap. Soc. Plant Physiol. (ed.), Studies on Microalgae and Photosynthetic Bacteria. The University of Tokyo Press, pp. 627-636.

Siebertz, H.P. & Heinz, E. 1977. Labelling experiments on the origin of hexa- and octadecatrienoic acids in galactolipids from leaves. Z. Naturforsch. 32c: 193-205.

Siebertz, H.P., Heinz, E., Linscheid, M., Joyard, J. & Douce, R. 1979. Characterization of lipids from chloroplast envelopes. Eur. J. Biochem. 101: 429-438.

Siebertz. H.P., Heinz, E., Joyard, J. & Douce, R. 1980. Labelling *in vivo* and *in vitro* of molecular species of lipids from chloroplast envelopes and thylakoids. Eur. J. Biochem. 108: 177-185.

Sigrist, M., Zwillenberg, C., Giroud, C., Eichenberger, W. & Boschetti, A. 1988. Sulfolipid associated with the light-harvesting complex associated with photosystem II apoproteins of *Chlamydomonas reinhardii*. Plant Sci. 58: 15-23.

Smith, K.L. & Harwood, J.L. 1983. The effect of trace metals on lipid metabolism in the brown alga *Fucus serratus*. Biochem. Soc. Transact. 11: 394-395.

Soll, J. & Roughan, P.G. 1982. Acyl-acyl carrier protein pool sizes during steady-state fatty acid synthesis by isolated spinach chloroplasts. FEBS Lett. 146: 189-192.

Spencer, D., Rerie, W.G., Randall, P.J. & Higgins, T.J.V. 1990. The regulation of pea seed storage protein genes by sulfur stress. Aust. J. Plant Physiol. 17: 355-363.

Stolyhwo, A., Martin, M. & Guiochon, G. 1987. Analysis of lipid classes by HPLC with the evaporative light scattering detector. J. Liquid Chromatogr. 10: 1237-1253.

Stuiver, C.E.E., De Kok, L.J., Santens, J.M.O., Kuiper & P.J.C. 1984. The effect of Na_2SO_4 on the lipid composition of sugar beet plants. Z. Pflanzenphysiol. 114: 187-191.

Takamura, H., Narita, H., Urade, R. & Kito, M. 1986. Quantitative analysis of polyenoic phospholipid molecular species by high performance liquid chromatography. Lipids 21: 356-361.

Toone, E.J., Simon, E.S. Bednarski, M.D. & Whitesides, G.M. 1989. Enzyme-catalyzed synthesis of carbohydrates. Tetrahedron 45: 5365-5422.

Trebst, A. & Schmidt, A. 1969. Photosynthetic sulfate and sulfite reduction by chloroplasts. In: Metzner, H. (ed.), Progress in Photosynthesis Research, Vol. III. H. Laupp jr., Tübingen, pp. 1510-1516.

Tulloch, A.P., Heinz, E. & Fischer, W. 1973. Combination and positional distribution of fatty acids in plant sulfolipids. Hoppe-Seyler's Z. Physiol. Chem. 354: 879-889.

Varin, L., DeLuca, V., Ibrahim, R.K. & Brisson, N. 1992. Molecular characterization of two flavonol sulfotransferases. Proc. Natl. Acad. Sci. USA 89: 1286-1290.

Voss, R., Radunz, A. & Schmid, G.H. 1992. Binding of lipids onto polypeptides of the thylakoid membrane. I. Galactolipids and sulpholipid as prosthetic groups of core peptides of the photosystem II complex. Z. Naturforsch. 47c: 406-415.

Weenink, R.O. 1963. Reaction of sulphoquinovose from red-clover lipids with anthrone. Nature 197: 62-63.

Wood, B.J.B., Nichols, B.W. & James, A.T. 1965. The lipids and fatty acid metabolism of photosynthetic bacteria. Biochim. Biophys. Acta 106: 261-273.

PHYSIOLOGICAL FUNCTIONS AND ENVIRONMENTAL RELEVANCE OF SULFUR-CONTAINING SECONDARY METABOLITES

Ewald Schnug
Institute of Plant Nutrition and Soil Science, Federal Agricultural Research Centre Braunschweig-Völkenrode (FAL), Bundesallee 50, D-3300 Braunschweig, F.R.G.

Introduction

An exciting characteristic of plant metabolism is the ability of certain species to synthesize metabolic by-products. The term 'secondary' is a poor one as it implies a degree of uselessness. Bell & Charlwood wrote in 1980 that 'It is a pity that the term "secondary" should have ever been applied to these compounds as the word gives the impression that they are all relatively unimportant.' Comprehensive discussion of the terminology and history of this term as well as the basic understanding of their environmental significance is given by Bell (1980) and Mothes (1980).

Secondary compounds can be understood as chemical tools that compensate for plants' immobility by enabling very complex reactions with their abiotic environment, and by facilitating a 'quiet communication' with their biotic environment. According to Bell (1980) plants synthesize a great array of secondary compounds because they cannot rely on physical mobility; the presence of secondary compounds is a form of chemical defense. According to the basic definition of the term 'secondary', compounds like phytochelatins (Grill *et al.* 1990; Rauser, this volume) and thionins (Bohlmann, this volume) need to be included in a treatise like this. However, this paper will not deal with those products as they are already well covered in this book. The major aim of this contribution is to present an overview of the most recent knowledge concerning the sulfur-containing members of this exciting part of phytochemistry, to provide a guide to the major literature of the biochemical background, and to provide an evaluation of their physiological functions and environmental significance from the viewpoint of applied agricultural biochemistry and ecology.

Sulfur-containing secondary compounds in fungal species

Certain fungal species are able to synthesize fairly extraordinary sulfur-containing compounds in their secondary metabolism. The first of three selected examples presented here is lenthionine (1,2,3,5,6-pentathiepane (Fig. 1); $C_2H_4S_5$)) with a heterocyclic structure consisting of 85.11% sulfur (Morita & Kobayashi 1966). The biochemical pathway by which the compound is synthesized is not yet clear. Lenthionine is the odorous principle from the edible mushroom *Lentinus edodes* ('Shiitake'). With its remarkable contribution to the odour, lenthionine is responsible for both the ecological and economic features of the fungi. Gliotoxin (2,3,5a,6 tetrahydro-

Sulfur Nutrition and Assimilation in Higher Plants
edited by L.J. De Kok et al., pp. 179–190

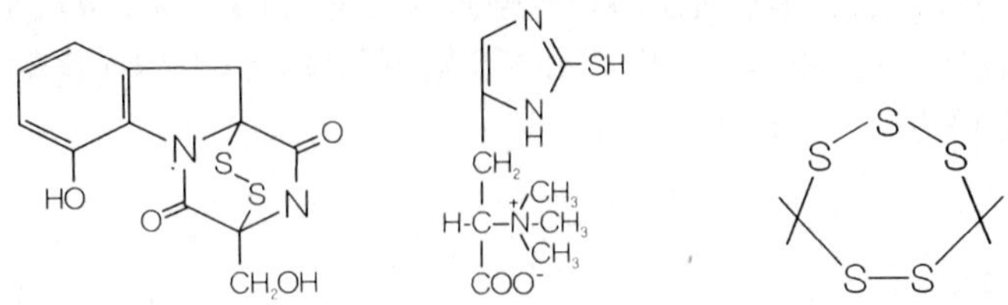

Gliotoxin from *Aspergillus fumigatus* | Ergothioneine from *Claviceps purpurea* | Lenthionine from *Lentius edodes*

Fig. 1. Molecular structures of some sulfur-containing secondary compounds found in fungal species.

6-hydroxy-3-(hydroxymethyl)-2-methyl-1OH-3,10a-epidithiopyrazino[1,2a]-indole-1,4-dione; Fig. 1) is a sulfur containing-compound of the secondary metabolism of various subspecies of *Penicillium*, *Trichoderma, Gladiocladium fimbriatum* and *Aspergillus fumigatus* (Beecham *et al.* 1966; Spenser 1968). Biochemically the compound originate from phenylalanine and serine by reaction with *S*-adenosyl-methionine (Suhadolnik & Chenoweth 1958). The compound is antimicrobial and thus part of the ecomonical system used by the fungi for establishment and survival in an ecosystem. However, gliotoxin is also of value in the production of pharmaceuticals.

Ergothioneine ((S)-α-carboxy-2,3-dihydro-N,N,N-trimethyl-thioxo-1H-imidazole-4-ethanaminium hydroxide, inner salt; Fig. 1) is an indol-alkaloid belonging to the more than 30 different members of the so called 'ergot'-alkaloids found in *Neurospora crassa* and *Claviceps purpurea*. The amino group of a histidin is methylated by reaction with methionine yielding an intermediate in the synthesis of ergothioneine known as hercynine. Hercynine is an alkaloid occurring in the fly-agaric *Amanita muscaria*. Ergothioneine is formed by reaction of hercynine with cysteine in which serine is released. Ergothioneine is not solely characteristic for fungal species. The substance has also been detected in tissue of higher plants and animals, namely in the king crab *Linulus polyphenus*. At the present stage of knowledge, it seems to be difficult to evaluate the environmental relevance of ergothioneine. Its toxicity for certain mammals might be interpreted as a mechanism for protecting mature sclerotia in the environment.

Sulfur-containing secondary compounds in higher plants

Thiols

Thiols are well known due to their unpleasant odour associated with the off-flavour of food. They are derived mostly from the degradation of cysteine or other larger sulfur containing molecules. The grapefruit aroma is largely due to p-mentha-8-thiol. The compound is a rare example of a thiol occurring in the intact plant. It is derived from the reaction of limonene (Simonsen 1947) with H_2S (Fig. 2). This reaction might be important in the detoxification of H_2S derived from sulfate reduction. The addition of the thiol group increases the vapour tension and thus the ecological impact of the compound. The detection limit for the odour of p-mentha-8-thiol is a mil-

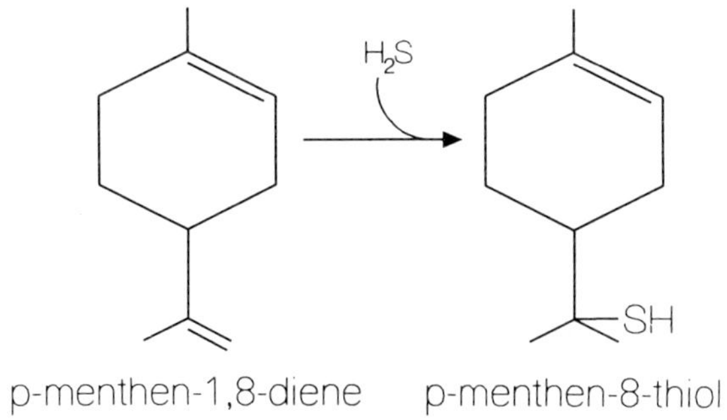

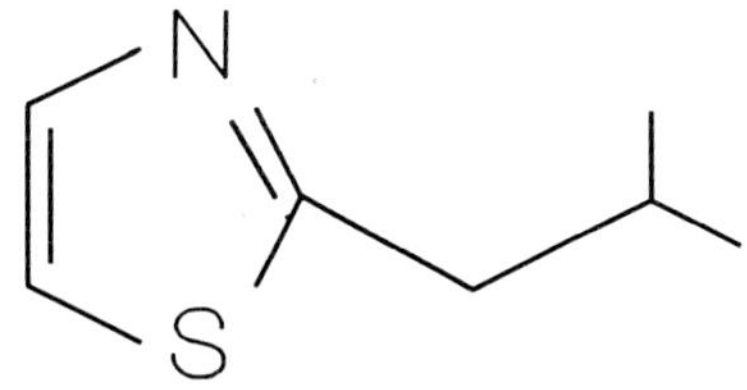

Fig. 2. Formation of p-mentha-8-thiol from p-mentha-1,8-diene.

Fig. 3. Molecular structure of 2-Isobutylthiazole from *Lycopersicon lycopersicon*.

lion times lower than that one of the original limonene and a thousand times lower than that one of methylmercaptane. As p-mentha-8-thiol has no phytotoxic features, the ecological relevance of the compound is that of an ecomone with a distinct function for the coevolution of animals.

Thiazoles

Thiazoles are compounds which are mainly developed during food processing providing a 'nutty' taste to boiled potatoes or cooked meat. 2-Isobutylthiazole is a thiazole that occurs as a by-product in the metabolism of an intact plant (Fig. 3) and is found in *Lycopersicon lycopersicon*. This thiazole is responsible for the 'green' winy taste of tomatoes. The compound is derived from the secondary metabolism of leucine and cysteine through decarboxylation of the two amino acids followed by deamination of leucine and cyclization of the remaining bodies via dehydration and oxidation (Belitz & Grosch 1983). As with p-mentha-8-thiol, thiazoles are volatile compounds with no phytotoxic features and their ecological relevance might be a factor for the coevolution of animals.

Cyclic disulfides

Asparagusic acid (1,2-dithiolane-4-carboxylic acid) is the precursor of the odorous principle of asparagus occurring after cooking of the vegetable, or in the urine of asparagus consumers. Although the molecular structure of both compounds involves a 1,2-dithiolane ring, the biochemical pathways of asparagusic acid and lipoic acid have nothing in common (Fenwick & Hanley 1989). According to Parry (1982), the biosynthesis of asparagusic acid (Metzner 1973; Fig. 4) involves also the formation of methacrylate as an important intermediate. Asparagusic acid is an allelopathic substance which inhibits the growth of seedlings to an extent comparable with abscisic acid (Kitahara *et al.* 1972). However, the compound also exhibits a strong positive effect on *Streptococcus faecalis* and thus might be a factor for coevolution of microorganisms.

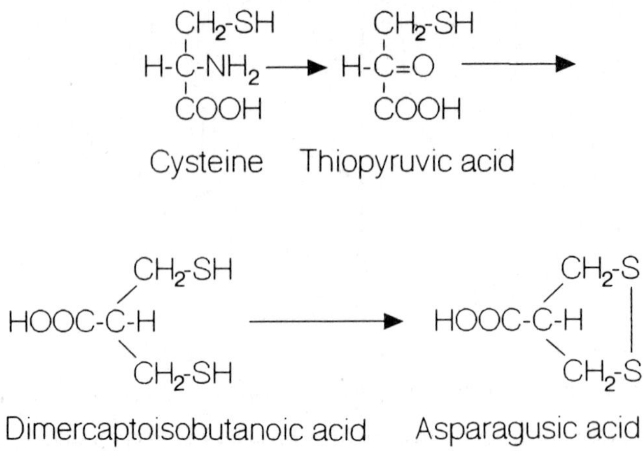

Fig. 4. Major steps in the biosynthesis of asparagusic acid.

Table 1. Comparison of major alk(en)yl radicals in *Allium* cepa and *Allium* sativum. *Relative proportion of the radical (R) in the volatile fraction (according to Freeman & Whenham 1975); **lachrymatory principle of *Allium* cepa.

R =	Principal sources	Final Alliin	Relative occurrence (%)*	
			A. cepa	*A. sativum*
$-CH=CH-CH_3$	cysteine + valine	*S*-(prop-1-enyl)cysteine sulfoxide**	6	–
$-CH_2-CH=CH_2$	serine + prop-2-enylmercaptane	*S*-(prop-2-enyl)cysteine sulfoxide	–	85
$-CH_2-CH-CH_3$	serine + propylmercaptane	*S*-propyl-cysteine sulfoxide	88	2

Alliins

Alliin is the common name for *S*-alk(en)ylcysteine sulfoxides which are characteristic sulfur compounds of the secondary metabolism of members of the genus *Allium*. Alliins are major contributors to odour and flavour of food products derived from these plants and also may provide pharmaceutical ingredients (Block 1992). The specific secondary biochemical pathway for alliins in onions starts with γ-glutamyl-*S*-2-carboxypropylcysteine. Specific to onions, this compound of the normal intermediary pathway is stepwise decarboxylated, oxidized and finally, *S*-(prop-1-enyl)cysteine sulfoxide is formed after cleavage of the intermediary product by γ-glutamyl-transpeptidase. The reactions for other *S*-alk(en)ylcysteine sulfoxides differ from this one only by the origin of the precursors involved in the metabolic pathway preceding the oxidation step (Block 1992; Fenwick & Hanley 1989).

More than 80% of the total sulfur in *Allium* species is bound to secondary compounds. In resting seeds and bulbs of *Allium* species where the dry matter contains up to more than one percent total sulfur, these compounds are stored as γ-glutamyl-peptides until germination or sprouting activates a γ-glutamyl-peptidase which releases the appropriate *S*-alk(en)ylcysteine sulfoxide. The alliinase system has no access to these storage compounds. The physical and biological features of the

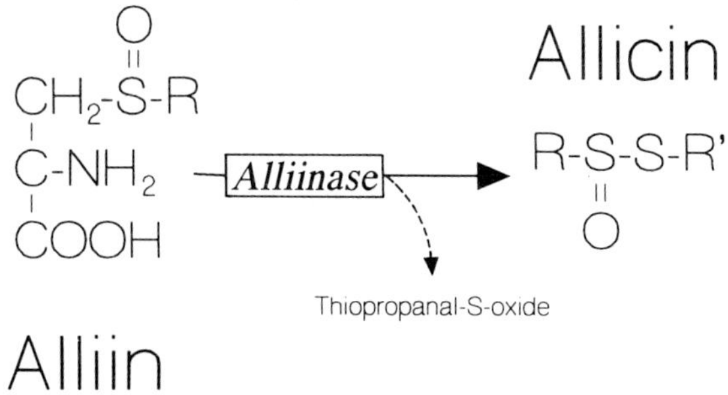

Fig. 5. Principal structure of alliins and allicins (for 'R' refer to Table 1).

alliins become active after cellular disruption and enzymatic cleavage of the compound by the *Allium* specific enzyme alliinase (EC 4.4.1.4). The reaction yields pyruvate and thiosulfinates (allicins) which again are the precursors of a large number of volatile substances like mono-, di- and trisulfides (Block 1992; Fenwick & Hanley, 1989). The pattern of individual *S*-alk(en)ylcysteine sulfoxides is specific for each *Allium* species (Table 1) and has been used for chemotaxonomic classification of members (Fig. 5) of the genus (Freeman & Whenham 1975).

Although many efforts have been made in investigating the pharmacological features of alliins, allicins and derivate products (Block 1992; Burger *et al.* 1992; Fenwick & Hanley 1989; Kensler 1992; Yang *et al.* 1992), little knowledge exists concerning their physiological functions in plants and environmental significance. In contrast to glucosinolates (see below), alliins are metabolically inert end-products of this biochemical pathway and the sulfur bound in these compounds is not reusable on the occasion of sulfur deficiency caused by insufficient supply. Consequently, the ratio between protein-sulfur and sulfur bound in secondary compounds in the *Allium* species is much wider (1:4-1:6) than in members of the *Brassica* family (1:0.3-1:2; see below). Alliins are a strong factor for coevolution resulting in some highly specialized *Allium* parasites, for instance *Hylemyia antiqua*.

Glucosinolates

Glucosinolates are characteristic sulfur containing compounds of the secondary metabolism of at least 15 dicotyledonous taxa. Within this group, the *Brassicaceae* account for most species of agricultural relevance. A basic treatise of the fundamental, environmental and agricultural aspects of glucosinolates is given in Schnug (1990a). Additional aspects of the analysis, food-chemistry and toxicology, are provided by Fenwick & Hanley (1989), Schnug (1991b), Schnug & Haneklaus (1991a) and Sørensen (1990).

One of the most exciting findings in recent agroecological research is the detection of the biorecyclization of glucosinolates through which glucosinolates provide the plant with a sulfur storage which can be utilized for the synthesis of primary products in the event of sulfur starvation (Schnug 1988a). After enzymatic cleavage, glucosinolates yield thiocyanates and sulfate (Srivastava & Hill 1974) which are precursors for the biosynthesis of amino acids (Fig. 6; Machev & Schraudolf 1977,1978;

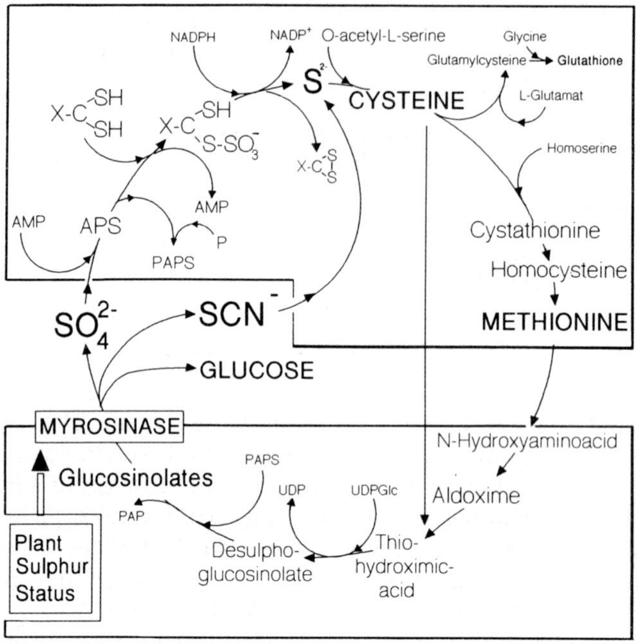

Fig. 6. Biosynthesis and biorecyclization of alkenyl-glucosinolates in *Brassica napus*.

Marquardt *et al.* 1968). Myrosinase has been localized in 'myrosin cell' labeled idioblasts (Bones & Iversen 1985; Iversen *et al.* 1979; Lüthy & Matile 1984; Tangstad *et al.* 1990). However, myrosinase activity in intact plants results in the occurrence of isothiocyanates among the headspace volatiles of *Brassica napus* (Seaton *et al.* 1990; Tollsten & Bergström 1988). The very fast reaction of glucosinolate concentrations in vegetative tissue in response to changes in the sulfur supply (Schnug 1990a) contributes to the evidence that, in vegetative tissue of *Brassica* species, glucosinolates and myrosinase are not totally compartmentalized in the vegetative tissue (see also: Höglund *et al.* 1991; Iversen 1970; Reiner 1987). A proposed way for the regulation of the glucosinolate/myrosinase system given in Fig. 7, also explains why compartmentalization of glucosinolate and myrosinase is not necessary.

The activity of the enzyme myrosinase is governed by the concentrations of ascorbic acid (lower left box in Fig. 7; Ettlinger *et al.* 1961; Björkman 1976). Vegetative tissue of *Brassica* species are well known for ascorbic acid concentrations in the mmol level (Horbowicz & Bakowski 1988; Reiner 1987) and thus in the range where increasing ascorbate concentrations are related to decreasing myrosinase activities (Björkman 1976). However, in *Brassica* seeds, ascorbate concentrations are too low to inhibit myrosinase activity. The fact that seed glucosinolate concentrations and their pattern of individuals are very stable within time (Schnug & Haneklaus 1988, 1991b; Schnug *et al.* 1992) proves that in seeds there is no myrosinase activity outside the myrosin cells. Again, the ascorbic acid concentrations are dependent upon the regeneration of dehydroascorbate in the ascorbate-glutathione cycle (box in middle of Fig. 7; Halliwell & Gutteridge 1989). The link with the sulfur nutritional status

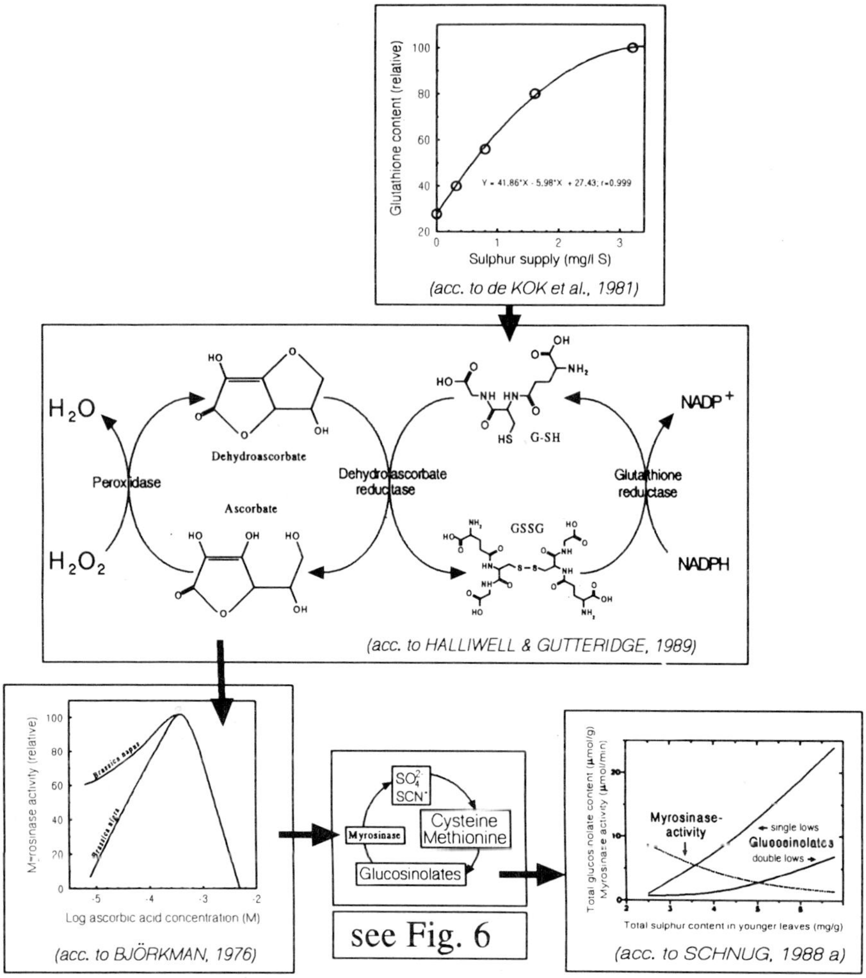

Fig. 7. Regulation of the biorecyclization of glucosinolates.

is evident from the strong dependency of the glutathione levels on the external sulfur supply (upper box in Fig. 7; De Kok *et al.* 1981).

This model for regulation explains some of the glucosinolate related phenomenons affecting plant resistance against pests and diseases (Ernst 1990; Schnug & Ceynowa 1990). The biologically active part of glucosinolates is the aglucone which is released from the molecule by enzyme activity. Wounding or infestation of vegetative tissue of *Brassica napus* results in a higher oxidation potential in the tissue causing increased consumption of ascorbate (Halliwell & Gutteridge 1989), followed by increased myrosinase activity resulting again in a decrease of the amount of intact alkenyl-glucosinolates (Koritsas *et al.* 1991).

The sulfur-supply dependent biorecyclization of glucosinolates is a fundamental part of the ecological survival strategy of *Brassica* species since the dissemination of

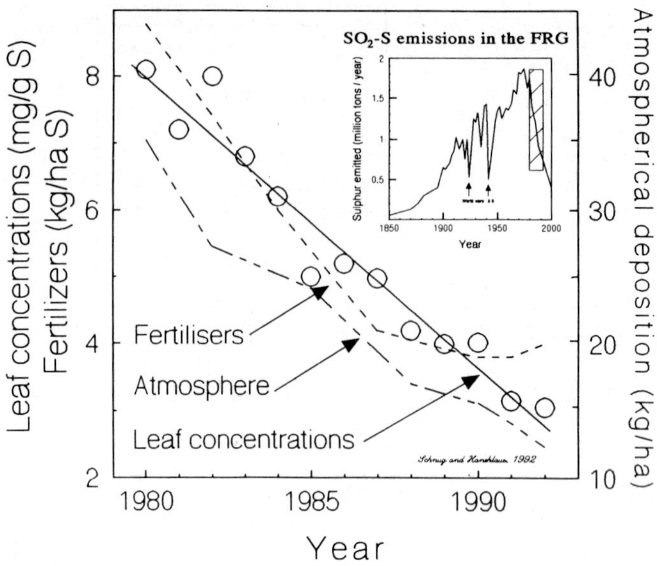

Fig. 8.

these plants relies on a large number of small seeds with high concentrations of sulfur containing amino acids. Thus, these plants have an extraordinary high demand for sulfur during seed production. The above ground biomass of a rapeseed crop yielding at 4t ha^{-1} of seed takes up 40 kg ha^{-1} sulfur in the first month from sowing. This is more than three times the amount of a high yielding cereal crop or the actual average annual supply by atmospheric deposition in the areas of Northern Europe (Schnug & Evans 1992; Schnug 1991b).

The poor performance of low glucosinolate containing *Brassica napus* breeds ('double lows'; Schnug 1990a, 1991a) in low sulfur input agroecosystems evident from visible symptoms of severe sulfur starvation being the most widespread visible symptom of nutrient deficiency in *Brassica* species (Haneklaus & Schnug 1992c) is impressive proof of the importance of the glucosinolate-biorecyclization in affecting plant vigour. By trying to meet human demands without fully understanding the physiological and ecological background, a synanthropic organism was established which failed after changes in its environment have took place.

The physiological background to this phenomenon is the fact that in the alkenyl- and sulfinyl-glucosinolate pathway of double low cultivar, a metabolic block exists prior to the formation of the aldoxim (Fig. 6). The myrosinase system has no access to those intermediary compounds and a remobilization of the sulfur into basic compounds is disabled. In the case of generative parts of *Brassica napus*, these intermediary compounds are enriched in the pod walls during seed development because they are not transported from the place of synthesis, which is the pod wall, into the seed. Later, as a consequence of senescence, these intermediates are susceptible to oxidative cleavage resulting in enrichments of sulfate as an unspecific secondary process which has nothing to do with the biochemical background of the 'double low' phenomenon. This sulfate is not usable for the organism as seed filling has finished and local protein biosynthesis is stopped at that stage of development.

Concluding remarks

Secondary plant products and especially the sulfur containing ones have never been treated very intensively by agricultural chemistry. However, after having satisfied the basic needs for nutrients in the community, agricultural production research now focuses more on the aspects of food quality and on environmentally sustainable systems for plant production.

Sulfur containing metabolites are important factors affecting plant resistance (Schnug & Ceynowa 1990), crop technology (Haneklaus & Schnug 1992 a,b) and the nutritional quality (Schnug 1990b) of crops. Most of these compounds are also valuable sources of pharmacological products (Block 1992; Burger *et al.* 1992; Cho *et al.* 1992; Chung 1992; Fenwick & Hanley 1989; Kensler 1992; Reddy & Rao 1992; Stoner 1992; Yang 1992).

The content of secondary metabolites is highly dependent on environmental factors (Luckner 1980), and of these, the sulfur nutritional status of the plants is one of the most important factors affecting the sulfur containing secondary compounds (Booth *et al.* 1991; Schnug 1988b, 1989). Thus, the great changes in the sulfur balance of agroecosystems in Northern Europe due to the reduction of industrial emissions (Schnug & Evans 1992; Schnug 1991b) urgently needs to be considered in future techniques of plant production if significant losses of natural plant resistance and food quality are to be avoided.

The example of reduced vigour in double low oilseed rape cultivars in which the ability for the biorecyclization of glucosinolates is reduced should remind both traditional plant breeders and modern molecular biologists that a contemplative understanding of the physiological functions and environmental significance of these compounds rather than short term human demands should be prerequisite of their work.

Acknowledgement

The author sincerely thanks Donal Murphy for assisting with the translation of this contribution.

References

Beecham, A.F., Friedrichson, J. & Mathieson, A.L. 1966. The structure and absolute configuration of gliotoxin and the absolute configuration of sporidesmin. Tetrahedron Lett. 27: 3131-3138.
Belitz, H.-D. & Grosch, W. 1983. Lehrbuch der Lebensmittelchemie. 2. Auflage. Springer-Verlag, Berlin.
Bell, E.A. 1980. The possible significance of secondary compounds in plants. In: Bell, E. A. & Charlwood, B. V. (ed.), Secondary Plant Products. Springer-Verlag, Berlin, pp. 11-21.
Bell, E.A. & Charlwood, B.V. 1980. Preface to: Secondary Plant Products, Springer-Verlag, Berlin.
Björkman, R. 1976. Properties and function of plant myrosinases. In: Vaughhan *et al.* (ed.), The Biology and Chemistry of the *Cruciferae*. Acad. Press, London, pp. 191-206.

Block, E. 1992. Die Organoschwefelchemie der Gattung *Allium* und ihre Bedeutung für die organische Chemie des Schwefels. Angew. Chem. 1158-1203.

Bones, A. & Iversen, T.-H. 1985. Myrosin cells and myrosinase. Isr. J. Bot. 34: 351-375.

Booth, E.J., Walker, K.C. and Schnug, E. 1991. The effect of site, foliar sulphur and nitrogen application on glucosinolate content and yield of oilseed rape (*Brassica napus*). Proc. 8th Int. Rapeseed Congress 2: 567-572.

Burger, R.A., Warren, L.D., Hughes, B.G. & Lawson, L.D. 1992. Aqueous garlic extract, polar and thiosulphinate fractions enhance *in vitro* human immune function. Proc. 204th Amer. Chem. Soc. Nat. Meeting, Div. Agric. Food Chem., Washington, D.C. August 23-28, *in press*.

Cho, C.G., Posner, G.H., Talalay, P. & Zhang, Y. 1992. Isothiocyanates and protection against chemical carcinogenesis. Proc. 204th Amer. Chem. Soc. Nat. Meeting, Div. Agric. Food Chem., Washington, D.C. August 23-28, *in press*.

Chung, F-L. 1992. Aromatic isothiocyanates in cancer chemoprevention. Proc. 204th Amer. Chem. Soc. Nat. Meeting, Div. Agric. Food Chem., Washington, D.C. August 23-28, *in press*.

De Kok, L.J., Kan, J.L. de, Tánczos, O.G. & Kuiper, P. J. C. 1981. Sulphate induced accumulation of glutathione and frost-tolerance of spinach leaf tissue. Physiol. Plant. 53: 435-438.

Ettlinger, M.G., Datev, G.P. Harrison, F.J. & Mabry, T.J. 1961. Vitamin C as a coenzyme: the hydrolysis of mustard oil glucosides. Proc. Nat. Acad. Sci. U.S.A. 47: 1875-1880.

Ernst, W.H.O. 1990. Ecological aspects of sulfur metabolism. In: Rennenberg, H., Brunold, C., De Kok, L.J. & Stulen, I. (eds.), Sulfur Nutrition and Sulfur Assimilation in Higher Plants; Fundamental, Environmental and Agricultural Aspects. SPB Academic Publishing, The Hague,pp. 131-144.

Fenwick, G.R. & Hanley, A.B. 1989. Glucosinolates, alliins and cyclic disulphides: sulphur containing secondary metabolites. In: Damani, L. A. (ed.), Sulphur-containing Drugs and Related Organic compounds – Chemistry, Biochemistry and Toxicology Vol. 1 Part B. E. Horwood Series in Biochemical Pharmacology, pp. 273-315.

Freeman, G.G. & Whenham, R.J. 1975. A survey of the volatile components of some *Allium* species in terms of *S*-alk(en)ylcysteine sulphoxides present as flavour precursors. J. Sci. Food Agric. 26: 1869-1886.

Grill, E., Winnacker, E.-L. & M.H. Zenk. 1990. Phytochelatins, the heavy metal binding peptides of the plant kingdom. In: Rennenberg, H., Brunold, C., De Kok, L.J. & Stulen, I. (eds.), Sulfur Nutrition and Sulfur Assimilation in Higher Plants; Fundamental, Environmental and Agricultural Aspects. SPB Academic Publishing, The Hague, pp. 89-96.

Halliwell, B. & Gutteridge, J.M.C. 1989. Free radicals in biology and medicine. 2 ed., Claredon Press Oxford, 277-289.

Haneklaus, S. & Schnug, E. 1992a. Baking quality and sulphur content of wheat. I. Relations between sulphur- and protein content and loaf volume. Sulphur in Agriculture, *in press*.

Haneklaus, S. & Schnug, E. 1992b. Baking quality and sulphur content of wheat. II. Evaluation of the relative importance of genetics and environment including sulphur fertilisation. Sulphur in Agriculture, *in press*.

Haneklaus, S. & Schnug, E. 1992c. Maçroskopic symptomatology of sulphur deficiency symptoms in *Brassica napus*. Phyton 32: 53-56.

Höglund, A.-S., Lenman, M., Falk, A. & Rask, L. 1991. Distribution of myrosinase in rapeseed tissues. Plant Physiol. 95: 213-221.

Horbowicz, M. & Bakowski, J. 1988. Effects of biological factors on the chemical composition of brussels sprouts. Biuletin Warzywywniczy XXXII: 161-169.

Iversen, T.H. 1970. Cytochemical localization of myrosinase (β-thioglucosidase) in root tips of *Sinapis alba*. Protoplasma 71: 451-466.

Iversen, T.-H., Baggerud, C., Beisvaag, T. 1979. Myrosin cells in *Brasicaceae* roots. Z. Pflanzenphysiol. 94: 143-145.

Kensler, T.W. 1992. Chemoprotection by dithiolethiones. Proc. 204th Amer. Chem. Soc. Nat. Meeting, Div. Agric. Food Chem., Washington, D.C. August 23-28, *in press*.

Kitahara, Y., Yanagawa, H., Kato, T. & Takahashi, N. 1972. Asparagusic acid, a new plant growth inhibitor in etiolated young *asparagus* shoots. Plant Cell Physiol. 13: 923-925.

Koritsas, V.M., Lewis, J.A. & Fenwick, G.R. 1991. Glucosinolate responses of oilsee rape, mustard and kale to mechanical wounding and infestation by cabbage stem flea beetle (*Psylliodes chrysocephala*). Ann. Appl. Biol. 118: 209-221.

Luckner, M. 1980. Expression and control of secondary metabolism. In: Bell, E.A. & Charlwood, B.V. (ed.), Secondary Plant Products, Springer-Verlag Berlin, pp. 24-63.

Lüthy, B. & Matile, P. 1984. The mustard oil bomb: Rectified analysis of the subcellular organisation of the myrosinase system. Biochem. Physiol. Pflanzen 179: 5-12.

Machev, N.P. & Schraudolf, H. 1977. Biochemical relation between thiocyanide assimilation and indole-glucosinolate biosynthesis in *Sinapis alba* L.. Proc. sec. Int. Symp. on Plant Growth Regulators, pp. 110-116.

Machev, N.P. & Schraudolf, H. 1978. Thiocyanate as predecessor of asparagine in *Sinapis alba* L. Plant Physiol. (Bulgaria): 26-33.

Marquard, R., Kühn, H. & Linser, H. 1968. Über die Verwendung dampfförmigen Allylsenföls durch Pflanzen. Atompraxis 14: 1-3.

Metzner, H. 1973. Stoffwechsel S-haltiger Verbindungen. In: Biochemie der Pflanzen. Enke-Verlag, Stuttgart, pp. 268-271.

Morita, K. & Kobayashi, S. 1966. Isolation and synthesis of lenthionine, an odourus substance of Shiitake an edible mushroom. Tetrahedron Lett. 6: 579-577.

Mothes, K. 1980. Historical introduction. In: Bell, E. A. & Charlwood, B. V. (ed.), Secondary Plant Products, Springer-Verlag Berlin, pp. 1-10.

Parry, R.J., Mizusawa, A.E. & Ricciardone, M. 1982. Biosynthesis of sulphur compounds. Investigation of the biosynthesis of asparagusic acid. J. Amer. Chem. Soc. 104: 1442-1448.

Reddy, B.S. & Rao, C.V. 1992. Chemoprevention of colon cancer by thiol compounds. Proc. 204th Amer. Chem. Soc. Nat. Meeting, Div. Agric. Food Chem., Washington, D.C. August 23-28, *in press*.

Reiner, H. 1987. Elektrophoretische Untersuchungen zum Glucosinolase-Glucosinolat-System in *Brassicaceae* (Cruciferen). Diplomarbeit am Institut für Chemie, Universität für Bodenkultur Wien.

Schnug, E. 1988a. Quantitative und qualitative Aspekte der Diagnose und Therapie der Schwefelversorgung von Raps (*Brassica napus* L.) unter besonderer Beruecksichtigung glucosinolatarmer Sorten. Habilitationsschrift, Agrarwiss. Fak. University of Kiel, FRG.

Schnug, E. 1988b. Fluctuations in the glucosinolate content in seeds of 0- and 00-oilseed rape. Proc. NIAB Double Low Forum London-Heathrow, pp. 24-30.

Schnug, E. 1989. Double low oilseed rape in West Germany: sulphur nutrition and levels. In: Production and Protection of Oilseed Rape and other *Brassica* Crops. Aspects Appl. Biol. 23: 67-82.

Schnug, E. 1990a. Glucosinolates – fundamental, environmental and agricultural aspects. In: Rennenberg, H., Brunold, C., De Kok, L.J. & Stulen, I. (eds.), Sulfur Nutrition and Sulfur Assimilation in Higher Plants; Fundamental, Environmental and Agricultural Aspects. SPB Academic Publishing, The Hague, pp. 97-106.

Schnug, E. 1990b. Sulphur nutrition and quality of vegetables. Sulphur in Agriculture 14: 3-7.

Schnug, E. 1991a. Sulphur metabolism in oilseed rape plants with particular reference to double low varieties. Proc. 8th Int. Rapeseed Congress Saskatoon 3: 695-700.

Schnug, E. 1991b. Sulphur nutritional status of European crops and consequences for agriculture. Sulphur in Agriculture 15: 7-12.

Schnug, E. & Ceynowa, J. 1990. Phytopathological aspects of glucosinolates in oilseed rape. J. Agron. Crop Sci. 165: 319-328.

Schnug, E. & Evans, E. 1992. Monitoring of the sulphur supply of agricultural crops in Northern Europe. Phyton 32: 121-124.

Schnug, E. & Haneklaus, S. 1988. Veränderungen des Gesamtglucosinolatgehaltes von Rapssaat während der Abreife nach Analysen mit Glucose-, HPLC-, H1- (Sulfatase-) und RFA-Methode. Fett Wiss. Technol. 90: 329-331.

Schnug, E. & Haneklaus, S. 1991a. Progress in the X-RF method for the determination of the total glucosinolate content in oilseed rape and processed meal. Proc. 8th Int. Rapeseed Congress Saskatoon 3: 837-844.

Schnug, E. & Haneklaus, S. 1991b. Study on the stability of sulphur in three rapeseed standard reference materials by X-RF. Report EEC Contract 5117/1/5/307/89/5-BCR-D(10), ILLIT Report 4.

Schnug, E., Boenke, A., Wagstaffe, P.J. & Lindsey, A.S. 1992. The characterisation of three rapeseed reference materials and their utilisation in the indirect determination of total glucosinolates in rapeseed by X-RF. Fat Sci. Technol. 94: 297 301.

Srivatava, V.K. & Hill, D.C. 1974. Glucosinolate hydrolytic products given by *Sinapis alba* and *Brassica napus*. Phytochemistry 13: 1053-1046.

Seaton, A., Harker, C., Cherrie, B., Brooke, M. & Robertson, A. 1990. Respiratory hazards from oilseed rape: experiments and hypotheses. Proc. Oilseed Rape/Allergy Meeting, Scottish Crop Research Institute Invergowrie Dundee, pp. 29-32.

Simonsen, J.L. 1947. The terpenes. Vol. 1, Univ. Press, Cambridge.

Spenser, I.D. 1968. The biosynthesis of alkaloids and other nitrogenous secondary metabolites. In: Florkin, M. & Stotz, E. H. (ed.), Comprehensive Biochemistry. Vol. 20. Elsevier Publ., Amsterdam.

Sfrensen, H. 1990. Glucosinolates: Structure – Properties – Function. In. Shahidy, F. (ed.), Rapeseed/Canola: Production, Chemistry, Nutrition and Processing Technology. Van Nostrand Reinhold Publ., pp. 149-172.

Stoner, G.D. 1992. Inhibition of esophageal tumorigenesis by ellagic acid and isothiocyanate. Proc. 204th Amer. Chem. Soc. Nat. Meeting, Div. Agric. Food Chem., Washington, D.C. August 23-28, *in press*.

Suhadolnik, R.J. & Chenoweth, R.G. 1958. Biosynthesis of gliotoxin. I. Incorporation of phenylalanin-1- and -2-C^{14}. J. Amer. Chem. Soc. 80: 4391-4392.

Thangstad, O.P., Iversen, T.-H., Slupphaug, G. & Bones, A. 1990. Immunocytochemical localisation in *Brassica napus*. Planta 189: 245-248.

Tollsten, L. & Bergström, G. 1988. Headspace volatiles of whole plants and macerated plant parts of *brassica* and *sinapis* species. Phytochemistry 27: 4013-4018.

Yang, C.S., Hong, J-Y. & Wang, Z.Y. 1992. Inhibition of chemical toxicity and carcinogenesis by garlic components. Proc. 204th Amer. Chem. Soc. Nat. Meeting, Div. Agric. Food Chem., Washington, D.C. August 23-28, *in press*.

Section 3.
Agricultural and environmental aspects of sulfur metabolism

Contents

ELEMENTAL SULFUR IN AGRICULTURE

Pascale Jolivet
Laboratoire de Chimie Biologique, Institut National Agronomique Paris-Grignon, F-78850 Thiverval-Grignon, France

Introduction

The sulfur requirement of crops has traditionally been met by using sulfur-containing fertilizers and sulfur in rainfall. With the reduction of these sources, sulfur deficiencies are becoming more widespread and occurring throughout the world. Since the development of environmental controls on industrial aerial pollution, atmospheric sulfur deposition has fallen by over 30% in the United Kingdom (Roberts & Fisher 1985). Since the introduction of sulfur free urea and triple superphosphate in 1965, sulfur contents of fertilizers have become negligible (1-2%). Sulfur deficiency was also promoted by use of high yielding varieties, intensive cropping, removal of crop residues and decrease in use of sulfur-containing pesticides (Tabatabai 1984).

Some soils contain less than 10 mg SO_4^{2-}-S kg^{-1}, a value considered inadequate to sustain optimum crop growth (Wainwright 1984; Scott 1985). Cereals use 10-15 kg S ha^{-1} and grass and cruciferous crops 20-30 kg S ha^{-1} each year. If sulfur supply from the soil is below this level, then the plants may become sulfur deficient. For most field crops, an N/S ratio of between 14-16/1 is required to insure maximum production of both dry matter and protein, elevated ratios are reflected in yield and quality depressions. The overall appearance of sulfur deficient plants is characterized by a chlorotic yellowing, although in white clover and in rapeseed an intense purple color may develop, and a poor leaf development which is difficult to distinguish from N deficiency.

Plants can absorb three forms of sulfur:

* Sulfate in the soil which is formed through organic matter mineralization and through oxidation of elemental sulfur, S^o.
* SO_2 which penetrates into leaves through stomata and is then metabolized (De Cormis 1968).
* S^o of antifungal treatments applied to foliage: a small part is metabolized into sulfate and sulfur amino acids (Legris-Delaporte *et al.* 1987; Pezet 1987) and the rest settles on the soil.

Up to 20 years ago, the traditional sulfur sources for correction of imbalance were gypsum and single superphosphate (>11% sulfur). Increased cost has led to the evaluation of a cheaper alternative such as the fungicidal preparation Thiovit, a micronized form of S^o in wettable powder form, with potential for both foliar and soil application. Greater opportunities for aerial application render foliar spraying attractive, though in such cases it is difficult to discriminate between fungicidal and nutrient effects (Martin *et al.* 1987).

First, in this review, the role of S^o in fertilization will be considered followed by its use as a fungicide and an acaricide.

Sulfur Nutrition and Assimilation in Higher Plants
edited by L.J. De Kok et al., pp. 193–206

Oxidation of elemental sulfur in the soil

S^o has been widely promoted as a fertilizer source but its effectiveness has been somewhat inconsistent, in part because S^o oxidation to plant-available sulfate is influenced by a number of fertilizer and soil variables. It is known that the rate of S^o uptake from the soil is slow because it first has to be converted to ionic oxysulfur forms by microbiological processes. This slow rate of uptake coupled with the low water solubility gives S^o a persistence and a slow-release nutrient character that may be of particular value.

The factors that affect the rate of S^o microbiological oxidation are numerous, soil temperature, soil moisture, the S^o-oxidizing biota present and the exposed surface area of S^o available for microbial attack for example.

So-oxidizers microorganisms

It is often assumed that autotrophic bacteria of the genus *Thiobacillus* are the most important organisms involved in sulfur oxidation in agricultural soils, but a wide range of heterotrophic microorganisms can also oxidize sulfur (Vitolins & Swaby 1969) amongst them bacteria (for example *Pseudomonas* species, Wainwright 1984) and fungi: *Penicillium* species (Wainwright 1978), *Fusarium solani* (Wainwright & Killham 1980), *Aureobasidium pullulans* (Killham *et al.* 1981). Microorganisms oxidize S^o to thiosulfate and sulfate, often with tetrathionate as an intermediate or by-product. It is likely that sulfur-oxidizing microorganism communities live in soils in mutualistic associations actively colonizing S^o and that the process involves a succession of microbial groups (Wainwright 1984; Lawrence & Germida 1991). Detection of rhodanese activity in soils, an enzyme which catalyzes the intermediate reaction $S_2O_3^{2-} \rightarrow SO_3^{2-}$ suggests that it has a role in S^o oxidation (Deng & Dick 1990; Dick & Deng 1991).

The size and activity of the microbial biomass determine the rate of S^o oxidation in agricultural soils (Lawrence & Germida 1988). Autotrophic thiosulfate-oxidizing communities were shown to increase in response to S^o additions and to decrease in response to SO_4^{2-} additions (Lluch & Olivares 1979; Hebert 1987; Lawrence & Germida 1991). The indigenous population generally suffices for S^o oxidation but an inoculation of Australian soils with *Thiobacilli* was reported to increase the rate of S^o oxidation (Wainwright 1984).

Forms of fertilizer sulfur

S^o is obviously the most concentrated sulfur carrier. It may be used in the solid form (screened S^o) or as various modifications: S^o/bentonite pills, S^o/anhydrite granules, miscellaneous S^o (popcorn, Thiovit). Normal and concentrated superphosphates can be fortified with S^o (18 to 35% S^o). Biosuper, produced by mixing apatite rock phosphate with S^o and *Thiobacilli* bacteria, was reported to be an effective fertilizer (Pathiratna *et al.* 1989). It is possible to coat granules of urea with amorphous and crystalline allotropes of S^o to provide a mantle which is impermeable to water (Jarrell & Boersma 1980). Sulfur-coated urea (SCU) appeared to be agronomically and economically feasible for use as a slow release nitrogen source (Gascho & Snyder

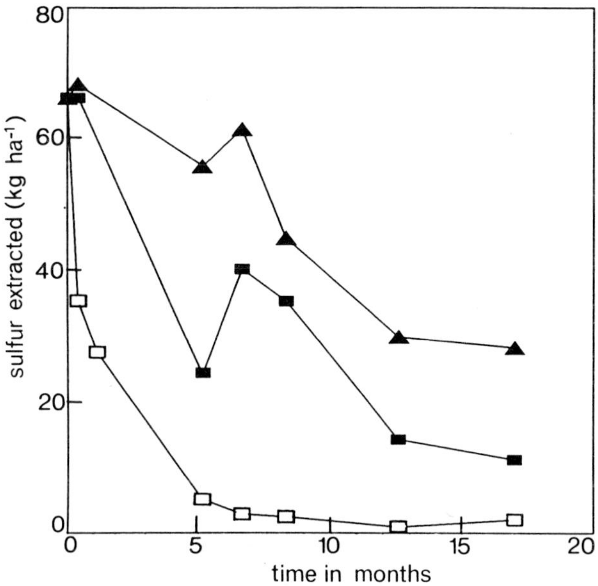

Fig. 1. Effect of particle size on recovery of extractable S° through two growing seasons after S° fertilization. From Barrow (1970) with permission of author and CSIRO Publications. (□), < 200 mesh; (■), 60-100 mesh; (▲), 20-40 mesh.

1976; Waddington & Turner 1980), moreover, less nitrogen leaching resulted from application of SCU (El Wali *et al.* 1980).

As S° has a very low water solubility, the rate of oxidation is dependent on the degree of sulfur particle dispersion within the soil (Fig. 1 from Barrow 1970; Bryant *et al.* 1987; McCaskill & Blair 1989; Janzen 1990). Particle sizes of 80-1000 mesh or smaller are usually required if S° is to be immediately effective. The microcrystalline structure and the molecular type of S° may be important factors to bacteria availability.

Soil pH

Soil pH is an important parameter in relation to S° oxidation; heterotrophs are the primary sulfur oxidizers between pH 7.5 and 6 (Vitolins & Swaby 1969; Lawrence & Germida 1988; Dick & Deng 1991). The acidity produced may create an unfavorable microenvironment for them but, conversely, *Thiobacilli* species and organisms with rhodanese activity become dominant below pH 5. The classical reduction in soil pH following S° oxidation is shown in Fig. 2 (from Wainwright 1984). The pH decrease is proportional to the amount of applied S°. It is sometimes necessary to maintain neutrality with liming: neutralization of 1 kg S° needs 3 kg $CaCO_3$ (Hebert 1987). Furthermore, sulfur may be used to decrease the alkalinity of the soil: treatment of calcareous soil having pH 8.7 with 1 ton S° ha^{-1} decreases the pH of the soil to a depth of 20-40 cm (1 to 2 pH units). A low percentage of seed germination and delay in emergence of seedling appeared clearly in calcareous soils and addition of

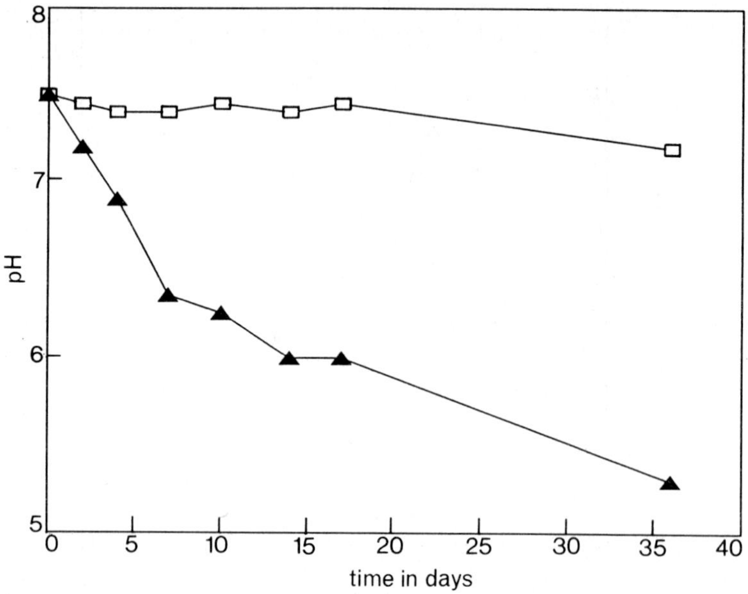

Fig. 2. Effect of S° oxidation on the pH of a fertile loam soil with or without added S°. From Wainwright (1984) with permission of author and Kluwer Academic Publishers. (▲), with S°; (□), without S°.

sulfur increased both the percentage and capacity of cotton seed germination (Abou-Khadrah 1987). In Pakistan, application of S° to calcareous soils helps to solubilize insoluble Ca and to make it available for the reclamation process (Rahmatullah & Salim 1987).

Soil temperature

Soil temperature is a critical factor in microbiological oxidation of S°. Oxidation rate (measured as acidity) is slow at 5 °C at pH 7 but changes by four orders of magnitude at 25 °C (Bryant *et al.* 1987). Rhodanese activity is less affected by temperature (Dick & Deng 1991). The rapid response of oxidation rate to the change in temperature from 5 to 25 °C (Fig. 3 from Bryant *et al.* 1987) indicates that seasonal temperature changes are likely to cause equally rapid changes in the rate of soil assimilation of S° to nutrient form. If S° is present in the soil and colonized with *Thiobacilli* even at lower temperatures, no delay occurs in microbiological oxidation when temperature increases. Thus, beneficial effects from S° application may be noted in field crops over long periods.

Soil moisture and aeration

Optimal soil moisture is about field water holding capacity. A decrease in oxidation rate was observed at lower or higher moisture tension (Kittams & Attoe 1965; Hebert 1987).

Since the autotrophic microbiological oxidizers are essentially aerobes, their ability

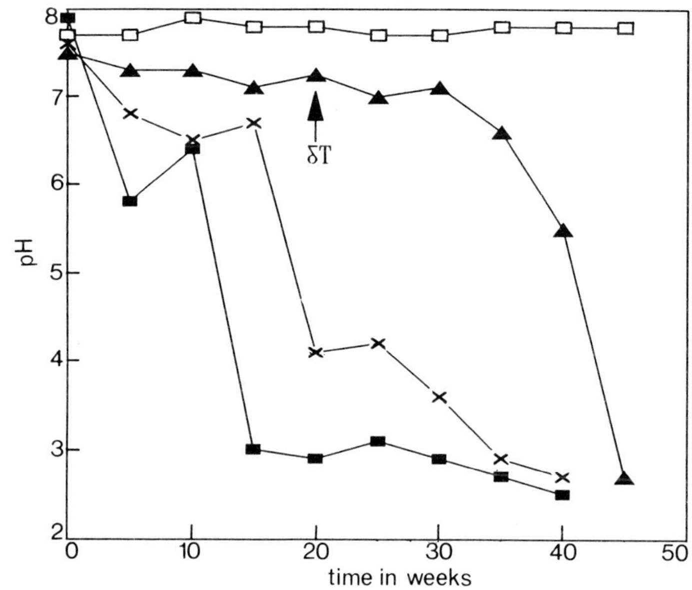

Fig. 3. Effect of temperature on the microbiological oxidation of S° in a soil environment. From Bryant *et al.* (1987) with permission of authors and Syndicat Français du Soufre publisher. (□), control soil; (▲), 5 °C then 25 °C; (✕), 18 °C; (■), 25 °C.

to convert S° into plant oxysulfur nutrient ions is sensitive to O_2 availability and is controlled by the rate of air diffusion from the surface, thus making the process depth dependent (Bryant *et al.* 1987). Oxidation is severely retarded by water saturation in the soil matrix which limits ingress of air. If a rapid assimilation of S° amendments is desired, the particulates should be applied near the surface. Deep ploughing of S° amendments, especially in cooler soils, will provide a long lasting source of nutrient due to lower microbiological oxidation rates. Furthermore, application of S° to potentially waterlogged soils should be avoided (Bryant *et al.* 1987).

So relation with other elements

It has been shown that NaCl (9%) reduced the rate of S° oxidation in soil and completely inhibited the process when added at a concentration of 11% (Wainwright 1984). Blair (1987) reported that sulfate production increased by up to 57% when phosphorus was added to a range of soils. A limitation of essential nutrients will reduce *Thiobacilli* growth and hence oxidation rates. Some fungicides inhibit S° oxidation in soil (Ray 1991); the effect is lower for insecticides such as lindane and benomyl and for 2-4 D herbicide (Wainwright 1984; Hebert 1987).

S°-dependent soil acidification may be useful for mobilizing soil nutrient reserves (Kittams & Attoe 1965; Schnug 1987; Mahler & Maples 1987). Solubilization of Mn and Fe is then ameliorated. The better Zn-supply of the plants grown on S° fertilized soils is due to an increased activity of the rhizoflora. Finally, SO_4^{2-}, derived from S° oxidation, can decrease anion uptake by the plants due to antagonistic effects, as is

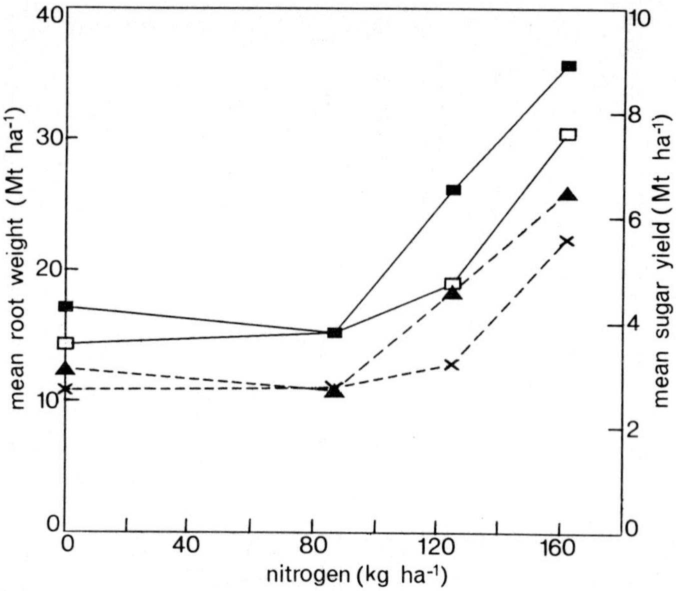

Fig. 4. Effect of sulfur and nitrogen fertilization on root weight (——) and sugar yield (– – –) of sugar beet. From Martin *et al.* (1987) with permission of authors and Syndicat Français du Soufre publisher. (□, ×), nil S°; (■, ▲), 10 kg S° ha^{-1}.

the case with arsenic, antinomy or bromine. The use of sulfur-containing fertilizers on soils contaminated with these toxic elements has been suggested to reduce their contents in plants (Schnug & Schnier 1986; Schnug 1987).

Elemental sulfur as a fertilizer

Crop response to S°-containing fertilizers has been largely studied. When applied in a finely divided form, the effectiveness of S° fertilizers in the year of application approaches that of sulfate. Because of their slow release characteristics, S° fertilizers exhibit significant residual effects on crops grown after the year of initial application when sulfate has to be applied each year (Karamanos & Janzen 1991).

Fertilization with S° appears to increase the yield of cereals particularly if the plants have sulfur contents lower than the critical values (Schnug 1987). The response to sulfur fertilizer is expressed by an increase in the number of ears m^{-2} or in the number of grains ear^{-1} (Recalde Manrique & Gomez Ortega 1979; Kettlewell *et al.* 1987; Thevenet & Taureau 1987). Scott *et al.* (1984) suggested that S° decreased spikelet mortality. On the other hand, Garcia del Moral & Ramos (1987) observed, on a culture of spring barley treated with S° during tillering, that the number of tillers per plant continued to increase until anthesis and that tiller survival was significantly higher than that of the control. These authors suggested that S° could increase the endogenous level of methionine, precursor to ethylene biosynthesis. Ethylene, as an inhibitor of auxin synthesis, would reduce apical dominance and could lead to a higher number of tillers. The application of sulfur on rice had a positive effect on grain yield, in increasing number of panicles per hill, number of grains per panicle, grain weight and reducing the percentage of empty grains (Ismunadji 1987).

In the case of oilseed rape, the application of S° was, as expected, more effective than on cereals. The seed yield increased significantly to about 7% (Schnug 1987). Joshi *et al.* (1991) observed that sulfur application markedly increased the oil content compared with no sulfur (34.69% with 75 kg S ha^{-1}). In the case of sugar beet, S° promoted a rise in yield of tops, sugar content of roots and crude protein yield per unit of sown area (Khomenko & Sinel'nik 1982; Martin *et al.* 1987). In particular, under high nitrogen dressings, insufficient sulfur may impose serious yield constraints (Fig. 4).

Sulfur influences also crop quality. It has been shown that a restricted sulfur supply seriously affected wheat grain quality: the grain was harder (higher pearling resistance), the dough had a greater resistance to extension and a lower extensibility, the panification index was significantly decreased (Moss *et al.* 1981; Camblin 1987). Wheat flour with low cysteine content might have poor disulfide bonding capacity and hence be unsuitable for breadmaking (Byers 1985). Sulfur starvation of barley resulted in a decrease in total hordein, in levels of lysine and threonine in hordein and then affected malting (Shewry *et al.* 1985). It would seem that more sulfur is needed for adequate grain quality than for a yield response (Byers 1985).

The economical reality of oilseed rape production depends on the level of yields as well as on the quality of crops, *i.e.* content of glucosinolates in seeds. A positive correlation of the glucosinolate contents of the seed with the raw-protein contents of rape was observed by Werteker (1991). These results indicate an influence of the nitrogen nutrition on the formation of glucosinolates and suggest that agricultural measures as well as environmental influences may be responsible for the increase of the protein level as well as the increase of unwished metabolites. A fertilization with 100 kg S ha^{-1} can increase the glucosinolate content in seeds by about 5-6 μmol g^{-1} defatted meal, but, in some cases, a slower action of S° in comparison with sulfate may lead to a lesser increase in the glucosinolate content of oilseed rape (Schnug 1987).

Sulfur is a nutrient essential for the maintenance of animal feed quality. Sulfur deficiency can alter the degradability of grass by either changing plant composition or by reducing sulfur supply to the rumen microbes (Millard *et al.* 1987). Experiments with sheep have shown that an intake of 1.95 g S day^{-1} is necessary to support maximal rumen microbial protein production (Hume & Bird 1970). Feeding sulfur deficient grass to ruminants decreases their liveweight gains (Jones *et al.* 1982).

Physiological effects of elemental sulfur on higher plants

Effects of S° on protein concentration in higher plants indicate that it penetrates into the plants and is involved with cellular metabolism.

Incorporation of S° into leaf tissues was followed by treating leaves with crystalline radioactive sulfur. A little part of this sulfur may cross the cuticular boundary and/or the stomata: 1 to 3% of the applied sulfur according to the different authors (Dreze *et al.* 1987; Legendre & Marty 1987; Legris-Delaporte *et al.* 1987; Mc Grath & Till 1987). It seems very important to enclose the treated leaves in plastic bags to

Table 1. Production of S-compounds (nmol ml^{-1}) from S^o (initial concentration: 65 μg ml^{-1}) by wheat chloroplasts (0.1 g chlorophyll l^{-1}) maintained for 2 h under different light and atmosphere conditions.

	Oxygen		Nitrogen	
	light	dark	light	dark
SO_4^{2-}	29.8	20.0	21.5	19.0
SO_3^{2-}	0.5	0	2.5	0.5
$S_2O_3^{2-}$	1.0	0.4	11.0	1.2

limit S^o losses by sublimation, to avoid subsequently S^o washing off the leaves onto the soil and to amplify the phenomenon related to foliar S^o uptake. Mc Grath & Till (1987) have observed that sulfur can enter the plant through the roots from the large proportion of spray application that falls onto the soil. The subcellular localization of exogenous S^o was studied by X-ray microanalysis and high resolution autoradiography. S^o was found in several subcellular compartments of the photosynthetic tissue: vacuoles, plastids, cytoplasm, mitochondria. Translocation of the radioactive tracer from the treated leaf to the other leaves in the same plant was demonstrated in the case of rape (Legendre & Marty 1987) and wheat (Legris-Delaporte *et al.* 1987). In winter wheat, radioactive sulfur compounds were translocated into ears (Dreze *et al.* 1987). In *Chlorella vulgaris*, an autotrophic green alga, S^o was metabolized as show by its entry into sulfur amino acids (cysteine, homocysteine, methionine) and incorporation into proteins and lipids (Pezet & Zuccaroni 1987). The biological reduction of S^o would seem to be connected to the mitochondria only. In the case of wheat, labeled sulfur was absorbed and metabolized into sulfate, cysteine, methionine and glutathione (Legris-Delaporte *et al.* 1987). The high intracellular sulfate content, the close correlation between the excess of oxygen uptake and oxygen needs for sulfur oxidation and the absence of H_2S released by treated leaves supported the idea of direct and fast oxidation of sulfur into sulfate. In fact, we have observed that S^o was metabolized by intact but not by boiled wheat chloroplasts (Jolivet & Kien 1992b). Although weak (transformation of 7% S^o), the process was significant and included oxidation: after two hours 10 to 15% of the metabolized S^o was transformed into sulfate (Table 1). When ^{35}S was added to S^o, $S_2O_3^{2-}$ and SO_3^{2-} (separated by HPLC and then analyzed by fluorimetry and continuous radioactivity detection) and SO_4^{2-} (precipitated by $BaCl_2$) were labeled. The effect of light was not significant on S^o metabolism by chloroplasts and it was verified that H_2O_2 could not oxidize S^o spontaneously. The operation of chloroplastic enzymatic mechanisms is assumed as a working hypothesis (Jolivet & Kien 1992a) and their presence sought.

Elemental sulfur as a fungicide

The fungicidal activity of S^o has been known for several centuries although the precise nature of that activity was not understood until recently. The aim of a good fungicide is to maintain the parasite population as low as possible and to prevent any modification of its structure. Using modern application systems and special formulations (dusts or wettable powders), the fungicide distribution is improved.

Mode of action of S^o as a fungicide

Mc Callan & Wilcoxon (1931) and Miller *et al.* (1953) suggested that the toxicity of S^o in *Fusaria* species is due to transfer of hydrogen atoms to S^o, instead of O_2, in the terminal part of the mitochondrial electron transport chain and to the toxicity of the hydrogen sulfide produced. In *Monilia fructicola*, Tweedy & Turner (1966) suspected an interference of S^o with the mitochondrial electron transfer at the level of cytochrome b and c. Beffa *et al.* (1987) have shown in *Phomopsis viticola* that exogenous S^o can quickly and non-enzymatically oxidize sulfhydryl groups implicated in many respiratory functions of mitochondria and is able also to reduce itself at the level of cytochrome c. These phenomena could produce a modification of the oxidation state of the respiratory complexes, disturb the electron flux in the mitochondrial respiratory chain and consequently alter the oxidative phosphorylation.

Uses of S^o as a fungicide

Sulfur as a fungicide for the protection of French vineyards against powdery mildew (*Uncinula necator*) was used for the first time in the middle of the last century and still insures, after more than one hundred years, the protection of most of vineyards against several parasites. It was observed that S^o fungicidal action was better when S^o was in the vicinity of mildew (Bessis 1987). Therefore the quality of treatment is important to obtain the best efficiency.

A fungicidal formulation containing 70% S^o has been compared with other anti-botrytis fungicides. Only S^o gave no marks on grapes and was recommended for the treatment of dessert grapes (Bourdier & Agulhon 1987). Late S^o spraying did not interfere with alcoholic fermentation and did not cause any change among the main wine analytical parameters. H_2S was only detected in non-settled musts of white wine (Barillère *et al.* 1987). However, it is often desirable that fungicidal treatments are stopped 3 or 4 weeks before grape gathering.

After the second world war, an extension of fruit growing in France favored the development of various diseases, among them, powdery mildew (*Podosphaera* and *Sphaerotheca* species). Sulfur products (dusting or wettable S^o) were then used and are still widely used in all French fruit production areas. S^o, which is not dangerous for the environment and soil microorganisms, can play a primary role in an integrated control in fruit tree crops, combining ecological, biological and chemical ways of protection (Manguin 1987).

In the case of cereal crops, sulfur is a less effective fungicide than modern systemic materials which are efficient at a much lower ratio of active ingredient. However, S^o is active against mildew and other fungal diseases when applications are frequent and the epidemic is not severe (Cook 1987; Falisse & Bodson 1987). S^o can be used alone, associated with a dithiocarbamate or in combination with other more sophisticated chemicals avoiding coating of active material by sulfur. Adding S^o to the most efficient fungicides often leads to an improvement of the treatments effectiveness and an excellent efficacy/cost ratio (Falisse & Bodson 1987). Furthermore, in the light of resistance to many mildew fungicides, a possible use of S^o might be in mixture with existing products in order to reduce the likelihood of resistance: the juxtaposition of fungicides having different sites of action limits the risks of breeding resistant strains (Falisse & Bodson 1987).

Elemental sulfur as an acaricide and an insecticide

S° was essentially the only acaricide used to control mites until about 1920 because of its many advantages: low toxicity to the applicator or crop consumers, low cost, availability, formulation with other pesticides, and relative specificity. Disadvantages include the fact that it is phytotoxic when applied during hot weather and ineffective during cool weather.

The mechanism of action of sulfur in mites and insects remains obscure. Hydrogen sulfide, known to develop from sulfur applied to leaves, was thought by some workers to be the factor responsible for the toxic action of sulfur. On the contrary, Hanks *et al.* (1992) observed with the ectoparasitic mite *Pyemotes tritici* that sulfur may physically impede the dispersal of immature mites by adhering to the cuticle, but that sulfur vapor did not act as a toxin.

Very few studies have compared the efficiency of different sulfur formulations as acaricides and especially whether differential toxicity might occur between pest mites and beneficial mite species. Spider mites and many pest insects are notorious for being able to develop resistances to nearly all chemical insecticides or acaricides applied for their control. However, there are no documented instances in which spider mites and eriophyid mites, which comprise most of the species feeding on higher plants, have acquired resistance to S°. Various laboratory and field trials have reported that S° was not toxic to phytoseiid mites which are effective biological control agents of spider mites and eriophyid (Fauvel *et al.* 1987; Strapazzon *et al.* 1987). It was suggested that long term exposure in the field resulted in the evolution of resistance to S° in the population of *Metaseiulus occidentalis* (Phytoseiidae), a beneficial species (Hoy 1987).

It has been noted that pest control programs would not be based on hypothetical stimulatory or repressive effects of fungicides on spider mites but would account for the differential toxicity of pesticides on phytophagous and predatory mites, particularly on Phytoseiides. In this view, the use of wettable S° is compatible with the programs of Tetranychidae biological control (Duso *et al.* 1987).

Other sulfur applications

Russeting

Russeting of apples and pears results from the secondary development of a phelloderm instead of epiderm. It is a normal anatomical phenomenon for grey varieties but it is considered as a defect in certain varieties of smooth and bright apples like Golden delicious which represents 2/3 of French fruit production. Russeting leads to a reduction in grade of fruit and some considerable economic consequence.

The factors inducing russeting are numerous and interactive. The factors involved most consistently are the genetic characters of varieties (a precocious development of a phelloderm in grey varieties or a particular epidermal fragility as in Golden delicious), nutritional effects (asphyxia, fruit position on the tree), climatic components (low temperature, high relative humidity) and external factors like treatments (cupric fungicides), diseases and mechanical injuries.

Though current research is directed towards growth substances and gibberellin use, S° remains the chief active ingredient recommended against russeting in association with micronutrients. S° has an action against microorganisms present on the fruit surface and could act via the synthesis of growth substances on epiderm development and cuticle formation (Bondoux 1987).

Pigment contents

For many years, farmers know the beneficial action of S° on the green color of leaves. In fact, foliar applications of exogenous S° have trophic effects. They are able to compensate for a lack of sulfate in hydroponic culture media for plants and they increase the chlorophyll content of non-deficient plants (Pezet 1987).

The use of micronized wettable S° is old in Japan and leads to increased citrus fruit quality through natural pigmentation and preservation resistance. S° improves fruit conservation by decreasing mold development on fruits and emphasizes natural coloring. Fruit pigmentation is controlled by ethylene (Gall 1986). The effect of S° is perhaps explained by its role on methionine synthesis, an amino acid implicated in ethylene biosynthesis.

Concluding remarks

This survey of different S° uses shows that much is still unknown about the involvement of this element into the cellular metabolism of higher plants and fungi. An understanding of the mechanism of toxicity of S° to insects and mites is also lacking and must be developed using modern techniques. However, custom led to the use of S° very widely and it appeared that the advantages of S° were numerous. In an attempt to improve the availability of S° to crops, sulfur-oxidizing microorganisms isolated and cultured in the laboratory have been added to soils. It was shown that the addition of these microorganisms increased both sulfate formation and the yield of crops (Buffat *et al.* 1987). This approach could be very interesting, particularly the isolation of high performance sulfur-oxidizing microorganisms able to transform S° applied as a fungicide or that originating from atmospheric pollution.

References

Abou-Khadrah, S. 1987. Application of sulphur for growing cotton in calcareous soil. In: Syndicat Français du Soufre (ed.), Proc. Int. Symp. Elemental Sulphur in Agriculture, Nice, France, pp. 307-311.

Barillère, J.M., Benard, P., Baraillа, R. & Vernet, S. 1987. Influence of late sulphur treatments on white wine flavours. In: Syndicat Français du Soufre (ed.), Proc. Int. Symp. Elemental Sulphur in Agriculture, Nice, France, pp. 205-211.

Barrow, N.J. 1970. Slowly available sulphur fertilizers in south-western Australia. Aust. J. Exp. Agric. Animal Husbandry 11: 211-216.

Beffa, T., Pezet, R. & Turian, G. 1987. Multiple-site inhibition by colloidal elemental sulfur (S°) of respiration by mitochondria from young dormant spores of *Phomopsis viticola*. Physiol. Plant. 69: 443-450.

Bessis, R. 1987. Sulfur effect upon powdery mildew of grapevine. Interest of early treatments based on the biology of the fungus and its mode of overwintering. In: Syndicat Français du Soufre (ed.), Proc. Int. Symp. Elemental Sulphur in Agriculture, Nice, France, pp. 71-81.

Blair, G.J. 1987. Sulfur supply to crops and pastures from elemental S. In: Syndicat Français du Soufre (ed.), Proc. Int. Symp. Elemental Sulphur in Agriculture, Nice, France, pp. 483-491.
Bondoux, P. 1987. Russeting. In: Syndicat Français du Soufre (ed.), Proc. Int. Symp. Elemental Sulphur in Agriculture, Nice, France, pp. 243-252.
Bourdier, L. & Agulhon, R. 1987. Sulphur in French Vineyard protection. Action polyvalency and safety. In: Syndicat Français du Soufre (ed.), Proc. Int. Symp. Elemental Sulphur in Agriculture, Nice, France, pp. 83-89.
Bryant, R.D., Laishley, E.J. & Hyne, J.B. 1987. The microbiological oxidation of elemental sulphur applied as a soil nutrient. In: Syndicat Français du Soufre (ed.), Proc. Int. Symp. Elemental Sulphur in Agriculture, Nice, France, pp. 709-716.
Buffat, P., Jadart, J.C., Paul, X., Ballester, J.M. & Ranchin, V. 1987. Interest of an association between sulfo-oxydant organisms and elemental sulphur for sulphur fertilization on crops. In: Syndicat Français du Soufre (ed.), Proc. Int. Symp. Elemental Sulphur in Agriculture, Nice, France, pp. 717-724.
Byers, M. 1985. The relationship between nitrogen and sulphur nutrition of wheat, grain protein composition and baking quality. J. Sci. Food Agric. 36: 268.
Camblin, P. 1987. Cultural techniques and cereal quality. In: Syndicat Français du Soufre (ed.), Proc. Int. Symp. Elemental Sulphur in Agriculture, Nice, France, pp. 459-467.
Cook, R.J. 1987. Sulphur as a cereal fungicide. In: Syndicat Français du Soufre (ed.), Proc. Int. Symp. Elemental Sulphur in Agriculture, Nice, France, pp. 417-425.
De Cormis, L. 1968. Contribution à l'étude de l'absorption du soufre par les plantes soumises à une atmosphère contenant du dioxyde de soufre. Annales Agronomiques 10: 99-112.
Deng, S. & Dick, R.P. 1990. Sulfur oxidation and rhodanese activity in soils. Soil Sci. 150: 552-560.
Dick, R.P. & Deng, S. 1991. Multivariate factor analysis of sulfur oxidation and rhodanese activity in soils. Biogeochem. 12: 87-101.
Dreze, P., Bodson, B., Decocq, G., Falisse, A. & Gasia, M.C. 1987. Incorporation of sulfur after treatment of winter wheat with elemental formulated sulfur. In: Syndicat Français du Soufre (ed.), Proc. Int. Symp. Elemental Sulphur in Agriculture, Nice, France, pp. 395-409.
Duso, C., Mazzaro, S. & Girolami, V. 1987. The influence of sulphur and dinocap on *Eotetranychus carpini* Oud. (acarina tetranychidae) and *Kampimodromus aberrans* Oud. (acarina phytoseiidae) in viticulture. In: Syndicat Français du Soufre (ed.), Proc. Int. Symp. Elemental Sulphur in Agriculture, Nice, France, pp. 145-152.
El Wali, A.M.O., Le Grand, F. & Gascho, G.J. 1980. Nitrogen leaching from soil and uptake by sugarcane from various urea-based fertilizers. Soil Sci. Soc. Am. J. 44: 119-122.
Falisse, A. & Bodson, B. 1987. Use of elemental sulphur in cereal crops. In: Syndicat Français du Soufre (ed.), Proc. Int. Symp. Elemental Sulphur in Agriculture, Nice, France, pp. 427-432.
Fauvel, G., Cotton, D., Audemard, H. & Fremond, J.C. 1987. Influence of sulphur wetting powder upon the evolution of phytophagous and beneficial mites in an apple orchard of southeastern France. In: Syndicat Français du Soufre (ed.), Proc. Int. Symp. Elemental Sulphur in Agriculture, Nice, France, pp. 233-242.
Gall, M. 1986. Utilisation du soufre élémentaire en agrumiculture au Japon. La Défense des Végétaux 239: 35-39.
Garcia Del Moral, L.F. & Ramos, J.M. 1987. Effects of foliar sulphur and ethrel on evolution and survival of tillers in spring barley. In: Syndicat Français du Soufre (ed.), Proc. Int. Symp. Elemental Sulphur in Agriculture, Nice, France, pp. 545-552.
Gascho, G.J. & Snyder, G.H. 1976. Sulfur-coated fertilizers for sugarcane: I. Plant response to sulfur-coated urea. Soil Sci. Soc. Am. J. 40: 119-122.
Hanks, L.M., McCelfresh, J.S., Millar, J.G. & Paine, T.D. 1992. Control of the straw itch mite (acari: *Pyemotidae*) with sulfur in an insect rearing facility. J. Econ. Entomol. 85: 683-686.
Hebert, J. 1987. Elemental sulphur in fertilization. In: Syndicat Français du Soufre (ed.), Proc. Int. Symp. Elemental Sulphur in Agriculture, Nice, France, pp. 17-29.
Hoy, M.A. 1987. Resistance to sulfur in the spider mite predator: *Metaseiulus occidentalis*. In: Syndicat Français du Soufre (ed.), Proc. Int. Symp. Elemental Sulphur in Agriculture, Nice, France, pp. 131-144.
Hume, I.D. & Bird, P.R. 1970. Synthesis of microbial protein in the rumen. IV. The influence of the level and form of dietary sulphur. Aust. J. Agric. Res. 21: 315-322.
Ismunadji, M. 1987. Sulphur and rice production. In: Syndicat Français du Soufre (ed.), Proc. Int. Symp. Elemental Sulphur in Agriculture, Nice, France, pp. 313-327.

Janzen, H.H. 1990. Elemental sulfur oxidation as influenced by plant growth and degree of dispersion within soil. Can. J. Soil Sci. 70: 499-502.
Jarrell, W.M. & Boersma, L. 1980. Release of urea by granules of sulfur-coated urea. Soil Sci. Soc. Am. J. 44: 418-422.
Jolivet, P. & Kien, P. 1992a. Sulphite metabolization by wheat chloroplasts. C. R. Acad. Sci. Paris 314: 179-184.
Jolivet, P. & Kien, P. 1992b. Elemental sulphur metabolization by wheat chloroplasts. C. R. Acad. Sci. Paris 314: 231-236.
Jones, M.B., Rendig, V.V., Torell, D.T. & Inouye, T.S. 1982. Forage quality for sheep and chemical composition associated with sulfur fertilization on a sulfur deficient site. Agronomy J. 74: 775-780.
Joshi, A.J., Ahlawat, R.P.S. & Thivedi, S.J. 1991. Effect of nitrogen and sulphur fertilization on growth and yield of mustard (*Brassica juncea*). Indian J. Agron. 36: 606-607.
Karamanos, R.E. & Janzen, H.H. 1991. Crop response to elemental sulfur fertilizers in central Alberta. Can. J. Soil Sci. 71: 213-225.
Kettlewell, P.S., Griffiths, M.W., Smith, J.J. & Hocking, T.J. 1987. The effects of late season foliar sulphur and nitrogen on yield and grain quality of winter wheat. In: Syndicat Français du Soufre (ed.), Proc. Int. Symp. Elemental Sulphur in Agriculture, Nice, France, pp. 451-457.
Khomenko, A.D. & Sinel'nik, A.D. 1982. Sulphur effects on yield and feeding value of sugar beets. Fiziol. Biokhim. Kul't Rast. 14: 362-365.
Killham, K, Lindley, N.D. & Wainwright, W. 1981. Inorganic sulfur oxidation by *Aureobasidium pullulans*. Appl. Environ. Microbiol. 42: 629-631.
Kittams, H.A. & Attoe, O.J. 1965. Availability of phosphorus in rock phosphate-sulfur fusions. Agronomy J. 57: 331-334.
Lawrence, J.R. & Germida, J.J. 1988. Relationship between microbial biomass and elemental sulfur oxidation in agricultural soils. Soil Sci. Soc. Am. J. 52: 672-677.
Lawrence, J.R. & Germida, J.J. 1991. Microbial and chemical characteristics of elemental sulfur beads in agricultural soils. Soil Biol. Biochem. 23: 617-622.
Legendre, S. & Marty, F. 1987. Incorporation and cellular localization of elemental sulphur after foliar spray on oil seed rape. In: Syndicat Français du Soufre (ed), Proc. Int. Symp. Elemental Sulphur in Agriculture, Nice, France, pp. 689-700.
Legris-Delaporte, S., Ferron, F., Landry, J. & Costes, C. 1987. Metabolization of elemental sulfur in wheat leaves consecutive to its foliar application. Plant Physiol. 85: 1026-1030.
Lluch, C. & Olivares, J. 1979. Effect of elemental sulphur and sulfate on soil microflora. Agrochimica 23: 273-278.
Mahler, R.J. & Maples, R.L. 1987. Effect of sulfur additions on soil and the nutrition of wheat. Commun. in Soil Sci. Plant Anal. 18: 653-673.
Manguin, J.P. 1987. Integrated fruit tree production: role of sulphur for fruit valorization. In: Syndicat Français du Soufre (ed.), Proc. Int. Symp. Elemental Sulphur in Agriculture, Nice, France, pp. 271-274.
Martin, A.D.E., Hailes, N.S.J. & Reynolds, S.B. 1987. Effects of foliar applied micronised elemental sulphur upon the growth and yield of sugar beet (*Beta vulgaris* cv. Primahill). In: Syndicat Français du Soufre (ed.), Proc. Int. Symp. Elemental Sulphur in Agriculture, Nice, France, pp. 639-647.
Mc Callan, S.E.A. & Wilcoxon, F. 1931. The fungicidal action of sulphur. II. The production of hydrogen sulphide by sulphured leaves and spores and its toxicity to spores. Contrib. Boyce Thompson Inst. 3: 13-38.
McCaskill, M.R. & Blair, G.J. 1989. A model for the release of sulfur from elemental S and superphosphate. Fertilizer Res. 19: 77-84.
Mc Grath, S.P. & Till, A.R. 1987. Uptake of sulphur following foliar applications of elemental sulphur. In: Syndicat Français du Soufre (ed.), Proc. Int. Symp. Elemental Sulphur in Agriculture, Nice, France, pp. 369-370.
Millard, P., Sharp, G.S. & Scott, N.M. 1987. Sulphur as a nutrient for plants and animals. In: Syndicat Français du Soufre (ed.), Proc. Int. Symp. Elemental Sulphur in Agriculture, Nice, France, pp. 509-515.
Miller, L.P., Mc Callan, S.E.A. & Weed, R.M. 1953. Quantitative studies on the role of hydrogen sulfide formation in the toxic action of sulfur to fungus spores. Contrib. Boyce Thompson Inst. 17: 151-171.
Moss, H.J., Wrigley, C.W., MacRitchie, F. & Randall, P.J. 1981. Sulfur and nitrogen fertilizer effects on wheat. II.Influence on grain quality. Aust. J. Agric. Res. 32: 213-226.

Pathiratna, L.S.S., Waidyanatha, U.P.De S. & Peries, O.S. 1989. The effects of apatite and elemental sulphur mixtures on growth and P content of *Centrocema pubescens*. Fertilizer Res. 21: 37-43.

Pezet, R. 1987. Physiological effects of elemental sulphur on arable crops: quantitative aspects, qualitative aspects. In: Syndicat Français du Soufre (ed.), Proc. Int. Symp. Elemental Sulphur in Agriculture, Nice, France, pp. 39-50.

Pezet, R. & Zuccaroni, P. 1987. Physiological effects of elemental sulphur applied on *Chlorella vulgaris*. In: Syndicat Français du Soufre (ed.), Proc. Int. Symp. Elemental Sulphur in Agriculture, Nice, France, pp. 657-670.

Rahmatullah & Salim, M. 1987. Sulphur responses of oilseed rape on problem soils of arid regions amended with elemental sulphur. In: Syndicat Français du Soufre (ed.), Proc. Int. Symp. Elemental Sulphur in Agriculture, Nice, France, pp. 591-598.

Ray, R.C. 1991. Inhibition of sulphur oxidation in soil by some fungicides. J. Environ. Biol. 12: 279-286.

Recalde Manrique, L. & Gomez Ortega, M. 1979. Investigations on some physiological factors that determine the productive capacity of wheat plants.I.Interaction between applications of nitrogen fertilizers and dusted elemental sulphur at tillering stage. Anales de Edafologia y Agrobiologia 38: 1839-1853.

Roberts, T.M. & Fisher, B.E.A. 1985. Estimated trends in dry and wet sulphur deposition over the UK between 1970 and 1982 and effects on crop growth. J. Sci. Food Agric. 36: 262-263.

Schnug, E. 1987. Response of crop yield and quality to fertilization with elemental sulfur in Northern Germany. In: Syndicat Français du Soufre (ed.), Proc. Int. Symp. Elemental Sulphur in Agriculture, Nice, France, pp. 553-561.

Schnug, E. & Schnier, C. 1986. The effect of sulfur-containing nitrogen fertilizers on the elemental composition of celery (*Apium graveolens*) grown on a polluted marsh soil. Plant Soil 91: 273-278.

Scott, N.M. 1985. Crop response to sulphur in the north of Scotland. J. Sci. Food Agric. 36: 263-264.

Scott, N.M., Dyson, P.W., Ross, J. & Sharp, G.S. 1984. The effect of sulphur on the yield and chemical composition of winter barley. J. Agric. Sci., Camb. 103: 699-702.

Shewry, P.R., Rahman, S., Bunce, N., Franklin, J. & Miflin, B.J. 1985. Control of storage protein synthesis in cereals by sulphur availability and its relationship to grain quality. J. Sci. Food Agric. 36: 264.

Strapazzon, A., Cappello, S. & Girolami, V. 1987. Influence of sulphur on *Amblyseius andersoni* (chant) apple. In: Syndicat Français du Soufre (ed.), Proc. Int. Symp. Elemental Sulphur in Agriculture, Nice, France, pp. 225-232.

Tabatabai, M.A. 1984. Importance of sulphur in crop production. Biogeochem. 1: 45-62.

Thevenet, G. & Taureau, J.C. 1987. Sulfur supply on winter wheat. Effects of sulphur fertilization on fields. Study on diagnosis and prognosis possibilities. In: Syndicat Français du Soufre (ed.), Proc. Int. Symp. Elemental Sulphur in Agriculture, Nice, France, pp. 469-482.

Tweedy, B.G. & Turner, N. 1966. The mechanism of sulfur reduction by conidia of *Monilia fructicola*. Contrib. Boyce Thompson Inst. 23: 255-265.

Vitolins, M.I. & Swaby, R.J. 1969. Activity of sulphur-oxidizing microorganisms in some Australian soils. Aust. J. Soil Res. 7: 171-183.

Waddington, D.V. & Turner, T.R. 1980. Evaluation of sulfur-coated urea on Merion Kentucky bluegrass. Soil Sci. Soc. Am. J. 44: 413-417.

Wainwright, M. 1978. Sulphur-oxidising micro-organisms on vegetation and in soils exposed to atmospheric pollution. Environ. Pollut. 17: 167-174.

Wainwright, M. 1984. Sulfur oxidation in soils. Adv. Agronomy 37: 349-396.

Wainwright, M. & Killham, K. 1980. Sulphur oxidation by *Fusarium solani*. Soil Biol. Biochem. 12: 555-558.

Werteker, M. 1991. Einflüsse auf den Glucosinolatgehalt des Körnerrapses durch die Schwefel- und Stickstoffernährung der Planze. Bodenkultur 42: 243-252.

CONSEQUENCES OF SULFUR DEFICIENCY

Konrad Mengel
Institute of Plant Nutrition of the Justus-Liebig University, Südanlage 6, D-6300 Giessen, F.R.G.

Introduction

The objective of this contribution is evident in its title; which measurements should be taken if sulfur deficiencies occur in crops. The ultimate answer on this question is to fertilize crops with sulfur fertilizers. How this can be done in a most rational way is discussed in the following.

Situation in Europe

Earlier investigations of Pfaff (1963) showed that the sulfur balance of arable crops is characterized by a high leaching rate (130 kg S year^{-1}) and by a relatively low crop demand (23 kg S year^{-1}). About 1/3 of the quantity required for the maintainance of the balance sheet was provided by sulfur in fertilizers and almost as high a quantity by wet deposition. Similar high sulfur leaching rates were reported from Northern France under arable crops (Ballif & Muller 1985). The sulfur deficit resulting from the high leaching losses was largely balanced by the entry of sulfur in the form of wet and dry deposition (Saalbach 1984). In the last years this sulfur source was drastically reduced in Central Europe and hence the hazard of sulfur deficiency in crops increased (Schnug 1991).

Recent, still unpublished field trials with rape, carried out by the Hanninghof Research Station from 1989 until 1991 showed that from the 17 sites tested about half were responsive to sulfur fertilizer application and gave a clear rape seed yield increase which in several cases was higher than 10%. Such an increase is profitable for farmers and hence has consequences for the farmers' fertilizer strategy. It is of utmost practical relevance to predict whether a crop yield is attained or not by sulfur fertilizer application. This is of particular importance for rape since this crop species has a high sulfur demand. Generally a reliable prediction of sulfur deficiency and a precise fertilizer application requires knowledge of the soil, and frequently a plant tissue test. Soil tests and plant tissue tests are not alternative techniques but may complement each other. Leaf analysis as proposed by Schnug (1989) is useful for the identification of sulfur deficiency. A reliable soil test not only may provide information whether sulfur fertilizer should be applied but also a quantification of the sulfur fertilizer rate required. Hence if sulfur fertilizer application becomes a common practice, at least for rape cultivation, there is a need for a reliable soil sulfur test.

Sulfur Nutrition and Assimilation in Higher Plants
edited by L.J. De Kok et al., pp. 207–209

Sulfur in soils

Soils of the temperate, humid climate contain about 0.1 to 15 g S kg dry weight $soil^{-1}$ of which only a small percentage is soluble in dilute acids and mainly consists of sulfate which is directly available to plant roots (Wainwright 1984). A high proportion of soil sulfur is in organic form and in forest soils the organic sulfur may amount to more than 90% of the total soil sulfur (Nodvin *et al.* 1986). Organic soil sulfur is divided into carbon-bound sulfur (reduced sulfur) and sulfate ester sulfur. According to experimental data of Maynard *et al.* (1985) inorganic sulfate added to soils can be easily converted to sulfate esters and this in turn may be quickly hydrolyzed if plants need sulfur. Hence sulfate esters are a potential sulfur reserve. This statement is in line with the fact that many soils contain relatively high concentrations of sulfohydrolases (Martens *et al.* 1992). Also oxidation of elemental and carbon-bound sulfur contributes to the formation of sulfate and thus to a form which can be directly taken up by plant roots. Nevell & Wainwright (1987) reported that the sulfur oxidation may be hampered in dry soils because of lack of water. Sulfate can also be adsorbed to sesquioxides and clay minerals (Turner & Kramer 1991). The adsorption is particularly strong at low soil pH (Novbin *et al.* 1986) and sulfate availability increases with soil pH (Martini & Mutters 1984).

Soil test for sulfur and sulfur fertilizers

In order to elaborate a soil test the most important soil characteristics must be considered. It is proposed to elaborate a sulfur test in field trials with rape which takes into consideration the sulfate and organic sulfur fractions (sulfate esters, carbon-bound sulfate), soil pH and soil texture of the upper soil layer analogous to a model recently developed for available soil nitrogen by Ziegler *et al.* (1992). Ngugen & Goh (1992) found that adsorbed sulfur and sulfur extracted by a $CaCl_2$ solution are a reliable indicator for sulfur mineralization.

The efficiency of sulfur fertilizers depends much on the capability of plant roots to exploit the soil sulfate. Still unpublished field trials of the Hanninghof research group found that after the harvest of rape high amounts of sulfate may be present in the rooting depth of a soil. The reason for these high sulfur amounts still needs further investigations. It is feasible that the soil sulfate results from rape leaves dropped during the process of maturation.

Sulfur fertilizer may be applied as elemental sulfur or as sulfate. The oxidation of elemental sulfur in the soil is associated with a soil acidification which on many soils is not desirable and may even favour sulfate fixation of soil colloids (Martini & Mutters 1984). Sulfate is present in various standard fertilizers. Depending on the nutrient status of soils and the fertilizer strategy of the farmer these fertilizers may well be used to supply the crop with sulfate. Gypsum is known as an appropiate sulfur fertilizer which supplies the soil also with Ca^{2+}, a plant nutrient may improve plant growth (Yan & Mengel 1992) and soil structure (Shainberg *et al.* 1989). In acid soils sulfate may complex aluminum cation species and thus reduce aluminum toxicity (Alva *et al.* 1991).

Concluding remarks

Total sulfur content in most soils is high as compared with the sulfur demand of crops. Most of soil sulfur, however, is not directly available to plants. Hence crops with a high sulfur demand, such as rape, may suffer from insufficient sulfur supply. In such cases application of sulfur fertilizers is recommended. For prognozing the rate of sulfur fertilizer application a reliable soil test for available soil sulfur is needed.

References

Alva, C., Kerven, G.L., Edwards, D.G. & Asher, C.J. 1991. Reduction in toxic aluminum to plants by sulfate complexation. Soil Sci. 152: 351-359.

Ballif, J.L. & Muller, J.C. 1985. Contibution à l'ètude de la dynamique des sulfates en sol de craie; dix années des mesures en lysimètre. C.R. Agric France 71: 1385-1397.

Martens, D.A., Johanson, J.B. & Frankenberger, W.T. 1992. Production and persistence of soil enzymes with repeated addition of organic residues. Soil Sci. 153: 53-61.

Martini, J.A. & Mutters, R.G. 1984. Effect of liming and fertilization on sulfur availability, mobility, and uptake in cultivated soils of South Carolina. Soil Sci. 138: 403-410.

Maynard, D.G., Steward, J.W.B. & Bettany, J.R. 1985. The effect of plants on soil transformation. Soil Biol. Biochem. 17: 127-134.

Nevell, W. & Wainwright, M. 1987. Influence of soil moisture on sulphur oxidation in brown earth soils exposed to atmospheric pollution. Biol Fertil. Soils 5: 209-214.

Ngugen, M.L. & Goh, K.M. 1992. Sulphur mineralization and release of soluble organic sulphur from camp and non-camp soils of grazed pastures receiving longterm superphosphate applications. Biol. Fertil. Soils 14: 272-279.

Nodvin, S.C., Driscoll, C.T. & Likens, G.E. 1986. The effect of pH on sulfate adsorption by forest soil. Soil Sci. 142: 69-75.

Pfaff, C. 1963. Über die Auswaschung von Calcium, Magnesium, Chlorid und Sulfat aus dem Boden (Lysimeterversuch). Z. Acker u. Pflanzenbau 117: 117-128.

Saalbach, E. 1984. Die Bedeuting der in der Atmosphäre enthaltenen Schwefelverbindungen für die Schwefelversorgung landwirtschaflicher Nutzpflanzen. Angew. Botanik 58: 147-156.

Schnug, E. 1989. Quantitative und qualitative Aspekte der Diagnose und Therapie der Schwefelversorgung von Raps (*Brassica napus* L.) unter besonderer Berücksichtigung glucosinolatarmer Sorten. Habilitation Thesis, Agric. Fac. Kiel.

Schnug, E. 1991. Sulphur nutritional status of European crops and consequences for agriculture. Sulphur in Agriculture 15: 7-12.

Shainberg, J., Sumner, M.E., Miller, W.P., Farina, M.P.W., Pavan, M.A. & Fey, M.V. 1989. Use of gypsum on soils: A review. Adv. Soil Sci. 9: 1-111.

Turner, L.J. & Kramer, J.R. 1991. Sulfate ion binding on goethite and hematite. Soil Sci. 152: 226-230.

Wainwright, M. 1984. Sulfur oxidation in soils. Adv. Agron. 37: 349-396.

Yan, F. & Mengel, K. 1992. Determination of lime requirement by electro-ultrafiltration (EUF). Eur. J. Agron. 1: 71-77.

Ziegler, K., Nemeth, K. & Mengel, K. 1992. Relationship between electroultrafiltration (EUF) extractable nitrogen, grain yield, and optimum nitrogen fertilizer rates for winter wheat. Fertil Res. 32. 37-43.

SIGNIFICANCE OF SULFUR-RICH PROTEINS IN SEEDS AND LEAVES

Holger Bohlmann
Institut für Pflanzenwissenschaften, ETH Zürich, Universitätsstrasse 2, LFW D.58, CH-8092 Zürich, Switzerland

Introduction

Proteins have two sulfur-containing amino acids, methionine and cysteine. Cysteine in particular has several specific functions:
* Cysteines can function as a redox system.
* Cysteines can bind metal ions.
* Cysteines can stabilize proteins by forming disulfide bridges.
As the first two functions are discussed by other authors in this volume (see papers by Schürmann and Rauser) I will give an overview of sulfur-rich seed storage proteins and cysteine-rich proteins of seeds and leaves whose common feature is that their cysteine residues form disulfide bridges to stabilize the protein. Since they are not methionine-rich, as in the case of the sulfur-rich seed storage proteins, their primary function does not seem to be sulfur storage. The primary function of most of these cysteine-rich proteins seems to be the defense against predators and pathogens.

Seed storage proteins

Seed storage proteins are considered mainly as a reserve source of nitrogen for seedling development. Consistent with this role they are usually rich in asparagine, glutamine, and arginine. Some storage proteins are, however, rich in sulfur-containing amino acids. Among the 2S protein family, the proteins from sunflower seed (Kortt *et al.* 1991) and Brazil nut (Sun *et al.* 1987) are very sulfur-rich. Both proteins are synthesized as preproproteins and, after processing, the mature proteins are deposited in protein bodies which is common for seed storage proteins. It is the mature proteins that are rich in sulfur-containing amino acids: 8% cysteine and 16% methionine (sunflower), 8% cysteine and 18% methionine (Brazil nut). Other examples of sulfur-rich storage proteins with high levels of methionine are the 10- and 15-kDa zeins from maize endosperm (22.5% and 11% methionine, respectively) (Kirihara *et al.* 1988). So far, no *in vivo* or *in vitro* function has been assigned to these sulfur-rich seed proteins in addition to a storage function. The high methionine contents of these proteins indicate their probable function as sulfur storage proteins.

Enzyme inhibitors

Both seeds and leaves of higher plants contain a variety of sulfur-rich proteins that have been shown *in vitro* to function as inhibitors of different enzymes, especially

Sulfur Nutrition and Assimilation in Higher Plants
edited by L.J. De Kok et al., pp. 211–219

```
                                                      1                 11
Soybean CII  (Glycine max)                                   S D H S S S D D E S
Adzuki bean II (Phaseolus angularis)                  S G H H E D T T D E P S E S
Peanut AII   (Arachis hypogaea)                              E A S S S S D D N V

15          21                   31                   41
S K P C C D L C M C T A S M P P - - Q C H C A D I R L N S C H S A C D R
S K P C C D Q C - C T K S M P P - - K C R C S D I R L D S C H S A C K S
- - - C C N G C L C D R R A P P Y F E C V C V D T F D H - C P A S C N S

51                  61                  71                  81
C A C T R S M P G Q C R C L D T T D F - C Y K P - C K S S D E D D D
C A C T Y S I P A K C F C T D I N D F - C Y E P - C K S S R D D D W D N
C V C T R S N P P Q C R C T D K T Q G R C P V T E C R S
```

Fig. 1. Amino acid sequences of 3 members of the Bowman-Birk family of enzyme inhibitors: Soybean (*Glycine max*) inhibitor CII (Odani & Ikenaka 1977), Adzuki bean (*Phaseolus angularis*) inhibitor II (Yoshikawa *et al.* 1979), peanut (*Arachis hypogaea*) inhibitor AII (Norioka and Ikenaka 1983). Cysteine residues are boxed. The arrowheads indicate the reactive site peptide bonds.

serine-proteinases such as trypsin and chymotrypsin, and α-amylases. Investigations of the reaction mechanisms of enzyme inhibitors have mostly dealt with proteinase inhibitors, especially of serine proteinases (see Richardson 1991). The interaction seems to be lock-and-key-like, with the inhibitors acting as highly specific substrates that are cleaved at the reactive site peptide bond. This cleavage, however, is not complete and the products do not dissociate. Most of the known enzyme inhibitors are cysteine-rich and it has been shown that the cysteines form disulfide bridges which help to stabilize the inhibitor proteins. As a consequence, most of these proteins are very resistant against denaturation. Their heat stability has sometimes been used in isolation procedures. The main function of the cysteine residues in the enzyme inhibitor proteins thus seems to be stabilization of the protein, although, in the case of the seed enzyme inhibitors, it cannot be excluded that they might, in addition, be a source of sulfur for the developing seedling.

The group of enzyme inhibitors is phylogenetically divergent and only functionally defined. Sequence comparisons, however, allow the enzyme inhibitors to be grouped into families. A good example is the Bowman-Birk family of enzyme inhibitors (for a complete list see Richardson 1991). Most are found in seeds of legumes, some in cereals, but there is also one inhibitor from alfalfa leaves. Fig. 1 shows 3 examples. The members of this family inhibit primarily trypsin and chymotrypsin. Their size is usually 8-9 kDa with up to 14 cysteine residues forming disulfide bridges.

The *in vitro* activity of some of these proteins as inhibitors of digestive enzymes from insects has led researchers to assume that these enzyme inhibitors might protect the plant against predators. Although this hypothesis is convincing, supporting experimental evidence is still scarce. The fact that a protein acts as an enzyme inhibitor under *in vitro* assay conditions does not prove that it would act the same way under *in vivo* conditions in, for example, an insect digestive tract. One example that has been investigated thoroughly are the proteinase inhibitors from tomato leaves (Ryan 1990). After wounding of a tomato leaf (for instance, by insects) two different proteinase inhibitors are induced systemically in the plant. Inhibitor I is active against chymotrypsin, inhibitor II is active against chymotrypsin and trypsin. Both inhibitors are synthesized as precursors and undergo posttranslational modification before the mature proteinase inhibitors are sequestered in the vacuole. Constitutive expression of inhibitor II in transgenic tobacco at a level above 100 $\mu g\ g^{-}$ tissue protects the plant against the larvae of the tobacco hornworm (Johnson *et al.* 1989). The presence of inhibitor I had little effect on the growth of the larvae, indicating that the trypsin inhibitory activity of inhibitor II was mainly responsible for the inhibition of larval growth.

The *in vivo* function of the enzyme inhibitors might not only be directed against predators but also against microbial pathogens. In this regard the recent observation of homology between the maize 22 kDa inhibitor, the PR5 protein from tobacco, and other stress-related proteins by Richardson *et al.* (1987) should be mentioned. PR (pathogenesis related) proteins, among them glucanases and chitinases, accumulate in plants after attack by pathogens and are assumed to be part of the plant's defense system (Linthorst 1991). It has therefore been speculated that the maize 22 kDa inhibitor might also be a defense protein, which protects the seed against microbial pathogens. This view is supported by recent findings that proteins of the PR5 family from tomato, corn, and other plant species have antifungal activities (Vigers *et al.* 1991, Woloshuk *et al.* 1991). An activity against microbial proteinases has been demonstrated several times for enzyme inhibitors *in vitro* (Richardson 1991) but the *in vivo* significance of these enzyme inhibitory effects remains unknown. There are only a few reports showing an increase in enzyme inhibitor activity after attack by a microbial pathogen. Peng & Black (1976), for instance, found increased levels of proteinase inhibitor activities (chymotrypsin and trypsin) in tomato leaves after attack by the phytopathogenic fungus *Phytophthora infestans*, but only in incompatible interactions. In compatible interactions the levels of proteinase inhibitor activity actually declined, even after an initial increase. Whether the increased levels of proteinase inhibitor activity in the incompatible interaction are inhibitory to growth of the fungus remains to be seen. An increase in proteinase inhibitor activity in melon plants after attack by *Colletotrichum lagenarium* was demonstrated by Roby *et al.* (1987). One should keep in mind, however, that seed-borne enzyme inhibitors and other resistance factors are usually pre-synthesized, while resistance proteins (and also phytoalexins) in leaves are, in many cases, inducible. The above-mentioned enzyme inhibitors from tomato and melon, and the PR proteins fall into the last category.

The emerging view about enzyme inhibitors from seeds and leaves is that many of them are part of the defense system of plants. There is now good evidence for a protective role against insects, but the evidence for a role as a defense factor against microbial pathogens is still very limited. In the case of the seed-borne enzyme inhibitors, these might play a double role, on one hand to protect the seeds and on the other hand to be utilized as storage proteins when the seeds germinate. Another function of some enzyme inhibitors seems to be the inhibition of endogenous enzymes, which has been demonstrated especially for the α-amylase inhibitors from the endosperm of cereals (for instance Mundy *et al.* 1984).

Thionins

Thionins are cysteine-rich proteins found in seeds and leaves of several systematically diverse seed plants. They were first isolated from the endosperm of wheat and other

```
                                                                  -28           -21
Precursor DB4 (Hordeum vulgare)               Leaf               M A P S K S I K S V
Precursor mu16 (Hordeum murinum)              Leaf               M A T N K T I K S V
Precursor HTH1 (Hordeum vulgare)              Endosperm          M G L - K G - - - V
Precursor Viscotoxin A3 (Viscum album)        Leaf + Stem        M E V V R G - S S L

          -11                  -1 1                  11
V I C V L I L G L V L E Q V Q V E G K S C C K A T L A R N C Y N T C H F A
V I C V L I L G L V Q E H V Q V E G K S C C K N T T A R N C Y N V C R L T
M V C L L I L G L V L E Q V Q V E G K S C C R S T L G R N C Y N L C R V R
V L L V L L L G A - L L V S Q V E S K S C C P N T T G R N I Y N A C R L T

20                  31                  41                  51
G G S R P V C A G A C R C K I I S G P K C P S D Y P K L N L L P E S G E P
G A P R Q V C A G G C G C K I I S G P T C P R D Y P K L N L L P E S G E P
G A Q K - L C A G V C R C K L T S S G K C P T G F P K L A L V S N S D E P
G A P R P T C A K L S G C K I I S G S T C P S D Y P K F - - - - - - - - -

56      61                  71                  81                  91
D V T Q Y C T I G C R N S V C D - N M D N V F R G Q E M K F D M G L C S N
D A T E Y C T I G C K T S V C D - N M D N V F R G Q Q M K I D M G L C S N
D T V K Y C N L G C R A S N C D Y M V N A A A D D E E M K L Y L E N C G D
- - - - Y C T M G C E S S Q C - - - A T N S - - - - N G D A E A V R C K T

93              101
A C A R F C N D G A V I Q S V E A
A C I R F C N D G A V I Q S V A A
A C V N F C N G D A G L T S L T A
A C S D L C Q F V - - - - - D D A
```

Fig. 2. Amino acid sequences for thionin precursors deduced from cDNA or genomic DNA clones. The HTH1 gene encodes the α-hordothionin. DB4: Bohlmann & Apel (1987), mu16: Bunge (1991), HTH1: Rodriguez-Palenzuela *et al.* (1988), Viscotoxin A3: Schrader & Apel (1991). Cysteine residues are boxed. Arrowheads indicate the processing steps necessary to liberate the mature thionins.

cereals, and from mistletoes (Garcia-Olmedo *et al.* 1989, Bohlmann & Apel 1991). Another group of thionins are the leaf-specific thionins from barley (Bohlmann & Apel 1987, Gausing 1987). Barley thus contains two distinct groups of thionins, one in the endosperm and one in leaves. The thionin from seeds of *Pyrularia pubera* was discovered quite recently, while the crambin (which is actually a mixture of at least 2 proteins) from the seeds of *Crambe abyssinica* (Abyssinian cabbage) has been known for some time. A comparison of the sequences shows that all thionins are homologous and that the 6 or 8 cysteine residues are always conserved. Fig. 2 shows several sequences of different thionins. With the exception of crambin, all thionins are highly basic. Another prominent feature of the thionins, crambin excepted, is their toxicity against microorganisms (fungi – including yeast, and bacteria), mammals and insects. In the latter cases this toxicity is only exerted if the proteins are administered intraperitoneally (LD50 for viscotoxin in mice, 0,5 mg kg^{-1} body weight) or intravenously (LD50 for viscotoxin in cats, 0,1 mg kg^{-1} body weight) or injected into the hemocoel. Fortunately, thionins are not toxic when eaten. Toxicity on various cell cultures has also been reported (for citations see Bohlmann & Apel 1991). The basic feature of the toxicity is the destruction of membranes although the exact mechanism is not yet known. One possibility is that thionins act like detergents to destroy membranes. In agreement with this is the known amphipathic structure of these proteins. As in the case of enzyme inhibitors, the cysteine residues seem to stabilize the three-dimensional structure through disulfide bridges. Viscotoxins, for example, have been reported to be very heat stable. For the *Pyrularia* thionin a different mechanism of toxicity was proposed which includes the activation of ion channels, resulting in an increase in phospholipase activity which leads to a destruction of membranes (Evans *et al.* 1989).

Barley leaf thionins have a toxic effect on plant cells. The regeneration of tobacco protoplasts cultivated 2 weeks under dim light was inhibited if isolated barley leaf thionins were added to the medium (Reimann-Philipp *et al.* 1989). A similar inhibition has also been shown for barley protoplasts (Dae-Won Lee *et al.*, manuscript in preparation). This indicates that plants which produce these toxins seem to require a protective mechanism against their own toxins. At least part of this mechanism might be to produce these toxins as non-toxic precursors (Fig. 2). Thionins are synthesized as preproproteins and the sequences shown were deduced from DNA clones. N-terminal to the thionin domain is a typical leader peptide which directs the proproteins into the endoplasmic reticulum. C-terminal to the thionin is the so-called acid-domain, which contains a large number of acidic residues. The 6 cysteine residues in this acid-domain are absolutely conserved as are several other amino acids. The homology of this acid-domain from diverse thionin precursors indicates an important role for these domains, especially for the cysteine residues. A similar distribution between the cysteine residues in the mature protein and a C-terminal extension is also found in hevein and win precursors (see "Other proteins"). So far, this acid protein has not been isolated as such, but it possibly neutralizes the basic thionin domain, thereby protecting the cell against the toxic action of the thionin while it is produced.

In the past, several different functions have been assigned to the thionins based mainly on *in vitro* experiments (see Bohlmann & Apel 1991). There is one short report about purothionin having α-amylase inhibiting activity (Jones & Meredith 1982) but the inhibition was only partial. A thorough investigation of the potential of thionins as enzyme inhibitors is still lacking. The thionins which are localized in seeds, especially crambin, which has no known toxicity, could function as storage proteins for sulfur. Experimental evidence for such a function is thus far lacking, however. On the other hand, research on leaf-specific thionins of barley has provided several independent lines of evidence suggesting that these proteins are part of the resistance mechanism of barley against microbial pathogens. Leaf-specific thionins of barley are coded for by a multigene family of about 50 genes or more. No genes for leaf-specific thionins could be detected in related cereals by Southern blotting, indicating that this multigene family has no closely related counterparts in species outside the genus *Hordeum*. Apparently, this gene family is evolving rapidly with the thionin domain showing a higher variability than the leader sequence or the acid domain (Bunge *et al.* 1992). The reason for this could be the selective pressure of genetically variable pathogens trying to overcome the toxic action of these proteins. The leaf-specific thionins of barley are distributed in vacuoles and cell walls of barley leaves. Isolated leaf thionins showed a toxic action against yeast and two phytopathogenic fungi, namely *Thielaviopsis paradoxa*, a pathogen of sugarcane, and *Pyrenophora* (*Drechslera*) *teres*, a pathogen of barley (Bohlmann *et al.* 1988). Several other investigators have in addition demonstrated a toxicity of endosperm thionins against phytopathogenic bacteria and fungi. For example, Cammue *et al.* (1992) showed very recently, that β-purothionin has an antifungal activity against several phytopathogenic fungi in liquid media when applied in the range of about 1-5 μg ml^{-1}.

Perhaps the most important pathogen of barley is mildew (*Erisyphe graminis f.* sp.

```
        1                   11                  21

SI1     R V C M G K S Q H H S F P C I S D R L C S N E C V K E E G G
SI2     R V C M G K S A G F K G L C M R D Q N C A Q V C L Q E - - G
SI3     R V C R R R S A G F K G L C M S D H N C A Q V C L Q E - - G

G1P     K I C R R R S A G F K G P C M S N K N C A Q V C Q Q E - - G
G2P     K V C R Q R S A G F K G P C V S D K N C A Q V C L Q E - - G
GH      R I C R R R S A G F K G P C V S N K N C A Q V C M Q E - - G

FST     R E C K T E S N T F P G I C I T K P P C R K A C I S E - - K
p322    R H C E S L S H R F K G P C T R D S N C A S V C E T E - - R
pSAS10  K T C E N L V D T Y R G P C F T T G S C D D H C K N K E - H


        31                  41

SI1     W T A G R C H - - L R Y C R C Q K A C
SI2     W G G G N C D G V M R Q C K C I R Q C W
SI3     W G G G N C D G V I R Q C K C I R Q C W

G1P     W G G G N C D G P F R R C K C I R Q C
G2P     W G G G N C D G P F R R C K C I R Q C
GH      W G G G N C D G P L R R C K C M R R C

FST     F T D G H C S K L L R R C L C T K P C
p322    F S G G N C H G F R R R C F C T K P C
pSAS10  L L S G R C R D D V R - C W C T R N C
```

Fig. 3. Amino acid sequences for γ-thionins and homologous proteins. SI1 – SI3: Sorghum inhibitors 1 – 3 (Bloch and Richardson 1991), G1P and G2P: γ1 – purothionin and γ2 – purothionin (Colilla *et al.* 1990), GH: γ – hordothionin (Mendez *et al.* 1990), FST: Flower specific thionin from tobacco deduced from cDNA (Gu *et al.* 1992), p322: protein from potato tubers deduced from cDNA (Stiekema *et al.* 1988), pSAS10: protein deduced from cDNA for stored mRNA in cotyledons of cowpea (Ishibashi *et al.* 1990). Cysteines are boxed.

hordei). Inoculation of barley with this fungus leads to a higher level of thionin transcripts, an increase which seems to be most pronounced in an incompatible interaction (Bohlmann *et al.* 1988). In another series of experiments the distribution of thionins at the infection site was investigated by Ebrahim-Nesbat *et al.* (1989) with the technique of immunogold-labeling using an antibody obtained against a fusion protein between β-galactosidase and the leaf-specific thionin precursor DB4 (Bohlmann & Apel 1987). The results show that the distribution of thionins in cell walls from cells not infected by mildew is quite different from that of cell walls of infected cells. Thionins seem to be newly synthesized after infection by mildew and incorporated into the papillae, at least in incompatible interactions. On the other hand, in compatible interactions, the fungus seems to have a mechanism to protect itself against the toxic thionins by somehow masking or digesting the proteins so that they are no longer recognized by the antibody.

In conclusion, leaf-specific thionins of barley are toxic against phytopathogenic fungi, are induced by mildew and seem to be incorporated into papillae, are distributed in vacuoles and cell walls, and have a high degree of variability. Leaf-specific thionins of barley and probably other thionins as well, with the possible exception of crambin, seem to be part of the plant's resistance mechanism against microorganisms.

γ-Thionins

A group of so-called γ-thionins has recently been found in barley and wheat endosperm and in tobacco flowers (Fig. 3). γ-thionins also have homology with proteins from potato and soybean and, interestingly, with enzyme inhibitors from sorghum seeds which have been shown to inhibit α-amylases from insects. Although called γ-thionins, the sequence homology of these proteins with the former thionins is extremely weak, so that I will not group them together with classical thionins. They are, however, very basic (except pSAS10 from cowpea) and cysteine-rich. The sequences of the tobacco, potato, and cowpea proteins are derived from cDNAs and all have leader peptides, indicating that the mature proteins might be found in protein bodies or vacuoles, or extracellularly. The tobacco flower protein has an acidic extension just as the classical thionins, but this extension contains no cysteine residues and there is no homology with the acid-domain of thionins. In addition to the enzyme inhibitory function of the sorghum proteins, the only other *in vitro* function reported thus far is the inhibition of protein synthesis in eukaryotic cell-free systems by the γ-hordothionin (Mendez *et al.* 1990). It remains to be seen if proteins of this group other than the sorghum proteins have enzyme inhibitory functions or if any of these proteins have toxic activities.

Other proteins

Lectins are a large group of diverse proteins functionally defined as carbohydrate binding proteins of non-immune origin. Several chitin-binding lectins have a homologous domain of about 40 amino acids (Chrispeels & Raikhel 1991) which is cysteine-rich. This chitin-binding domain is also found in chitinases. The 43 amino acid protein hevein from the laticifers of rubber trees is another protein which is homologous to this cysteine-rich domain and binds chitin. Hevein is probably produced as a much larger preproprotein with a 144 amino acid C-terminal extension which contains 6 cysteine residues (Broekaert *et al.* 1990). This precursor has homology with proteins deduced from wound-induced genes (win1, win2) of potato (Stanford *et al.* 1989). All of these chitin-binding lectins and chitinases are thought to play a role in the defense of plants against insects and against fungi which contain chitin in their cell walls.

Last, but perhaps not least, a recent report describes two related peptides (36, and 37 amino acids, respectively) from the seeds of *Mirabilis jalapa*, which have no sequence homology with thionins but which are also cysteine-rich (6 residues), basic and have antimicrobial activities (Cammue *et al.* 1992). The mechanism of their toxicity is not known. These proteins are another example of the growing list of small, cysteine-rich, basic proteins with antimicrobial activities which includes thionins, perhaps γ-thionins, *Mirabilis jalapa*-antimicrobial peptides, and the defensins from animals (Lehrer *et al.* 1991).

Concluding remarks

Several groups of cysteine-rich plant proteins in seeds, leaves and perhaps also in other plant organs seem to be part of the plant's defense system. These proteins are usually produced as precursors which are secreted so that the mature proteins are extracellular or are found in vacuoles and protein bodies. The main function of the cysteine residues is probably to stabilize these proteins with disulfide bridges, giving most of them a pronounced stability against denaturation. While the defense proteins discussed in this chapter have this feature in common, the mechanisms by which they exert their activities are quite different.

Acknowledgment

I thank Drs. Francesco Campos and Gregory Armstrong for reading the manuscript.

References

Bloch, C., Jr. & Richardson, M. 1991. A new family of small (5 kDa) protein inhibitors of insect α-amylases from seeds of sorghum (*Sorghum bicolor* L. Moench) have sequence homologies with wheat γ-purothionins. FEBS Lett. 279: 101-104.

Bohlmann, H. & Apel, K. 1987. Isolation and characterization of cDNAs coding for leaf-specific thionins closely related to the endosperm-specific hordothionin of barley (*Hordeum vulgare* L.). Mol. Gen. Genet. 207, 446-454.

Bohlmann, H. & Apel, K. 1991. Thionins. Annu. Rev. Plant Physiol. Plant Mol. Biol. 42: 227-240.

Bohlmann, H., Clausen, S., Behnke, S., Giese, H., Hiller, C., Reimann-Philipp, U., Schrader, G., Barkholt, V. & Apel, K. 1988. Leaf-specific thionins of barley – a novel class of cell wall proteins toxic to plant-pathogenic fungi and possibly involved in the defense mechanism of plants. EMBO J. 7: 1559-1565.

Broekaert, W., Lee, H.-I., Kush, A., Chua, N.-H. & Raikhel, N. 1990. Wound-induced accumulation of mRNA containing a hevein sequence in laticifers of rubber tree (*Hevea brasiliensis*). Proc. Natl. Acad. Sci. USA 87: 7633-7637.

Bunge, S. 1991. Vergleichende Untersuchungen an Blatt-Thionin-Genen in der Gattung Hordeum. PhD Thesis, Kiel.

Bunge, S., Wolters, J. & Apel, K. 1992. A comparison of leaf thionin sequences of barley cultivars and wild barley species. Mol. Gen. Genet. 231: 460-468.

Cammue, B.P.A., De Bolle, M.F.C., Terras, F.R.G., Proost, P., Van Damme, J., Rees, S.B., Vanderleyden, J. & Broekaert, W.F. 1992. Isolation and characterization of a novel class of plant antimicrobial peptides from *Mirabilis jalapa* L. seeds. J. Biol. Chem. 267: 2228-2233.

Chrispeels, M.J. & Raikhel, N.V. 1991. Lectins, lectin genes, and their role in plant defense. Plant Cell 3: 1-9.

Colilla, F.J., Rocher, A. & Mendez, E. 1990. γ-Purothionins: amino acid sequence of two polypeptides of a new family of thionins from wheat endosperm. FEBS Lett. 270: 191-194.

Ebrahim-Nesbat, F., Behnke, S., Kleinhofs, A. & Apel, K. 1989. Cultivar-related differences in the distribution of cell-wall-bound thionins in compatible and incompatible interactions between barley and powdery mildew. Planta 179: 203-210.

Evans, J., Wang, Y., Shaw, K.-P. & Vernon, L.P. 1989. Cellular responses to *Pyrularia* thionin are mediated by Ca^{2+} influx and phospholipase A_2 activation and are inhibited by thionin tyrosine iodination. Proc. Natl. Acad. Sci. USA 86: 5849-5853.

Garcia-Olmedo, F., Rodriguez-Palenzuela, P., Hernandez-Lucas, C., Ponz, F., Marana, C., Carmona, M.J., Lopez-Fando, J., Fernandez, J.A. & Carbonero, P. 1989. The thionins: a protein family that includes purothionins, viscotoxins and crambins. Oxford Surv. Plant Mol. Cell Biol. 6: 31-60.

Gausing, K. 1987. Thionin genes specifically expressed in barley leaves. Planta 171: 241-246.

Gu, Q., Kawata, E.E., Morse, M.-J., Wu, H.-M. & Cheung, A.Y. 1992. A flower-specific cDNA encoding a novel thionin in tobacco. Mol. Gen. Genet. 234: 89-96.

Ishibashi, N., Yamauchi, D. & Minamikawa, T. 1990. Stored mRNA in cotyledons of *Vigna unguiculata* seeds: nucleotide sequence of cloned cDNA for a stored mRNA and induction of its synthesis by precocious germination. Plant Mol. Biol. 15: 59-64.

Jones, B.L. & Meredith, P. 1982. Inactivation of alpha-amylase activity by purothionins. Cereal Chem. 59: 321.

Johnson, R., Narvaez, J., An, G., Ryan, C. 1989. Expression of proteinase inhibitors I and II in transgenic tobacco plants: Effects on natural defense against *Manduca sexta* larvae. Proc. Natl. Acad. Sci. USA 86: 9871-9875.

Kirihara, J.A., Petri, J.B. & Messing, J. 1988. Isolation and sequence of a gene encoding a methionine-rich 10-kDa zein protein from maize. Gene 71: 359-370.

Kortt, A.A., Caldwell, J.B., Lilley, G.G. & Higgins, T.J.V. 1991. Amino acid and cDNA sequences of a methionine-rich 2S protein from sunflower seed (*Helianthus annuus* L.). Eur. J. Biochem. 195: 329-334.

Lehrer, R.I., Ganz, T. & Selsted, M.E. 1991. Defensins: Endogenous antibiotic peptides of animal cells. Cell 64: 229-230.

Linthorst, H.J.M. 1991. Pathogenesis-related proteins of plants. Crit. Rev. Plant Sciences 10: 123-150.

Mendez, E., Moreno, A., Colilla, F., Pelaez, F., Limas, G.G., Mendez, R., Soriano, F., Salinas, M. & de Haro, C. 1990. Primary structure and inhibition of protein synthesis in eukaryotic cell-free system of a novel thionin, γ-hordothionin, from barley endosperm. Eur. J. Biochem. 194: 533-539.

Mundy, J., Hejgaard, J. & Svendsen, I. 1984. Characterization of a bifunctional wheat inhibitor of endogenous α-amylase and subtilisin. FEBS L. 167: 210-214.

Norioka, S. & Ikenaka, T. 1983. Amino acid sequences of trypsin-chymotrypsin inhibitors (A-I, A-II, B-I, and B-II) from peanut (*Arachis hypogaea*): A discussion on the molecular evolution of legume Bowman-Birk type inhibitors. J. Biochem. 94: 589-599.

Odani, S.& Ikenaka, T. 1977. Studies on soybean trypsin inhibitors. XI. Complete amino acid sequence of a soybean trypsin-chymotrypsin-elastase inhibitor, C-II. J. Biochem. 82: 1523-1531.

Peng, J.H. & Black, L.L. 1976. Increased proteinase inhibitor activity in response to infection of resistant tomato plants by *Phytophthora infestans*. Phytopathology 66: 958-963.

Reimann-Philipp, U., Schrader, G., Martinoia, E., Barkholt, V. & Apel, K. 1989. Intracellular thionins of barley. A second group of leaf thionins closely related to but distinct from cell wall-bound thionins. J. Biol. Chem. 264: 8978-8984.

Richardson, M. 1991. Seed storage proteins: the enzyme inhibitors. Methods in Plant Biochemistry 5: 259-305.

Richardson, M., Valdes-Rodriguez, S. & Blanco-Labra, A. 1987. A possible function for thaumatin and a TMV-induced protein suggested by homology to a maize inhibitor. Nature 327 432-434.

Roby, D., Toppan, A. & Esquerre-Tugaye, M.-T. 1987. Cell surfaces in plant micro-organism interactions. VIII. Increased proteinase inhibitor activity in melon plants in response to infection by Colletotrichum lagenarium or to treatment with an elicitor fraction from this fungus. Physiol. Mol. Plant Pathol. 30: 453-460.

Rodriguez-Palenzuela, P., Pintor-Toro, J., Carbonero, P. & Garcia-Olmedo, F. 1988. Nucleotide sequence and endosperm specific expression of the structural gene for the toxin α-hordothionin in barley (*Hordeum vulgare* L.). Gene 70: 271-281.

Ryan, C.A. 1990. Protease inhibitors in plants: Genes for improving defenses against insects and pathogens. Annu. Rev. Phytopathol. 28: 425-449.

Schrader, G. & Apel, K. 1991. Isolation and characterization of cDNAs encoding viscotoxins of mistletoe (*Viscum album*). Eur. J. Biochem. 198: 549-553.

Stanford, A., Bevan, M. Y Northcote, D. 1989. Differential expression within a family of novel wound-induced genes in potato. Mol. Gen. Genet. 215: 200-208.

Stiekema, W.J., Heidekamp, F., Dirkse, W.G., van Beckum, J., de Haan, P., ten Bosch, C. & Louwerse, J.D. 1988. Molecular cloning and analysis of four potato tuber mRNAs. Plant Mol. Biol. 11: 255-269.

Sun, S.S.M., Altenbach, S.B. & Leung, F.W. 1987. Properties, biosynthesis and processing of a sulfur-rich protein in Brazil nut (*Bertholletia excelsa* H.B.K.). Eur. J. Biochem. 162: 477-483.

Vigers, A.J., Roberts, W.K. & Selitrennikoff, C.P. 1991. A new family of plant antifungal proteins. Mol. Plant-Microbe Interact. 4: 315-323.

Woloshuk, C.P., Meulenhoff, J.S., Sela-Buurlage, M., Van den Elzen, P.J.M. & Cornelissen, B.J.C. 1991. Pathogen-induced proteins with inhibitory activity toward *Phytophthora infestans*. Plant Cell, 3, 619-628.

Yoshikawa, M., Kiyohara, T., Iwasaki, T., Ishii, Y. & Kimura, N. 1979. Amino acid sequences of proteinase inhibitors II and II' from adzuki beans. Agric. Biol. Chem. 43: 787-796.

GLUTATHIONE IN THE METABOLISM AND DETOXIFICATION OF XENOBIOTICS IN PLANTS

Gerald L. Lamoureux & Donald G. Rusness
USDA, Agricultural Research Service, Biosciences Research Laboratory, Fargo, North Dakota 58105, U.S.A.

Introduction

Much of our current knowledge concerning the metabolism of xenobiotics in plants has been obtained from pesticide metabolism studies conducted during the last 25 years (Lamoureux & Rusness 1986a; Hatzios 1991; Aizawa 1989). Although attempts to generalize observations made with pesticides to a broader range of xenobiotics may result in some distortions, many valid conclusions regarding xenobiotic metabolism can be reached from these studies. When xenobiotics are detoxified by metabolism, detoxification usually occurs during the first or second step, and in plants, only seven chemical reactions are utilized with frequency during these first two steps. Listed in approximate order of frequency of occurrence, these reactions are: oxidation, hydrolysis, glucoside conjugation, glutathione (GSH) conjugation, amino acid conjugation, malonyl conjugation, and reduction. Although conjugation with GSH is not the most commonly observed of these reactions, it is extremely important because it is utilized to detoxify electrophilic alkylating agents. The following aspects of xenobiotic GSH conjugation in plants are considered in this review: (a) the basic GSH conjugation reactions; (b) the role of GSH conjugation in detoxification and toxification of xenobiotics; (c) the metabolism and break-down of GSH conjugates in plants; and (d) the glutathione transferase enzymes-sources, basic properties, isozymes, isolation, induction, inhibition, safeners, molecular biology, and possible natural functions.

Glutathione conjugation

The GSH conjugation reaction

The primary requirements for conjugation of a xenobiotic with GSH are: the xenobiotic must contain an electrophilic site or be metabolized to an intermediate that contains an electrophilic site, an adequate supply of GSH or an active homolog such as homoglutathione (hGSH) must be available, and a glutathione *S*-transferase (GST) enzyme may be necessary to catalyze the reaction.

Nucleophilic displacement

In plants, the most commonly observed GSH conjugation reaction has been the nucleophilic displacement of a halogen from an electrophilic site on an aromatic ring, a heterocyclic ring, or an alkyl group (Fig. 1). Among the pesticides metabolized in

Sulfur Nutrition and Assimilation in Higher Plants
edited by L.J. De Kok et al., pp. 221–237

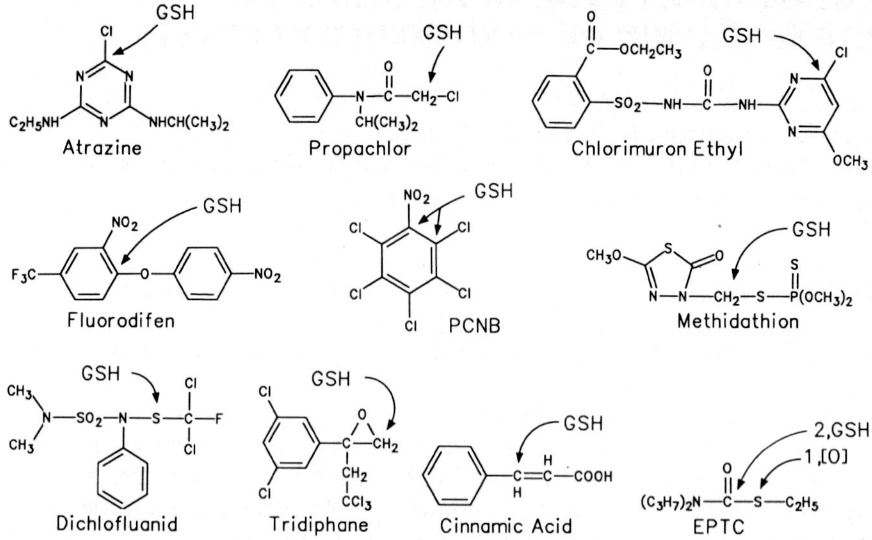

Fig. 1. Structures of compounds known to be metabolized by conjugation with glutathione or homoglutathione in plants.

this manner are the chloro-s-triazine herbicides such as atrazine, the chloroacetanilides such as propachlor, and a sulfonylurea herbicide, chlorimuron ethyl (Lamoureux & Rusness 1986a; Brown 1990). Nucleophilic displacements of phenols from diphenylether herbicides such as fluorodifen, the displacement of a nitro group from PCNB, and the displacement of an alkyl group from an organophosphorothioate insecticide, methidathion, also have been reported (Lamoureux & Rusness 1986a) (Fig. 1).

GSH conjugation at noncarbon sites

Disulfide glutathione conjugates have been proposed as products of the nucleophilic attack of GSH at sulfur centers in the metabolism of fungicides such as dichlofluanid in strawberry (Schuphan *et al.* 1981) (Fig. 1) and thiocyanate insecticides in insects (Fukami 1984). GSH conjugation also occurs by attack at nitrogen centers, as in the metabolism of aromatic nitroso compounds in animals (Eyer & Schneller 1983).

Nucleophilic addition

In plants, GSH conjugation by nucleophilic addition has been reported less frequently than GSH conjugation by nucleophilic displacement, but this may be a consequence of the types of xenobiotics that have been studied rather than a true indication of the ability of plants to catalyze these various reactions. The metabolism of tridiphane in corn and giant foxtail is an example of an addition reaction between GSH and an epoxide, and the metabolism of *trans*-cinnamic acid in pea is an example of an addition reaction between GSH and a double bond (Diesperger & Sandermann 1979; Lamoureux & Rusness 1986b) (Fig. 1).

Activation prior to GSH conjugation

Some xenobiotics must be activated before GSH conjugation can occur. The asymmetrical triazine herbicides substituted with an alkylsulfide and the thiocarbamate herbicides must be activated by oxidation of a sulfur to a sulfoxide (Lay & Casida 1976; Frear *et al.* 1985; Lamoureux & Rusness 1986a) (Fig. 1). Conjugation then occurs by nucleophilic displacement of the alkylsulfinyl group with GSH. Recent evidence indicates that GSH conjugation of a sulfonylurea herbicide may occur after oxidation of an aromatic ring; however, the mechanism of this reaction has not been determined (Lamoureux *et al.* 1991b). Although GSH conjugation following the formation of an epoxide occurs with some frequency in mammals (Sies & Ketterer 1988), there is less evidence that this is a common route of metabolism in plants.

GSH conjugation, a detoxification process

GSH conjugation is usually a detoxification process. This was demonstrated in metabolism/photosynthesis studies with atrazine in tolerant and susceptible inbred maize lines and in tolerant maize and susceptible pea (Shimabukuro *et al.* 1978). In soybean, chlorimuron ethyl is detoxified by conjugation with homoglutathione (hGSH). Chlorimuron ethyl is an inhibitor of acetolactate synthase, but the hGSH conjugate is inactive (Brown 1990). Based on the relationship between metabolism and tolerance, GSH conjugation also appears to be a detoxification process in the metabolism of the chloroacetanilide, diphenylether, and thiocarbamate herbicides (Lamoureux & Rusness 1989a; Lamoureux *et al.* 1991c).

Toxic products can be produced as a result of GSH conjugation

In strawberry, dichlofluanid is metabolized to a thiophosgene derivative in a process that appears to involve two GSH conjugation steps (Schuphan *et al.* 1981). In insects, hydrogen cyanide is produced during GSH conjugation of thiocyanate insecticides (Fukami 1984), and in the rat a highly reactive nephrotoxic GSH conjugate is produced in the metabolism of dibromoethane (Inskeep *et al.* 1986). However, in spite of such exceptions, GSH conjugation usually results in the production of relatively nontoxic products.

The metabolism of GSH conjugates

Glutathione conjugates are commonly metabolized via intermediary cysteine conjugates to a variety of other metabolites in plants, mammals, fish, birds, and insects (Lamoureux & Bakke 1984). In plants, the half-life of a GSH conjugate may be only a few hours (Lamoureux & Rusness 1989b); or in isolated cases, *e.g.*, the GSH conjugate of tridiphane in giant foxtail, the conjugate may be moderately stable (G.L. Lamoureux, unpublished results). Since metabolites of GSH conjugates may be chromatographically very similar to the parent GSH conjugates, considerable care must be exercised to correctly identify these metabolites (Lamoureux & Rusness 1989b).

The various metabolites and reactions which have been detected in the metabolism

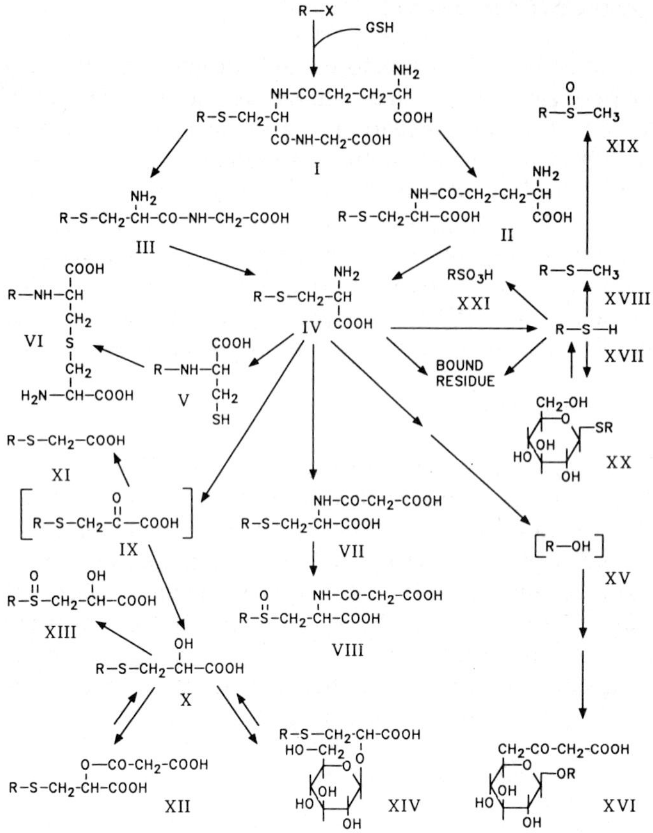

Fig. 2. Pathways utilized in the metabolism of glutathione conjugates in plants.

of xenobiotic GSH conjugates in plants are shown (Fig. 2). The metabolism of GSH conjugates in plants has been discussed in several reviews (Lamoureux & Rusness 1981, 1986a, 1989a). In plants, GSH conjugates (I) are usually metabolized *via* γ-glutamylcysteine conjugates (II) to *S*-cysteine conjugates (IV) (Lamoureux & Rusness 1986a, 1989a). The *S*-cysteine conjugates occupy a key pivotal position in the pathway. Occasionally, *S*-cysteine conjugates may persist in some plant tissues, but more frequently they are metabolized to other products (Lamoureux & Rusness 1983). The *S*-cysteine conjugate of atrazine undergoes a nonenzymatic rearrangement to an *N*-cysteine conjugate (V) in sorghum and it is subsequently transformed into a lanthionine conjugate (VI). The lanthionine conjugate is a major terminal metabolite of atrazine in sorghum (Lamoureux *et al.* 1973). This appears to be the only xenobiotic lanthionine conjugate reported; however, the nonenzymatic rearrangement of an *S*-cysteine conjugate to an *N*-cysteine conjugate was observed in the metabolism of dimethametryn in rice (Mayer *et al.* 1981). The rearrangment of *S*- to *N*-cysteine conjugates may be more common than indicated by these two isolated reports.

N-Malonylcysteine conjugates (VII) are commonly formed during the metabolism of a variety of pesticide GSH conjugates in many different plant species. The *N*-malonylcysteine conjugates frequently represent major terminal metabolites, comparable to the mercapturic acids formed from GSH conjugates in animals (Lamoureux & Rusness 1983). In some cases, the *N*-malonylcysteine conjugates are metabolized further by oxidation of the sulfur to yield the corresponding sulfinyl conjugates (VIII). Some sulfinyl conjugates are unstable (Lamoureux & Rusness 1983).

Derivatives of *S*-thiolactic acid conjugates (X) are commonly formed during the metabolism of xenobiotic GSH conjugates in plants. In animals, *S*-thiolactic acid conjugates are formed from *S*-thiopyruvate intermediates (IX) (Tomisawa *et al.* 1992) (Fig. 2), but *S*-thiopyruvate intermediates have not been isolated from plants. An *S*-thioacetic acid conjugate (XI) is a minor metabolite of PCNB in peanut (Lamoureux & Rusness 1989a). This metabolite is probably formed by decarboxylation of the corresponding *S*-thiopyruvate intermediate.

Xenobiotic *S*-thiolactic acid conjugates can undergo further metabolism in plants. In corn and cotton, *S*-(2-*O*-malonyl)-3-thiolactic acid conjugates (XII) of chloroacetamide and thiocarbamate herbicides account for up to 40% of the herbicide dose one to four weeks following treatment (Lamoureux & Rusness 1987b, 1989b; Khalifa & Lamoureux 1990). The *S*-(2-*O*-malonyl)-3-thiolactic acid conjugates of the chloroacetamides appear to be be further metabolized to simple *SO*-thiolactic acid conjugates (XIII) that are the most abundant products of chloroacetamide metabolism in corn after 30 days (Khalifa & Lamoureux 1990). A variation in the metabolism of the *S*-thiolactic acid conjugates was observed in a Norway spruce cell suspension culture. In that tissue, an *S*-(2-*O*-glucosyl)-3-thiolactic acid conjugate (XIV) accounted for nearly 21% of the fluorodifen herbicide dose after 6 days (Lamoureux *et al.* 1991a). This metabolite is not stable and undergoes metabolism to other unidentified products. Norway spruce was the first member of the gymnosperm subdivision of the plant kingdom shown to metabolize xenobiotics by GSH conjugation (Schröder *et al.* 1990).

O-Glucosides and *O*-(6-*O*-malonyl)glucosides (XVI) can be formed as major metabolites derived from the GSH pathway, ie., in the metabolism of chloroacetamide herbicides in maize and other species (Blattman *et al.* 1986; Lamoureux & Rusness 1989b). The mechanism of this transformation is uncertain, but it probably involves a hydroxylated intermediate (XV) and therefore it can be concluded that hydroxylated metabolites and glucoside conjugates of xenobiotics can be formed by mechanisms that do not involve simple oxidative routes of metabolism. Some hydroxylated derivatives of atrazine isolated from sorghum may actually have been produced via this mechanism (Lamoureux *et al.* 1973).

Cysteine lyase enzymes capable of hydrolyzing xenobiotic cysteine conjugates to xenobiotic thiols (XVII) have been detected in several plant species (Lamoureux & Rusness 1986a, 1989a; Lamoureux *et al.* 1992). Although xenobiotic thiols have not been commonly reported as metabolites in plants, they are relatively reactive and there is indirect and enzymatic evidence that xenobiotic thiols are intermediates in the metabolism of some xenobiotic cysteine conjugates.

S-CH_3 derivatives (XVIII) can be formed from xenobiotic thiol intermediates (Fig. 2). In a Norway spruce cell suspension culture, a volatile *S*-CH_3 derivative of fluorodifen was produced by this route and accounted for ca. 25% of the herbicide dose. These reactions were demonstrated *in vitro* with cysteine lyase and *S*-adenosylmethionine methyl transferase enzymes from spruce (Lamoureux *et al.* 1992). The formation of *S*-CH_3 derivatives by this route has also been observed in the metabolism of PCNB in several plant species and the selectivity of the *S*-adenosylmethionine methyl transferase enzyme from onion has been examined (Lamoureux & Rusness 1980, 1981). The *S*-CH_3 derivatives can be further oxidized to sulfoxides (XIX) and perhaps to sulfones.

Xenobiotic thiols liberated by cysteine lyase or other enzymes can also serve as intermediates to other metabolites. An *S*-glucoside (XX) was the most abundant metabolite produced during the metabolism of fluorodifen by a Norway spruce cell suspension culture (Lamoureux *et al.* 1992). This appears to be the only documented report of the formation of an *S*-glucoside from the GSH pathway in plants. A sulfonic acid derivative (XXI) of fluorodifen was also detected in that study. Both Fluorodifen and the *S*-glucoside of fluorodifen served as precursors of the sulfonic acid conjugate. Although sulfonic acid derivatives have been reported in the metabolism of GSH conjugates in soil (Lamoureux & Rusness 1989b), this is the first demonstration that plant tissues can also form these products as metabolites from the GSH pathway.

Bound residues can be important terminal products of pesticides that are initially metabolized to GSH conjugates in plants (Lamoureux & Rusness 1981, 1986a). Little is known about the mechanisms by which bound residues are formed from the GSH pathway; however, a cysteine conjugate and a thiophenol derived from the metabolism of PCNB by the GSH pathway were both converted in high yield to bound residues in a peanut cell suspension culture (Lamoureux & Rusness 1981). It is likely that other intermediary metabolites from the pathway shown in Fig. 2 may also be precursors to bound residues.

The glutathione *S*-transferase (GST) enzymes

Distribution of GST enzymes and general considerations

Glutathione *S*-transferase (GST) enzymes in plants were discovered in maize and related species in 1970 (Frear & Swanson 1970). Before 1986, only one or two papers on plant GST enzymes were published per year; however, in 1986 papers dealing with plant GST enzymes began to increase and at least 13 papers were published on this subject in 1991. Glutathione *S*-transferase enzymes have now been detected in at least 33 species in the Plant Kingdom, including members of the Pteridophyta Division and members of both the Gymnospermae and Angiospermae Subdivisions of the Spermatophyta Division (Table 1). Only the soluble plant GST enzymes have been studied, but microsomal GST activity has been detected in pea (Diesperger & Sanderman 1979) and corn (Komives *et al.* 1985). Microsomal GST also occurs in mammals (Sies & Ketterer 1988). In corn, GST accounts for 1 to 2% of the soluble protein (Mozer *et al.* 1983; Edwards & Owen 1988). Although the plant GST enzymes are assumed to be cytoplasmic, their intracellular location has not been studied. It is possible that GST enzymes are present in the chloroplasts since the GSH conjugate of atrazine is localized around the bundle sheath chloroplasts (Huber & Sautter 1987)

Table 1. List of plant species in which glutathione *S*-transferase (GST) activity has been demonstrated and a literature reference for each species.

Abutilon theophrasti (velvetleaf)	Anderson & Gronwald (1991)
Allium cepa L. (onion)	Lamoureux & Rusness (1980)
Amaranthus retroflexus (redroot pigweed)	Burkholder (1977)
Arachis hypogaea L. (peanut)	Frear & Swanson (1973)
Avena sativa L. (oat)	Singh & Shaw (1988)
Chenopodium album (common lambsquarters)	Burkholder (1977)
Cicer arietinum L. (chickpea)	Hunaiti & Ali (1990)
Cucumis sativus L. (cucumber)	Frear & Swanson (1973)
Cucurbita maxima (squash)	Frear & Swanson (1973)
Dianthus caryophyllus (carnation)	Meyer *et al.* (1991)
Digitaria ischaemum L. (large crabgrass)	Burkholder (1977)
Digitaria sanguinalis L. (smooth crabgrass)	Burkholder (1977)
Glycine max L. (soybean)	Frear & Swanson (1973)
Gossypium hirsutum L. (cotton)	Frear & Swanson (1973)
Hevea brasiliensis (rubber tree)	Balabaskaran & Muniandy (1984)
Hibiscus esculentus (okra)	Frear & Swanson (1973)
Hordeum vulgare L. (barley)	Price *et al.* (1990)
Lycopersicon esculentum (tomato)	Frear & Swanson (1973)
Medicago sativa (alfalfa)	Edwards & Dixon (1991)
Nicotiana tabacum L. (tobacco)	Takahasi & Nagata (1992)
Oryza sativa L. (rice)	Komives & Dutka (1989)
Phaseolus vulgaris (French bean)	Edwards & Dixon (1991)
Picea abies (Norway spruce)	Schröder *et al.* (1990)
Picea glauca (white spruce)	Schröder *et al.* (1990)
Pisum sativum L. (pea)	Frear & Swanson (1973)
Saccharum officianarum (sugarcane)	Frear & Swanson (1970)
Setaria lutescens L. (yellow foxtail)	Burkholder (1977)
Sorghum bicolor L. (sorghum)	Dean *et al.* (1990)
Sorghum halepense L. (Johnson grass)	Frear & Swanson (1970)
Sorghum sudanenese (Sudan grass)	Frear & Swanson (1970)
Sorghum vulgare L. (sorghum)	Frear & Swanson (1970)
Tortula ruralis (moss)	Dhindsa (1991)
Triticum aestivum L. (wheat)	Jablonkai & Hatzios (1991)
Zea mays L. (corn)	Frear & Swanson (1970)

and a high percent of the GSH is in the chloroplasts (Rennenberg & Lamoureux 1990). The GST enzymes in mammalian tissues have been localized by immunological techniques (Oberley *et al.* 1991); therefore, it may be possible to localize the GST enzymes in plants without isolating organelles.

The soluble GST enzymes from maize share some amino acid homology and/or immunological similarity with GST enzymes from other plant species and with GST enzymes from insects and mammals (Toung *et al.* 1990; Hunaiti & Ali 1991; Reinemer *et al.* 1991; Singhal *et al.* 1991; Takahashi & Nagata 1992). The molecular weight of these dimeric proteins is usually about 54 kDa; the monomers have molecular weights of 26 to 29 kDa and consist of approximately 214 amino acid residues (Timmerman 1989).

GST isolation and general considerations

GST enzymes from plants have been isolated from roots (Mozer *et al.* 1983), seeds (Williamson & Beverley 1988), green and etiolated tissues (Mozer *et al.* 1983), cell cultures (Edwards & Owen 1986), protoplasts (Takahashi & Nagata 1992), and needles (Schröder *et al.* 1990). GST enzymes are frequently extracted from tissue that is frozen and ground under liquid nitrogen. The ground tissue is extracted with 0.2 M potassium phosphate (pH 6.8-7.8) (Dean *et al.* 1990) or Tris-HCl (pH 7.8) (O'Connell *et al.* 1988) buffer that may contain 1 mM EDTA, 1 mM GSH, 5 to 7.5% polyvinylpyrrolidone, and aprotinin (O'Connell *et al.* 1988; Williamson & Beverley 1988; Dean *et al.* 1990). Nucleic acids may be precipitated with protamine sulfate and the GST preparation is frequently fractionated by ammonium sulfate precipitation (Mozer *et al.* 1983). After dialysis or gel filtration, the enzyme preparation frequently is subjected to procedures such as DEAE chromatography (Mozer *et al.* 1983), affinity chromatography (O'Connel *et al.* 1988), phenyl Sepharose CL-4B chromatography (Williamson & Beverely 1988), chromatography on Mono Q (Dean *et al.* 1991), or electrophoresis (O'Connel *et al.* 1988). GST enzymes from plants are frequently stable, but some variation in the stability of different isozymes may occur (Timmerman 1989). Plants contain natural inhibitors of GST (Singh & Shaw 1988; Lee 1991) and a number of precautions may be required to isolate GST activity from certain organisms (Kumagai *et al.* 1988).

Plant GST isozymes display considerable specificity for the xenobiotic substrate (Lamoureux & Rusness 1989a; Dean *et al.* 1990, 1991), but very little is known about their specificity for the nucleophilic substrate (GSH, homoGSH, *etc.*). A crude GST system from maize did not utilize L-cysteine (Frear & Swanson 1970), but this may not be the case with all plant GST isozymes (Still & Rusness 1977). Some important leguminous species such as soybean contain hGSH rather than GSH, but very little has been published regarding the selectivity of these isozymes for different γ-glutamylcysteine tripeptides. Significant selectivity towards GSH or analogs, has been observed among mammalian GST isozymes (Adang *et al.* 1988).

GST Isozymes

Pea was one of the first plant species shown to contain multiple forms of GST activity. Two isozymes active with *trans*-cinnamic acid were separated from a third isozyme that was active with fluorodifen (Diesperger & Sandermann 1979). The most detailed plant GST isozyme studies have been conducted with maize. In the first study dealing with isozymes from maize, a constitutive GST (GST I) and an inducible GST (GST II) that utilize CDNB and metolachlor as substrates were reported (Mozer *et al.* 1983). More recently, maize was shown to contain a complex mixture of GST isozymes active with the following substrates: atrazine (3 isozymes), CDNB (1 isozyme), *trans*-cinnamic acid (1 isozyme), EPTC sulfoxide (2 isozymes), and metolachlor (2 isozymes) (Dean *et al.* 1991). These isozymes vary significantly in their selectivity towards different substrates. Sorghum contains a constitutive isozyme active with CDNB and another that is active with metolachlor (Dean *et al.* 1990). Sorghum also contains activity for atrazine (Frear & Swanson 1970), but it was not determined whether GST activity for atrazine was due to additional isozymes. Determining the number of isozymes in a species or biotype may be difficult because isozymes vary signficantly in their substrate selectivity and isozyme composition may be dependent upon the stage of plant development or prior chemical exposure (Edwards & Owen 1986; Dean *et al.* 1990).

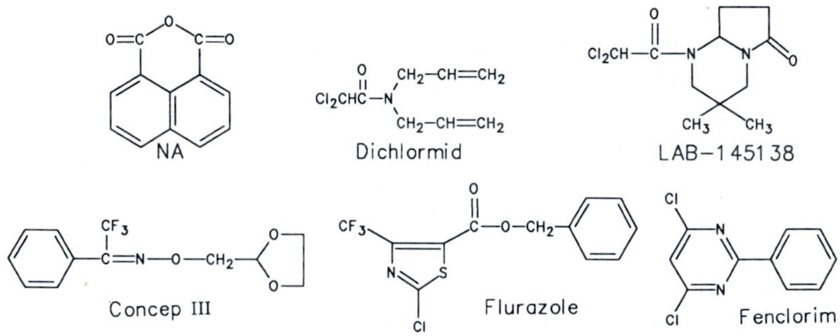

Fig. 3. Structures of some typical herbicide safeners.

Induction of GST activity with herbicide safeners

The GST enzymes from plants (Hatzios & Hoagland 1989) animals (Sies & Ketterer 1988), insects (Lamoureux & Rusness 1989a), and microorganisms (Tamaki *et al.* 1990) can be induced by chemical or environmental factors. The induction of plant GST enzymes is an active area of research and many compounds that induce GST activity in plants have proven effective as herbicide safeners. Safeners are compounds used in combination with herbicides to decrease the toxic effects of herbicides on plants. Some safeners are selective; *i.e.*, they reduce herbicide toxicity to only a few plant species. Other safeners reduce herbicide toxicity to many species and are considered nonselective.

There are several successful safeners and/or families of safeners and all or most of these cause an elevation of GST activity in some plant species (Fig. 3). Naphthalic anhydride (NA) and related anhydrides are nonselective safeners that elevate GST activity in sorghum and corn (Pallos & Casida 1978; Hatzios & Hoagland 1989). The dichloroacetanilides (dichlormid and LAB-145138/BAS 145 138) are selective safeners that induce GST activity in both corn and sorghum (Hatzios & Hoagland 1989; Dean *et al.* 1990, 1991). The oxime safeners (Concept III) and flurazole have been most extensively studied in sorghum and corn, but they induce GST in several other species as well (Mozer *et al.* 1983; Hatzios & Hoagland 1989; Dean *et al.* 1990). Fenclorim also induces GST and it is an effective safener in rice (Komives & Dutka 1989). These safeners are effective with either or both the chloroacetamide or thiocarbamate herbicides which are metabolized by GSH conjugation.

In corn, 0.3 ppm dichlormid causes a 2- to 3-fold elevation in the enzymatic rate of EPTC sulfoxide conjugation with GSH, and at 3 ppm it also causes a 2-fold elevation in GSH levels (Lay & Casida 1976). Dichlormid and other dichloroacetamide safeners have been consistently shown to cause an increase in the rate of GSH conjugation (Hatzios & Hoagland 1989; Fuerst & Lamoureux 1992; Dean *et al.* 1991). In corn, induction of GST activity by dichlormid and related compounds begins within 12 h and reaches a maximum within 48 h (Komives *et al.* 1986). In sorghum, induction of GST by an oxime ether begins within 12 h and reaches a maximum in 36 h; the induction is blocked with either a transcription inhibitor (cordycepin) or a translation inhibitor (cycloheximide) (Dean *et al.* 1990). Two constitutive GST isozymes and five new GST isozymes are induced with different safeners or metolachlor herbicide, but the induction patterns vary somewhat depending upon what safener or herbicide is used (Dean *et al.* 1990). In corn, flurazole causes a 3- to 9-fold increase in the steady state level of mRNA (Wiegand *et al.* 1986; Edwards & Owen 1988), and a dichloroacetamide safener (CGA-154281) selectively induces several GST isozymes (Dean *et al.* 1991). An isozyme active with cinnamic acid and three isozymes active with atrazine are not affected by CGA-154281, but the isozymes active with CDNB, metolachlor, and EPTC sulfoxide are elevated. A new isozyme active with CDNB and a new isozyme active with metolachlor also are induced (Dean *et al.* 1991).

GST activity also can be induced by herbicides. In corn, atrazine metabolism by GSH conjugation is accelerated by pretreatment with a low dose of atrazine (Jachetta & Radosevich 1981). The pretreatment results in a more rapid recovery from inhibition of photosynthesis caused by atrazine. In corn, CDAA metabolism by GSH conjugation is accelerated by pretreatment with a low dose of CDAA which increases tolerance for CDAA (Ezra *et al.* 1985). In sorghum, pretreatment with metolachlor causes an induction of GST isozymes and accelerates the *in vitro* rate of metolachlor GSH conjugation. The GST isozyme pattern produced by metolachlor in sorghum is slightly different from those produced by safeners (Dean *et al.* 1990). In chickpea (*Cicer arietinum*), a 24-h exposure to 10 ppm oxadiazon causes a 3.7-fold elevation in GST activity in the roots. The elevation in GST activity is observed within 6 h and occurs in the seedling, roots, stem, leaves, and shoots. Oxadiazon also causes the induction of GST in *Vicia faba* and other species (Hunaiti & Ali 1990, 1991).

The auxinic compounds 2,4-D, IAA, and 1-napthalene acetic acid cause the induction of GST activity in tobacco mesophyll protoplasts (Takahashi & Nagata 1992). Although GST activity with CDNB was elevated only about 40% by 2,4-D, CDNB may not be a good substrate for this enzyme (Dean *et al.* 1991). The auxin-regulated cDNA responsible for this activity was isolated and expression was demonstrated during transition of the cultured protoplasts from the G_O to the S phase. Induction of mRNA was observed within 10 to 20 min of treatment and was maximal in 4 h. The amino acid sequence deduced from the cDNA shows significant homology to maize GST III. The cDNA with an expression vector was successfuly used to transform *E. coli* which then produced the expected GST (Takahashi & Nagata 1992).

The mechanism of GST induction by safeners has not been determined, but likely mechanisms proposed are: (a) the safener may combine with a gene repressor in a negative control mechanism, (b) the safener may combine with a gene activator in a postive control mechanism, or (c) an indirect method might involve an activator-repressor complex that becomes activated when the safener combines with the repressor which releases the activator to fit on the gene and stimulate the gene to produce mRNA (Hatzios & Hoagland 1989). Since a large number of new proteins with GST activity are produced as a result of certain herbicide safener treatments, it is possible that other processes, including post-translational ones, may also be involved. The mechanism of GST induction is an area that needs considerable research.

One known mechanism of action of herbicide safeners is to cause an increase in the rate of herbicide detoxification by GSH conjugation. This is true with safeners used with the chloroacetanilide and thiocarbamate herbicides. However, safeners also can elevate monooxygenase activity involved in the metabolism of some herbicides (Frear *et al.* 1991), glutathione reductase and GSH levels (Komives *et al.* 1986), glucosyltransferase activity (Lamoureux & Rusness 1991), and the uptake and assimilation of sulfur (Lamoureux & Rusness 1989a). More than one of these effects may occur at the same time and the protective mechanism may involve several processes, including some not related to pesticide detoxification. Clearly, the effects of herbicide safeners on plants are complex.

Since herbicide safeners elevate the levels of several metabolic enzymes, they might be expected to cause an alteration in the chemical composition of herbicide residues present in the plant. The effect of a dichloroacetanilide safener, LAB-145138 (BAS 145 138) on the metabolism of two chloroacetanilide herbicides (metolachlor and propachlor) in corn was examined. In neither case was the qualitative nature of the metabolites altered by the safener (Khalifa & Lamoureux 1990). Although these safeners did not affect the herbicide residues detected in these studies, they may have some impact on the metabolites produced with other classes of herbicides (Lamoureux & Rusness 1991).

Molecular biology and genetic engineering

A number of studies on the molecular biology of GST isozymes from maize were reported in the mid and late 1980s (Moore *et al.* 1986; Shah *et al.* 1986; Wiegand *et al.* 1986; Grove *et al.* 1988; Timmerman 1989; Wosnick *et al.* 1989). Genes for GST activity in maize have been cloned, inserted into shuttle vectors and GST activity has been expressed in both *E. coli* and yeast.

Inhibition of GST activity

Crude GST preparations from corn and pea are inhibited by the common inhibitors of mammalian GST activity, including triphenyl tin chloride, bromocresol green, and sulfobromophthalein (Lamoureux & Rusness 1986b, 1989a). In addition, the natural plant constituents quercetin, ellagic acid, juglone, hemin, and chlorophyllin are also inhibitors of GST (Singh & Shaw 1988; Lee 1991).

The glutathione conjugate and the glutathione sulfoxide conjugate of tridiphane are among the most potent inhibitors of plant GST activity. They inhibit plant GST enzymes (corn, pea, giant foxtail) and also GST enzymes from equine liver and housefly. The I_{50} values range from 0.7 to 19 μM when assayed with atrazine, fluorodifen, PCNB, propachlor, diazinon, and CDNB as the substrates. Inhibition of GST from giant foxtail by the tridiphane GSH conjugate is competitive with respect to GSH (K_i = 2.2 μM) and noncompetitive with respect to CDNB (K_i = 12.2 μM). The GSH conjugate of tridiphane is 20- to 1000-times more effective as an inhibitor of GST than several other xenobiotic GSH conjugates tested (Lamoureux & Rusness 1986b, 1989a).

Tridiphane is metabolized to a GSH conjugate in giant foxtail and is a potent synergist of atrazine which is also metabolized by GSH conjugation in this species. The synergism of atrazine toxicity appears to be at least partly related to inhibition of atrazine metabolism (Lamoureux 1989). Tridiphane also synergizes diazinon toxicity in the housefly, apparently by inhibition of GSH conjugation of diazinon (Lamoureux & Rusness 1987a). Some caution is needed in the interpretation of *in vivo* results with tridiphane since tridiphane also inhibits monooxygenase and GSH reductase activities (Lamoureux & Rusness 1989a; Moreland *et al.* 1989). The insecticide, fenitrothion is synergized in the housefly by IBP fungicide (*O,O*-diisopropyl *S*-benzyl phosphorothiolate). It appears that the fungicide is metabolized to a GSH conjugate that inhibits GST-mediated demethylation of fenitrothion (Shiotsuki 1991).

Lipophilic GSH conjugates are excellent inhibitors of mammalian GST enzymes and they are usually competitive with GSH at the active site (Mannervik *et al.* 1989). A variety of halogenated benzoquinones are irreversible inhibitors of mammalian GST activity. Conversion of these benzoquinones to GSH conjugates greatly increases the rate at which irreversible inhibition of GST activity occurs (Van Ommen *et al.* 1991). It is uncertain whether these compounds could be effectively used *in vivo* as specific inhibitors of GST. Glutathione sulfonate also inhibits GST activity and it has been used to study the structure of mammalian GST enzymes (Reinemer *et al.* 1991).

Natural function of GST isozymes in plants

In addition to protecting plants against toxic xenobiotics, plant GST enzymes may play other important roles. They probably are involved in the metabolism of endogenous chemicals or natural plant hormones. The findings that several plant species have GST activity for cinnamic acid (Edwards & Dixon 1991), the occurrence of a GST conjugate of caftaric acid in wine (Cheynier *et al.* 1986), and the identification of gibberthione as a natural plant constituent that could be synthesized from a GSH conjugate of 3-keto GA_3 are all consistent with this hypothesis (Lamoureux & Frear 1987).

GST enzymes may play a role in the metabolism and detoxification of lipid peroxidation products. A GST preparation from pea catalyzes the detoxification of linoleic acid hydroperoxide to an alcohol (Williamson & Beverley 1987). All three major classes of GST enzymes from humans catalyze the GSH conjugation of 9,10-epoxystearic acid (Sharma *et al.* 1991), and it is likely that plant GST isozymes could catalyze similar detoxification reactions. Toxic aldehydes, such as 3-nonenal and 4-hydroxy-2-nonenal, are produced during lipid peroxidation. These aldehydes can be detoxified by GSH conjugation (Mannervik *et al.* 1989; Spitz *et al.* 1991) and GST enzymes increase the rate of this detoxification reaction (Spitz *et al.* 1991). It is possible that this class of lipid peroxidation products may be detoxified by plant GST enzymes.

A GST preparation from oat is inhibited by the tetrapyrroles, chlorophyllin and hemin (Singh & Shaw 1988). This may suggest that GST enzymes play a role in the transport of these compounds between subcellular compartments within the plant cell. A similar proposal has been made for mammalian GST isozymes (Boyer & Olsen 1991). In wheat, a protein with GST activity is induced by a fungal strain that is not pathogenic in wheat. This wheat is protected from subsequent invasion by a pathogenic form of the fungus. This may suggest that GST enzymes play a role in protecting plants from pathogenic infection (Dudler *et al.* 1991; Mauch *et al.* 1991). It is not known whether GST activity with cinnamic acid is involved in this protective effect (Edwards & Dixon 1991). Herbicide safeners possibly could be used to help protect plants against some fungi. The recent finding that auxinic compounds induce GST activity may point to a natural function for these enzymes in the proliferative activity of cells (Takahashi & Nagata 1992). GST induction in carnation in response to ethylene also has been reported and this suggests a natural function for GST (Meyer *et al.* 1991); however, the induction of GST by ethylene was not detected in tobacco protoplasts (Takahashi & Nagata 1992). GST enzymes may also play a role in the protection of plants during drought since these enzymes can be induced by drought (Dhindsa 1991).

Concluding remarks

Most reported examples of GSH conjugation of xenobiotics in plants involve nucleophilic displacement reactions at a carbon site, but other classes of GSH conjugation reactions also have been observed. Few if any new classes of xenobiotics have been shown to be metabolized by GSH conjugation in plants, but several new pathways have been discovered in the metabolism and breakdown of GSH conjugates. GST activity has been demonstrated in over 33 plant species, including members of the *Pteridophyta* division and in gymnosperms and angiosperms in the *Spermatophyta* division. Our knowledge of the GST isozymes and the induction of these isozymes has been expanded considerably in recent years. The molecular biology of GST isozymes has been advanced to the point where transgenic yeast and bacteria that contain plant GST isozymes have been produced. Knowledge of the active site and kinetics of these enzymes has not been appreciably advanced. However, studies with auxinic compounds, plant pathogens, lipid oxidation products, natural plant constituents, *etc.*, have yielded results that indicate important natural functions for GST enzymes in addition to the detoxification of xenobiotics.

References

Adang, E.P., Brussee, J., Meyer, D.J., Coles, B., Ketterer, B., Van Der Gen, A. & Mulder, G.J. 1988. Substrate specificity of rat liver glutathione *S*-transferase isoenzymes for a series of glutathione analogues, modified at the γ-glutamyl moiety. Biochem. J. 255: 721-724.

Aizawa, H. 1989. Metabolic Maps of Pesticides, Vol. 2. Academic Press, New York, pp. 1-272.

Anderson, M.P. & Gronwald, J.W. 1991. Atrazine resistance in a velvetleaf (*Abutilon theophrasti*) biotype due to enhanced glutathione *S*-transferase activity. Plant Physiol. 96: 104-109.

Balabaskaran, S. & Muniandy, N. 1984. Glutathione *S*-transferase from *Hevea brasiliensis*. Phytochemistry 23: 251-256.

Blattman, P., Gross, D., Kriemler, H.P. & Ramsteiner, K. 1986. Identification of thiolactic acid type conjugates as major degradation products in glutathione dependent metabolism of the 2-chloroacetamide herbicides metolachlor (Dual), dimetachlor (Teridox) and Pretilachlor (Kifit). Abstract, 6th International Congress of Pesticide Chemistry (IUPAC), Ottawa, Canada.

Boyer, T.D. & Olsen, E. 1991. Role of glutathione *S*-transferases in heme transport. Biochem. Pharmacol. 42: 188-190.

Brown, H.M. 1990. Mode of action, crop selectivity, and soil relations of the sulfonylurea herbicides. Pestic. Sci. 29: 263-281.

Burkholder, R.R.S. 1977. A survey of several important agronomic plant species for the presence of glutathione-*S*-transferase activity. Thesis for Master of Science in Chemistry, North Dakota State University, College of Chemistry, Fargo, North Dakota.

Cheynier, V.F., Trousdale, E.K., Singleton, V.L., Salgues, M.J. & Wylde, R. 1986. Characterization of 2-*S*-glutathionyl-caftaric acid and its hydrolysis in relation to grape wines. J. Agric. Food Chem. 34: 217-221.

Dean, J.V., Gronwald J.W. & Eberlein, C.V. 1990. Induction of glutathione *S*-transferase isozymes in sorghum by herbicide antidotes. Plant Physiol. 92: 467-473.

Dean, J.V., Gronwald, J.W. & Anderson, M.P. 1991. Glutathione *S*-transferase activity in nontreated and CGA-154281-treated maize shoots. Z. Naturforsch. 46c: 850-856.

Dhindsa, R.S. 1991. Drought stress enzymes of glutathione metabolism oxidation injury and protein synthesis in *tortula-ruralis*. Plant Physiol. 95: 648-651.

Diesperger, H. & Sandermann, H. 1979. Soluble and microsomal glutathione *S*-transferase activities in pea seedlings. Planta 146: 643-648.

Dudler, R., Hertig, C., Rebmann, G., Bull, J. & Mauch, F. 1991. A pathogen-induced wheat gene encodes a protein homologous to glutathione-*S*-transferase. Molec. Plant-Microbe Inter. 4: 14-18.

Edwards, R. & Owen, W. 1986. Comparison of glutathione *S*-transferase of *Zea mays* responsible for herbicide detoxification in plants and suspension-cultured cells. Planta 169: 208-215.

Edwards, R. & Owen, W.J. 1988. Regulation of glutathione *S*-transferases of *Zea mays* in plants and cell cultures. Planta 175: 99-106.

Edwards, R. & Dixon, R.A. 1991. Glutathione *S*-cinnamoyl transferases in plants. Phytochemistry 30: 79-84.

Eyer, P. & Schneller, M. 1983. Reactions of the nitroso analogue of chloramphenicol with reduced glutathione. Biochem. Pharmacol. 32: 1029-1036.

Ezra, G., Rusness, D.G., Lamoureux, G.L. & Stephenson, G.R. 1985. The effect of CDAA (N,N,-diallyl-2-chloroacetamide) pretreatments on subsequent CDAA injury to corn (*Zea mays* L.). Pestic. Biochem. Physiol. 23: 108-115.

Frear, D.S. & Swanson, H.R. 1970. Biosynthesis of *S*-(4-ethylamino-6-isopropylamino-2-s-triazino) glutathione. Partial purification and properties of glutathione *S*-transferase from corn. Phytochemistry 9: 2123-2132.

Frear, D.S. & Swanson, H.R. 1973. Metabolism of substituted diphenylether herbicides in plants. Enzymatic cleavage of fluorodifen in peas (*Pisum sativum* L.). Pestic. Biochem. Physiol. 3: 473-482.

Frear, D.S., Swanson, H.R. & Mansager, E.R. 1985. Alternate pathways of metribuzin metabolism in soybean: Formation of *N*-glucoside and homoglutathione conjugates. Pestic. Biochem. Physiol. 23: 56-65.

Frear, D.S., Swanson, H.R. & Thalacker, F.W. 1991. Induced microsomal oxidation of diclofop, triasulfuron, chlorsulfuron, and linuron in wheat. Pestic. Biochem. Physiol. 41: 274-287.

Fuerst, E.P. & Lamoureux, G.L. 1992. Mode of action of the dichloroacetamide antidote BAS 145 138 in corn. 2. Effects on metabolism, absorption, and mobility of metazachlor. Pestic. Biochem. Physiol. 42: 78-87.

Fukami, J. 1984. Metabolism of several insecticides by glutathione *S*-transferase. In: Matsumura, F. (ed.). Internatl. Encycl. Pharmacol. Ther. Pergamon Press, Oxford, pp. 223-264.

Grove, G., Zarlengo, R.P., Timmerman, K.P., Li, N-Q., Tam, M.F. & Tu, C-P. 1988. Characterization and heterospecific expression of cDNA clones of genes in the maize GSH *S*-transferase multigene family. Nucl. Acids Res. 16: 425-438.

Hatzios, K.H. 1991. Biotransformations of herbicides in plants. In: Grover, R. & Cessna A.J. (eds.). Environmental Chemistry of Herbicides, Vol. II. CRC Press, Boca Raton, pp. 141-187.

Hatzios, K.K. & Hoagland, R.E. 1989. Crop Safeners for Herbicides. Academic Press Inc., New York.

Huber, S.J. & Sautter, C. 1987. Immunofluorescence localization of conjugated atrazine in leaf pieces of corn (*Zea mays*). Z. Pflanzenkr. Pflanzenschutz 93: 608-613.

Hunaiti, A.A. & Ali, B.R. 1990. Glutathione *S*-transferase from oxadiazon treated chickpea. Phytochemistry 29: 2431-2435.

Hunaiti, A.A. & Ali, B.R. 1991. The induction of chickpea glutathione *S*-transferase by oxadiazon. Phytochemistry 30: 2131-2134.

Inskeep, P.B., Koga, N., Cmarik, J.L. & Guengessich, F.P. 1986. Covalent binding of 1,2-dihaloalkanes to DNA and stability of the major DNA adduct, *S*-[2-(N^7-guanyl)ethyl]glutathione. Cancer Res. 46: 2839-2844.

Jablonkai, I. & Hatzios, K.K. 1991. Role of glutathione and glutathione *S*-transferase in selectivity of acetochlor in maize and wheat. Pestic. Biochem. Physiol. 41: 221-231.

Jachetta, J.J. & Radosevich, S.R. 1981. Enhanced degradation of atrazine by corn. Weed Sci. 29: 37-44.

Khalifa, M.A. & Lamoureux, G.L. 1990. The effect of BAS 145 138 safener on the secondary/tertiary steps of propachlor metabolism in corn. Abstract, 7th International Congress of Pesticide Chemistry (IUPAC), Hamburg, Germany.

Komives, T. & Dutka, F. 1989. Biochemical mode of action of the antidote fenclorim in rice. Biochem. Physiol. Pflanz. 184: 475-477.

Komives, T., Komives, V.A., Balazs, M. & Dutka, F. 1985. Role of glutathione-related enzymes in the mode of action of herbicide antidotes. British Crop Protection Conference-Weeds B9-2: 1155-1160.

Komives, A.V., Komives, T. & Dutka, F. 1986. Effects of thiocarbamate herbicides on the activity of glutathione *S*-transferase in maize (Zea-mays). Cereal Res. Comm. 13: 253-258.

Kumagai, H., Tamake, H., Koshino, Y., Suzuki, H. & Tochikura, T. 1988. Distribution formation and stabilization of yeast glutathione *S*-transferase. Agric. Biol. Chem. 52: 1377-1382.

Lamoureux G.L. & Rusness, D.G. 1980. In vitro metabolism of pentachloronitrobenzene to pentachloromethylthiobenzene by onion: Characterization of glutathione *S*-transferease, cysteine C-S lyase, and *S*-adenosylmethionine methyl transferase activities. Pestic. Biochem. Physiol. 14: 50-61.

Lamoureux, G.L. & Rusness, D.G. 1981. Catabolism of glutathione conjugates of pesticides in plants. In: Rosen, J., Magee, P. & Casida, J. (eds.). Sulfur in Pesticide Action and Metabolism. 158. Am. Chem. Soc., Washington, DC, pp. 133-164.

Lamoureux, G.L. & Rusness, D.G. 1983. Malonylcysteine conjugates as end-products of glutathione conjugate metabolism in plants. In: Miyamoto, J. & Kearney, P.C. (eds.). Pesticide Chemistry Human Welfare and the Environment Vol. 3. Pergaman Press, Oxford, pp. 295-300.

Lamoureux, G.L. & Bakke, J.E. 1984. Formation and metabolism of xenobiotic glutathione conjugates in various life forms. In: Caldwell, J. & Paulson, G.D. (eds.). Foreign Compound Metabolism. Taylor and Francis, London, pp. 185-199.

Lamoureux, G.L. & Rusness, D.G. 1986a. Xenobiotic conjugation in higher plants. In: Paulson, G.D., Caldwell, J., Hutson, D.H. & Menn, J.J. (eds.). Xenobiotic Conjugation Chemistry. 299, Am. Chem. Soc., Washington, DC, pp. 62-105.

Lamoureux, G.L. & Rusness, D.G. 1986b. Tridiphane [2-(3,5-dichlorophenyl)-2-(2,2,2-trichloroethyl)oxirane] an atrazine synergist: enzymatic conversion to a potent glutathione *S*-transferase inhibitor. Pestic. Biochem. Physiol. 26: 323-342.

Lamoureux, G.L. & Frear, D.S. 1987. Current problems, trends, and developments in pesticide metabolism in plants. In: Greenhalgh, R. & Roberts, T.R. (eds.). Pesticide Science and Biotechnology. Blackwell Scientific Publications, Oxford, pp. 455-463.

Lamoureux, G.L. & Rusness, D.G. 1987a. Synergism of diazinon toxicity and inhibition of diazinon metabolism in the house fly by tridiphane: Inhibition of glutathione *S*-transferase activity. Pestic. Biochem. Physiol. 27: 318-329.

Lamoureux, G.L. & Rusness, D.G. 1987b. EPTC metabolism in corn, cotton, and soybean: Identification of a novel metabolite derived from the metabolism of a glutathione conjugate. J. Agric. Food Chem. 35: 1-7.

Lamoureux, G.L. & Rusness, D.G. 1989a. The role of glutathione and glutathione *S*-transferases in pesticide metabolism, selectivity, and mode of action in plants and insects. In: Dolphin, D., Poulson, R. & Avramovic, O. (eds.), Glutathione: Chemical, Biochemical, and Medical Aspects. Part B, pp. 153-196.

Lamoureux, G.L. & Rusness, D.G. 1989b. Propachlor metabolism in soybean plants, excised soybean tissues, and soil. Pestic. Biochem. Physiol. 34: 187-204.

Lamoureux, G.L. & Rusness, D.G. 1991. The effect of BAS 145 138 on chlorimuron ethyl metabolism and toxicity in corn. Z. Naturforsch. 46c: 96-100.

Lamoureux, G.L., Stafford, L.E., Shimabukuro, R.H. & Zaylskie, R.G. 1973. Atrazine metabolism in sorghum: catabolism of the glutathione conjugate of atrazine. J. Agric. Food Chem. 21: 1020-1030.

Lamoureux, G.L., Rusness, D.G., Schröder, P. & Rennenberg, H. 1991a. Diphenyl ether herbicide metabolism in a spruce cell suspension culture: The identification of two novel metabolites derived from a glutathione conjugate. Pestic. Biochem. Physiol. 39: 291-301.

Lamoureux, G.L., Rusness, D.G. & Tanaka, F.S. 1991b. Chlorimuron ethyl metabolism in corn. Pestic. Biochem. Physiol. 41: 66-81.

Lamoureux, G.L., Shimabukuro, R.H. & Frear, D. S. 1991c. Glutathione and glucoside conjugation in herbicide selectivity. In: Caseley, J.C., Cussans, G.W. & Atkin R.K. (eds.). Herbicide Resistance in Weeds and Crops. Butterworth-Heinemann, Oxford, pp. 227-261.

Lamoureux, G.L., Rusness, D.G. & Schröder, P. 1992. Metabolism of a diphenylether herbicide to a volatile thioanisole and a polar sulfonic acid metabolite in spruce (*Picea*). Pestic. Biochem. Physiol. (submitted, 1992).

Lay, M.-M. & Casida, J.E. 1976. Dichloroacetamide antidotes enhance thiocarbamate sulfoxide detoxifi cation by elevating corn root glutathione content and glutathione *S*-transferase activity. Pestic. Biochem. Physiol. 6: 442-456.

Lee, K.W. 1991. Glutathione *S*-transferase activities in phytophagous insects – induction and inhibition by plant phototoxins and phenols. Insect Biochemistry 21: 353-361.

Mannervik, B., Carlberg, I. & Larson, K. 1989. Glutathione: General review of mechanism of action. In: Coenzymes and Cofactors. John Wiley & Sons, New York, Vol. III, Part A, pp. 475-517.

Mauch, F., Hertig, C., Rebmann, G., Bull, J. & Dudler, R. 1991. A wheat glutathione-*S*-transferase gene with transposon-like sequences in the promoter region. Plant Molec. Biol. 16: 1089-1091.

Mayer, P., Kriemler, H. & Laanio, T. 1981. *N*-(1',2'-dimethylpropyl)-N'-ethyl-6-methylthio-1,3,5-triazine-2,4-diamine (C 18 898) in paddy rice. Agric. Biol. Chem. 45: 361-368.

Meyer, R.C., Goldsbrough, P.B. & Woodson, W.R. 1991. An ethylene-responsive flower senescence-related gene from carnation encodes a protein homologous to glutathione *S*-transferases. Plant Molec. Biol. 17: 277-281.

Moore, R.E., Davies, M.S., O'Connell, K.M., Harding, E.I., Wiegand, R.C. & Tiemeier, D.C. 1986. Cloning and expression of a cDNA encoding a maize glutathione *S*-transferase in *E. coli*. Nucl. Acid Res. 14: 7227-7235.

Moreland, D.E., Novitzky, W.P. & Levi, P.E. 1989. Selective inhibition of cytochrome P-450 isozymes by the herbicide synergist tridiphane. Pestic. Biochem. Physiol. 35: 42-50.

Mozer, T., Tiemeier, D. & Jaworski, E. 1983. Purification and characterization of corn glutathione *S*-transferase. Biochemisty 22: 1068-1072.

O'Connell, K.M., Breaux, E.J. & Fraley, R.T. 1988. Different rates of metabolism of two chloroacetanilide herbicides in Pioneer 3320 corn. Plant Physiol. 86: 359-363.

Oberley, T.D., Oberley, L.W., Slattery, A.F. & Elwell, J.H. 1991. Immunohistochemical localization of glutathione-*S*-transferase and glutathione peroxidase in adult Syrian hamster tissues and during kidney development. Amer. J. Pathol. 139: 355-369.

Pallos, F.M. & Casida, J.E. 1978. Chemistry and Action of Herbicide Antidotes. Academic Press, New York.

Price, A., Lucas, P.W. & Lea, P.J. 1990. Age dependent damage and glutathione metabolism in ozone fumigated barley: A leaf section approach. J. Exp. Bot. 41: 1309-1317.

Reinemer. P., Dirr, H.W., Ladenstein, R., Schaffer, J., Gallay, O. & Huber R. 1991. The three-dimensional structure of class-pi glutathione *S*-transferase in complex with glutathione sulfonate at 2.3 A resolution. EMBO J. 10: 1997-2005.

Rennenberg, H. & Lamoureux, G.L. 1990. Physiological processes that modulate the concentration of glutathione in plant cells. In: Rennenberg, H., Brunold, C.H., De Kok, L.J. & Stulen, I., (eds.). Sulfur Nutrition and Sulfur Assimilation in Higher Plants. SPB Academic Publishing, The Hague, pp. 53-67.

Schröder, P., Lamoureux, G.L., Rusness, D.G. & Rennenberg, H. 1990. Glutathione *S*-transferase activity in spruce needles. Pestic. Biochem. Physiol. 37: 211-218.

Schuphan, I., Westphal, D., Hague, A. & Ebing, W. 1981. Biological and chemical behavior of perhalogenmethylmercapto fungicides: Metabolism and *in vitro* reactions of dichlofluanid in comparison with captan sulfur. In: Rosen, J.D., Magee, P.S. & Casida, J.E. (eds.). Pesticide Action and Metabolism. 158, Am. Chem. Soc., Washington, DC, pp. 65-85.

Shah, D.M., Hironaka, C.M., Wiegand, R.C., Harding, E.I., Krivi, G.G. & Tiemeier, D.C. 1986. Stuctural analysis of a maize gene coding for glutathione *S*-transferase involved in herbicide detoxification. Plant Molec. Biol. 6: 203-211.

Sharma, R., Gupta, S., Singhal, S.S., Ansari, G.A.S. & Awasthi, Y.C. 1991. Glutathione *S*-transferase catalyzed conjugation of 9,10-epoxystearic acid with glutathione. J. Biochem. Toxicol. 6: 147-153.

Shimabukuro, R.H., Lamoureux, G.L. & Frear, D.S. 1978. Glutathione conjugation: A mechanism of herbicide detoxification and selectivity in plants. In: Pallos, F.M. & Casida, J.E. (eds.). Chemistry and Action of Herbicide Antidotes. Academic Press, New York, pp. 133-150.

Shiotsuki, T. 1991. Mode of action of saligenin cyclic phosphates on organophosphate-resistant houseflies. J. Pestic. Sci. 16: 523-531.

Sies, H. & Ketterer, B. 1988. Glutathione Conjugation: Mechanisms and Biological Significance, Academic Press, New York, pp. 1-480.

Singh, B.R. & Shaw, R.W. 1988. Selective inhibition of oat glutathione-*S*-transferase activity by tetrapyrroles. FEBS Lett. 234: 379-382.

Singhal, S.S., Tiwari, N.K., Ahmad, H., Srivastava, S.K. & Awasthi, Y.C. 1991. Purification and characterization of glutathione *S*-transferase from sugarcane leaves. Phytochemistry 30: 1409-1414.

Sptiz, D.R., Sullivan, S.J., Malcolm, R.R. & Roberts, R.J. 1991. Glutathione dependent metabolism and detoxification of 4-hydroxy-2-nonenal. Free Radical Biol. & Med. 11: 415-423.

Still, G.G. & Rusness, D.G. 1977. *S*-cysteinyl-hydroxychlorpropham: Formation of the *S*-cysteinyl conjugate of isopropyl-3-chloro-4-hydroxy-carbanilate in oat (*Avena sativa* L.). Pestic. Biochem. Physiol. 7: 210-219.

Takahashi, Y. & Nagata, T. 1992. Part B: An auxin-regulated gene encoding glutathione *S*-transferase. Proc. Natl. Acad. Sci. USA 89: 56-59.

Tamaki, H., Kumagai, H. & Tochikura, T. 1990. Glutathione *S*-transferase in yeast induction of messenger RNA complementary DNA cloning and expression in *Escherichia coli*. Biochem. Biophys. Res. Commun. 172: 669-675.

Timmerman, K.P. 1989. Molecular characterization of corn glutathione *S*-transferase isozymes involved in herbicide detoxication. Physiol. Plant. 77: 465-471.

Tomisawa, H., Hayashi, M., Uda, F., Okamoto, A., Hattori, K., Ozawa, N. & Tateishi, M. 1992. Purification and characterization of 3-mercaptolactic acid *S*-conjugate oxidases. Biochem. Pharmacol. 44: 703-714.

Toung Y-PS., Hsieh, T-S. & Tu, C-PD. 1990. *Drosophila* glutathione *S*-transferase 1-1 shares a region of sequence homology with the maize glutathione *S*-transferase III. Proc. Natl. Acad. Sci. USA 87: 31-35.

Van Ommen, B., Ploemen, J.H.T.M., Bogaards, J.J.P., Monks, T.J., Gau, S.S. & Van Bladeren, P.J. 1991. Irreversible inhibition of rat glutathione *S*-transferase-1-1 by quinones and their glutathione conjugates – structure activity relationship and mechanism. Biochem. J. 276: 661-666.

Wiegand, R.C., Shah, D.M., Mozer, T.J., Harding, E.I., Diaz-Collier, J., Saunders, C., Jaworski, E.G. & Teimemeier, D.C. 1986. Messenger RNA encoding a glutathione-*S*-transferase responsible for herbicide tolerance in maize is induced in response to safener treatment. Plant Molec. Biol. 7: 235-243.

Williamson, G. & Beverley, M.C. 1987. The purification of acidic glutathione *S*-transferases from pea seeds and their activity with lipid peroxidation products. Biochem. Soc. Trans. 15: 1103-1104.

Williamson, G. & Beverley, M.C. 1988. Wheat glutathione *S*-transferase purification and properties. J. Cereal Sci. 8: 155-164.

Wosnick, M.A., Barnett, R.W. & Carlson, J.E. 1989. Total chemical synthesis and expression in *Escherichia coli* of a maize glutathione-transferase (GST) gene. Gene 76: 153-160.

METAL-BINDING PEPTIDES IN PLANTS

Wilfried E. Rauser
Department of Botany, University of Guelph, Guelph, Ontario N1G 2W1, Canada

Introduction

Plants may use various mechanisms to deal with excess metals acquired from the environment (Tomsett & Thurman 1988). Intracellular chelation of metal has received the most attention. Amongst the potential metal complexors, emphasis has been given to the subset involving sulfur-rich polypeptides named cadystins (Kondo *et al.* 1984) and phytochelatins (Grill *et al.* 1985). These compounds were discussed in 1989 at the first workshop in this series with the following generalizations (Grill *et al.* 1990). The primary structure of the polypeptides was (γ-glutamyl-cysteinyl)$_n$-glycine or $(\gamma EC)_nG$ with the number of γEC pairs, n, ranging from 2 to 8 depending on the organism. Low amounts of the smaller peptides occurred in normal plant cells, the synthesis of the range of peptides occurred with certain metals. Cadmium caused the appearance of $(\gamma EC)_nG$ polypeptides in a wide variety of organisms including two members of the Kingdom Fungi and many members of the Kingdom *Phyta* (*Plantae*) ranging from algae through to vascular plants. The polypeptides were not synthesized by ribosomes glutathione was required for their synthesis. Plants containing homoglutathione produced the homologous molecules (γ-glutamyl-cysteinyl)$_n$-β-alanine or homophytochelatins. Cadmium was bound through the thiol of the polypeptides. Complexes between Cd, a heterogeneous population of sulfur-rich polypeptides, and sulfide were extractable from cells. The complexes could account for high percentages of intracellular Cd.

The purpose of this essay is to describe and assess developments in the field of metal-binding peptides that have occurred since the first workshop on sulfur metabolism in plants. The pertinent information available prior to and shortly after the 1989 workshop is considered extensively in reviews (Robinson 1989, 1990; Steffens 1990; Reddy & Prasad 1990; Rauser 1990) and is addressed selectively here. No consensus is yet available on the nomenclature of the thiol-rich polypeptides. The trivial names cadystins, phytochelatins, and homophytochelatins are restrictive. The name γ-glutamyl-cysteinyl isopeptide or γEC isopeptide will be used. This name accommodates the range of peptides occurring in metal-binding complexes and applies irrespective of the phylogenetic classification of the organism producing them.

Synthesis of (γ-glutamyl-cysteinyl)$_n$-glycines

In vitro studies

Involvement of glutathione (GSH) in the biosynthesis of $(\gamma EC)_nG$ was suggested by two lines of evidence. Kinetic analyses showed that, at early times, the γEC iso-

Sulfur Nutrition and Assimilation in Higher Plants
edited by L.J. De Kok et al., pp. 239–251

peptides were formed at the direct expense of GSH (Grill *et al.* 1987; Berger *et al.* 1989; Mendum *et al.* 1990), and inhibition of γ-glutamyl-cysteine synthetase with buthionine sulfoximine to limit formation of γEC, a precursor of GSH biosynthesis, inhibited formation of $(\gamma EC)_nG$ (Grill *et al.* 1987; Scheller *et al.* 1987; Reese & Wagner 1987).

Direct evidence of GSH involvement in the biosynthesis of $(\gamma EC)_nG$ was demonstrated with an enzyme preparation from cell cultures of *Silene cucubalus* (Grill *et al.* 1989; Loeffler *et al.* 1989). The enzyme was a γ-glutamyl cysteine dipeptidyl transpeptidase. It catalysed the transfer of the γEC moiety from a donor molecule of GSH to an acceptor molecule of GSH producing the pentapeptide $(\gamma EC)_2G$ and glycine. Once the pentapeptide accumulated and was in competition with GSH an an acceptor molecule, the γEC moiety from a GSH donor molecule was transferred to the pentapeptide to form the heptapeptide $(\gamma EC)_3G$ and glycine. In the absence of GSH the enzyme transferred γEC from the pentapeptide as donor molecule to the pentapeptide as acceptor molecule producing $(\gamma EC)_3G$ and GSH. The general reaction catalyzed by the dipeptidyl transpeptidase was

$$(\gamma EC)_nG + (\gamma EC)_nG \rightarrow (\gamma EC)_{n-1}G + (\gamma EC)_{n+1}G$$

where n = 1, 2, 3, ... (Grill *et al.* 1989).

The dipeptidyl transpeptidase enzyme was extracted from cultured cells not previously exposed to Cd, hence it was designated a constitutive enzyme. The enzyme activity was also found in extracts from cell cultures of *Podophyllum peltatum*, *Eschscholtzia california*, *Beta vulgaris* and *Equisetum giganteum*, but could not be shown in extracts of other cell cultures or from differentiated plant tissues (Grill *et al.* 1989).

Hayashi *et al.* (1991) confirmed the γEC dipeptidyl transpeptidase reaction using an enzyme preparation from the fission yeast *Schizosaccharomyces pombe*. Furthermore, they discovered that at low GSH concentration (12 μM instead of 1 mM) an enzyme activity transferred γEC from GSH to γEC, $(\gamma EC)_2$ or $(\gamma EC)_3$ to produce the respective $(\gamma EC)_{n+1}$ peptide. In the absence of GSH the enzyme preparation directly polymerized γEC into $(\gamma EC)_{2\text{-}3}$. These data provide the first indication that the glycine-less (desGly or $(\gamma EC)_n$) variants of $(\gamma EC)_nG$ can be products of synthesis rather than products of degradation as suggested in the past (Gekeler *et al.* 1989). Hayashi *et al.* (1991) also showed that by providing purified glutathione synthetase and glycine the $(\gamma EC)_{2\text{-}3}$ polymers could be converted to $(\gamma EC)_{2\text{-}3}G$. From these data it could be deduced that the $(\gamma EC)_nG$ found commonly in metal-binding complexes was formed from GSH when GSH was abundant and from γEC when GSH was low. It remains unclear whether the various transpeptidase reactions are catalyzed by a single enzyme or a number of separate enzymes present in the preparations.

The dipeptidyl transpeptidase activity depended absolutely on Cd or certain other metals (Grill *et al.* 1989). The transfer of γEC ceased when sufficient $(\gamma EC)_nG$ had been formed to complex Cd at a thiol:Cd ratio of about 2:1 (Loeffler *et al.* 1989). The Cd-activated enzyme could be stopped prematurely by chelating free Cd with NaEDTA or a complement of metal-free $(\gamma EC)_nG$. It was suggested that the dipeptidyl transpeptidase activity was self-regulated in that free Cd activated the enzyme and once the products of the reaction chelated the free Cd catalysis ceased (Grill *et al.* 1989, Loeffler *et al.* 1989). This interpretation was, however, not supported by Hayashi *et al.* (1991) who found transpeptidase activity with or without Cd present. Impure enzymes were used by both groups. The status of metal activation is unclear until highly purified enzymes are evaluated.

In vivo studies

Experiments with cultured cells of *Rauvolfia serpentina* (Grill *et al.* 1987) and *Datura innoxia* (Robinson *et al.* 1988; Berger *et al.* 1989) showed a rapid accumulation of $(\gamma EC)_2G$ after addition of Cd. The $n = 3$, 4 and 5 isopeptides appeared successively later and in turn accumulated more slowly. These data indicated that the shorter isopeptide was the precursor to the larger one. This pattern was matched by the kinetic studies with γEC dipeptidyl transpeptidase *in vitro* (Grill *et al.* 1989).

Most of the studies on induction of (γEC) isopeptides have been with Cd. Grill *et al.* (1987) showed that Cd, Pb, Zn, Sb, Ag, Ni, Hg, arsenate, Cu, Sn, selenate, Au, Bi, Te and W induced the peptides with decreasing effectiveness. The concentrations of metals and anions used seem to be arbitrary. Without knowledge of the local concentration at the controlling site it is difficult to justify ranking metals according to their effectiveness as peptide inducers. The γEC isopeptides appeared in the algae *Scenedesmus acutiformis* and *Chlorella fusca* on exposure to excess Ag, Cd, Cu, Hg, Pb and Zn (Gekeler *et al.* 1988). Responses to a wide variety of metals in the same organism have not been verified independently. In fission yeast Cd and Cu induced γEC isopeptides (Reese *et al.* 1988) as did Cd in tobacco cells (Reese & Wagner 1987), yet in both cases Zn was without effect. Lack of response may be related to too low concentration of metal at the control site or genuine differences between species in ability to respond.

In some cultured plant cells addition of Cd caused the concentration of GSH to decline rapidly followed by some increase towards the end of 9 and 24 hour experiments (Grill *et al.* 1987; Scheller *et al.* 1987; Mendum *et al.* 1990). For *Datura innoxia* cells, however, Cd exposure for 2 hours caused little change in GSH despite synthesis of γEC isopeptides (Berger *et al.* 1989). The concentration of GSH in Cd-exposed tomato cells increased by the second day and the higher levels were maintained for 12 to 16 days (Gupta & Goldsbrough 1991). The concentration of GSH in roots of maize seedlings also declined rapidly after exposure to Cd (Rauser 1987; Tukendorf & Rauser 1990; Rauser *et al.* 1991; Meuwly & Rauser 1992), however, unlike the situation in cultured cells, the concentration of GSH did not recover to that in controls within 4 to 7 days (Meuwly & Rauser 1992; Rüegsegger & Brunold 1992). The lack of recovery only in roots of intact plants was puzzling because Cd had no adverse effect on GSH flux to maize roots at least over the initial 32 hours (Rauser *et al.* 1991) and thiols were present to participate in some binding of Cd (Meuwly & Rauser 1992). It is uncertain whether the lower concentrations of GSH are themselves of some biological strain to the root tissues.

The normally low concentration of γEC in maize roots rose 6 hours after exposure to Cd, at a time when the greatest loss of GSH had occurred (Rauser *et al.* 1991, Meuwly & Rauser 1992). The concentration of γEC continued to increase to about 12-fold over control in 4 days (Rüegsegger & Brunold 1992) and 28-fold after 6 days (Meuwly & Rauser 1992). Significant accumulations of γEC in shoots started after

1 or 2 days and continued to 4- and 8-fold over controls. Rüegsegger and Brunold (1992) found that roots of maize incorporated more [^{35}S]sulfate into Cys, γEC and GSH when seedlings were exposed to Cd. The extractable activity of γ-glutamyl-cysteine synthetase increased more than 2-fold, whereas that of GSH synthetase was 1.6-fold that of controls. The observed accumulation of γEC in Cd-exposed maize roots was suggested to be due to a combination of higher contents of Cys, greater activity of γ-glutamyl cysteine synthetase and elimination of feed-back inhibition of γ-glutamyl cysteine synthetase by low GSH content. A combination of γEC availability and dipeptidyl transpeptidase activity (as per Hayashi *et al.* 1991) would explain the presence of a Glu + Cys peptide (Bernhard & Kägi 1987) and of $(\gamma EC)_2$ (see later) in Cd-binding complexes from maize roots. DesGly variants have also been identified, along with $(\gamma EC)_nG$ peptides, in Cd-binding complexes from two yeasts (Mehra *et al.* 1988) and Cu-binding complex from fission yeast (Mehra & Winge 1988). These data support the view that the $(\gamma EC)_n$ peptides found in plants and yeasts are products of biosynthesis; that they are products of degradation remains possible.

Acid-soluble thiols

Once it became clear that Cd-binding complexes contained sulfur-rich peptides like $(\gamma EC)_nG$, assays of just the peptides in small samples became prevalent. In one procedure tissues or cultured cells were extracted with NaOH followed by reduction of thiols with $NaBH_4$ prior to acidification (Grill *et al.* 1991). An alternative was to simply homogenize the plant material in acid such as 5-sulfosalicylic acid (Gupta & Goldsbrough 1991) or HCl (Tukendorf & Rauser 1990). These methods largely deproteinized the extracts leaving amino acids and polypeptides in solution. It is not known how quantitative these methods are for γEC isopeptides, a recovery of approximately 90% was reported for GSH (Scheller *et al.* 1987). In the author's hands the $NaOH/NaBH_4$ extraction of maize tissues resulted in unsatisfactory thiol profiles. Constituents could be separated by reverse-phase HPLC and the thiols quantitated (Grill *et al.* 1991; Rauser 1991). A major limitation, however, was the displacement of metal by the acid making it impossible to directly ascertain which thiols bound metal *in vivo*.

The quantitation of acid-soluble thiols following separation by HPLC is based on their reaction with the thiol-specific reagent 5,5'-dithiobis(2-nitrobenzoic acid) (DTNB, Ellman's reagent) added to the HPLC effluent at a constant rate. The time allowed for the yellow color to develop was not specified in the original application of the method (Grill *et al.* 1985), from a later publication it could be calculated to be 1.25 min (Gekeler *et al.* 1989). A reaction time of 1.2 minutes, derived independently, gave nearly complete color development with the various thiols from a Cd-binding complex of maize roots and GSH was a satisfactory standard for all the thiols detected (Rauser 1991). A reaction time of 6 seconds gave equivalent thiol responses for GSH and $(\gamma EC)_2G$, but for the larger peptides (n = 3 to 6) the response factors were reduced which necessitated calibration with the respective individual peptides (Gupta & Goldsbrough 1991).

Plant cell cultures are a popular model system to study the dynamics of metal-

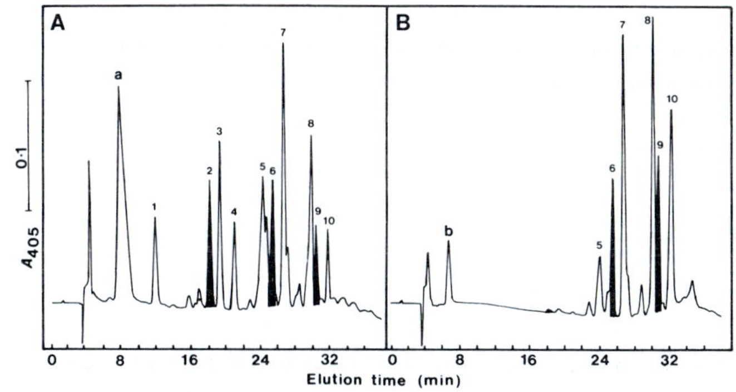

Fig. 1. Thiol profiles from roots of maize exposed to 3 μM $CdSO_4$ for five days. Acid-soluble materials in an extract equivalent to 0.30 g fresh weight of roots (A) and from a Cd-binding complex equivalent to 1.69 g fresh weight of roots (B) were separated by reverse-phase HPLC and thiols determined by post-column reaction with Ellman's reagent (Rauser 1991). The thiols were eluted by a linear gradient of 0-20% acetonitrite in 0.1% trifluoroacetic acid over 40 minutes. GSH and γEC occurred together in peak a, peak b was for sulfide.

binding peptides, few studies are available for roots and shoots of intact plants. The thiol profiles of acid extracts of cultured cells are usually simple, showing Cys, GSH and 4 to 8 $(\gamma EC)_nG$ isopeptides (Grill *et al.* 1987; Fujita & Izumi 1990; Reese *et al.* 1992). Other acid-soluble thiols are not evident or appear as small peaks in published thiol profiles. Some workers mention that minor thiol peaks eluted between the predominant $(\gamma EC)_nG$ isopeptides without showing a thiol profile (Gupta & Goldsbrough 1991). In roots of maize seedlings exposed to 3 μM Cd for 5 days the thiol profile was complicated (Fig. 1A). In this example 18 acid-soluble thiols eluted after the emergence of GSH and γEC. Identity of the thiols in the ten major peaks was sought in the absence of standard γEC isopeptides. Peaks 2, 6 and 9 were previously designated $(\gamma EC)_2G$, $(\gamma EC)_3G$ and $(\gamma EC)_4G$, respectively on the basis of amino acid analyses (Tukendorf & Rauser 1990). The thionitrobenzoate derivatives of peaks 1 through 10 were collected, purified by HPLC and the amino acid compositions determined (Meuwly, P. & Rauser, W.E.; unpublished results). The peptides were composed primarily of Glu and Cys or Glu, Cys and Gly. The amino acids in purified thionitrobenzoate derivatives were sequenced by mass spectrometry. Thiol peak 6 was the pentapeptide $(\gamma EC)_2G$, the polymer based on GSH (γECG). At the time of writing the γ linkage in all compounds is inferred on the basis of parallels in fragmentation patterns with standard derivatized γECG and γEC. Thiol peak 7 was $(\gamma EC)_2$, the polymer presumably based on the abnormally accumulated γEC (see above). Thiol peak 8 was $(\gamma EC)_2E$, a novel polymer not based on γECG (glutathione) nor γEC but presumably on thiol peak 1, the novel tripeptide γECE. Structural identification of the other thiol peaks in Fig. 1A is in progress. This incomplete study indicated that previous identifications of thiols in maize roots based on amino acid compositions alone were incorrect (Tukendorf & Rauser 1989; Rauser 1990). Comparison of the thiol complement in Fig. 1A with that in Fig. 1B indicated that only some of the acid-soluble thiols participated in forming Cd-binding complex. Peak 7, the desGly $(\gamma EC)_2$, was the single most abundant thiol in the acid extract 5 days after exposure to Cd. Furthermore, in roots, the combined concentration of γEC and other thiols containing Glu and Cys increased after the pool of GSH was significantly depleted (Meuwly & Rauser 1992). From 24 hours of exposure onward the concentration of these thiols exceeded the accumulating $(\gamma EC)_nG$ peptides by 1.2 to 2.4-fold.

Peptide function

Metal transfer to apoenzymes

One aspect of metal homeostasis in cells is the transfer of metal from a site of chelation or storage to an apoform of a metalloenzyme. Thumann *et al.* (1991) reconstituted separate Cu and Zn complexes with individual $(\gamma EC)_nG$ peptides. The apoform of the Cu-requiring enzyme diamino oxidase from spinach was reactivated by $Cu(\gamma EC)_2G$ nearly as effectively as by an equal amount of Cu from $CuSO_4$. The $Cu(\gamma EC)_4G$ and $Cu(\gamma EC)_5G$ complexes transferred metal less efficiently, perhaps due to strong metal binding in the complex. The Zn-requiring enzyme carbonic anhydrase from bovine erythrocytes was reactivated by $Zn(\gamma EC)_2G$ and $Zn(\gamma EC)_7G$ complexes. The rates of reactivation with the complexes were 30% and 68% slower, respectively than with an equal amount of Zn from $ZnSO_4$. These are the first data to show that metal-binding peptides from plants activate apoforms of metalloenzymes, a role consistent with metal homeostasis. The relative effectiveness of γEC isopeptides in this role needs to be compared to other metal chelators such as the organic acids prevalent in plant tissues.

Chelating metal in complexes

The major putative function of the sulfur-rich polypeptides in plants is metal binding. This function *in vivo* can currently only be assessed by isolating the combined entity of metal and polypeptides as metal-binding complex. Vascular plants through to algae and certain yeasts can have a small to large proportion of certain cellular metals present as metal-binding complexes. The most rigorous data are for the Cd complexes, the Cu complexes from plants are poorly understood. Complexes with the range of other metals that induced the appearance of $(\gamma EC)_nG$ peptides (Gekeler *et al.* 1989; Grill *et al.* 1990) remain to be isolated. The actual form in which the low amounts of $(\gamma EC)_nG$ peptides occur in control cells and tissues is unknown.

The Cd-binding complexes from plant cells and tissues contain Cd, a complement of γEC isopeptides and some acid-labile sulfur. The stoichiometry between the three types of constituents is not available for plant Cd-binding complexes, most characterizations are partial and qualitative. The complex from cultured cells of tomato contained $(\gamma EC)_nG$ peptides of $n = 3$, 4, 5 and 6, the $n=2$ peptide was minimal, acid-labile sulfur was not measured (Gupta & Goldsbrough 1991). Complexes from tomato roots contained the same peptides, however, the $n = 4$ peptide predominated in a high sulfide-containing complex while $n = 3$ and 4 peptides dominated in a low sulfide-containing complex (Reese *et al.* 1992). Ultraviolet spectroscopy of tomato complex, particularly the high sulfide form, showed transitions typical of CdS crystallite as described for the Cd complex from fission yeast (Dameron *et al.* 1989). In this model, CdS crystallite occurs as particles of 20 Å diameter capped by about 30 peptides of predominantly $(\gamma EC)_2G$ and $(\gamma EC)_2$. A Cd-binding complex from maize roots contained only some of the thiols and in a different ratio (Fig. 1B) to those found in the acid extract of comparable roots (Fig. 1A), illustrating the danger of equating amounts of thiolate peptides in cell or tissue extracts with their participation in binding metal *in vivo*. The sulfide evident in Fig. 1B was part of that in the complex, an unknown loss occurred during preparation of the sample for chromatography.

The Cd-binding complex from *Rauvolfia serpentina* contained the following $(\gamma EC)_nG$ peptides in μmol: n = 1, 3.9 (GSH); n = 2, 1.7; n = 3, 71.0; n = 4, 22.9; n = 5, 3.9; n = 6, 1.1; a low amount of sulfide (1.0 μmol) and 89.4 μmol Cd (Strasdeit *et al.* 1991). This description probably reflects an aggregate or composite of several complexes rather than a stoichiometric composition of an individual complex. Extended X-ray absorption fine structure analysis showed that each Cd was coordinated by the S of four cysteines at a Cd-S bond length of 2.52 ± 0.02 Å. The various $(\gamma EC)_nG$ molecules participated in forming the discrete $Cd(SCys)_4$ units. The carboxylate groups of the peptides were non-coordinating. The location of these groups on the surface explained the high negative charge so characteristic of the Cd-binding complexes and the tendency of the complex to increase in apparent molecular size at low ionic strengths (Grill *et al.* 1987, Rauser 1990). It is hoped that the extended X-ray analysis will be applied to a complex containing high amounts of acid-labile sulfide.

The importance of the thiol-rich peptides in binding metal has been evaluated by estimating the amount of cellular metal that occurred as a complex. Extracts are prepared and chromatographed by gel filtration, then the amount of Cd in the major Cd-binding peak known to contain $(\gamma EC)_nG$ peptides is determined. Values of greater than 90% were reported for Cd from cultured plant cells (Grill *et al.* 1990, Gupta & Goldsbrough 1991) and up to 70% from the alga *Chlamydomonas reinhardtii* (Howe & Merchant 1992). These reports lack information on 1) removal of extracellular Cd by customary exchange methods and 2) partitioning of intracellular Cd between the cell debris and the soluble phase. The percentage of Cd occurring as Cd-binding complex in gel filtration of soluble phase material does not represent the proportion of cellular Cd in the complex. An extreme example of the problem is the situation in the freshwater moss *Rhynchostegium riparioides* where most metal was bound by the cell walls leaving about 0.6% of the total Cu and 2% of the total Cd extractable, yet most of the extractable Cu and Cd occurred as metal-binding complexes (Jackson *et al.* 1991). Washed cells of *Datura innoxia* bound 77% of the soluble Cd as Cd-binding complex (Jackson *et al.* 1984). In *Silene vulgaris* plants exposed to 40 μM Cd for 21 days, 52% of the Cd in washed roots occurred as Cd-binding complex (Verkleij *et al.* 1990). In maize seedlings 37% of the Cd in desorbed roots occurred as complex after 2 days and 57% after 5 days of exposure to 3 μM Cd (Rauser, unpublished data). In maize shoots only 12% of the Cd was bound as a complex after 5 days. The importance of Cd-binding complexes in roots increases because roots generally contain more of the plant Cd than do shoots. For example, roots of maize contained about 60% of the plant Cd, about 80% was in tomato roots and up to 87% in the roots of the grass *Agrostis gigantea* (Rauser 1986). Bean plants exposed to the very high concentration of 2.5 mM Cd for 21 days contained about 85% of the plant Cd in roots (Leita *et al.* 1991). The bean roots had 24% of their Cd in a Cd-binding complex, the leaves 17%.

The ready solubilization of Cd-binding complexes is consistent with their putative location in the aqueous phase of cells. Vögeli-Lange and Wagner (1990) showed that purified vacuoles contained virtually all of the $(\gamma EC)_nG$ peptides and Cd present in tobacco leaf protoplasts. GSH was found in the protoplasts but not in the vacuoles. Since the dipeptidyl transpeptidase was deemed a soluble enzyme (Grill *et al.* 1989), it was highly probable that the $(\gamma EC)_nG$ peptides were synthesized extravacuolarly where the substrate was available. Transport of the peptides into the vacuoles was postulated (Vögeli-Lange & Wagner 1990). This process and whether Cd accompanies the peptides requires direct evaluation.

If one accepts the evidence that only some of the cellular Cd is associated with γEC isopeptides (30% in the case of cultured tobacco cells, Krotz *et al.* 1989; 57% in maize roots, Rauser unpublished data), the question of the speciation of the remaining Cd arises. More than enough organic acids occurred in vacuoles of tobacco cells to chelate the remaining Cd, and all the Zn which in these cells did not elicit (γEC) isopeptides (Krotz *et al.* 1989). Computer calculations of the ion species in tobacco vacuoles predicted that γEC isopeptides and the organic acids malate, oxalate and citrate formed soluble Cd complexes (Wang *et al.* 1991). Citrate had a particularly high potential for complexing Cd over a wide range of vacuolar pH. These studies indicated that considering γEC isopeptides as the principal metal chelators in plants (*i.e.* Grill *et al.* 1990) may apply to high and very high level Cd exposures and not to low level Cd exposures generally occurring in agricultural production.

Two low-molecular weight Cu-binding complexes isolated from pea leaves contained about 30% of the total Cu in the tissue (Palma *et al.* 1990). Challenge of Cu-sensitive and Cu-tolerant cultivars with excess Cu caused an increase of one of the complexes in the Cu-tolerant pea. The protein in the complex contained little Cys, excluding the consideration of γEC isopeptides, but was rich in isoleucine (30%). A poly-isoleucine protein was suggested as participating in binding some of the Cu in leaves of the Cu-tolerant pea.

Metal tolerance

A role for γEC isopeptides in metal tolerance was supported for Cd with cultured plant cells and one intact plant. Inhibition of peptide accumulation by reducing available glutathione and γEC with buthionine sulfoximine sensitized cultured cells to Cd (refs. in Grill *et al.* 1990). A comparable effect was reported for Cu in intact plants of *Mimulus guttatus* where buthionine sulfoximine markedly reduced root growth when an otherwise Cu-tolerant clone was exposed to excess Cu (Salt *et al.* 1989). Unfortunately no details were provided on peptide levels with the various treatments. An entirely opposite effect was demonstrated for *Festuca rubra* where buthionine sulfoximine did not increase, rather it decreased, the inhibitory effect of excess Zn on root growth of Zn-tolerant and Zn-sensitive cultivars (Davies *et al.* 1991). The suggestion that γEC isopeptides did not have a key role in Zn tolerance was made in the absence of any peptide measurements. Tolerance of Cd by cultured cells of *Datura innoxia* was not due to overproduction of γEC isopeptides, only rapid production of Cd-γEC isopeptide complex correlated with tolerance (Delhaize *et al.* 1989). After a 21 day exposure to Cd no correlation was found between the amount of Cd-binding complex and Cd tolerance in roots of *Silene vulgaris*, incorporation of acid-soluble sulfide might be of importance (Verkleij *et al.* 1990). Both Cu-tolerant and non-tolerant root apecies of *Silene vulgaris* had equal γEC isopeptide contents supporting the conclusion that the peptides were not involved in differential Cu tolerance (Schat & Kalff 1992). This study is exemplary in that the target tissue, root apecies, was analyzed and comparisons were made between plants with equivalent growth responses. The amount of Cu-γEC isopeptide complex was not measured which precluded assessment of this component in Cu tolerance. Insufficient specific data are at hand for a variey of plant species, and of metals, to definitively rule out metal complexation by γEC isopeptides in the phenomenon of differential metal tolerance.

Metallothionein

Identification of γEC isopeptides in plants arose from the search for the classical metallothionein (MT), *i.e.* the class I MT closely related to equine renal cortex MT. Since a considerable portion of metal, notably Cd, can be accounted for through binding to γEC isopeptides in plant cells and tissues, the challenge to isolate and characterize the classical MTs in plants has been neglected. The finding that mature wheat embryos contained a Cys-rich Zn-binding protein with amino acid sequence homology to rat liver MT and crab MT (Lane *et al.* 1987) has not been carried further. A new impetus came from the discovery of two genes similar to those of MTs amongst the Cu-regulated genes in Cu-tolerant *Mimulus guttatus* (de Miranda *et al.* 1990). The strong homology was due to two sequences coding for 14 or 15 amino acids in two separate domains each containing 6 Cys residues. The genes encoded a putative polypeptide 72 amino acids long. Similar genes were found in peas (Evans *et al.* 1990), barley (Okumura *et al.* 1991), soybean (Kawashima *et al.* 1991) and maize (de Framond 1991). The protein from these plant MT-like genes needs to be isolated from plants and characterized. The MT-like gene from pea was fused to the glutathione-S-transferase fusion-protein expression vector and expressed in *E. coli* (Tommey *et al.* 1991). After growth in 500 μM Cu, Cd or Zn the *E. coli* produced fusion protein with the MT portion binding each of the metals. The pH of half-dissociation of Cu, Cd or Zn from the fusion protein was 1.45, 3.95 and 5.35 respectively, values similar to those found for equine renal MT. Kille *et al.* (1991) reisolated the MT-like gene from peas and expressed it via a plasmid in *E. coli*. Only degraded protein could be isolated from *E. coli* as if the recombinant protein was attacked by proteolytic enzymes *in vivo*. If the same situation were to occur inside plants cells it would explain previous difficulties in isolating classical MTs from plants (Kille *et al.* 1991).

Antibodies to rat MT were used in a radioimmunoassay to show that soybean roots exposed to 2.5 μM Cu increased the MT concentration by 6-fold over the control (Chongpraditnun *et al.* 1991; Nakajima *et al.* 1991). Immunohistochemical techniques showed staining for MT in selected parts of the soybean: in root cap and cotyledonary epidermal cells of hydrated embryos; cells of the shoot apex below and to the side of the tunica-corpus and in root cap cells of seedlings. Perhaps the putative plant MTs are involved particularly with the essential elements Cu and Zn in highly regenerative parts of plants. The mature cells in the bulk of the tissues respond rapidly, certainly to Cd, with the synthesis of γEC isopeptides.

Concluding remarks

Considerable progress has been made in the biochemistry and physiology of the ubiquitous sulfur-rich peptides identified as (γ-glutamyl-cysteinyl)$_n$-glycine. Characterization of an enzyme involved in biosynthesis of the peptides, γ-glutamylcysteine dipeptidyl transpeptidase, provides a key element to enable studies of the gene for the enzyme and its regulation. The possibility of biosynthesis of related thiolate peptides directly from γ-glutamyl cysteine emphasizes the diversity of mechanisms operating in plants and the complexity we seek to understand. Exposure to excess Cd particularly enhances the metabolism of sulfur through the accentuated demand for organic sulfur and even sulfide to form complexes.

The demonstration that Cu- and Zn (γEC)$_n$G peptides activate apoforms of metalloenzymes indicates that the peptides can function in metal homeostasis. Their role in complexing a large portion of intracellular metal in situations of high external doses is clear for Cd in cultured plant cells yet appears more restrained in roots and shoots of intact plants. The comparable situation for other metals known to induce the peptides remains unclear. Other chelators such as organic acids may sequester metals in those situations where external doses are more like the metal concentrations found in lightly to moderately contaminated soils. The role of γEC isopeptides in the phenomenon of differential metal tolerance in intact plants remains unclear. Much of the knowledge gained for γEC isopeptides has been for the accentuated reaction of plants to Cd, a non-essential element. For other metals, including the essential micronutrients Cu and Zn, it is necessary to consider that the reactions of plants may differ in kind and degree from those elicited by Cd. Plants may well use an array of different, yet functionally parallel, mechanisms to deal with metals including proteins closely related to the classical mammalian metallothioneins.

Acknowledgement

Work in the author's laboratory has been supported by grants from the Natural Sciences and Engineering Research Council of Canada.

References

Berger, J.M., Jackson, P.J., Robinson, N.J., Lujan, L.D. & Delhaize, E. 1989. Precursor-product relationships of poly(γ-glutamylcysteinyl)glycine biosynthesis in *Datura innoxia*. Plant Cell Rep. 7: 632-635.

Bernhard, W.R. & Kägi, J.H.R. 1987. Purification and characterization of atypical cadmium-binding polypeptides in *Zea mays*. Experientia Suppl. 52: 309-315.

Chongpraditnun, P., Suzuki, K., Kawaharada, U., Nakajima, K. & Chino, M. 1991. Immunohistochemical localization of metallothionein in plant tissues. Water Air Soil Pollut. 57-58: 545-553.

Dameron, C., Reese, R.N., Mehra, R.K., Kortan, A.R., Carroll, P.J., Steigerwald, M.L., Brus, L.E. & Winge, D.R. 1989. Biosynthesis of CdS quantum crystallites. Nature 338: 596-597.

Davies, K.L., Davies, M.S. & Francis, M. 1991. The influence of an inhibitor of phytochelatin synthesis on root growth and root meristematic activity in *Festuca rubra* L. in response to zinc. New Phytol. 118: 565-570.

de Framond, A.J. 1991. A metallothionein-like gene from maize (*Zea mays*). Cloning and characterization. FEBS Lett. 290: 103-106.

de Miranda, J.R., Thomas, M.A., Thurman, D.A. & Tomsett, A.B. 1990. Metallothionein genes from the flowering plant *Mimulus guttatus*. FEBS Lett. 260: 277-280.

Delhaize, E., Jackson, P.J., Lujan, L.D. & Robinson, N.J. 1989. Poly(γ-glutamylcysteinyl)glycine synthesis in *Datura innoxia* and binding with cadmium. Role in metal tolerance. Plant Physiol. 89: 700-706.

Evans, I.M., Gatehouse, L.N., Gatehouse, J.A., Robinson, N.J. & Cray, R.R.D. 1990. A gene from pea (*Pisum sativum* L.) with homology to metallothionein genes. FEBS Lett. 262: 29-32.

Fujita, M. & Izumi, K. 1990. Induction of phytochelatins and variation of glutathione level in fruits and callus cultures of pumpkin treated with heavy metals. J. Food Hyg. Soc. Japan 31: 404-408.

Gekeler, W., Grill, E., Winnacker, E.-L. & Zenk, M.H. 1988. Algae sequester heavy metals via synthesis of phytochelatin complexes. Arch. Microbiol. 150: 197-202.

Gekeler, W., Grill, E., Winnacker, E.-L. & Zenk, M.H. 1989. Survey of the plant kingdom for the ability to bind heavy metals through phytochelatins. Z. Naturforsch. 44C: 361-369.

Grill, E., Winnacker, E.-L. & Zenk, M.H. 1985. Phytochelatins: the principal heavy-metal complexing peptides of higher plants. Science 230: 674-676.

Grill, E., Winnacker, E.-L. & Zenk, M.H. 1987. Phytochelatins, a class of heavy-metal-binding peptides from plants, are functionally analogous to metallothioneins. Proc. Natl. Acad. Sci. 84: 439-443.

Grill, E., Löffler, S., Winnacker, E.-L. & Zenk, M.H. 1989. Phytochelatins, the heavy-metal-binding peptides of plants, are synthesized from glutathione by a specific γ-glutamylcysteine dipeptidyl transpeptidase (phytochelatin synthase). Proc. Natl. Acad. Sci. 86: 6338-6342.

Grill, E., Winnacker, E.-L. & Zenk, M.H. 1990. Phytochelatins, the heavy metal chelating peptides of the plant kingdom. In: Rennenberg, H., Brunold, C., De Kok, L.J. & Stulen, I. (eds.), Sulfur Nutrition and Sulfur Assimilation in Higher Plants; Fundamental, Environmental and Agricultural Aspects. SPB Academic Publishing, The Hague, pp. 89-95.

Grill, E., Winnacker, E.-L. & Zenk, M.H. 1991. Phytochelatins. In: Methods in Enzymology Vol. 205. Academic Press, New York, pp. 333-341.

Gupta, S.C. & Goldsbrough, P.B. 1991. Phytochelatin accumulation and cadmium tolerance in selected tomato cell lines. Plant Physiol. 97: 306-312.

Hayashi, Y., Nakagawa, C.W., Mutoh, N., Isobe, M. & Goto, T. 1991. Two pathways in the biosynthesis of cadystins $(\gamma EC)_nG$ in the cell-free system of the fission yeast. Biochem. Cell Biol. 69: 115-121.

Howe, G. & Marchant, S. 1992 Heavy metal-activated synthesis of peptides in *Chlamydomonas reinhardtii*. Plant Physiol. 98: 127-136.

Jackson, P.J., Roth, E.J., McClure, P.R. & Naranjo, C.M. 1984. Selection, isolation, and characterization of cadmium-resistant *Datura innoxia* suspension cultures. Plant Physiol. 75: 914-918.

Jackson, P.P., Robinson, N.J. & Whitton, B.A. 1991. Low molecular weight metal complexes in the freshwater moss *Rhynchostegium riparioides* exposed to elevated concentrations of Zn, Cu, Cd and Pb in the laboratory and field. Environ. Exptl. Bot. 31: 359-366.

Kawashima, I., Inokuchi, Y., Chino, M., Kimura, M. & Shimizu, N. 1991. Isolation of a gene for a metallothionein-like protein from soybean. Plant Cell Physiol. 32: 913-916.

Kille, P., Winge, D.R., Harwood, J.L. & Kay, J. 1991. A plant metallothionein produced in *E. coli*. FEBS Lett. 295: 171-175.

Kondo, N., Imai, K., Isobe, M., Goto, T., Murasugi, A., Wada-Nakagawa, C. & Hayashi, Y. 1984. Cadystin A and B, major unit peptides comprising cadmium-binding peptides induced in fission yeast – separation, revision of structure and synthesis. Tetrahedron Lett. 25: 3869-3872.

Krotz, R.M., Evangelou, B.P. & Wagner, G.J. 1989. Relationship between cadmium, zinc, Cd-peptide, and organic acid in tobacco suspension cells. Plant Physiol. 91: 780-787.

Lane, B.L., Kajioka, R. & Kennedy, T. 1987. The wheat germ E_c protein is a zinc-containing metallothionein. Biochem. Cell Biol. 65: 1001-1005.

Leita, L., Contin, M. & Maggioni, A. 1991. Distribution of cadmium and induced Cd-binding proteins in roots, stems and leaves of *Phaseolus vulgaris*. Plant Sci. 77: 139-147.

Loeffler, S., Hochberger, A., Grill, E., Winnacker, E.-L. & Zenk, M.H. 1989. Termination of the phytochelatin synthase reaction through sequestration of heavy metals by the reaction product. FEBS Lett. 258: 42-46.

Mehra, R.K. & Winge, D.R. 1988. Cu(I) binding to the *Schizosaccharomyces pombe* γ-glutamyl peptides varying in chain lengths. Arch. Biochem. Biophys. 265: 381-389.

Mehra, R.K., Tarbet, E.B., Gray, W.R. & Winge, D.R. 1988. Metal-specific synthesis of two metallothioneins and γ-glutamyl peptides in *Candida glabrata*. Proc. Natl. Acad. Sci. 85: 8815-8819.

Mendum, M.L., Gupta, S.C. & Goldsbrough, P.B. 1990. Effect of glutathione on phytochelatin synthesis in tomato cells. Plant Physiol. 93: 484-485.

Meuwly, P. & Rauser, W.E. 1992. Alteration of thiol pools in roots and shoots of maize seedlings exposed to cadmium: adaptation and developmental cost. Plant Physiol. 99: 8-15.

Nakajima, K., Suzuki, K., Otaki, N. & Kimura, M. 1991. Epitope mapping of metallothionein antibodies. In: Methods in Enzymology. Vol. 205. Academic Press, New York, pp. 174-189.

Okumura, N., Nishizawa, N.-K., Umehara, Y. & Mori, S. 1991. An iron deficiency-specific cDNA from barley roots having two homologous cysteine-rich MT domains. Plant Mol. Biol. 17: 531-533.

Palma, J.M., Yáñez, J., Gómez, M. & del Rio, L.A. 1990. Copper-binding proteins and copper tolerance in *Pisum sativum* L. Characterization of low-molecular-weight metalloproteins from plants with different sensitivity to copper. Planta 181: 487-495.

Rauser, W.E. 1986. The amount of cadmium associated with Cd-binding protein in roots of *Agrostis gigantea*, maize and tomato. Plant Sci. 43: 85-91.

Rauser, W.E. 1987. Changes in glutathione content of maize seedlings exposed to cadmium. Plant Sci. 51: 171-175.

Rauser, W.E. 1990. Phytochelatins. Annu. Rev. Biochem. 59: 61-86.

Rauser, W.E. 1991. Cadmium-binding peptides from plants. In: Methods in Enzymology. Vol. 205. Academic Press, New York, pp. 319-333.

Rauser, W.E., Schupp, R. & Rennenberg, H. 1991. Cysteine, γ-glutamylcysteine, and glutathione levels in maize seedlings. Distribution and translocation in normal and cadmium-exposed plants. Plant Physiol. 97: 128-138.

Reddy, G.N. & Prasad, M.N.V. 1990. Heavy metal binding proteins/peptides: Occurrence, structure, synthesis and function. A review. Environ. Exp. Bot. 30: 251-264.

Reese, R.N. & Wagner, G.J. 1987. Effects of buthionine sulfoximine on Cd-binding peptide levels in suspension-cultured tobacco cells treated with Cd, Zn, or Cu. Plant Physiol. 84: 574-577.

Reese, R.N., Mehra, R.K., Tarbet, E.B. & Winge, D.R. 1988. Studies on the γ-glutamyl Cu-binding peptide from *Schizosaccharomyces pombe*. J. Biol. Chem. 263: 4186-4192.

Reese, R.N., White, C.A. & Winge, D.R. 1992. Cadmium-sulfide crystallites in Cd-$(\gamma EC)_nG$ peptide complexes from tomato. Plant Physiol. 98: 225-229.

Robinson, N.J. 1989. Algal metallothioneins: secondary metabolites and proteins. J. Appl. Phycol. 1: 5-18.

Robinson, N.J. 1990. Metal-binding polypeptides in plants. In: Shaw, A.J. (ed.), Heavy Metal Tolerance in Plants: Evolutionary Aspects. CRC Press, Boca Raton, pp. 195-214.

Rüegsegger, A. & Brunold, C. 1992. Effect of cadmium on γ-glutamylcysteine synthesis in maize seedlings. Plant Physiol. 99: 428-433.

Salt, D.E., Thurmann, D.A., Tomsett, D.A. & Sewell, A.K. 1989. Copper phytochelatins of *Mimulus guttatus*. Proc. Royal Soc. London B236: 79-89.

Schat, H. & Kalff, M.M.A. 1992. Are phytochelatins involved in differential metal tolerance or do they merely reflect metal-imposed strain? Plant Physiol. 99: 1475-1480.

Scheller, H.W., Huang, B., Hatch, E. & Goldsbrough, P.B. 1987. Phytochelatin synthesis and glutathione levels in response to heavy metals in tomato cells. Plant Physiol. 85: 1031-1035.

Steffens, J.C. 1990. The heavy metal-binding peptides of plants. Annu. Rev. Plant Physiol. Plant Mol. Biol. 41: 553-575.

Strasdeit, H., Duhme, A.-K., Kneer, R., Zenk, M.H., Hermes, C. & Nolting, H.-F. 1991. Evidence for discrete $Cd(Cys)_4$ units in cadmium phytochelatin complexes from EXAFS spectroscopy. J. Chem. Soc., Chem. Comm. issue 16: 1129-1130.

Thumann, J., Grill, E., Winnacker, E.-L. & Zenk, M.H. 1991. Reactivation of metal-requiring apoenzymes by phytochelatin-metal complexes. FEBS Lett. 284: 66-69.

Tommey, A.M., Shi, J., Lindsay, W.P., Urwin, P.E. & Robinson, N.J. 1991. Expression of the pea gene $PsMT_A$ in *E. coli*. Metal-binding properties of the expressed protein. FEBS Lett. 292: 48-52.

Tomsett, A.B. & Thurman, D.A. 1988. Molecular biology of metal tolerances of plants. Plant Cell Enivron. 11: 383-394.

Tukendorf, A. & Rauser, W.E. 1990. Changes in glutathione and phytochelatins in roots of maize seedlings exposed to cadmium. Plant Sci. 70: 155-166.

Verkleij, J.A.C., Koevoets, P., Van't Riet, J., Bank, R., Nijdam, Y. & Ernst, W.H.O. 1990. Poly(γ-glutamylcysteinyl)glycines or phytochelatins and their role in cadmium tolerance of *Silene vulgaris*. Plant Cell Environ. 13: 913-921.

Vögeli-Lange, R. & Wagner, G.J. 1990. Subcellular localization of cadmium and cadmium-binding peptides in tobacco leaves. Implication of a transport function for cadmium-binding peptides. Plant Physiol. 92: 1086-1093.

Wang, J., Evangelou, D.A., Nielsen, M.T. & Wagner, G.J. 1991. Computer-simulated evaluation of possible mechanisms for quenching heavy metal ion activity in plant vacuoles. Plant Physiol. 97: 1154-1160.

PLANTS AS SOURCES OF ATMOSPHERIC SULFUR

Peter Schröder
Fraunhofer Institute for Atmospheric Environmental Research, Kreuzeckbahnstrasse 19, D-8100 Garmisch-Partenkirchen, F.R.G.

Introduction

Sulfur emissions from the biosphere

Atmospheric sulfur originates from numerous sources, both anthropogenic and natural. Amongst the natural sources, biogenic emissions are presently thought to constitute about half of the total sulfur burden to the atmosphere (Aneja & Cooper 1989; Aneja 1990; Kesselmeier 1991; Rennenberg 1991), thus representing an important variable in the total biogeochemical sulfur cycle.

However, present estimates of the global sulfur budget and of emission rates from ecosystems have a high uncertainty (Table 1). This problem has recently been addressed by several authors (*e.g.* Rennenberg 1991; Kesselmeier 1991). It may be noteworthy in this context to realize that the sparseness of reliable data for concentrations and fluxes may be as much a result of analytical problems in the successful quantification of sulfur compounds as in their rapid atmospheric reactions by photolysis (Crutzen 1976) or after reactions with hydroxyl radicals (Ravishankara *et al.* 1980; Adewuyi 1989) and NO_3^- (Jensen *et al.* 1991). Both these processes finally yield SO_2 or SO_4^{2-} which are scavenged out of the atmosphere by dry and wet deposition. A compilation of approximate atmospheric lifetimes of some important sulfur containing gases is given in Table 1, indicating that reliable emission data may, at present, only be obtained for stable compounds.

Nevertheless, several volatile reduced sulfur compounds from various plant species growing in different ecosystems have been identified, and emission is a phenomenon recognized in the literature. H_2S, COS, DMS, CS_2 and methylmercaptan seem to be frequently released into the atmosphere by plants. The distribution of emissions amongst plant species has been reviewed by Aneja & Cooper (1989) and Rennenberg (1991). Comprehensive information on the distribution of these emissions from different ecosystems is summarized in the work of Adams *et al.* (1981), Goldan *et al.* (1987), Aneja (1990) and Kesselmeier *et al.* (1991). Ecosystem fluxes estimated by several authors still are afflicted with a large uncertainty, because of the large biological diversity within these systems, and the technical prerequisites for exact sulfur measurements have only recently been established (MacTaggart *et al.* 1987, Haunold *et al.* 1992). Ten years ago, estimates for global biogenic sulfur emissions fluctuated between 270 Tg S $year^{-1}$ to about 35 Tg S $year^{-1}$ (Freney *et al.* 1982), nowadays they are calculated to be in the range between 170 and 73 Tg $year^{-1}$ (Berresheim *et al.* 1989; Andreae 1990). But even these data sets can be used to distinguish typical low and high emitting biotopes. Except for ocean salt spray, which is not considered in this review, several authors report the highest biogenic source strengths from salt marshes (3.5 up to 650 g S m^{-2} $year^{-1}$), followed by humid and dry forests (0.88 and

Sulfur Nutrition and Assimilation in Higher Plants
edited by L.J. De Kok et al., pp. 253–270

Table 1. Atmospheric concentrations and lifetimes of reduced sulfur gases. Compilation of concentrations by: [a], Meszaros (1981), [b], Huber *et al.* (1992). Lifetimes as estimated by Andreae *et al.* (1982); n.d, no data available.

Species	Common name	Atmospheric	
		Conc. [pptv]	Lifetime
H_2S	Hydrogen sulfide	0[a] – 230[b]	<1 day
DMS	Dimethyl sulfide	3 – 36	2 days
DMDS	Dimethyl disulfide	n.d.	<1 day
COS	Carbonyl sulfide	510 – 1120	>365 days
CS_2	Carbon disulfide	190 – 1470	14 days
CH_3SH	Methyl mercaptan	n.d.	<1 day

0.023 g S m^{-2} $year^{-1}$, respectively) and crop fields (with maximum emissions from corn, 0.273 g S m^{-2} $year^{-1}$, data from Adams *et al.* 1981 and Aneya 1990). Studies with single coniferous trees showed that they may generally be classified as low emitters (0.007 g S m^{-2} $year^{-1}$, Aneya & Cooper 1989; 0.001 g S m^{-2} $year^{-1}$, Rennenberg *et al.* 1990).

In all cases, the main sulfur species reported to be emitted by the ecosystems are H_2S and DMS. Flux data for CH_3SH and CS_2 from ecosystems are scarce; for the former gas, this might be due to rapid atmospheric reactions, and for the latter gas this may be due to its low source strength and high atmospheric stability. Methyl mercaptan has been identified in emissions from rape seed (Kesselmeier *et al.* 1991), clover, wheat (Fall *et al.* 1988), and from the marsh grass *Spartina alternifolia* (Steudler & Peterson 1984, 1985; Goldan *et al.* 1987). Only a few reports identify CS_2 emissions as plant related. The tropical tree *Stryphnodendron excelsum* seems to emit CS_2 in high amounts from its roots (0.15 -1.5 g S m^{-2} $year^{-1}$; Haines *et al.* 1989), and *Spartina alternifolia* has been shown to emit 0.2 g S m^{-2} $year^{-1}$ (Aneja *et al.* 1979).

It is interesting to note that the emissions from soil samples consistently differ from those from total ecosystems. Alfisols, inceptisols and spodosols are low sulfur emitters (Adams *et al.* 1981), whereas histosols and mollisols are soil types which emit reduced sulfur gases most abundantly, with equal amounts of H_2S and DMS, followed by CS_2, COS and DMDS (Adams *et al.* 1981; Goldan *et al.* 1987; Aneja 1990). It seems to be a fact that large amounts of the sulfur emitted by soils is taken up by the plants covering the system and may be re-emitted into the atmosphere in a different form. Nice examples for this interaction and the role of the soil can be compiled using data from laboratory experiments with wheat plants (Fall *et al.* 1988) and also from field studies (Goldan *et al.* 1987). The soil-plant system as a whole shows a net emission of DMS, methyl mercaptan and low fluxes of COS and CS_2, whereas the soil system alone emits only COS and CS_2 (Table 2). This observation was confirmed during field measurements in a wheat field by B. Huber (personal communication). Similar results for the source of H_2S from a French lawn were obtained by Delmas *et al.* (1980); the lawn was calculated to emit 0.24 g S m^{-2} $year^{-1}$, whereas the lawn soil was found to emit only 0.07 g S m^{-2} $year^{-1}$.

During recent years the role of the biosphere as a sink for atmospheric sulfur has been investigated in some depth (De Cormis 1968; Freer-Smith 1985; Taylor & Tingey

Table 2. Sulfur fluxes from canopies, showing the distinction between bare soil and vegetation covered sites; together with identification of the sulfur source. Measurements were performed after cryogenic sampling and analysis by FPD-GC. Data are compiled from (a) Goldan *et al.* (1987), field measurements; (b) Fall *et al.* (1988), laboratory studies.

Species	Soil + Plant	Soil – Plant
	flux (ng S m^{-2} min^{-1})	
H_2S (mollisol)[a]	1 – 3	0.3 – 0.9
H_2S (histosol)[a]	4 – 8	4.3 – 4.9
CH_3SH[b]	3 – 4	< 0.2
DMS (mollisol)[a]	2 – 4	0.7 – 1.1
DMS (histosol)[a]	2 – 8	0.3 – 1.0
DMS[b]	7 – 8	< 0.2
COS (mollisol)[a]	1 – 2	2.4 – 3.2
COS (histosol)[a]	1 – 2.2	5.4 – 8.4
COS[b]	1	2.0 – 3.0
CS_2 (mollisol)[a]	0.3 – 0.7	0.4 – 0.8
CS_2 (histosol)[a]	0.8 – 1.6	1.1 – 1.7
CS_2[b]	0.5 – 1	> 4

1983; De Kok *et al.* 1989; De Kok 1989, 1990; Protoschill-Krebs & Kesselmeier 1992), however, information on the factors governing the emission of sulfur compounds from plants is scarce. The present paper presents a compilation and evaluation of the processes leading to the formation of volatile sulfur compounds in plants and the factors governing their release into the atmosphere.

Physiological conditions for the emission of sulfur gases

Single plant emission studies reflect a complex relationship between the ontogenetic status of the whole plant, the plant part under consideration, nutritional effects, the amount and form of sulfur applied, and the plant's physiological parameters, as well as the meteorological situation. The influence of these factors on the emission of reduced sulfur compounds has been reviewed by Rennenberg (1989, 1991) and Anderson (1990). As all higher plants seem to be able to emit H_2S, the formation of this substance has been investigated in numerous studies. It has been suggested in these studies that hydrogen sulfide may be emitted in order to regulate homeostatically the size of the cysteine pool and maintain it at a low level because of its cytotoxicity (Rennenberg *et al.* 1982; Filner *et al.* 1984; Rennenberg 1984).

Emission as reaction to excess pedospheric sulfur

Plants may be exposed to excess inorganic sulfur by fertilization or by growing them in soils with a high organic sulfur content because of incorporated decaying organic matter. Under these circumstances, the roots absorb various sulfur compounds, such as sulfite, sulfide, sulfate, and also sulfur containing amino acids like methionine or cysteine (Spaleny 1977; Filner *et al.* 1984; Rennenberg 1989). Although some data are available on the emission of methyl mercaptan after amino acid uptake (Schmidt *et al.* 1985), H_2S is the most abundant gas emitted by various plant species in response to excess sulfur (Rennenberg 1991). Hydrogen sulfide emissions in response to high sulfate loads occur in correlation with the amount of sulfur in the fertilizer, but are also highly dependent on the plant's stage of development (Rennenberg & Filner 1983; Rennenberg 1984). Sulfate uptake into the plant is regulated by the roots (Rennenberg & Thoene 1987; Rennenberg 1991) due to inhibitory concentrations of glutathione. This regulatory mechanism may prevent the accumulation of high burdens of inorganic sulfur in the tissue under circumstances where H_2S emission is not possible or not wanted. Detached spruce branches, for example, emit large amounts of H_2S during winter after sulfite/sulfate donations to the transpiration water. The whole tree, however, does not emit any H_2S, even if the concentrations of sulfate in the soil are increased to 30 mM. After trimming the roots, H_2S emission rates similar to those with detached branches are found. This experiment demonstrates the immense barrier potential of the roots (B. Huber, personal communication). The present database is, however, scarce and needs accurate uptake and emission studies under controlled conditions.

The emission of sulfur compounds under the influence of high concentrations of sulfur containing amino acids has been studied with leaf discs or suspension cultured cells rather than with whole plants. Donations of L- as well as D-cysteine to cell cultures resulted in the emission of H_2S (Sekiya *et al.* 1982b), and surplus amounts of methionine led to the volatilization of CH_3SH from the cells. The production of these gases from amino acids was always found to be light dependent and under developmental control (Rennenberg & Filner 1983). Although it has been postulated in the past decades, L-cysteine does not seem to be an intermediate in the formation of H_2S from sulfate (Rennenberg 1989). It is obscure whether the emission of H_2S under the influence of L- and D-cysteine is mediated by cysteine-desulfhydrase, because this enzyme's activity is not coupled to the emission of the gas (Schütz *et al.* 1991). In the reverse reaction however, cysteine desulfhydrase has been shown to participate in the assimilation of H_2S from atmospheric sources (Schütz *et al.* 1991). At present there is no clear cut evidence available to show if the hydrogen sulfide precursor D-cysteine is formed from L-cysteine by a racemase in higher plants (Rennenberg *et al.* 1992).

Emission after fumigation

The effects of plant fumigation with SO_2 have attracted interest because of a possible connection to pollution damage and forest dieback. Uptake of sulfur compounds from the atmosphere may, however, also be beneficial for plants. Under conditions of low sulfur availability from the soil, atmospheric SO_2 can support plant growth and prevent deficiency symptoms (*e.g.* Thomas *et al.* 1943, 1944; Olsen 1957; Faller *et al.* 1970; Faller 1972; Anderson 1990). However, when certain threshold values for the intake are exceeded, other symptoms might arise including leaf lesions, effects on electron transport, and malfunctions of the chloroplast (for review see Thomas 1951; Anderson 1990). This fact is of special importance because, apparently, the uptake of SO_2 even in injurious amounts can not be avoided by the plants (Rennenberg 1984). The existence of such threshold values has been shown in ornamental plants, where the deposition of SO_2 was found to be proportional to the atmospheric concentration but decreasing with time (Elkiey & Ormrod 1981). Of the many studies investigating adverse effects of SO_2 fumigation (Winner *et al.* 1985), the work of Pfanz and coworkers emphasize the cellular acidification and inhibitory effects of SO_2 on the metabolism of plant cells (Pfanz & Heber 1985; 1986; Heber *et al.* 1987). Growth reductions in trees under the influence of SO_2 of about 20 to 50%, depending on the species under consideration, have been reported by Freer-Smith (1985) and Whitmore & Freer-Smith (1982). The latter authors did not observe re-emission of sulfur in any form. This is contrary to earlier results obtained by Materna (1966) and De Cormis (1968) which indicated a positive correlation between SO_2 fumigation and H_2S emission in several plants as a detoxification reaction. Forest canopies have been identified as sinks as well as sources of volatile sulfur by Hicks *et al.* (1982). However, only a small amount of the sulfur taken up by the trees seemed to be re-emitted. Sekiya *et al.* (1982a) reported an emission of 15% of the SO_2-sulfur in reduced form, some 60% seemed to be further oxidized and stored in the vacuole. Increasing amounts of sulfur-containing amino acids in treated cells indicate that plants are able to utilize atmospheric SO_2 or H_2S in low concentrations to fill or maintain their sulfur pool. In this context the role of glutathione (γ-glutamyl-cysteinyl-glycine) as a storage peptide and its formation (Brunold 1990; Bergmann & Hell 1990) or degradation (Anderson 1990) has been reviewed extensively (Rennenberg & Lamoureux 1990; Rennenberg 1991).

COS has been shown to be deposited in plants very effectively (Fall *et al.* 1988; Goldan *et al.* 1988; Hofmann *et al.* 1992), obviously being utilized as a carbon source for photosynthetic C-fixation (Protoschill-Krebs & Kesselmeier 1992). In an excellent review of the database on the topic, Goldan *et al.* (1988) estimated the annual plant uptake of COS to be 0.2-0.6 Tg which would balance known global sources and sinks for the gas.

Unlike CO_2, COS is not an activator of ribulose-1,5-bisphosphate-carboxylase, which might limit its importance for photosynthesis (Lorimer & Pierce 1989). Whereas these early investigations claimed a direct uptake of COS by rubisco, recent studies have elucidated that COS is first metabolized by the action of a carbonic anhydrase, the affinity of which is 1000 times higher for COS than for CO_2. The sulfur group seems to be rapidly cleaved, reduced and emitted in the form of H_2S. About 60% of the S taken up as COS rapidly is re-emitted as H_2S (Protoschill-Krebs & Kesselmeier 1992).

Emission under normal conditions

One of the most important discoveries in recent years is the observation that plants emit sulfur compounds throughout the day even without excess sulfur from whatever sources (Goldan *et al.* 1987; Lamb *et al.* 1987; Fall *et al.* 1988; Rennenberg *et al.* 1990; Hofmann *et al.* 1992). These emissions of reduced sulfur compounds are now thought to occur as regulatory steps in order to level the sulfur pools in the plant. This maintenance of constant sulfur pools is important because the plant sulfur status and requirements are subject to changes during ontogenetic development. Whereas the proteins within the vegetative parts of the plants are generally found to be sulfur-rich, a significant shift to sulfur poor amino acids occurs in the fruits during ripening (Rennenberg 1991). Hence, the emission patterns are generally found to be independent of sulfur nutrition and modulated, on the one hand, by plant age as well as developmental stage, and, on the other hand by environmental factors, such as light intensity, temperature, humidity and photosynthetic CO_2 fixation.

Enzymatic background for the synthesis of sulfur gases

The synthesis of H_2S in higher plants has been investigated in numerous studies. Whereas the enzymatic degradation of cysteine by cysteine desulfhydrase action was favoured as source for H_2S from plant tissue, recent studies throw considerable doubt on this hypothesis. Instead, a reduction of sulfate or sulfite under the influence of light is favoured. This would include the removal of the thiol group from cysteine to yield *O*-acetylserine and sulfide, the latter being transformed to sulfite, sulfate, sulfide and finally being emitted as hydrogen sulfide. Critical pools are represented by sulfate and cysteine. Whereas sulfate uptake by the roots can be regulated by glutathione, the cysteine pool size is maintained by the cysteine desulfhydrase. As already mentioned above, it is not yet clear whether L-cysteine could be transformed to D-cysteine by racemase action and degraded by a D-cysteine specific cysteine desulfhydrase.

In higher plants methyl mercaptan and DMS are derived from the breakdown of methionine and *S*-methylmethionine, as has been shown by Schwenn *et al.* (1982) with *Catharanthus roseus* cells, and by Fall *et al.* (1988) with *Triticum aestivum*. No enzymatic pathways are presently known for the evolution of COS and CS_2 from plant tissues. Recent investigations by Kesselmeier and co-workers indicate a possible correlation between the formation COS and the photosynthetic and respiratory carbon cycle (Hofmann *et al* 1992; Protoschill-Krebs & Kesselmeier 1992). The sources of CS_2 remain obscure and might be connected to microbial metabolism rather than to higher plants. Giovanelli (1987) and Rennenberg (1991) suggest that COS and CS_2 are volatile intermediates in the metabolism of nonprotein sulfur amino acids and unusual sulfur compounds. However, the pathways to biosynthesis and degradation of such compounds are still only poorly understood.

Significance of sulfur gas emission as a regulatory factor

The significance of sulfur gas emission from plants as a regulatory factor in sulfur assimilation has been discussed by several authors (*e.g.* Filner *et al.* 1985; Rennenberg 1984, 1991; Ernst 1990). Especially Ernst (1990) points out that many experiments concerning this topic have been performed which neglect ecologically relevant concentrations, and that a lack of knowledge about the total sulfur budget of a plant makes it impossible to judge about the efficiency of the system. Taking into consideration that the emission of sulfur compounds from a plant may only be a small fraction of the annual sulfur burden from the atmosphere or pedospheric sources to the plant (Unsworth *et al.* 1985; Meyers *et al.* 1991), it may be useful to think of translocation, storage pools, or detoxification pathways for excess sulfur. In this context the reader should refer to the review of Rennenberg (1984) who discusses whether the minute amounts of sulfur emitted may be significant and sufficient to balance the internal sulfur budget of a plant cell.

Table 3. Thermodynamic constants for hydrogen sulfide (from Snoeyink & Jenkins 1980).

	$DH°_f$	$DG°_f$
	(kcal mol^{-1})	
$HS^-_{(aq)}$	-4.22	+3.01
$H_2S_{(g)}$	-4.815	-7.98
H_2S	-9.4	-6.54

The formation of volatile sulfur compounds and their escape from plants

Among the reduced sulfur gases, the largest data set is available for H_2S. Therefore the following paragraphs will focus predominantly on the formation, features and fate of this compound.

H_2S might be formed either by enzymatic action or, as discussed above, spontaneously and nonenzymatically in a chemical reaction between its precursors SH^- and H^+; only the formation of SH^- with a positive $DG°_f$ requires energy (Table 3). Given a pK_a of 7.02 and a negative DG_f value, it is obvious that hydrogen sulfide will form spontaneously in aqueous solutions, with a high tendency to escape into the gas phase. Hence, the energy dependent synthesis of the HS^- ion from its precursors seems to be the rate limiting step in the reaction chain, and might explain the observation of light (and energy) dependent H_2S emissions from plants.

Release of H_2S from the cell

Any gaseous pollutant entering or escaping from a leaf must diffuse through the apoplastic space filled with water. The diffusion process requires a dissolution of the respective substance in the aqueous phase. Cell wall pH values reported are in the range 4 to 7, with the majority between 5 and 6.5 (Grignon & Sentenac 1991). As the apoplast is acidified in the light, a different solubility of gases in the night as compared to that in the day must be expected. H_2S behaves like a multiprotic acid with transitional states as H_2S in the pH range between 4 and 7, dissociation to HS^- and H^+ in the pH range 8 to 12 and to S^- above pH 13. In the physiological pH range between 5 to 6 only 1 to 9% of the H_2S is dissociated. Equilibrium constants for the reactions are $K_{a1} = 10^{-7}$ for $H_2S \rightarrow HS^- + H^+$ and $K_{a2} = 10^{-14}$ for $HS^- + H^+ \rightarrow H_2S$. The dissociation constant, describing the transition from H_2S to SH^- and H^+ and *vice versa*, is given as $k = 4.3 \times 10^{-3}\ s^{-1}$ under these conditions (Snoeyink & Jenkins 1980). An uncharged acid like H_2S may then be able to permeate rapidly through organelle or cell membranes into the apoplast. The apoplast volume is only 1.5% of the total cell volume per unit leaf area (Pfanz & Dietz 1987) and will thus be rapidly saturated with $H_2S_{(aq)}$.

Henry's law states that the partial pressure of a particular gas above a liquid is proportional to the number of gas molecules dissolved in the liquid phase (Nobel 1991). Henry's constant for H_2S is given as 0.1 mol l^{-1} bar^{-1} (Snoeyink & Jenkins 1980). Under supersaturation conditions, hydrogen sulfide will, despite its high solubility in water (Table 4) be volatilized from the apoplastic liquid into the gas phase

Table 4. Selected physico-chemical and chemical properties of sulfur containing gases (from Taylor *et al.* 1983; Baldocci 1991).

	Molecular weight	Solubility in water, 30 °C	Molecular diameter	Diffusivity ratio	Activation energy	Q_{10}
		(μmoles cm^{-3})	(x 10^{-8} cm)		(kJ mol^{-1})	
SO_2	64	1218	8.9	0.53	–	–
H_2S	34	210	12.5	0.73	70	2.6
COS	60	37	11	0.49	75	2.75
DMS	62	0	18.2	0.54	70-86	2.8
CS_2	76	36	10.8	0.48	–	–
CH_3SH	48	259	22.9	0.61	58	2.2

of the intercellular cavities and escape from the leaves. At room temperature and a sulfur concentration of 10^{-3} M in an aqueous solution of pH 8.5, the partial pressure of the gas has been determined using Henry's constant to be approximately $10^{-3.5}$ atm (Snoeyink & Jenkins 1980).

Escape of sulfur gases from the leaf

From a physico-chemical point of view, the emission of sulfur gases from the mesophyll into the atmosphere is a simple process. The flux rate depends on (a) the production rate of the gas under consideration; (b) the solubility and viscosity of the gas; (c) the concentration gradients in the system, and (d) the conductivity of the system for the gas. The data summarized in Table 4 give some clues to the behaviour of the single gases and Fig. 1 tries to interpret these escape steps in an admittedly oversimplified, but useful scheme. A compilation of general factors influencing the trace gas emission from leaves and canopies and a critical evaluation of different models describing these emissions is presented by Baldocci (1991).

Production rates. Some attempts to parametrize the sulfur gas fluxes have recently been made, demonstrating that both the flux strength and direction are a function of temperature and photon flux (Fall *et al.* 1988). Studies with intact plants and detached leaves revealed that H_2S formation is a light dependent reaction (Wilson *et al.* 1978, Anderson 1990). For spruce trees (*Picea abies*) the influence of light intensity on H_2S emissions has recently been parametrized (Schröder *et al.* 1992). It was shown that H_2S fluxes increase with light intensity and are positively correlated to photosynthetic CO_2 uptake (Fig. 2). It is interesting to note that the emission of hydrogen sulfide from these plants does not start immediately with the beginning of the light period, but is delayed until the compensation point of photosynthesis is surpassed (Schröder *et al.* 1992). This indicates a close connection between the release of H_2S with basic physiological parameters like metabolic activity, transpiration and photosynthesis. Some comprehensive parametrization experiments investigating these processes have recently been performed by Rennenberg *et al.* (1991, 1992) and Rennenberg & Schröder (1991).

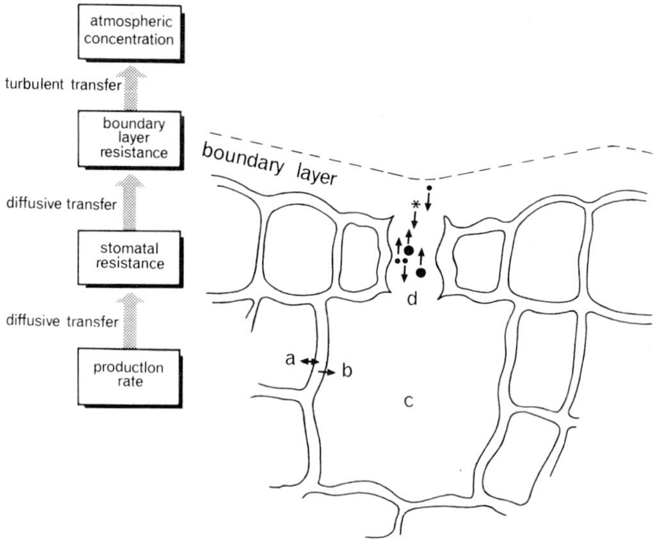

Fig. 1. A schematic outline of the factors influencing the escape of gas molecules from a leaf. (a) metabolic activity of the cells, (b) solubility of the gas in the apoplastic fluid, (c) concentration in substomatal cavities, (d) effusion through the stomates according to Bunsen-Graham's law (Jeans 1962) and (e) laminar transport through the boundary layer. •, *, ●, molecules with different sizes (see Baldocci 1991 for a comprehensive review).

Solubility. Among the sulfur gases, COS, DMS and CS_2, because of their low solubility in water, are more likely to escape or stay in the atmosphere. H_2S and methyl mercaptan have a higher solubility but are still more likely to be be found in the gas phase than SO_2 which will stay in the liquid phase. The diffusivity ratios given for the single species in Table 4 are derived from Bunsen-Graham's law of effusion

$$\frac{v_1}{v_2} = \frac{\sqrt{MW_2}}{\sqrt{MW_1}}$$

and give the relative velocity (v_1) of sulfur gases as compared to the velocity of water molecules (v_2) derived from their molecular weights ($MW_{1,2}$) when escaping through narrow openings like stomates.

H_2S is the only gas species with an efflux velocity close to water fluxes; the lowest velocities are calculated for COS and CS_2. The activation energies for the emission and the Q_{10} coefficients, describing the increase in biological reaction rates as temperatures rise by 10 °C, are almost equal for the gases considered (Baldocci 1991).

Stomatal control. Although the permeability of leaf and fruit cuticles for sulfur gases is three orders of magnitude higher than for water, the emission of sulfur gases occurs predominantly via the stomata (Rennenberg & Schröder 1991). The release of gaseous compounds from the mesophyll into the atmosphere depends on the stomatal aperture and is thus governed by the diurnal rhythms of plant transpiration as well as by environmental factors (*e.g.* temperature, relative humidity, water availability).

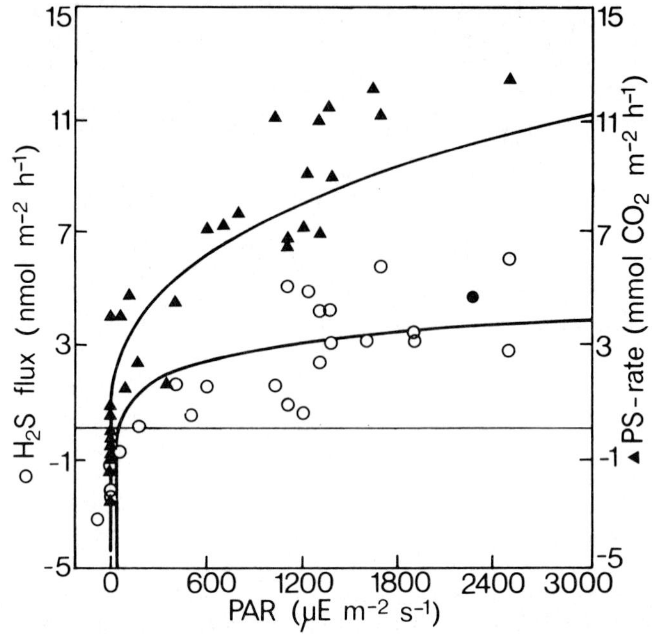

Fig. 2. H_2S emission from spruce plotted against CO_2 fixing rates. The study was conducted with branches attached to the whole tree using dynamic gas exchange cuvettes. Samples of 2 to 5 L of purified air were taken before and after rinsing the chamber. The sulfur gases were trapped at -186 °C. After desorption they were swept onto a GC column and detected by FPD (Rennenberg *et al.* 1990).

From the Bunsen-Graham relationship (*cf.* Table 4, Fig. 1), the stomatal conductance of any gas can easily be deduced in relation to the loss of water vapour through the stomates (Taylor *et al.* 1983; De Kok *et al.* 1989; Schröder & Weiß 1992). In experiments with young spruce trees, the diurnal emission patterns of H_2S from the branches correlated well with the transpiration rate (Rennenberg *et al.* 1990). In four independent chamber experiments with spruce, where the stomatal conductances for water were plotted against the H_2S emissions for the respective trees, a linear relationship was observed, indicating a strong stomatal control of the emissions (Fig. 3). Stomatal control of sulfur gas emissions has also been reported from chamber experiments with corn, soybean, wheat (Fall *et al.* 1988) and other agricultural plants (MacTaggart *et al.* 1987). Besides stomatal control, it is important to consider that gas escape from a canopy is also governed by laminar transport phenomena through the leaf boundary layer. This restriction becomes valid in cases where gradient techniques or the eddy correlation method are applied for flux quantitation instead of chamber techniques. For comprehensive information on this topic, the reader is referred to Müller *et al.* (1992).

Metabolic status of the plant. In chamber experiments conducted in the autumn, approximately 1 nanomole of H_2S was found to be emitted per mole of water. In the experiment conducted during the summer, up to 6 nmoles H_2S were emitted per mole of H_2O. A connection between H_2S emission and stomatal opening was also

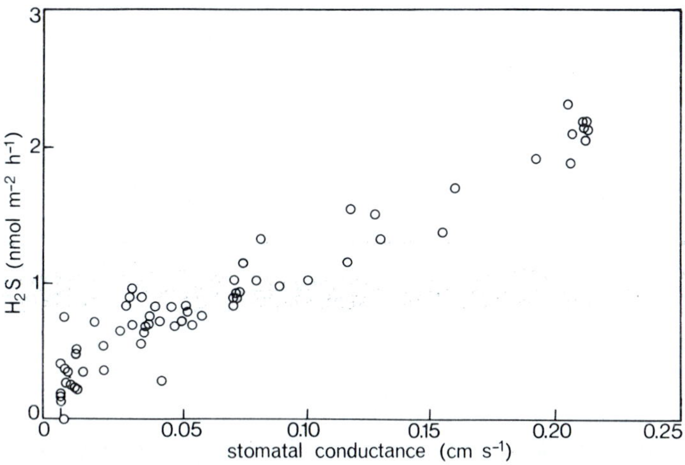

Fig. 3. H_2S emission from spruce trees versus the stomatal conductance for water vapour in four independent experiments. For experimental conditions see legend to Fig. 2.

reflected by the correlation between H_2S emissions and stomatal conductances for water (Rennenberg *et al.* 1992). Whereas this water vapour/hydrogen sulfide emission relationship fits well for spring measurements, considerable deviation from the ideal behaviour is found with plants in autumn (Table 5).

For the summer data, the calculated fluxes are in good agreement with the measurements. In autumn, however, the measured emission fluxes during daytime are lower than those calculated on the basis of transpiration data. Apparently, the emission of H_2S is limited due to some internal resistance as the plants prepare for hibernation. A similar but reciprocal effect was found for wheat plants. The emission of H_2S proceeded slower than calculated during spring but was up to predicted values during summer. In this case the metabolic background is thought to be the production of sulfur deprived storage proteins for the seeds (see above). Whether these two observations are different images of the same effect in perennial and crop plants remains to be elucidated.

Atmospheric concentrations. The rate and direction of the fluxes for sulfur gases depend not only on plant mediated processes and diurnal variations, but greatly on the atmospheric concentrations of the respective compounds. Emission and deposition of molecules from or to surfaces are two independent processes which, according to the kinetic theory of gases (Jeans 1962), occur simultaneously. Simplifying, the number (N) of gas molecules crossing an area (a) per unit time (t) is described as

$$dN/dt = a\ n\ \langle v \rangle/4$$

with n, the number of gas molecules per unit volume and $\langle v \rangle$, the mean of the absolute velocity of the molecules under consideration (see also Schröder 1989). Note in this context that $\langle v \rangle$ is a function of the temperature of the system, $\langle v \rangle = T^{1/2}$ and that rising temperatures as well as increasing molecule numbers (n) will increase the

Table 5. Flux of H_2S from spruce branches in summer and autumn as compared to water fluxes and calculated stomatal conductances. The measurements were conducted as described in the legend to Fig. 2. Data from Rennenberg *et al.* (1992).

	July		October	
	Night	Day	Night	Day
Stomatal conductance for H_2O (cm s^{-1})	0.015 ± 0.01	0.120 ± 0.01	0.074 ± 0.01	0.200 ± 0.02
Measures H_2S flux (nmoles m^{-2} h^{-1})	0.35 ± 0.1	6.71 ± 0.77	0.97 ± 0.1	2.1 ± 0.13
Calculated H_2S flux (nmoles m^{-2} h^{-1})	0.26 ± 0.14	7.78 ± 0.62	1.27 ± 0.19	7.8 ± 0.6

flux per unit time. Provided no temperature difference is observed and no sinks or additional sources are present, the flux in both directions is statistically equal, and therefore can be described by the equations

$$dN/dt\ (up) = a\ n\ \langle v \rangle /4$$
$$dN/dt\ (down) = a\ n\ \langle v \rangle /4,$$

or

$$a\ n\ \langle v \rangle /4\ (up) = a\ n\ \langle v \rangle /4\ (down),$$

which can be reduced to

$$n\ (up) = n\ (down),$$

or, under steady state conditions and because $c = n\ V$, where c, the concentration (mol l^{-1}) and V, the unit volume, can be expressed as concentrations in the mesophyll (c_m) and the atmosphere (c_a):

$$c_m = c_a$$

If a source or a sink for the gas was present, the molecular movements between plant and atmosphere would be highly dependent on the concentrations available in the system. In practice, *in vivo* plant gas exchange is measured adopting von Caemmerer & Farquhar's (1981) approach using an enclosure method. The movement of gases from the atmosphere to the plant and *vice versa* is determined by measuring the change in the respective gas concentrations of the air flowing through the chamber. This is obvious from the micrometeorological flux equation

$$F = Q \frac{c_o - c_i}{A}$$

where F represents the flux (ng $m^{-2}h^{-1}$), Q is the flow rate [m^3 h^{-1}], A is the projected leaf area (m^2) and $c_{i,o}$ are the respective concentrations at the inlet (i) and outlet (o)

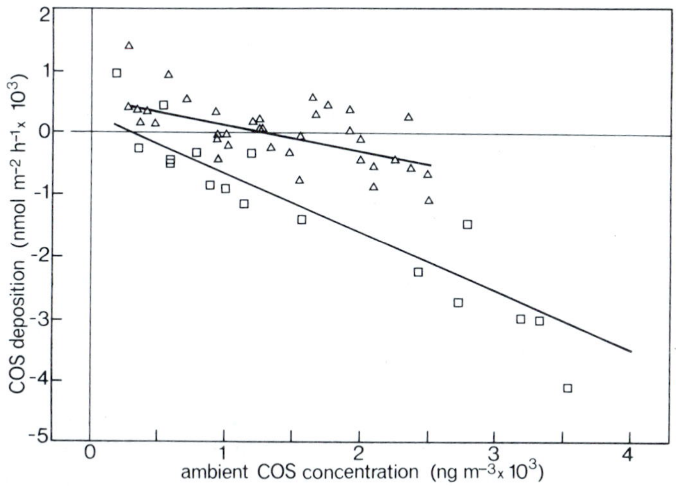

Fig. 4. COS fluxes in a wheat field during spring and summer measurements. The study was conducted with canopy plots in the field covered by dynamic gas exchange cuvettes. Experimental conditions are given in the legend to Fig. 2. Squares: measurement performed in May, vegetative stage; triangles: measurement performed in June, grain filling stage.

of the enclosure used for the measurements (ng m^{-3}). As long as outlet concentrations (c_o) are higher than inlet concentrations (c_i), F remains positive and the flux is directed upward (emission). Increasing inlet concentrations will compensate for the plant mediated emission and thus, at a certain concentration, the sign of F becomes negative and deposition occurs.

The critical concentration, the so called compensation point (see above), reflects the equilibrium between emission and deposition. As mentioned above, the plant's sulfur status could change during development and so the seasonal requirements for sulfur. Changes like this should lead to a modified flux pattern and also to a shift in the compensation point.

This phenomenon has been observed by Rennenberg *et al.* (1991, 1992) in field experiments, when sulfur fluxes were determined in a wheat canopy during two measurement campaigns in May (vegetative growth) and June/July (grain filling). Whereas the fluxes of H_2S and COS are generally directed towards the plants during the vegetative period, the deposition is lowered and fluxes are more frequently directed into the atmosphere during the ripening stage. Under ambient conditions, this decreasing ability to take up atmospheric sulfur is reflected in an increasing equilibrium concentration between uptake and emission (Fig. 4). For COS, the compensation point was altered from 300 ng m^{-3} in May to 1280 ng m^{-3} in June, which is almost equal to the atmospheric background concentration of the gas (Rennenberg *et al.* 1991, 1992).

The above discussion of relevant factors for the emission of sulfur gases from plant leaves reveals the close connection between the direction and magnitude of the fluxes from any ecosystem, the metabolic state of the inhabiting plants, and the atmospheric concentrations of the gases under consideration. It is not possible to obtain reliable emission rates for reduced sulfur gases if any of these factors is ignored.

SO_2-release from plants

Investigations of the flux of SO_2 to a pine forest (Hicks *et al.* 1982) and other plant species (Taylor *et al.* 1983) indicated small but significant sulfur fluxes directed upward from the canopy under certain meteorological conditions. The observation of an SO_2 emission from plants is rather curious, as deposition of this gas is generally observed. The experiments of Hicks *et al.* (1982) were conducted using a SO_2 specific flame photometric detector, which excludes reduced sulfur compounds. Therefore the source of the signal can only be interpreted as SO_2 or, possibly, a particulate sulfate liberating SO_2 during the analytical procedure. A similar observation of positive SO_2 fluxes from a forest was also reported from Whiteface Mountain during the summer months (V. Mohnen, personal communication). Although unlikely to occur, SO_2 release without any diurnal pattern was detected from 6 year old spruce trees during an investigation of the phenomenon under laboratory conditions (Rennenberg *et al.* 1990). The gas was emitted from the trees in single "outbursts" during the night and early morning hours, before changes in light intensities were measurable. The amounts of SO_2 emitted by the trees were not correlated with stomatal movements or with the H_2S released (Rennenberg *et al.* 1990). SO_2 was observed to be emitted independent of the sulfur status of the needles; the emission, however, increased with increasing sulfate content of the soil. The mechanism of this SO_2 evolution remains obscure. It has been suggested that either an SO_2 enrichment in the stomatal cavities during the night or a pH-shift in cell compartments might be responsible for the observed outburst (Rennenberg *et al.* 1990).

Concluding remarks

From the data presented, sulfur emissions are not restricted to certain species or ecosystems, but occur ubiquitously. Conifers from temperate forests seem to be among the lowest emitters, and plants from wetland ecosystems, particularly from the tropics, may be the most significant sources of biogenic sulfur. H_2S and DMS are the most important gases released by higher plants, followed by methyl mercaptan and then COS. CS_2 emission has only been observed from a few species and might therefore be somewhat restricted to wetland plants.

The amounts of sulfur gases produced by plants depend on the internal concentrations of the metabolic precursors governed by the developmental stage of the plant, and also on environmental factors such as temperature, light intensity and air humidity. Recent studies also indicate a correlation between the emission of certain sulfur species and their atmospheric concentration. In general, the emission of reduced sulfur compounds might be a significant detoxification step under conditions of excess sulfur or high atmospheric SO_2. It may also have a regulatory function during plant development under moderate sulfur supply.

Although the concentrations are low when compared to other trace gases, sulfur compounds may have considerable impact on the earth's climate. This is, on the one hand, due to the role of DMS as a molecule involved in the formation of cloud condensation nuclei, and, on the other hand, due to the rapid reactions with radicals and species like NH_3 (Jensen *et al.* 1991). The stable compounds COS and CS_2 might, furthermore, contribute significantly to the stratospheric sulfate budget and be involved in climate stability (Crutzen 1976). To overcome the uncertainties for global estimates of sulfur fluxes, further measurements will have to be conducted to determine physiological, ecological and meteorological prerequisites for the emission of these compounds from plants properly.

Acknowledgement

The author thanks Dr. P. Borrell for critically reading the manuscript and correcting the language.

References

Adams, D.F., Farell, S.O., Robinson, E., Pack, M.R. & Bamesberger, W.L. 1981. Biogenic sulfur source strengths. Environ. Sci. Technol. 15, 1493-1498.

Adewuji, Y.G. 1989. Oxidation of biogenic sulfur compounds in aqueous media: Kinetics and environmental implications. In: Saltzman, E.S. & Cooper, W.J. (eds.), Biogenic Sulfur in the Environment. ACS Symposium series, pp. 529-559.

Anderson J.W. 1990. Sulfur metabolism in Plants. In: Miflin B.J., Lea P.J. (eds.), The Biochemistry of Plants. Academic Press, New York, pp. 327-381.

Andreae, M.O. 1985. The emission of sulfur to the remote atmosphere; Background paper. In: Galloway, J.N., Charlson, R.J., Andreae, M.O. & Rohde, H. (eds.), The Biochemical Cycling of Sulfur and Nitrogen in the Remote Atmosphere. Reidel Publ., pp. 331-362.

Andreae, M.O. 1991. The global biogeochemical cycle: A review. In: Moore, B & Schimel, D.S. (eds.), Trace Gases and the Earth System. Natl. Ctr. for Atmosph. Res. Boulder, Colorado.

Andreae, M.O., Andersen, N.R., Balzer, W., Bingemer, H.G., Brinckman, F.E., Bruner, F., Ehhalt, D.H., Hallberg, R.O., Jorgenson, B.B., Liss, P.S., Lovelock, J.E. & Schmidt, U., 1982. Biogenic contributions to atmpsoheric chemistry. In: Goldberg, E.D. (ed.), Atmospheric Chemistry, Dahlem Konferenzen 1982, Springer, pp. 251-272.

Aneja, V.P. & Cooper, W.J. 1989. Biogenic sulfur emissions. In: Saltzman, E.S. & Cooper, W.J. (eds.), Biogenic Sulfur in the Environment. ACS Symposium series, pp. 2-13.

Aneja, V.P. 1990. Natural sulfur emissions into the atmosphere. J. Air and Waste Manag. Assoc. 40(4):469-476.

Aneja, V.P., Overton, J.H., Cupitt, L.T., Durham, J.L. & Wilson, W.E. 1979. Carbon disulfide and carbonyl sulfide from biogenic sources and their contribution to the global sulfur cycle. Nature 282: 493.

Baldocci, D.D. 1991. Canopy control of trace gas emissions. In: Sharkey, T.D., Holland, E.A. & Mooney, H.A. (eds.), Trace Gas Emissions by Plants. Academic Press, San Diego, pp. 293-334.

Bergmann, L. & Hell, R. 1990. Determination and characterization of γ-glutamylcysteine synthetase from higher plants. In: Rennenberg, H., Brunold, C., De Kok, L.J. & Stulen, I. (eds.), Sulfur Nutrition and Sulfur Assimilation in Higher Plants. SPB Publishing, The Hague, pp. 167-172.

Berresheim, H., Andreae, M.O., Ayers, A.P. & GIllet, R.W. 1989. Distribution of biogenic sulfur compounds in the remote southern hemisphere. In: Moore, B. & Schimel, D.S. (eds.), Trace Gases and the Earth System. Natl. Ctr. for Atmosph. Res. Boulder, Colorado.

Brunold, C. 1990. Reduction of sulfate to sulfide. In: Rennenberg, H., Brunold, C., De Kok, L.J. & Stulen, I. (eds.), Sulfur Nutrition and Sulfur Assimilation in Higher Plants. SPB Publishing, The Hague, pp. 13-31.

Crutzen, P.F. 1976. The possible importance of CSO for the sulphate layer of the stratosphere. Geophys. Res. Lett. 3: 73-76.

De Cormis, L. 1968. Release of hydrogen sulfide into an atmosphere containing sulfur dioxide. C.R. Acad. Sci. 266: 683-685.

De Kok, L.J. 1989. Responses of sulfur metabolism in plants to atmospheric hydrogen sulfide. Phyton 29: 189-201.

De Kok, L.J. 1990. Sulfur metabolism in plants exposed to atmospheric sulfur. In: Rennenberg, H., Brunold, C., De Kok, L.J. & Stulen, I. (eds.), Sulfur Nutrition and Sulfur Assimilation in Higher Plants. SPB Publishing, The Hague, pp. 111-130.

De Kok, L.J., Stahl, K. & Rennenberg, H. 1989. Fluxes of atmospheric hydrogen sulphide to plant shoots. New. Phytol. 112: 533-542.

Delmas, R., Baudet, J., Servant, J. & Baziard, Y. 1980. Emissions and concentrations of hydrogen sulfide in the air of the tropical forest of the Ivory Coast and of temperate regions in France. J. Geophys. Res. 85: 4468-4474.

Elkiey, T. & Ormrod, D.P. 1981. Absorption of ozone, sulphur dioxide, and nitrogen dioxide by petunia plants. Environ. Exp. Bot. 21: 63-70.

Ernst, W.H.O. 1991. Ecological aspects of sulfur metabolism. In: Rennenberg, H., Brunold, C., De Kok, L.J. & Stulen, I. (eds.), Sulfur Nutrition and Sulfur Assimilation in Higher Plants. SPB Publishing, The Hague, pp. 131-144.

Fall, R., Albritton, D.L., Fehsenfeld, F.C., Kuster, W.C. & Goldan, P.D. 1988. Laboratory studies of some environmental variables controlling sulfur emissions from plants. J. Atmos. Chem. 6: 341-362.

Faller, N. 1972. Schwefeldioxid, Schwefelwasserstoff, nitrose Gase und Ammoniak als ausschließliche S- bzw. N-Quellen der höheren Pflanze. Z. Pflanzenernährung 131: 120-130.

Faller, N., Herwig, K. & Kühn, H. 1970. Die Aufnahme von Schwefeldioxid ($S^{35}O_2$) aus der Luft. I. Einfluß auf den pflanzlichen Ertrag. Plant Soil 33: 177-190.

Filner, P., Rennenberg, H., Sekiya, J., Bressan, R.A., Wilson, L.G., Le Cureux, L. & Shimei, T. 1984. Biosynthesis and emission of hydrogen sulfide by higher plants. In: Koziol, M.J. & Whatey, F.R. (eds.), Gaseous Air Pollutants and Plant Metabolism. Butterworth, London, pp. 291-312.

Freer-Smith, P. 1985. The influence of SO_2 and NO_2 on the growth, development and gas exchange of *Betula pendula* Roth. New Phytol. 99: 417-430.

Freney, J.R., Jacq, V.A: & Baldensberger, J.F. 1982. The significance of the biological sulfur cycle in rice production. In: Dommerques, Y.R. & Diem, H.G. (eds.), Microbiology of Tropical Soils and Plant Productivity. Martinus Nijhof/Dr. W. Junk, The Hague, pp. 271-315.

Giovanelli, J. 1987. Sulfur amino acids of plants: an overview. In: Jacoby, W.B. & Griffith, O.W. (eds.), Methods in Enzymology 143, Academic Press, New York, pp. 419-426.

Goldan, P.D., Kuster, W.C., Albritton, D.L. & Fehsenfeld, F.C. 1987. The measurement of natural sulfur emissions from soils and vegetation: Three sites in the eastern United States revisited. J. Atm. Chem. 5: 439-367.

Grignon, C. & Sentenac, H. 1991. pH and ionic conditions in the apoplast. Annu. Rev. Plant Physiol. Plant Mol. Biol. 42: 103-128.

Haines, B., Black, M. & Bayer, C., 1989. Sulfur emission from roots of the rain forest tree *Stryphnodendron excelsum*. In: Saltzman, E.S. & Cooper, W.J. (eds.), Biogenic Sulfur in the Environment. ACS Symp. Ser. Washington, pp. 58-69.

Haunold, W., Georgii, H.-W. & Ockelmann, G. 1992. Gas chromatographic analysis of atmospheric sulfur dioxide and reduced sulfur compounds. LC-GC Intl. 5,10: 28-35.

Heber, U., Laisk, A., Pfanz, H. & Lange, O.L. 1987. Wann ist SO_2 Nährstoff und wann Schadstoff? AFZ 27/28: 700-705.

Hicks, B.B., Weseley, M.L., Durham, J.L. & Brown, M.A. 1982. Some direct measurements of atmospheric sulfur fluxes over a pine plantation. Atmos. Environ. 16: 2899-2903.

Hofmann, U., Hofmann, R. & Kesselmeier, J. 1992. Field measurements of reduced sulfur compounds over wheat during a growing season. In: Slinn, W.G.N. & Schwartz, S. (eds.), Precipitation Scavenging and Atmosphere-Surface Exchange Processes, Hemisphere Publ. Washington, pp. 954-963.

Hofmann, U., Protoschill-Krebs, G., Merk, L., Bingemer, H., Kesselmeier, J. & Andreae, M.O. 1992. Field measurements of reduced sulphur compounds and first physiological results on the uptake of COS by higher plants. In: Borrell, P. *et al.* (eds.) The Proceedings of the EUROTRAC Symposium '90. SPB Academix. Publishing, The Hague, pp. 149-154.

Huber, B., Schröder, P. & Rennenberg, H. 1992. Fluxes of sulfur compounds in a wheat field. Data published in Meixner, F.X. & Ludwig, J. 1992. A field experiment on surface exchange of trace gases: Ecosystem "wheatfield", Manndorf, Lower Bavaria. In: Angeletti, G., Beilke, S. & Slanina, J. (eds.), Field Measurements and Interpretation of Species related to Photooxidants and Acid Deposition. CEC report No. 39, Brussels, pp. 325-338.

Jeans, J. 1962. An introduction to the kinetic theory of gases. Cambridge University Press, Cambridge.

Jensen, N.R., Hjorth, J., Lohse, C., Skov, H. & Restelli, G. 1991. Products and mechanism of the reaction between NO_3 and Dimethylsulphide in air. Atmos. Environ. 25A: 1897-1904.

Kesselmeier, J. 1991. Emission of sulfur compounds from vegetation and global scale extrapolation. In: Sharkey, T.D., Holland, E.A. & Mooney, H.A. (eds.), Trace Gas Emissions by Plants. Academic Press, San Diego, pp. 261-266.

Kesselmeier, J., Hofmann, U., Protoschill-Krebs, G., Hofmann, R., Merk, L., Bartell, U. & Andreae, M.O. 1991. Exchange of reduced sulfur compounds between atmosphere and biosphere. EUROTRAC annual report 1990, Garmisch-Partenkirchen, pp. 162-167.

Lamb, B., Westberg, H., Allwine, G., Bamesberger, L. & Guenther, A. 1987. Measurements of biogenic sulfur emissions from soils and vegetation: Application of dynamic enclosure methods with Nautsch filter and GC/FPD analysis. J. Atmos. Chem. 5: 469-491.

Lorimer, G.H. & Pierce, J. 1989. Carbonyl sulfide: an alternate substrate for but not an activator of ribulose-1,5-bisphosphate carboxylase. J. Biol. Chem. 264: 2764-2772.

MacTaggart, D.L., Adams, D.F. & Farwell, O.S. 1987. Measurement of biogenic sulfur emissions from soils and vegetation using dynamic enclosure methods: Total sulfur gas emissions via MFC/FD/FPD determinations. J. Atmos. Chem. 5: 417-439.

Materna J. 1966. Die Ausscheidung des durch die Fichtennadeln absorbierten Schwefeldioxids. Archiv für Forstwesen, 15: 691-692.

Meszaros, E. 1981. Atmospheric Chemistry, Fundamental aspects. Studies in Environmental Sciences Vol. 11. Elsevier Scientific Publ., Amsterdam.

Meyers, T. P., Hicks, B.B., Hosker, R.P., Womack, J.D. & Satterfield, L.C. 1991. Dry deposition inferential measurement techniques – II. Seasonal and annual deposition rates of sulfur and nitrate. Atmos. Environ. 25A: 2361-2370.

Müller, H., Kramm, G., Meixner, F.X., Dollard, G.J., Fowler, D. and Possanzini, M. 1992. Determination of HNO_3 dry deposition by modified bowen ratio and aerodynamic profile techniques. Tellus, Series B, *in press*.

Nobel, P.S., 1991. Physicochemical and environmental plant physiology. Academic Press, San Diego.

Olsen, R.A. 1957. Absorption of sulfur dioxide from the atmosphere by cotton plants. Soil Sci. 85: 107-111.

Pfanz, H. & Dietz, K.J. 1987. A fluorescence method for the determination of the apoplastic proton concentration in intact leaf tissues. J. Plant Physiol. 129: 159-14.

Pfanz H. & Heber, U. 1985. Protonenflüsse und zelluläre Pufferkapazitäten in Blättern bei SO_2-Belastung. In: Proc. Intl. Workshop on Physiology and Biochemistry of Stressed Plants, GSF 44/85, pp. 103-111.

Pfanz H. & Heber, U. 1986. Buffer capacities of leaves, leaf cells and leaf cell organelles in relation to fluxes of potentially acidic gases. Plant Physiol. 81: 597-602.

Protoschill-Krebs, G. & Kesselmeier, J. 1992. Enzymatic pathways for the consumption of carbonyl sulphide (COS) by Higher plants. Botanica Acta 105: 206-212.

Ravishankara, A.R., Kreutter, N.M., Shah, R.C. & Wine, P.H. 1980. Rate reaction of OH with COS. Geoph. Res. Lett. 7: 861-864.

Rennenberg, H. 1984. The fate of excess sulfur in higher plants. Annu. Rev. Plant Physiol. 35: 121-153.

Rennenberg, H. & Filner, P. 1983. Developmental changes in the potential for H_2S emission in cucurbit plants. Plant Physiol. 71: 269-275.

Rennenberg, H. & Lamoureux, G.L. 1990. Physiological processes that modulate the concentration of glutathione in plant cells. In: Rennenberg, H., Brunold, C., De Kok, L.J. & Stulen, I. (eds.), Sulfur Nutrition and Sulfur Assimilation in Higher Plants. SPB Publishing, The Hague, pp. 53-65.

Rennenberg, H., 1989. Synthesis and emission of hydrogen sulfide by higher plants. In: Saltzman, E.S. & Cooper, W.J. (eds.), Biogenic Sulfur in the Environment. ACS Symposium Series, pp. 44-57.

Rennenberg, H., 1991. The significance of higher plants in the emission of sulfur compounds from terrestrial ecosystems. In: Sharkey, T.D., Holland, E.A. & Mooney, H.A. (eds.), Trace Gas Emissions by Plants. Academic Press, San Diego, pp. 217-260.

Rennenberg, H. & Schröder, P., 1991. Emission of reduced sulfur compounds from agricultural and forest ecosystems. In: Borrell, P., Borrell, P.M. & Seiler, W. (eds.), Proceedings of EUROTRAC-Symposium 90. SPB Academic Publishing,, The Hague, pp. 169-172.

Rennenberg, H. & Thoene, B. 1987. The mobility of sulphur in higher plants. In. Proceedings of the International Symposium on Elemental Sulphur in Agriculture, Nice, Vol. 2, pp. 701-707.

Rennenberg, H., Huber, B., Schröder, P., Stahl, K., Haunold, W., Georgii, H.-W., Slovik, S. & Pfanz, H. 1990. Emission of volatile sulfur compounds from spruce trees. Plant Physiol. 92: 560-564.

Rennenberg, H., Schröder, P. & Huber B. 1991. Emission of reduced sulfur compounds from agricultural and forest ecosystems. A contribution to the EUROTRAC subproject ONEF14 BIATEX.

EUROTRAC annual report 1990, Garmisch-Partenkirchen, pp. 156-162.
Rennenberg, H., Schröder, P. & Huber, B. 1992. Emission of sulfur compounds from spruce trees. In: Schwartz, S.E. & Slinn W.G.N. (eds.) Precipitation Scavenging and Atmosphere Aurface Exchange Processes. Hemisphere Publ. Washington, pp. 979-990.
Schmidt, A., Rennenberg, H., Wilson, L.G. & Filner, P. 1985. Formation of methanethiol from methionine by leaf tissue. Phytochemistry 24: 1181-1185.
Schröder, P. 1989. Characterization of a thermo-osmotic gas transport mechanism in *Alnus glutinosa* (L.) Gaertn. Trees 3: 38-44.
Schröder, P., Huber, B., Schütz, B. & Rennenberg, H. 1992. Emission of reduced sulfur compounds from agricultural and forest ecosystems – Annual report 1991. A contribution to the EUROTRAC subproject ONEF14 BIATAX. In: Biosphere Atmosphere Exchange of Pollutants, Annual Report 1991, Part 4. EUROTRAC ISS, Garmisch-Partenkirchen, pp. 189-193.
Schröder, P. & Weiss, A. 1992. Uptake and detoxification of chlorinated hydrocarbons by spruce trees. In: Schwartz, S.E. & Slinn W.G.N. (eds.), Precipitation Scavenging and Atmosphere Surface Exchange Processes. Hemisphere Publ. Washington, pp. 1011-1021.
Schütz, B., De Kok, L.J. & Rennenberg, H. 1991. Thiol accumulation and cysteine desulfhydrase activity in H_2S fumigated leaves and leaf homogenates of cucurbit plants. Plant Cell Physiol. 32: 733-736.
Schwenn, J.D., Schriek, U. & Kilz, H.H. 1982. Dissimilation of methionine in cell suspension cultures from *Catharanthus roseus* L. Planta 158: 540-549.
Sekiya, J., Schmidt, A., Rennenberg, H., Wilson, L.G. & Filner, P. 1982a. Hydrogen sulfide emission by cucumber leaves in response to sulfate in light and dark. Phytochemistry 21: 2173-2178.
Sekiya, J., Schmidt, A., Wilson, L.G. & Filner, P. 1982b. Emission of hydrogen sulfide by leaf tissue in response to L-cysteine. Plant Physiol. 70: 430-436.
Snoeyink, V.L. & Jenkins, D. 1980. Water Chemistry. John Wiley, New York, pp. 64 f.
Spaleny, J. 1977. Sulphate transformation to hydrogen sulphide in spruce seedlings. Plant Soil 48: 557-563.
Steudler, P.A. & Peterson, B.J. 1984. Contribution of the sulfur from salt marshes to the global sulfur cycle. Nature 311: 455-457.
Steudler, P.A. & Peterson, B.J. 1985. Annual cycle of gaseous sulfur emissions from a New England *Spartina alternifolia* marsh. Atmos. Environ. 19: 1411-1416.
Taylor, G.E. & Tingey, D.T. 1983. Sulfur dioxide flux into leaves of *Geranium carolinianum* L. Evidence for a nonstomatal or residual resistance. Plant Physiol. 72: 237-244.
Taylor, G.E., McLaughlin, S.B., Shriner, D.S. & Selvidge, W.J. 1983. The flux of sulfur-containing gases to vegetation. Atmos. Environ. 17: 789-796.
Thomas, M.D. 1951. Gas damage to plants. Annu. Rev. Plant Physiol. 2: 293-322.
Thomas, M.D., Hendricks, R.H., Collier, T. R. & Hill, G.R. 1943. The utilization of sulfate and sulfur dioxide for the sulfur nutrition of alfalfa. Plant Physiol. 18: 345-371.
Thomas, M.D., Hendricks, R.H., Bryner, R. & Hill, G.R. 1944. A study of the sulphur metabolism of wheat, barley and corn using radioactive sulphur. Plant Physiol. 19: 227-244.
Unsworth, M.H., Crawford, D.V., Gregsen, S.K. & Rowlatt, S.M. 1985. Pathways for sulfur from the atmosphere to plants and soil. In: Winner, W.E., Mooney, H.A. & Goldstein, R.A. (eds.), Sulfur Dioxide and Vegetation, Stanford Univ. Press, California, pp. 375-388.
von Caemmerer, S. & Farguhar, G.D. 1981. Some relationships between the biochemistry of photosynthesis and the gas exchange of leaves. Planta 153: 376-387.
Whitmore, M.E. & Freer-Smith, P. 1982. Growth effects of SO_2 and/or NO_2 on woody plants and grasses during spring and summer. Nature 300: 55-56.
Wilson, L.G., Bressan, R.A. & Filner, P. 1978. Light-dependent emission of hydrogen sulfide from plants. Plant Physiol. 61: 184-189.
Winner, W.E., Mooney, H.A. & Goldstein, R.A. 1985. Sulfur Dioxide and Vegetation. Stanford Univ. Press., California.

DEPOSITION OF GASEOUS SULFUR COMPOUNDS TO VEGETATION

Dennis D. Baldocchi
Atmospheric Turbulence and Diffusion Division/ARL, National Oceanic and Atmospheric Administration, P.O. Box 2456, Oak Ridge, TN 37831, U.S.A.

Introduction

Sulfur compounds play an important role in the chemistry, climate and biogeochemistry of the globe (Andreae 1985; Charlson *et al.* 1992; Schulze 1989; Schwartz 1989). Reduced sulfur compounds (H_2S, COS, dimethyl sulfide, CS_2) enter the atmosphere primarily via natural pathways. Natural sources include the decomposition of organic material, volcanos, sulfur springs and sea spray (Andreae 1985). Oxidized sulfur compounds, such as SO_2 and SO_4^{2-}, have anthropogenic origins. Key anthropogenic sources include fossil fuel combustion, petroleum refining and ore smelting (Semb 1978).

Once in the atmosphere, sulfur compounds are ultimately lost through chemical transformation, dry and wet deposition, cloud scavenging, and long-range transport (Schwartz 1989). In other words, sulfur compounds introduced into the atmosphere are subject to Newtonian physics: what goes up ultimately must come down. Precipitation of sulfur-containing droplets and the dry deposition of gaseous and particulate sulfur compounds are of particular interest to this audience because deposition of sulfur compounds affects the acidity and nutrient status of aquatic and terrestrial ecosystems (Schulze 1989). On the scale of individual plants, exposure to high sulfur concentrations promotes stomatal closure, inhibits photosynthesis and affects plant metabolism and bioenergetics (Wellburn 1982; Rennenberg 1984).

Whatever ones interest regarding sulfur compounds, we cannot assess their impact on the chemistry and climate of the atmosphere and biogeochemistry of natural ecosystems unless we understand what controls the deposition of gaseous sulfur compounds to vegetation and can quantify the rates of this transfer. Material in this paper discusses: 1) methods for measuring gaseous sulfur deposition; 2) results from field studies; 3) processes that control deposition rates; and 4) the models that predict rates of deposition. The material in this review can be supplemented by specific reviews on sulfur deposition by Garland (1977), Fowler (1981, 1985), Unsworth *et al.* (1985), and Murphy & Sigmon (1990) and by general reviews on deposition processes by Hosker & Lindberg (1982), Taylor *et al.* (1988), Davidson & Wu (1990), and Wesely & Hicks (1977). The deposition of particulate sulfate and cloud water are only briefly discussed here, since they are relatively minor contributors to sulfur deposition, and their discussion is beyond the scope of this essay. For more information on particle deposition the reader is referred to Davidson & Wu (1990), and Sehmel (1980). Cloud water deposition of sulfate is thoroughly discussed by Vong *et al.* (1991).

Sulfur Nutrition and Assimilation in Higher Plants
edited by L.J. De Kok et al., pp. 271–293

Methods of measuring sulfur deposition

Micrometeorological and surface analysis methods provide a means for evaluating dry deposition rates of sulfur compounds. Below, I discuss these techniques and instrumentation that is used to measure sulfur concentrations.

Micrometeorological measurement theory

Micrometeorological methods allow one to measure short-term flux densities (moles per unit area and time) of sulfur compounds to crop and forest ecosystems. The equation describing the conservation of mass provides the basic framework for applying micrometeorological methods to measure the vertical flux density of sulfur (F) between the surface and the atmosphere. The conservation equation describes the factors that control the time rate of change of a scalar mixing ratio in a controlled volume. To grasp an understanding of this relationship, let's consider the factors controlling the water level in a bath tube. The water level will remain the same if the amount of water flowing into the tub equals that removed through the drain. In the atmosphere, the concentration of a sulfur compound will remain unchanged if the mean and turbulent fluxes entering a controlled volume equal those leaving (the flux divergence is zero). On the contrary, concentrations will vary with time if the flux of sulfur entering the system differs from that leaving, as when plume impaction occurs.

How can we apply the conservation equation to measure fluxes? In the field, we measure fluxes at a given height above the surface, but we want to know the rate sulfur is deposited to the surface below. The vertical flux density of sulfur will remain unchanged with height if the underlying surface is: 1) homogeneous and extends upwind for a considerable distance (this requirement ensures the development of a surface boundary layer); 2) if scalar concentrations are steady with time; and 3) if no chemical reactions are occurring between the surface and the measurement height.

Condition one can be met easily through proper site selection. As a rule of thumb the site should be flat and horizontally homogeneous for a distance between 75 and 100 times the measurement height (Monteith & Unsworth 1990). Condition two is met often for many scalars. Non-steady conditions are most apt to occur during abrupt transitions between unstable and stable atmospheric thermal stratification, during the passage of a front or from the impaction of a sulfur plume from nearby power plants. It is our experience in Oak Ridge that non-steady SO_2 concentrations occur often, in comparison to other scalars (Matt *et al.* 1987). A rapid and large change in background concentrations occurs during the morning when the stable inversion layer breaks up and entrains elevated SO_2 levels from the planetary boundary layer into the surface layer. Flux densities measured during transient occurrences are difficult to interpret and should be viewed with caution.

In-air chemical reactions involving SO_2 are a potential sink. Key gas phase reactions with SO_2 involve OH, HO_2 and CH_3O_2 (Calvert *et al.* 1978; Schwartz 1989). Yet, these reactions are relatively slow compared to the time scales of turbulent mixing (100 to 200 s), so chemical reactions do not influence the short-term mass conservation budget; typical summertime oxidation rates in rural areas cause 0.1 to 1.5% loss of SO_2 per hour (Calvert *et al.* 1978).

Application of micrometerology theory

Micrometeorologists use a variety of techniques to measure or infer vertical flux densities of sulfur compounds. These methods are outlined below. For the curious reader, more detailed discussions on micrometeorological methods are provided by Businger (1986), Baldocchi *et al.* (1988), Fowler & Duyzer (1989), and Wesely *et al.* (1989).

Eddy correlation technique

The eddy correlation method is a direct method for measuring flux densities of sulfur compounds. It has been used to measure SO_2 deposition over conifer forests (Hicks et al. 1982; Fowler & Cape 1982; McMillen *et al.* 1987), over deciduous forests (Matt *et al.* 1987; Meyers & Baldocchi 1987), over crops (Hicks *et al.* 1989), and over grasslands (Hicks et al. 1982 1986; Neumann & den Hartog 1985; Wesely *et al.* 1983).

The vertical flux density is proportional to the covariance between vertical wind velocity (w) and scalar concentration fluctuations (c):

$$F \approx \overline{-w'c'} \tag{1}$$

Primes represent fluctuations from the mean and the overbar represents time averaging.

A wide range of turbulent eddies contribute to the turbulent transfer of material. Proper implementation of Eq. 1 requires that we sample across a spectrum of eddies. In frequency domain, eddies contributing to turbulent transfer having periods between 0.5 and 2000 s typically contribute to mass and energy exchange (Wesely *et al.* 1989). Hence, wind and chemical instrumentation must be capable of responding to high frequency fluctuations. And computer-controlled data acquisition systems must sample the instrumentation frequently to avoid aliasing and average the signals over a sufficiently long period to capture all the contributions to the transfer.

On applying Eq. 1, it is assumed implicitly that the mean vertical flux density is perpendicular to the streamlines of the mean horizontal wind flow. Consequently, the mean vertical velocity, perpendicular to the streamlines of the mean wind flow, equals zero. In practice, non-zero vertical velocities occur due to instrument misalignment, sloping terrain and density fluctuations. These effects must be removed when processing the data, otherwise mean mass flow can introduced a bias error (Businger 1986; Baldocchi *et al.* 1988).

Evaluating the accuracy of the eddy correlation method is complicated. Factors contributing to instrument errors include time response of the sensor, signal to noise ratio, sensor separation distance, height of the measurement, and signal attenuation due to path averaging and sampling through a tube. Natural variability is due to nonsteady conditions and surface inhomogeneities. Under ideal conditions natural variability exceeds about +/-10%, so it is desirable to design a system with an error approaching this metric.

Flux-gradient method

The flux-gradient methods have been used the most and longest to measure sulfur deposition flux densities. Their long history stems from the availability of simple and slow responding instruments. Sulfur deposition studies, based on flux-gradient methods, have been performed over grass and heather (Duyzer & Bosveld 1988; Garland *et al.* 1974; Davis & Wright 1985), cereal crops (wheat: Fowler 1978, Fowler & Unsworth 1979), pine forests (Lorenz & Murphy 1985; Garland 1977), bare soil and water (Garland 1977).

Flux-gradient, or K-theory, methods are inferential, not direct. Turbulent flux densities are computed using a relationship that is analogous to Fick's equation for molecular diffusion:

$$F = K \frac{\partial c}{\partial z} \tag{2}$$

where K is the eddy exchange coefficient. The concept of down-gradient transfer holds well over aerodynamically-smooth vegetation. On the other hand, use of flux-gradient methods can be problematic over forests. The great height of forests causes them to be aerodynamically rough and to be immersed in a turbulent field dominated by large eddies. These features causes turbulent mixing to be effective and vertical concentration gradients to be relatively small (and, hence, harder to measure). Flux-gradient theory is invalid over forests when the length scales of turbulent mixing exceeds those associated with the curvature of the concentration gradient; under this circumstance turbulent transfer is dominated by large scale transport, which causes counter-gradient transport (Raupach 1988).

Various methods are used to derive the eddy exchange coefficient, K. The energy balance and momentum techniques are most widely used. Momentum-based methods employ mean wind and temperature profile measurements. Eddy exchange coefficients, derived from this method, must be corrected for atmospheric stability effects and for the difference between the transfer of momentum and scalars. The energy balance method uses measurements of mean temperature and humidity profiles and net radiation and soil heat flux densities to calculate the eddy exchange coefficient for water vapor and heat transfer. The eddy exchange coefficient (K) for SO_2 can be assumed to equal that for water vapor and heat, since the transfer pathways are similar. Equations for applying K theory are discussed by Garland (1977) and Fowler & Duyzer (1989).

The accuracy of flux-gradient methods are linked to the accuracy by which one can measure concentration gradients and K. Early SO_2 deposition studies cite large errors in the measurements of concentration gradient (20 to 40%) and flux densities (up to 50%) (Garland 1977; Fowler & Unsworth 1979).

Other micrometeorological methods

Various alternative approaches have been proposed to measure turbulent fluxes. These include the mass balance, variance methods, eddy accumulation and dual tracer methods. It is beyond the scope of this paper to discuss each method. The curious reader is referred to the cited reviews for further detail.

Chemical instrumentation

The flame photometric detection (FPD) is an old and common method of measuring sulfur. The principle of the method follows. Sulfur compounds exposed to a flame are decomposed. Occasionally, two sulfur atoms bond, to form an excited S_2 molecule. Sulfur can be detected with a photon detector when the excited sulfur molecules radiate UV photons (Anderson *et al.* 1989). The method has inherent limitations. It is noisy and is non-linear at low sulfur concentrations (Anderson *et al.* 1989; McMillen *et al.* 1987; Neumann & den Hartog 1985). Non-linearity problems can be circumvented by adding known quantities of SF_6 to the hydrogen gas that fuels the flame. The FPD method has a detection limit between 0.5 and 1 ppb and a 0.5 s response time. Hence, it is not a reliable tool to measure the concentration of natural sulfur gases, whose concentrations are much lower than 0.5 ppb. When using the eddy correlation method, errors due to sensor noise can be minimized by averaging many runs (Wesely & Hart 1985). Furthermore, sensor noise is not critical if the noise is not correlated with vertical velocity fluctuations. For example, Neumann & den Hartog (1985) and McMillen *et al.* (1987) report that a noisy SO_2 power spectrum, measured with a FPD, yields a relatively clean co-spectrum, from which the flux covariance is derived.

A pulsed fluorescence analyzer measures SO_2, specifically. It photo-excites SO_2 with 214 nm radiation. The fluorescence from the excited molecule is detected with a photometer. The detection limit is 0.1 ppb and a 60 s sampling time (Kok et al. 1990). This method can suffer from interference, because UV photoexcitation can also cause hydrocarbons to fluoresce. The slow response time of the method does not lend it to eddy correlation applications, but it can be used in gradient applications.

Tunable diode laser (TDL) spectrometers detect SO_2 concentrations by means of absorption spectroscopy. For SO_2, the absorption wavenumber is in the 1350 cm^{-1} region (Ogram *et al.* 1988). A TDL can be used in eddy correlation experiments because it has a fast response time (0.3 s; Ogram *et al.* 1988) and a reasonable detection limit (1 ppb; Anderson *et al.* 1989).

Surface analysis methods

Surface analysis methods include the throughfall technique (and its analog, the leaf washing technique). The throughfall method estimates sulfate and SO_2 deposition to forests by measuring precipitation above and below a plant canopy. Rain captured below the canopy contains material that washed off the foliage, flowed down the stems and fell through the canopy. Ideally, the dry deposition flux is proportional to the difference between the amount of material measured in the rain below the canopy and in an opening. Complications occur when: 1) the volume of rainfall is not great enough to wash the canopy clean; 2) when sulfur compounds taken up through the stomata are irreversibly immobilized inside the leaf; 3) when sulfur is taken up from the soil by roots and is leached out as sulfate (internal cycling); and 4) if previously accumulated or translocated material is leached out of the leaves and washed off (Schaefer & Reiners 1990, Ivens *et al.* 1990, Lindberg & Lovett 1992).

With regard to sulfur deposition, internal cycling is a small component of the total sulfur budget; internal cycling accounts for less than 10% of total sulfur deposition in red maple, yellow poplar and loblolly pine (Garten *et al.* 1988; Garten 1990). There continues to be uncertainty about the magnitude of irreversible SO_2 uptake. Ivens *et al.* (1990) consider this component to be small. Garten (1990) reports that only half of deposited SO_2 onto forests is later leached as sulfate. Radioactive tracer studies by Garland & Branson (1977) suggest that less than 10% of absorbed sulfur is removed by washing. Throughfall methods also do not capture material that is transferred directly to the forest floor, by way of large scale eddies penetrating through the canopy (T.P. Meyers & D.D. Baldocchi, unpublished results).

Despite the complications cited above, the throughfall method has several appealing advantages. It is inexpensive. It requires limited instrumentation. Throughfall methods are also conducive for examining the role of edges or mountainous terrain on deposition, where micrometeorological methods have limited utility. In fact, throughfall studies suggest that deposition is enhanced at the edge of forests. For example, Ivens *et al.* (1990) report that sulfur deposition is 1.5 times greater to vegetation at the edge of a forest than to vegetation growing inside the forest.

SO_2 deposition to plant canopies

Most field studies on sulfur deposition report a quantity called the deposition velocity (V_d): the ratio between the vertical flux density (F) and the ambient concentration (c(z)). This parameter is in vogue because it is a quantity used by chemical transport and transformation models (Carmichael & Peters 1984; Baer & Nester 1987). Its use is also appealing because it normalizes space and time differences in fluxes on the basis of ambient SO_2 concentrations. This paradigm is valid for SO_2 because its flux is uni-directional (downward) and the surface is a perfect sink for SO_2 (Hicks & Matt 1988). It is not recommended to use the deposition velocity concept to study air-surface exchange of other sulfur species (*e.g.* COS, H_2S) because they are emitted from and deposit to plant canopies (Goldan *et al.* 1988; Taylor *et al.* 1983; Rennenberg 1991).

The goal of many early studies was to support modeling efforts by quantifying the magnitude and variability of V_d over a variety of surfaces. On examination of the scientific literature, one discovers that V_d ranges between 0.1 and 2 cm s^{-1} (conifer forests: Galbally 1979; Garland 1977; *Hicks et al.* 1982; Fowler & Cape 1983; McMillen *et al.* 1987; deciduous forests: Petit *et al.* 1977; Matt *et al.* 1987; Meyers & Baldocchi 1987; crops: Fowler 1978; Fowler & Unsworth 1979; Hicks *et al.* 1989; grasslands: Hicks *et al.* 1983, 1986; Neumann & den Hartog 1985; Duyzer & Bosveld 1988; Davis & Wright 1985; Garland 1977; water bodies: Whelpdale & Shaw 1974; Garland 1977; bare soil: Garland 1977). Typical daytime values over a variety of plant surfaces are on the order of 0.8 to 1 cm s^{-1} (Nicholson *et al.* 1988; Garland 1977, 1978; Wesely & Hicks 1977). Greater values (exceeding 2 cm s^{-1}) are observed over water surfaces (Whelpdale & Shaw 1974) and smaller ($<$ 0.13 cm s^{-1}) values are measured over snow (Davidson & Wu 1990; Valdez *et al.* 1987) and bare soil (Payrissat & Beilke 1974; Garland 1977).

Diurnal and seasonal trends

Hourly, daily, and seasonal variations are primarily due to alterations in surface conditions. V_d varies between 0.1 to 1.5 cm s^{-1} throughout the course of a day over a dry cereal crop (Fowler 1978, 1981). When the crop is vegetative, diurnal changes in V_d are explained by variations in stomatal aperture. Greatest values occur during the day when stomata are most open and lowest values occur at night when stomata are closed. Yet, at night SO_2 exchange does not go to zero, like transpiration, because significant amounts of material is deposited to leaf surfaces and the underlying soil.

Distinct diurnal trends have also been reported for deciduous (Matt *et al.* 1987) and conifer (Galbally *et al.* 1979) forest canopies and for conifer shoots (Hallgren 1982; Johansson *et al.* 1983). Over a deciduous forest, a maximum V_d (1.4 cm s^{-1}) was reported to occur around mid-morning. This high value was due to enhanced deposition to a canopy that had been wetted by previous precipitation. Through the remainder of the day, V_d decreased gradually, reaching a minimum during the night (< 0.1 cm s^{-1}). Factors accounting for the gradual daytime decrease in V_d include drying of the canopy and stomatal closure due to midday water stress and lowering light levels.

Few data exist on the seasonal variation in SO_2 deposition. Meyers *et al.* (1991) evaluated the seasonal variation in SO_2 deposition to a variety of surfaces (deciduous and evergreen forests, crops, and grasslands). They estimated deposition fluxes using monitored SO_2 concentrations and meteorological and surface variables to drive an inferential deposition model. Seasonal trends were evident over most sites surveyed. Maximum values occurred during the summer, minimum values occurred during the winter and spring and fall values were similar.

Seasonal trends in SO_2 deposition velocities are expected for annual crops and deciduous forests, which possess distinct seasonal changes in phenology. Wintertime death or loss of leaves constitute a reduction in the sink strength for SO_2. In contrast, the presence of foliage during the growing season constitute a strong sink for SO_2. For example, Fowler (1985) reports that leaf senescence causes V_d of a cereal crop to drop below detection level of the measurement system, even though atmospheric conditions are apt for transfer. Over a deciduous forest growing near Oak Ridge, TN mean summer fluxes over a 3 year period were 2.37 kg ha^{-1}, while mean fluxes during the winter were 1.96 kg ha^{-1} (Meyers *et al.* 1991).

Evergreens are capable of absorbing SO_2 throughout the year, even in subzero temperatures (Garsed 1985). Yet, greater V_d values occur in the summer than the winter. Greater stomatal resistance during the winter cause most sulfur to be deposited to the needle surface (Johansson *et al.* 1983), a less efficient pathway.

Seasonal trends in stomatal conductance are not the only factor controlling the seasonal trend of SO_2 deposition to conifer forests. In the Netherlands, maximum deposition values are observed during the winter and minimum values occur during the summer (Vermetten *et al.* 1991). The distinct north European climate contributes to this seasonal pattern. Forests are typically wet during the winter, a feature that exacerbates deposition (see below). During the summer, lower SO_2 deposition fluxes occur because the frequency of occurrence of a wet canopy is lower and higher evaporation rates in the summer allow it to dry faster.

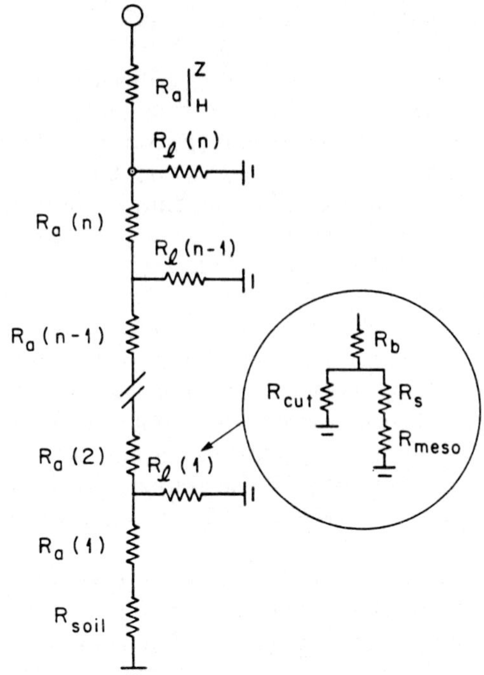

Fig. 1. Resistance network for deposition to a plant canopy.

Mechanisms for SO_2 deposition to plant canopies

Practitioners have attempted to study V_d in conjunction with controlling the meteorological, biological and physio-chemical processes. The big-leaf model, adapted from evaporation studies, has been invoked to interpret deposition flux measurements (Wesely & Hicks 1977; Fowler & Unsworth 1979; Fowler 1985; Baldocchi *et al.* 1987; Hicks *et al.* 1987; Murphy & Sigmon 1990; Fig. 1). The inverse deposition velocity equals the sum of the aerodynamic (R_a), quasi-laminar (R_b) and canopy resistances (R_c).

$$V_d = \frac{1}{R_a + R_b + R_c} \tag{3}$$

For canopies with the same leaf area indices, one can expect V_d of short C_3 grasses to be less than V_d for taller C_3 crops (wheat, barley, cotton), V_d of a C_4 crop to be less than that of a C_3 crop, and V_d of a tall C_3 broadleaf forest to be less than a shorter C_3 crop (Meyers & Hicks 1988; Fowler 1985). The factors controlling the resistances that comprise V_d are discussed next.

Turbulence and diffusion

Turbulent eddies are responsible for transporting material through the surface boundary layer. The aerodynamic resistance determines the rate that momentum, and other scalars, are transported between a given level in the atmosphere and the vegetation's effective surface sink. The aerodynamic resistance is expressed as:

$$R_a = \frac{1}{ku*} \ln \frac{z-d}{z_o} \psi_c \qquad (4)$$

where k is von Karman's constant (0.4), u* is friction velocity, z is height, d is the zero-plane displacement, z_o is the roughness parameter and ψ_c is a diabatic correction function. The reader is advised to consult Monteith & Unsworth (1990) and Wesely & Hicks (1977) for detailed discussions on these parameters.

The quasi-laminar resistance (R_b) is introduced because the sinks for momentum and SO_2 (and any other scalar) differ. In the immediate vicinity of the canopy, mass and energy exchange is controlled by the molecular properties of the fluid, while the rate of momentum transfer is affected by bluff-body pressure effects: a process that has no analog in the case of mass and heat transfer. R_b is formulated as:

$$R_b = \frac{1}{ku*} \ln \frac{z_o}{z_c} = \frac{const}{ku*} (Sc/Pr)^{2/3} \qquad (5)$$

where z_c is the scalar roughness length, Sc is the Schmidt number and Pr is the Prandtl number. The constant in Eq. 5 is often assumed to equal 2 over closed canopies, but can be much greater over rough incomplete canopies (Wesely & Hicks 1977).

The aerodynamic and quasi-laminar resistances are affected by wind speed, crop height, leaf size, and atmospheric stability (Wesely & Hicks 1977; Fowler 1985; Murphy & Sigmon 1990). In general, R_a plus R_b decreases with increasing wind speed and crop height. Hence, smaller resistances are expected over tall forests than over short grass and under unstable atmospheric thermal stratification, than under neutral and stable stratification. To give the reader some flavor for the magnitude of these resistances, let's examine typical boundary layer resistance (R_a) for several vegetation types. When wind speeds are 4 m s^{-1} theoretical boundary layer resistances over a 0.1 m tall grass, a 1.0 m crop and a 10 m conifer forest are about 60, 20 and 10 s m^{-1}, respectively (Fowler 1985). Experimental measurements by Matt *et al.* (1987) show that both R_a and R_b are less than 20 s m^{-1} during the day over a temperate deciduous forest. Greater R_a values (up to 150 s m^{-1}) occur at night when turbulent mixing is reduced.

Canopy resistance

Numerous field studies show that R_c is the major controller of SO_2 deposition to a plant canopy (Fowler & Unsworth 1979; Matt *et al.* 1987). The canopy resistance (R_c) is a function of the canopy stomatal resistance (R_{stom}), the canopy cuticle resistance ($R_{cuticle}$), and the soil resistance (R_{soil}). In turn, these resistances are affected by leaf area, stomatal physiology, soil pH, and the presence and chemistry of liquid drops and films. The stomatal, leaf surface (cuticle) and soil resistances act in parallel, causing R_c to be formulated as:

$$\frac{1}{R_c} = \frac{1}{R_{stom}} + \frac{1}{R_{soil}} + \frac{1}{R_{cuticle}} \tag{6}$$

Below we examine these components in more detail.

Physiological control of canopy resistance

Many physiological studies show that the main pathway of SO_2 to a plant is through the stomata (Garland & Branson 1977; Olszyk & Tingey 1985; Hallgren *et al.* 1982; Taylor & Tingey 1983; Johansson et al. 1983). Reviews by Garsed (1985) and Taylor *et al.* (1988) conclude that between 75-90% of SO_2 deposited to a leaf follows this route, while the residual 10-25% is deposited on the leaf surface. On the other hand, studies reporting SO_2 emission from leaves are rare and the episodes that have been observed are short-lived; emission of SO_2 from spruce leaves occurs as a burst just before a light period (Rennenberg *et al.* 1990).

As gas molecules enter the leaf, deposition occurs as molecules react with the moist cells in the sub-stomatal chamber and the mesophyll. Taylor & Tingey (1983) report that the stomatal resistance to SO_2 uptake, based on transpiration measurements, is less than the stomatal resistance derived from sulfur dioxide flux measurements. They surmise that the diffusion path of a SO_2 molecule is less than for water (29 μm for SO_2 *vs.* 60 μm for H_2O).

One exceptional study in the literature is worth discussing. Hällgren *et al.* (1982), studying deposition to Scots pine needles, noted a marked difference between diurnal variation of deposition velocity and stomatal conductance. Furthermore, they did not observe a one-one relation between needle surface conductance for SO_2 and stomatal conductance. They conclude that an additional resistance can be inferred because the leaf conductance to SO_2 uptake was less than the stomatal conductance. This result is controversial because Johansson *et al.* (1983) performed a similar study at the same site and found contrasting results; they observed that the total conductance for SO_2 uptake by needles (V_d) was greater than the stomatal conductance for SO_2 uptake. The controversy may be resolved by considering the SO_2 concentrations to which the needles were exposed. Hallgren *et al.* (1982) exposed conifer needles to high SO_2 levels (15 to 125 ppb) and hypothesized that their results were affected by sulfur emissions. On the other hand, Johansson *et al.* (1983) had exposure levels $<$ 10 ppb, a condition when sulfur emission are less apt to occur.

Since stomatal action exerts major control of SO_2 uptake, several conclusions can be drawn about environmental effects and interspecific differences on SO_2 deposition based on known stomatal physiology. In other words, factors that increase stomatal resistance should decrease the SO_2 deposition velocity (Baldocchi 1988). Stomatal physiology studies show that stomatal resistance decreases hyperbolically with increasing light and increases linearly with increasing vapor pressure deficits (Jarvis 1976). Soil water deficits cause stomata to close after some threshold deficit level is exceeded. Low and high temperatures cause stomatal closure and moderate temperatures promotes stomatal opening. Leaf age, nutrition and adaptation are other factors affecting stomatal resistance. Elevated exposure to SO_2 causes stomata to close.

A significant difference in V_d is expected between C_3 and C_4 species because leaves of C_3 species exert a lower stomatal resistance than do C_4 leaves (Pearcy & Ehleringer 1983). Among different C_3 species, V_d differences are not as well defined. Olsyzk & Tingey (1985) examined a spectrum of vegetation types (crop, native herbaceous, woody mediterranean). They reported that the same SO_2 flux was experienced to the leaves of most species (except for a mutant). They also concluded that internal resistances could be discounted, except for the mutant.

On the canopy scale, the stomatal component (R_{stom}) of the canopy resistance is proportional to the mean stomatal resistance (r_s) divided by the canopy's leaf area index. Care is needed when modeling or measuring R_{stom} because r_s is a function of light, which varies appreciably with depth into the canopy and on sunlit and shaded leaves (Baldocchi *et al.* 1987). Typical values of R_{stom}, under ideal conditions, range between 30 and 300 s m^{-1} for a range of herbaceous annuals and woody perennials (C_3 annuals < woody perennials < C_4 annuals) (Baldocchi *et al.* 1987; Fowler 1985; Matt *et al.* 1987).

The canopy cuticle resistance far exceeds the canopy stomatal resistance; $R_{cuticle}$ ranges between 3000 and 40000 s m^{-1} (Van Hove *et al.* 1989), the cuticle is not an impermeable barrier to SO_2. Nor is it a static quantity. Van Hove *et al.* (1989) envisions the cuticle as a sponge, which expands with humidity as the pores of the cutin matrix become water-filled. This swelling process provides external SO_2 molecules accessibility to aqueous sites inside the leaf, which have an affinity for SO_2. Hence, cuticle resistance decreases as relative humidity increases (Van Hove *et al.* 1989; Garsed 1985). SO_2 deposition to the surface also occurs via absorption and via chemical oxidation reactions with the surface.

Physio-chemical control of SO_2 uptake

Many authors report that SO_2 deposition is enhanced over wet vegetative surfaces (Fowler & Unsworth 1979; Fowler 1985; Garland & Branson 1977; Vermetten *et al.* 1991). Enhanced deposition occurs over wet vegetation because SO_2 is very soluble in water. The absorption of SO_2 is associated with a series of reactions (see Seinfeld 1985; Brimblecombe 1986):

$$\begin{aligned} &SO_2(g) + H_2O \Leftrightarrow SO_2{\cdot}H_2O && (a) \\ &SO_2{\cdot}H_2O \Leftrightarrow H^+ + HSO_3^- && (b) \\ &HSO_3^- \Leftrightarrow H^+ + SO_3^{2-} && (c) \end{aligned} \tag{7}$$

The total dissolved sulfur in solution is related to the partial pressure of SO_2 over the solution (p_{SO2}) and the pH of the solution:

$$[S(IV)] = K_H\, pSO_2\, (1 + K_1/[H^+] + K_1K_2/[H^+]^2) \tag{8}$$

K_H is the Henry coefficient for SO_2 (5.4 mol l^{-1} atm^{-1} at 15 °C) and K_1 and K_2 are the rate coefficients for reactions 7b and 7c, respectively. From Eq. 8 one can conclude that an increase in acidity will reduce the amount of dissolved SO_2. With regards to SO_2 chemistry, there are two sources of protons which affect the acidity of a solution and SO_2 solubility. One source of protons is the disassociation described in Eqs. 7b and 7c. Another source of protons comes from the oxidation of S(IV) to S(VI). Oxidation of S(IV) typically occurs by O_2, O_3 and H_2O_2. The oxidation of aqueous SO_2 is slow unless there is a metal catalyst (iron or manganese; Brimblecombe 1986).

The rate of SO_2 deposition to a wet surface depends on whether the surface as wetted by rain or dew (Fowler & Unsworth 1979). The pH of rain is less than pure water due to carbonate chemistry and the scavenging of acidic precursors and oxidants during the precipitation process (see Schwartz 1989). The chemistry of dew is affected by uptake of acidic precursors and oxidants, particles on the leaf surface, chemical lifetime, droplet size and the leaching of nutrients from the leaf (Brimblecombe 1978; Wesely *et al.* 1990; Chameides 1987).

Several field studies illustrate the effect of differential deposition to dew and rain wetted canopies. For example, Fowler (1978, 1981, 1985) reports that V_d, at night, rose from 0.3 to 0.6 cm s^{-1} as dew deposited on the canopy, even though stomata remained closed. If the crop was rain-wetted instead of dew wetted, V_d values drop by half (to 0.3 cm s^{-1}) on the rain wetted crop. Fowler (1985) also reports modest deposition rates to a rain wetted Scots pine canopy ($<$ 0.4 cm s^{-1}).

Wesely *et al.* (1990) modeled deposition velocities to dew wetted grass. V_d values ranged between 0.05 and 1.0 cm s^{-1}. The aqueous resistance to S(IV) uptake limits SO_2 uptake when V_d exceeds 0.5 cm s^{-1} (if O_3 is the only oxidant available). Low deposition velocities occur when atmospheric mixing is weak. Yet these are the conditions under which dew forms. When turbulent mixing increases, so does V_d. However, this effect is temporary because greater turbulence promotes evaporation and the disappearance of dew.

Vermetten *et al.* (1991) studied SO_2 deposition to a Douglas fir forest. They report that V_d was strongly enhanced by wetness; V_d at midday approached 2 cm s^{-1} over a dry canopy and 4 to 5 cm s^{-1} over a wet one. The amount of water storage on the canopy was a critical factor controlling the rate of deposition. V_d increased from near zero to 3 cm s^{-1}, as canopy water storage increases from 0 to 1 mm. At higher storage levels deposition velocities plateaued.

Only one study shows contradictory results: deposition is not enhanced over a wet canopy. Nicholson & Davies (1988) report a mean V_d of 1.55 cm s^{-1} over a dry agricultural region and a mean V_d of 0.75 cm s^{-1} when the region was wet. The low pH of the precipitation may explain the suppressed deposition to the region.

Co-deposition of SO_2 and NH_3

As discussed above the deposition of SO_2 to a wet surface will decrease as the pH of the solution decreases. The deposition of NH_3 to an acid solution neutralizes the solution. Thereby, the co-deposition of SO_2 and NH_3 should experience higher SO_2 deposition rates to water surfaces than what would be experienced by SO_2 alone (Wesely *et al.* 1990).

Van Hove *et al.* (1989) and Fowler *et al.* (1991) present evidence showing that the surface SO_2 uptake is, indeed, greatly enhanced in the presence of NH_3 emission. One site of reaction is in the water-filled pores of the cuticle. Another site is on the leaf surface, where reactions forming $(NH_4)_2SO_4$ and $(NH_4)_2SO_3$ act as a sink for SO_2.

Sulfur deposition to soils and litter

Deposition to canopy involves deposition to both vegetation and soil. Consequently, we can only assess the deposition flux to vegetation by subtracting the flux to the soil from the total canopy flux. Garland (1977) reports that the SO_2 deposition velocity to soil increases with pH. Payrissat & Beilke (1975) report that a pH increase from 4.5 to 7.6 corresponds to a 10 fold decrease in the soil surface resistance to SO_2 uptake. They also find that soil surface resistance, at a constant pH, decreases as relative humidity increases. Petit *et al.* (1977) report that R_{soil} is between 303 and 385 s m^{-1} at pH 3 and between 31 and 40 s m^{-1} at pH 7.

Early studies assumed that deposition to soils under vegetation was relatively small. For example, Fowler (1981) estimates that flux to the soil is only 5 to 10% of total flux to a cereal canopy. Yet, recent work shows that a significant amount of material is deposited on the soil below the vegetation. 20 to 30 % of SO_2 depositing on a deciduous forest is received at the forest floor (T.P. Meyers & D.D. Baldocchi, unpublished results). This substantial transfer occurs because large scale intermittent eddies are able to punch through the vegetation and deliver a significant amount of material to the soil.

Deposition of other sulfur compounds

H_2S, COS, methyl mercaptan and CS_2 are other sulfur compounds that have vegetative sinks (Goldan *et al.* 1988; Taylor *et al.* 1983). Taylor *et al.* (1983) report that the magnitude of the flux to a leaf is ranked in the following order: $SO_2 > H_2S >$ COS $>$ methyl mercaptan $> CS_2$. Water solubility and molecular size accounted for 73 and 87%, respectively, of the variation in surface fluxes among compounds. Uptake through the stomata was the major route. However, surface uptake was not insignificant; for *Glycine max* between 30 and 50% of the uptake was on the leaf surface.

Sulfate deposition has been studied by Wesely *et al.* (1985, 1982) and Hicks *et al.* (1982), among others. The deposition of particles is mainly dependent on physical processes and quantities, such as turbulent mixing, particle size distribution. Over grass sulfate V_d is dependent on friction velocity (u*) and atmospheric stability (Wesely *et al.* 1985). Mean values are on the order of 0.22 cm s^{-1}, but can approach 0.5 cm s^{-1} under very windy conditions.

Cloud water deposition of sulfur compounds

Cloud water deposition is a contributor to the sulfur budget of high elevation sites that are immersed in clouds or in valleys with high occurrence of fog (Lindberg & Lovett 1992). The deposition of SO_2 in clouds or fog is enhanced by up to a factor of three because the relatively large Henry constant, associated with SO_2, allows it to be readily transferred to the aqueous phase (Pandis & Seinfeld 1991).

Fowler *et al.* (1990) report that cloud water deposition may increase the amount of wet deposited SO_4^{2-} by 12%, and if the high elevation sites are forested this increase may approach 44 %. Vong *et al.* (1991) report that cloud water deposition fluxes at a variety of Eastern mountain sites ($>$ 1000 m) range between 0.9 to 11.5 kg ha^{-1} $month^{-1}$ of sulfate. Cloud water deposition contributes 25-50% of total sulfate input (ppt + dry + cloud) to mountain sites in the eastern US (Lindberg & Lovett 1992; Vong *et al.* 1991).

Sulfur budgets

To address ecological and biogeochemical questions, information is needed on the long-term, (*e.g.* annual) input of sulfur compounds to vegetation. Unsworth *et al.* (1985) performed a detailed study on the sulfur budget of a wheat crop in Great Britain. Over the course of a growing season, SO_2 deposition was 6.8 kg ha^{-1}. Of this total 5.0 kg ha^{-1} was deposited when the crop was growing and 1.8 kg ha^{-1} was deposited when the crop was senescent.

In the eastern US, annual deposition rates to a deciduous forest, a grassy meadow and an agricultural region are 8.7, 12.6 and 5.0 kg ha^{-1}, respectively (Meyers *et al.* 1991). Much smaller values (0.3 kg ha^{-1}) are detected over a grassland in the Rocky Mountain region of North America.

Modeling SO_2 deposition

Scaling SO_2 fluxes from the leaf to canopy scales involves linking a leaf-level, physiology and physio-chemical model to a canopy micrometeorology model. A physiology model is needed to evaluate the stomatal resistance of leaves. A physio-chemical model is needed to assess leaf boundary layer resistances, and gaseous uptake at the leaf surface and by water drops and films. A canopy micrometeorology model computes variables that control stomatal resistance (such as light, temperature, and vapor pressure deficits) at the local microclimate of the leaf. Micrometeorological models are also used to compute leaf wetness and turbulent mixing within and above the canopy. Below, the linkage between these model classes is discussed. For more information on linking physiological, physio-chemical and micrometeorological trace gas exchange models the reader is referred to Baldocchi (1991).

Leaf uptake of sulfur gases

The flux density of gaseous sulfur compounds to a dry leaf can be computed using a resistance-analog model. In other words, the flux density is assumed to be proportional to the potential (the difference between concentrations in the atmosphere and at the leaf surface) and is inversely proportional to the sum of resistances that restrict this transfer. The algorithm describing leaf pollutant uptake is:

$$F_l = \frac{(C(z) - C_i)}{r_b + r_c} \tag{9}$$

where C is the SO_2 concentration, r_b is the leaf boundary layer resistance and r_c is the surface resistance. Resistance algorithms for gaseous uptake by leaves have been used by O'Dell *et al.* (1973), Murphy *et al.* (1977), Baldocchi *et al.* (1987, 1988) and Meyers (1987) to extend computations of SO_2 deposition to leaves to the canopy scale.

The leaf surface resistance contains the parallel resistances exerted by the stomata and cuticle. Stomatal resistance can be computed using a multiplicative phenomenological model (Jarvis 1976; Baldocchi *et al.* 1987) or a photosynthesis-driven model (Collatz *et al.* 1991). The internal concentration (C_i) and mesophyll resistance (r_m) for SO_2 transfer are often zero (O'Dell *et al.* 1973; Taylor & Tingey 1983). Empirical relationships are usually used to specify the cuticle resistance.

Deposition to dry plant canopies

Big-leaf and multi-layer models are used to compute SO_2 deposition to plant canopies. A big-leaf model is a based on Eq. 3. Since this model was discussed above, its description will not be repeated here.

Multi-layer models are used to mechanistically scale leaf level fluxes to the canopy level. This approach has many advantages over the big-leaf method. For example, leaf to canopy scaling requires the evaluation of dependent, non-linear functions in terms of independent variables that vary in time and space. From a statistical standpoint, one should evaluate the expected value of a non-linear function instead of evaluating the function in terms of the mean independent variable. The application of a multi-layer model allows one to evaluate the spatial variability of controlling independent variables and, hence, an ability to properly evaluate non-linear dependent functions.

Estimates of the net turbulent flux of material between a plant canopy and the overlying atmosphere are determined by integrating the sink strength of discrete vegetative layers and the soil. The vegetative sink strength of a discrete layer can be modeled by multiplying Eq. 9 by the leaf area density of the layer. Resistance models are also used to specify the soil boundary conditions (Meyers 1987).

Two basic reference frames exist for computing turbulent fluxes over plant canopies. They are the Eulerian and Lagrangian frames. The principles behind these frameworks are explored in the following subsections.

Eulerian models

The Eulerian framework evaluates the scalar concentration at a fixed point and time, as occurs when measuring the concentration of a given scalar from a tower. However, the conservation budget equation cannot be readily solved because the conservation budget does not form a closed set of equations and unknowns. The equation defining the time rate of change in C contains a higher order moment which is also a function of C. This higher order moment is the vertical turbulent flux. The simplest, and earliest, Eulerian models on turbulent exchange in plant canopies adopted a first order closure scheme, called 'K-theory'. K-theory models have been used by Shreffler (1976), Murphy *et al.* (1977) and Baldocchi (1988) for modeling SO_2 deposition to plant canopies. The appeal of this model class is its simple reduction of the number of unknown variables. K-theory models assume that turbulent transfer and molecular diffusion are analogs, thereby the vertical velocity-scalar covariance is represented as the product of the scalar concentration gradient and a turbulent diffusivity (K), such as Eq. 2. An accumulating body of evidence shows that many of the assumptions supporting K-theory are false inside plant canopies. Turbulent transport is

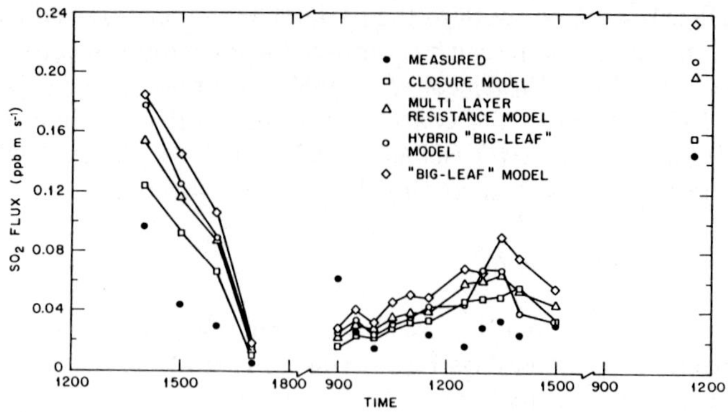

Fig. 2. Intercomparison of big-leaf and multi-layer Eulerian deposition models (after Meyers & Baldocchi 1988).

dominated by large scale and intermittent eddies, which can cause counter-gradient transfer (Raupach 1988).

Higher-order closure models have been proposed as a means of circumventing the inherent limitation of first order closure models (Meyers 1987). Higher order closure models introduce formal budget equations for higher order moments, such as the vertical turbulent flux. Equations that describe mean wind speed and turbulence are also introduced to evaluate dependent terms in the second moment equation and in the source-sink function (*i.e.* r_b and c(z)). The budget equations for the second order moments, unfortunately, include additional unknowns of the third order. Hence, an equal set of equations and unknowns can only be obtained through parameterizing the highest order moment with an 'effective' eddy exchange coefficient (Meyers 1987; Meyers & Baldocchi 1988).

A hierarchy of Eulerian closure models have been tested for their ability to simulate SO_2 fluxes above a deciduous forest canopy (Meyers 1987; Meyers & Baldocchi 1988). Model tests reveal that an increase in model sophistication yields an improvement in model accuracy. For calculating SO_2 deposition to a deciduous forest, a higher order closure model is better than a multi-layer K-theory model, which is better than a big-leaf resistance model (Meyers & Baldocchi 1988; Fig. 2). The more detailed models perform better because they are better able to assess the local environmental variables that control the resistances, r_b and r_s.

Lagrangian models

Deardorff (1978) criticized the use of 'effective' exchange coefficients to close Eulerian budget equations of higher order moments. Deardorff argues that 'effective' exchange coefficients are inadequate for near-field flows: which occur in the vicinity of sources and sinks. This is because any turbulent diffusivity, K, in the vicinity of a source or sink is linearly related to the time period that fluid parcels have travelled. Only after a long travel distance is the time independent, "far-field" limit of K reached.

The Lagrangian framework circumvents the closure problem ailing Eulerian

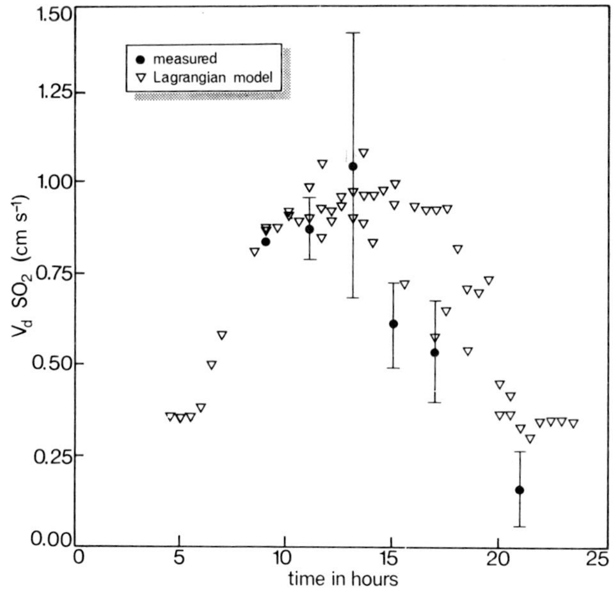

Fig. 3. A comparison of calculated SO_2 deposition velocities, derived from a multi-layer Lagrangian model, and eddy correlation measurements (D.R. Matt, personal communication). The experimental field conditions are described by Matt *et al.* (1987). Details of the Lagrangian model are described by Baldocchi (1992a,b).

models because Lagrangian models are able to explicitly differentiate between near and far field diffusion. The Lagrangian approach analyzes the conservation equation by following parcels of fluid as they move with the wind, much like the trajectory of a neutrally-buoyant balloon. In the Lagrangian frame, a concentration field and its turbulent flux are defined by the statistics of an ensemble of dispersing marked fluid parcels and the strength and spatial distribution of sources and sinks (Raupach 1988). The concentration of a scalar at a particular location is a function of the probability that a fluid parcel released from a point in space, z_o, at time, t_o, will be observed at another location and time (z,t).

The probability density function for the diffusion of fluid parcels depends only on the properties of the turbulent wind field, which must be prescribed. A difficulty arises when one tries to apply this method inside a plant canopy, where turbulence length, time and velocity scales of turbulence vary with height and the probability density functions of the three vector velocity components are skewed and kurtotic (Raupach 1988; Baldocchi & Meyers 1988). When turbulence is heterogeneous, analytical solutions cannot be used. Instead, Markovian "random-walk" approaches must be employed to compute the trajectory of a large number of fluid parcels.

Fluid parcel movement in a Markovian random-walk model is computed using the Langevin equation, a function that defines the acceleration of a fluid parcel as a function of the memory of its initial value and a random forcing (Raupach 1988). Inside plant canopies, the vertical gradient in the vertical velocity variance imposes a downward drift on a Markovian random flight model, and an artificial accumulation of matter. Heuristic models have been applied to remove this unrealistic accumulation

Table 1. Comparison of SO_2 deposition velocities computed with a multi-layer K-theory model (Baldocchi 1988) and the random-walk Lagrangian model that was discribed in this paper.

V_d (cm s^{-1})	@u = 1 m s^{-1}	@u = 2 m s^{-1}	@u = 4 m s^{-1}
K-theory	0.554	0.764	0.866
Lagrangian	0.716	0.818	0.915
relative difference	0.226	0.066	0.0535

of matter. One approach introduces an additional force term into the Langevin equation, yielding a mean upward drift velocity in the solution of the differential equation.

Theory on Lagrangian modeling has been developing rapidly over the last decade (Raupach 1988), but has only recently been used to calculate fluxes of trace gases in real forest and crop canopies (Baldocchi 1992, 1993), and until here has not been used to assess SO_2 deposition velocities.

Fig. 3 compares computations of V_d derived from a Lagrangian SO_2 deposition model against field measurements over a temperate deciduous forest during dry periods. The model computes a distinct diurnal course: a minimum occurs during dark periods and a maximum is reached during midday. Calculated and measured values are in relatively good agreement until noon. After this period, model values overestimate measured ones. The field flux data imply that the surface resistance increase in the afternoon, thereby lowering V_d (Matt *et al.* 1987). This version of the model does not allow stomata to close with increasing vapor pressure or soil moisture deficits. The omission of a stomatal response to vapor deficit is based on work Meyers & Baldocchi (1988) on *Quercus alba*. It is not possible to conclude whether or not the daily evaporation demand exceeded soil water uptake and caused the stomata to partially close.

How does the Lagrangian model compare with an Eulerian first-order K-theory model? Table 1 shows that at relatively high wind speeds ($u \geq 2$ m s^{-1}) deposition velocities calculated by the 'theoretically-flawed' K-theory model underestimate those calculated by the Lagrangian model by less than 6%, a trivial difference. Only at low wind speeds (≤ 1 m s^{-1}) is a divergence in calculated values observed. At low wind speeds, an artificially large drawn-down in the scalar concentration profiles is computed by a K-theory model (Baldocchi 1988; Meyers & Baldocchi 1988), which in turn causes the integrated sink term to be underestimated.

Deposition to wet canopies

The deposition rate of SO_2 to a dew-wetted canopy has been modeled by Chameides (1987) using a Big-leaf resistance approach, where V_d is:

$$V_d = \frac{1}{R_a + R_b + R_i + R_{sfc}} \qquad (10)$$

R_a and R_b retain the same definitions given above. R_i is an impactive resistance, defining the rate that a species impacts and sticks to a dew drop. R_i is inversely related to the square of the drop radius, the thermal velocity and the drop concentration and the accommodation coefficient; its value is typically three orders of magnitude smaller than $R_a + R_b$. The surface resistance for a wet surface (R_{sfc}) is distinct from that for a dry surface. The surface resistance is formulated as:

$$R_s = \frac{n_m F}{A H_I \times 10^{-3}} \left(\frac{W}{\gamma_I} + \frac{dW}{dt} \right)^{-1} \tag{11}$$

γ_I is the lifetime of chemical I in solution, W is dewfall concentration, dW/dt is its time rate of change, H_I is Henry's coefficient, A is Avogadro's number, n_M is surface number density and F defines a function that describes the relation between the ambient air concentration and that in the drop due to its aqueous phase molecular diffusion and chemical consumption. In general, the surface resistance to SO_2 uptake decreases as the solubility coefficient (H_I) increases. On the other hand, an increase in the chemical time scale or evaporation of the dew increases the concentration of material in the droplets, thereby increasing R_{sfc}.

Concluding remarks

The predominant pathway for gaseous sulfur uptake to dry vegetation is via turbulent transfer through the atmosphere surface boundary layer and molecular diffusion through the laminar boundary layer and the stomata. The soil surface is a significant, but weaker sink for sulfur. The appreciable solubility of SO_2 causes its uptake to be enhanced greatly in the presence of moisture on leaves. However, the aqueous uptake of SO_2 causes the pH of a solution decreases, producing a reduction in the solubility of SO_2.

Three classes of models have been developed to compute deposition fluxes. Big-leaf models are the simplest and are used routinely. Multi-layer Eulerian and Lagrangian models can describe the controlling abiotic factors better, hence they yield more accurate estimates of deposition fluxes.

Several areas of research can be identified from this overview. One, work is needed on scaling models of deposition of wet leaves to the canopy scale. This is a difficult endeavor because one must evaluate how wet the canopy is, where it is wet, how fast it is drying, the size and distribution of droplets, the chemistry of the droplets, the leaching of compounds from leaves and the deposition of oxidizing compounds into the droplets or films. Chameides (1987) provides an initial intellectual framework from which to advance, but more work needs to be done to develop an accurate application model.

The current parameterization of water stress and its consequential stomatal closure is weak (see Fig. 3). Incorporating a water balance into the canopy deposition model would be helpful.

Finally, we need to be able to calculate fluxes to heterogeneous land surfaces. Many forests and canopies are not closed, are in mountainous terrain or consist of small fields in a heterogeneous patchwork. These conditions are difficult to address with conventional micrometerology. Yet, we need to know how variations in topography and landscape components affect large scale average fluxes, which are required as inputs in global and regional atmospheric chemistry and biogeochemistry models.

Acknowledgements

This work was partially supported by the National Oceanic and Atmospheric Administration and the US Department of Energy. I thank Dr. D.R. Matt for providing me with some SO_2 deposition data. I am also grateful to Drs. D.R. Matt and Steve Lindberg for providing me with constructive comments on the manuscript.

References

Anderson, S.M., Delany, A.C., Fehsenfeld, F.C., Goldan, P.D., Kolb, C.E., Ridley, B.A. & Steele, L.P. 1989. Status of chemical sensors for flux measurements. In: Lenschow, D.H. & Hicks, B.B. (eds.), Global Tropospheric Chemistry: Chemical Fluxes in the Global Atmosphere. Natl. Center Atmos. Res., Colorado, p 47.

Andreae, M.O. 1985. The emmission of sulfur to the remote atmosphere: background paper. In: Galloway, J.N. *et al.* (eds.), The Biogeochemical Cycling of Sulfur and Nitrogen in the Remote Atmosphere. D. Reidel Publishing Co. Dordrecht, pp. 5-25.

Baer, M. & Nester K. 1987. Simulation of concentration, dry deposition and fluxes of SO_2 during TULLA. Annales Geophysicae, 5B: 497-502.

Baldocchi, D. 1988. A multi-layer model for estimating sulfur dioxide deposition to a deciduous oak forest canopy. Atmos. Environ. 22: 869-884.

Baldocchi, D.D. 1991. Canopy control of trace gas emissions. In: Sharkey, T.D., Holland, E.A. & Mooney, H.A. (eds.), Trace Gas Emissions by Plants. Academic Press, pp. 293-334.

Baldocchi, D.D. 1992. A Lagrangian random-walk model for simulating water vapor, CO_2 and sensible heat flux densities and scalar profiles over and within a soybean canopy. Boundary-Layer Meteorology. 61: 113-144.

Baldocchi, D.D. 1993. Scaling water vapor and carbon dioxide exchange from leaves to a canopy: rules and tools. In: Ehleringer, J. & Field, S.B., Scaling Processes between Leaf and Landscape Levels, 1. Academic Press. pp. 77-114.

Baldocchi, D.D., Hicks, B.B. & Camara, P. 1987. A canopy stomatal resistance model for gaseous deposition to vegetated surfaces. Atmos. Environ. 21: 91-101.

Baldocchi, D.D., Hicks, B.B. & Meyers, T.P. 1988. Measuring biosphere-atmosphere exchanges of biologically related gases with micrometeorological methods. Ecology 69: 1331-1340.

Brimblecombe, P. 1978. "Dew" as a sink for sulphur dioxide. Tellus 30: 151-157.

Brimblecombe, P. 1986. Air Composition and Chemistry. Cambridge University Press, Great Britain.

Businger, J. 1986. Evaluation of the accuracy with which dry deposition can be measured with current micrometeorological techniques. J. Climate Appl. Meteorology 25: 1100-1124.

Calvert, J.G., Fu Su, Bottenheim, J.W. & Strausz, O.P. 1978. Mechanism of the homogeneous oxidation of sulfur dioxide in the troposphere. Atmos. Environ. 12: 197-226.

Carmichael, G.R. & Peters, L.K. 1984. An eulerian transport/transformation/removal model for SO_2 and sulfate-I. model development. Atmos. Environ. 18: 937-951.

Chameides, W.L. 1987. Acid dew and the role of chemistry in the dry deposition of reactive gases to wetted surfaces. J. Geophys. Res. 92(D10): 11895-11908.

Charlson, R.J., Schwartz, S.E., Hales, J.M., Cess, R.D., Coakley, Jr., J.A., Hansen, J.E. & Hofmann, D.J. 1992. Climate forcing by anthropogenic aerosols. Science 255: 423-255.

Collatz, G.J., Ball, J.T., Grivet, C. & Berry, J.A. 1991. Regulation of stomatal conductance and transpiration: a physiological model of canopy processes. Agric. Forest Meteorology. 54: 107-136.

Davidson, C.I. & Wu, Y.L. 1990. Dry deposition of particles and vapors. In: Lindberg, S.E., Page, A.L. & Norton, S.A. (eds.), Acidic Precipitation: Sources, Deposition & Canopy Interactions. Springer-Verlag, New York, pp. 1-216.

Davis, C.S. & Wright, R.G. 1985. Sulfur dioxide deposition velocity by a concentration gradient measurement system. J. Geophys. Res. 90(D1): 2091-2095.

Deardorff, J.W. 1978. Closure of second- and third-moment rate equations for diffusion in homogeneous turbulence. Physics of Fluids 21: 525-530.

Duyzer, J. H. & Bosveld, F.C. 1988. Measurements of dry deposition fluxes of O_3, NO_x, SO_2 and parti-

cles over grass/heathland vegetation and the influence of surface inhomogeneity. The Netherlands.
Fowler, D. 1978. Dry deposition of SO_2 on agricultural crops. Atmos. Environ. 12: 369-373.
Fowler, D. 1981. Turbulent Transfer of sulphur dioxide to cereals: a case study. In: Plants and Flux. pp. 139-146.
Fowler, D. 1985. Deposition of SO_2 onto plant canopies. In: Winner, W.E., Mooney, H.A. & Goldstein, R.A. (eds.), Sulfur Dioxide and Vegetation. Stanford University Press, California, pp. 389-402.
Fowler, D. & Unsworth, M.H. 1979. Turbulent transfer of sulphur dioxide to a wheat crop. Quart. J. R. Met. Soc. 105: 767-783.
Fowler, D. & Cape, J.N. 1982. Dry deposition of SO_2 onto a scots pine forest. In: Pruppacher *et al.* (eds.), Precipitation Scavenging, Dry Deposition, and Resuspension. Elsevier Publ. Co., Inc., New York, p. 763.
Fowler D. & Duyzer, J.H. 1989. Micrometeorological techniques for the measurement of trace gas exchange. In: M.O. Andreae & Schimel, D.S. (eds.), Exchange of Trace Gases between Terrestrial Ecosystems and the Atmosphere. John Wiley & Sons Ltd, pp. 189-207.
Fowler, D., Morse, A.P. Gallagher, M.W. & Choularton, T.W. 1990. Measurements of cloud water deposition on vegetation using a lysimeter and a flux gradient technique. Tellus 42B: 285-293.
Fowler, D., Duyzer, J.H. & Baldocchi, D.D. 1991. Inputs of trace gases, particles and cloud droplets to terrestrial surfaces. Proc. Royal Soc. Edinburgh 97B: 35-59.
Gallbally, I. 1979. Sulphur uptake from the atmosphere by forest and farmland. Nature 280: 49-50.
Garland, J.A. 1977. The dry deposition of sulphur dioxide to land and water surfaces. Proc. Royal. Soc. Lond. A354: 245-268.
Garland, J.A. 1978. Dry and wet removal of sulphur from the atmosphere. Atmos. Environ. 12: 349-362.
Garland, J.A. & Branson, J.R. 1977. The deposition of sulphur dioxide to pine forest assessed by a radioactive tracer method. Tellus 29: 445-454.
Garland, J.A., Atkins, D.F. Readings, C.J. & Caughey, S.J. 1974. Deposition of gaseous sulphur dioxide to the ground. Atmos. Environ. 8: 75-79.
Garsed, S.G. 1985. SO_2 uptake and transport. In: Winner, W.E., Mooney, H.A. & Goldstein, R.A. (eds.), Sulfur Dioxide and Vegetation: Physiology, Ecology, and Policy Issues. Stanford Univ. Press, pp. 75-95.
Garten, C.T., Jr. 1990. Foliar leaching, translocation, and biogenic emission of ^{35}S in radiolabeled loblolly pines. Ecology 71: 239-251.
Garten, Jr., C.T., Bondietti, E.A. & Lomax, R.D. 1988. Contribution of foliar leaching and dry deposition to sulfate in net throughfall below deciduous trees. Atmos. Environ. 22: 1425-1432.
Goldan, P.D., Fall, R., Kuster, W.C. & Fehsenfield, F.C. 1988. Uptake of COS by growing vegetation: a major tropospheric sink. J. Geophys. Res. 93: 14186-14192.
Hällgren, J.E., Linder, S., Richter, A., Troeng, E. & Granat, L. 1982. Uptake of SO_2 in shoots of Scots pine: filed measurements of net flux of sulphur in relation to stomatal conductance. Plant Cell Environ. 5: 75-83.
Hicks, B.B. & Matt, D.R. 1988. Combining biology, chemistry, and meteorology in modeling and measuring dry deposition. J. Atmos. Chem. 6: 117-131.
Hicks, B.B., Wesely, M.L. Durham, J.L. & Brown, M.A. 1982. Some direct measurements of atmospheric sulfur fluxes over a pine plantation. Atmos. Environ. 16: 2899-2903.
Hicks, B.B., Wesely, M.L. Coulter, R.L. Hart, R.L. Durham, J.L. Speer, R.E. & Stedman, D.H. 1982. An experimental study of sulfur deposition to grassland. In: Pruppacher *et al.* (eds.), Precipitation Scavenging, Dry Deposition, and Resuspension. Elsevier Science Publishing Co., Inc., pp. 933-942.
Hicks, B.B., Baldocchi, D.D., Meyers, T.P. Hosker, R.P., Jr. & Matt, D.R. 1987. A preliminary multiple resistance routine for deriving dry deposition velocities from measured quantities. Water Air Soil Pollut. 36: 311-330.
Hicks, B.B., Matt, D.R. & McMillen, R.T. 1989a. A micrometeorological investigation of surface exchange of O_3, SO_2 and NO_2: A case study. Boundary-Layer Meteorol. 47: 321-336.
Hicks, B.B., Matt, D.R., McMillen, R.T., Womack, J.D., Wesely, M.L., Hart, R.L., Cook, D.R., Lindberg, S.E., de Pena, R.G. & Thomson, D.W. 1989b. A field investigation of sulfate fluxes to a deciduous forest. J. Geophys. Res. 94(D10): 13003-13011.
Hosker, R.P., Jr. & Lindberg, S.E. 1982. Review: Atmospheric deposition and plant assimilation of gases and particles. Atmos. Environ. 16. 889-910.
Ivens, W., Kauppi, P., Alcamo, J., Posch, M. 1990. Sulfur deposition onto European forests: throughfall data and model estimates. Tellus 42B: 294-303.
Jarvis, P.G. 1976. The interpretation of the variations in leaf water potential and stomatal conductance found in canopies in the field. Philosoph. Trans. Royal Soc. London 273: 593-610.

Johansson, C., A. Richter & L. Granat. 1982. Dry deposition on coniferous forest of SO_2 at ppb levels. In: Pruppacher *et al.* (eds.), Precipitation Scavenging, Dry Deposition, and Resuspension. Elsevier Science Pub. Co., Inc., New York, p. 775.

Kok, G.L, Schanot, A.J., Lindgren, P.F., Dasgupta, P., Hegg, D.A., Hobbs, P.V. & Boatman, J.F. 1990. An airborne test of three sulfur dioxide measurement techniques. Atmos. Environ. 24: 1903-1908.

Lindberg, S.E. & Lovett, G.M. 1992. Deposition and forest canopy interactions of airborne sulfur: results from the integrated forest study. Atmos. Environ. 26A: 1477-1492.

Lorenz R. & Murphy, C.H., Jr. 1985. Sulfur dioxide, carbon dioxide, and water vapor flux measurements utilizing a microprocessor-controlled data acquisition system in a pine plantation. In: Hutchison, B.A. & Hicks, B.B. (eds.), The Forest-Atmosphere Interaction. pp. 133-147.

Matt, D.R., McMillen, R.T., Womack, J.D. & Hicks, B.B. 1987. A comparison of estimated and measured SO_2 deposition velocities. Water Air Soil Pollut. 36: 331-347.

McMillen, R.T., Matt, D.R., Hicks, B.B. & Womack, J.D. 1987. Dry Deposition Measurements of Sulfur Dioxide to a Spruce-Fir Forest in the Black Forest: A Data Report. NOAA Technical Memorandum ERL ARL-152.

Meyers, T.P. 1987. The sensitivity of modeled SO_2 fluxes and profiles to stomatal and boundary layer resistances. Water Air Soil Poll. 35: 261-278.

Meyers, T.P. & Baldocchi, D.D. 1988. A comparison of models for deriving dry deposition fluxes of O_3 and SO_2 to a forest canopy. Tellus 40B: 270-284.

Meyers T.P. & Hicks, B.B. 1988. Dry deposition of O_3, SO_2, and HNO_3 to different vegetation in the same exposure environment. Environ. Pollut. 53: 13-25.

Meyers, T. P., Hicks, B.B., Hosker, R.P., Jr., Womack, J.D. & Satterfield, L.C. 1991. Dry deposition inferential measurement techniques. II. Seasonal and annual deposition rates of sulfur and nitrate. Atmos. Environ. 25A: 2361-2370.

Monteith, J. L. & Unsworth, M.H. 1990. Principals of Environmental Physics, Edward Arnold Publishing Co., New York.

Murphy, C.E., Jr. & Sigmon, J.T. 1990. Dry deposition of sulfur and nitrogen oxide gases to forest vegetation. In: Lindberg, S.E., Page, A.L. & Norton, S.A. (eds.), Acidic Precipitation: Sources, Deposition, and Canopy Interactions. Springer-Verlag, New York, pp. 217-240.

Murphy, C.E., Jr., Sinclair, T.R. & Knoerr, K.R. 1977. An assessment of the use of forests as sinks for the removal of atmospheric sulfur dioxide. J. Environ. Qual. 6: 388-396.

Neumann, H. H. & den Hartog, G.D. 1985. Eddy correlation measurements of atmospheric fluxes of ozone, sulphur, and particulates during the Champaign intercomparison study. J. Geophys. Res. 90(D1): 2097-2110.

Nicholson, K.W. & Davies, T.D. 1988. Letter to the editors: The dry deposition of sulphur dioxide at a rural site. Atmos. Environ. 22: 2885-2889.

O'Dell, R.A., Taheri M. & Kabel, R.L. 1977. A model for uptake of pollutants by vegetation. J. Air Pollut. Control Ass. 27: 1104-1109.

Ogram, G.L., Northrup, F.J. & Edwards, G.C. 1988. Fast time response tunable diode laser measurements of atmospheric trace gases for eddy correlation. J. Atmos. Oceanic Technol. 5: 521-527.

Olszyk, D.M. & Tingey, D.T. 1985. Interspecific variation in SO_2 flux: Leaf surface versus internal flux, and components of leaf conductance. Plant Physiol. 79: 949-956.

Pandis, S.N. & Seinfeld, J.H. 1990. On the interaction between equilibration processes and wet or dry deposition. Atmos. Environ. 24A: 2313-2327.

Payrissat, M. & Beilke, S. 1975. Laboratory measurements of the uptake of sulphur dioxide by different european soils. Atmos. Environ. 9: 211-217.

Pearcy, R.W. & Ehleringer, J. 1984. Comparative ecophysiology of C_3 and C_4 plants. Plant Cell Environ. 7: 1-13.

Petit, C., Ledoux, M. & Trinite, M 1977. Transfer resistances to SO_2 capture by some plane surfaces, water and leaves. Atmos. Environ. 11: 1123-1126.

Raupach, M.R. 1988. Canopy transport processes. In: Steffen, W.L. & Denmead, O.T., Flow and Transport in the Natural Environment. Springer-Verlag, Berlin.

Rennenberg, H. 1984. The fate of excess sulfur in higher plants. Ann. Rev. Plant Physiol. 35: 121-53.

Rennenberg, H. 1991. The significance of higher plants in the emission of sulfur compounds from terrestrial ecosystems. In: Sharkey, T.D., Holland, E.A. & Mooney, H.A. (eds.), Trace Gas Emissions by Plants. Academic Press, Inc., New York, p. 217.

Rennenberg, H., Huber, B., Schröder, P., Stahl, K., Haunold, W., Georgii, H.W., Slovik, S. & Pfanz, H. 1990. Emission of volatile sulfur compounds from spruce trees. Plant Physiol. 92: 560-564.

Schaefer, D.A. & Reiners, W.A. 1990. Throughfall chemistry and canopy processing mechanisms. In:

Lindberg, S.E., Page, A.L. & Norton, S.A. (eds.), Acidic Precipitation: Sources, Deposition, and Canopy Interactions. Springer-Verlag, New York, pp. 241-284.

Schulze, E.D. 1989. Air pollution and forest decline in a spruce (*Picea abies*) forest. Science 244: 776-783.

Schwartz, S.E. 1989. Acid Deposition: unraveling a regional phenomenon. Science 243: 753-763.

Sehmel, G.A. 1980. Particle and gas dry deposition: a review. Atmos. Environ. 14: 983-1011.

Seinfeld, J. 1985. Atmospheric Chemistry and Physics of Air Pollution. John Wiley, New York.

Semb, A. 1978. Sulphur emissions in Europe. Atmos. Environ. 12: 455-460.

Shreffler, J.H. 1976. A model for the transfer of gaseous pollutants to a vegetational surface. J. Appl. Meteorology 15: 744-746.

Taylor, G.E. & Tingey, D.T. 1983. Sulfur dioxide flux into leaves of *Geranium carolinianum* L. Plant Physiol. 72: 237-244.

Taylor, G.E., McLaughlin, S.B., Jr., Shriner, D.S. & Selvidge, W.J. 1983. The flux of sulfur-containing gases to vegetation. Atmos. Environ. 17: 789-796.

Taylor, G.E., Hanson, P.J. & Baldocchi, D.D. 1988. Pollutant deposition to individual leaves and plant canopies: sites of regulation and relationship to injury. In: Heck, W.W., Taylor, O.C. & Tingey, D.T. (eds.), Assessment of Crop Loss From Air Pollutants. Elsevier Applied Science, New York, pp. 227-257.

Unsworth, M.H., Crawford, D.V., Gregson, S.K. & Rowlatt, S.M. 1985. Pathways for sulfur from the atmosphere to plants and soil. In: Winner, W.E., Mooney, H.A. & Goldstein, R.A. (eds.), Sulfur Dioxide and Vegetation. Stanford Univ. Press, California, pp. 375-388.

Valdez, M.P., Bales, R.C., Stanley, D.A. & Dawson, G.A. 1987. Gaseous deposition to snow. 1. Experimental study of SO_2 and NO_2 deposition. J. Geophys. Res. 92(D8): 9779-9787.

Van Hove, L.W.A., Adema, E.H., Vredenberg, W.J. & Pieters, G.A. 1989. A study of the absorption of NH_3 and SO_2 on leaf surfaces. Atmos. Environ. 23: 1479-1486.

Van Hove, L.W.A., Vredenberg, W.J. & Adema, E.H. 1990. The effect of wind velocity, air temperature and humidity on NH_3 and SO_2 transfer into bean leaves (*Phaseolus vulgaris* L.). Atmos. Environ. 24A: 1263-1270.

Vermetten, A.W.M., Hofschreuder, P., Duyzer, J.H., Diederen, J.S.M.A., Bosveld, F.C. & Bouten, W. 1991. Dry deposition of SO_2 onto a stand of douglas fir: the influence of canopy wetness. IPSASEP: 1-12.

Vong, R.J., Sigmon, J.T. & Mueller, S.F. 1991. Cloud water deposition to appalachian forests. Environ. Sci. Technol. 25: 1014-1021.

Wellburn, A.R. 1982. Effects of SO_2 and NO_2 on metabolic function. In: Unsworth, M.H. & Ormrod, D.P. (eds.), Effects of Gaseous Air Pollution in Agriculture and Horticulture. Butterworth Scientific, Boston, pp. 169-187.

Wesely, M.L. & Hicks, B.B. 1977. Some factors that affect the deposition rates of sulfur dioxide and similar gases on vegetation. APCA Journal 27: 1110-1116.

Wesely, M.L. & Hart, R.L. 1985. Variability of short term eddy-correlation estimates of mass exchange. In: Hutchison, B.A. & Hicks, B.B. (eds.), Forest-Atmosphere Interactions. Dordrecht, The Netherlands, pp. 591-612.

Wesely, M.L., Cook, D.R., Hart, R.L., Hicks, B.B., Durham, J.L., Speer, R.E., Stedman, D.H. & Tropp, R.J. 1982. Eddy-correlation measurements of the dry deposition of particulate sulfur and submicron particles. In: Pruppacher (ed.), Precipitation Scavenging, Dry Deposition & Resuspension. Elsevier Science Publishing Co., pp. 943-952.

Wesely, M. L., Cook, D.R. & Hart, R.L. 1983. Fluxes of gases and particles above a deciduous forest in wintertime. Boundary-Layer Meteorology 27: 237-255.

Wesely, M.L., Cook, D.R., Hart, R.L. & Speer, R.E. 1985. Measurements and parameterization of particulate sulfur dry deposition over grass. J. Geophys. Res. 90(D1): 2131-2143.

Wesely, M.L., Lenschow, D.H. & Denmead, O.T. 1989. Flux meaurement techniques. In: Lenschow, D.H., Hicks, B.B. (eds.), Global Tropospheric Chemistry. Chemical Fluxes in the Global Atmosphere. Natl. Center for Atmos. Res., Colorado, p. 31.

Wesely, M.L., Sisterson, D.L. & Jastrow, J.D. 1990. Observations of the chemical properties of dew on vegetation that affect the dry deposition of SO_2. J. of Geophys. Res. 95(D6): 7501-7514.

Whelpdale, D.M. & Shaw, R.W. 1974. Sulphur dioxide removal by turbulent transfer over grass, snow, and water surfaces. Tellus XXVI: 1-2.

ECOLOGICAL ASPECTS OF SULFUR IN HIGHER PLANTS: THE IMPACT OF SO_2 AND THE EVOLUTION OF THE BIOSYNTHESIS OF ORGANIC SULFUR COMPOUNDS ON POPOLATIONS AND ECOSYSTEMS

Wilfried H.O. Ernst
Department of Ecology and Ecotoxicology, Faculty of Biology, Vrije Universiteit, De Boelelaan 1087, 1081 HV Amsterdam, The Netherlands

Introduction

Sulfur supply to plants depends on its distribution within the three environmental compartments soil, water and atmosphere. In these environmental compartments sulfur occurs in a lot of chemical forms, ranging from very reduced to very oxidized sulfur, and from inorganic to organic sulfur compounds. Adaptation of higher plants to the various habitats will depend on the chemical sulfur speciation and on their environmental concentrations in space and time. In the case of a sulfur surplus, adaptation will ultimately be determined by the physiological potential of a diminished uptake, quantitative and qualitative changes of the biosynthesis of sulfur-containing compounds and their storage capacity, either at the cellular and/or tissue level. With regard to populations, communities and ecosystems, the success of the modified sulfur compounds may affect intra- and interspecific competition at the same trophic level and the plant-herbivore and plant-pathogen relationships.

Due to geological differentiation of the lithosphere and hydrosphere the horizontal pattern of sulfur concentration ranges from areas of low supply *e.g.* grey wooded soils in Canada (Bentley *et al.* 1956) to areas with sulfur toxicity to non-adapted plants, *e.g.* gypsum soils (Heinze *et al.* 1982; Meyer 1986). The ecophysiological aspects of the adaptation to high sulfur supply have recently been reviewed (Ernst 1990a); the small progress in this research area does not justify an update.

Atmospheric sulfur compounds have received increasing interest, because of anthropogenic emissions of SO_2 and H_2S (De Kok 1990, for a review) and consequently their deposition and impact on terrestrial ecosystems (Autri & Fitzgerald 1991). Biogenic sulfur emission is of special interest due to the impact of CS_2 (Schröder, this volume) and of dimethyl sulfide (DMS) and its atmospheric oxidation product, methanesulfonate, on the global sulfur cycle and on global climate by controlling global albedo (Kiene & Bates 1990; Ayers *et al.* 1991). It may represent 30 to 50% of the global sulfur flux to the atmosphere (the hypothesis of Charlson *et al.* 1987). With regard to higher plants, the production of dimethyl sulfide and its release to the atmosphere is restricted to *Zostera* (Gorham *et al.* 1980) and *Spartina* (Van Diggelen *et al.* 1986), whereas the greatest mass of DMS is produced by marine phytoplankton and macroalgae (Gibson *et al.* 1990; Karsten *et al.* 1991). The deposition of methanesulfonate in the antarctic ice core (Legrand *et al.* 1991) demonstrates that at least plants in coastal ecosystems, *e.g.* mangroves and salt marshes, may be affected by the volatile thioether DMA. The interaction between DMS and higher

Sulfur Nutrition and Assimilation in Higher Plants
edited by L.J. De Kok et al., pp. 295–313

plants has not received any further attention during the last years (Ernst 1990a) so this aspect is not considered any further.

My intention is to show the ecological aspects of the anthropogenic enrichment of the environment with sulfur compounds, emphasizing changes at the population and community level and to look to the ecological role of organic sulfur compounds in plant-heterotroph relationships.

The impact of anthropogenic sulfur emission on ecosystems

As soon as man affects nature, there are two possibilities of biological responses: a negative one varying from plant injury to extinction of species (Korneck 1984) or even disappearance of ecosystems; a positive one, selecting resistant organisms which can build up new communities and ecosystems (Ernst 1990b). In the case of anthropogenic sulfur emission, the negative aspects are fully recognized (Guderian 1977), but I will also elaborate some positive effects.

The damage

Anthropogenic sulfur emission will return from the atmosphere and deposit on components of the ecosystem and change their functioning. In sites with a high level of SO_2 pollution, damage of the vegetation, especially of coniferous trees, agricultural and horticultural crops occurs (Unsworth & Omrod 1982). These processes are well documented as part of the cascade of causes and effects of forest decline in the Northern hemisphere (Guderian 1977; Schulze *et al.* 1989; Kozlowski *et al.* 1991). The effects of anthropogenic sulfur emission on ecosystems are very dependent on the vegetation type. As demonstrated in the Solling project, a beech (*Fagus sylvatica*) forest intercepts 50 kg S ha^{-1} yr^{-1} whereas a spruce (*Picea abies*) forest has an interception of 85 kg S ha^{-1} yr^{-1} compared to the annual sulfur deposition of 23 kg ha^{-1} via precipitation (Ellenberg *et al.* 1986). Once the deposition has reached the soil surface, either directly or indirectly, it will affect the herb-, grass- and moss layer and the biological processes in the humus layer. Only recently more emphasis has been given to the chemical speciation of sulfur in soils, which is predominantly present as organic sulfur (Johnson *et al.* 1982; Strickland *et al.* 1987; Autri & Fitzgerald 1991) and thus not directly available to plants. Organic sulfur, however, provides a pool of readily mineralized available sulfur in ecosystems.

As a consequence of vegetation-specific interception of sulfur aerosols, sulfur fixation and the synthesis of humus compounds, the passage of sulfur through the soil profile will also be related to the vegetation type as demonstrated by a long-lasting lysimeter experiment. Discharge of sulfate from 1947 to 1983 was studied in four lysimeters (625 m^2 each), one with bare sand, the others vegetated with a shrub vegetation of buckthorn (*Hippophae rhamnoides*), with an oak (*Quercus robur*) forest and with an Austrian pine (*Pinus nigra* var. *nigra*) forest (Table 1). Whereas the lysimeter without any vegetation had a nearly constant discharge quality, the sulfate concentration in the discharge of the vegetated lysimeters increased steadily, the strongest under the pine forest (Stuyfzand 1984). This latter result is in good agreement with the sulfate discharged under a beech and spruce forest in the Solling

Table 1. Sulfate deposition rate (mg m^{-2} day^{-1}) and sulfate in drainage water (mg l^{-1}) from four lysimeters (625 m^2 surface each), filled with calcareous dune sand and vegetated from 1941 onwards, at the Dutch coast. Data are combined from Minderman & Leeflang (1968) and Stuyfzand (1984). n.d. = not determined.

			Lysimeter with		
Discharge period	Sulfate deposition rate	Bare sand	*Hippophae rhamnoides*	*Quercus robur*	*Pinus nigra*
1912-1956	14.0	16.8	39.5	32.1	64.6
1957-1961	19.3	16.1	35.3	40.5	98.8
1966-1968	42.0	n.d.	n.d.	n.d.	n.d.
1980-1983	12.0	16.5	50.8	47.4	304.0

(Ellenberg *et al.* 1986) and with other European and North American forests (Likens *et al.* 1977). The sulfate concentration of the discharge is determined by four components, the deposited sulfur, its allocation in the living biomass, its retention in the soil, and the water consumption by the vegetation. The high sulfate concentration of the lysimeter discharge under Austrian pine is partly caused by the high water demands of this coniferous trees, being a factor of two higher than on the two other vegetated lysimeters (Stuyfzand 1989). The increased availability of sulfur together with changes of the mobility of other chemical soil constituents (Smith 1981; Wookey & Ineson 1991) will affect plant growth. The crucial metabolic aspect for plants at SO_2 exposure is the storage of the additional sulfate if other sulfur compounds are not synthesized. SO_2 injury of Scotch pine and spruce was associated with higher sulfate concentration in needles of SO_2-sensitive lines than of SO_2-resistant trees (Rohmeder *et al.* 1962). Experiments with increasing supply of sulfate fertilizers have confirmed that in four-year old spruce needles 50% of the total needle sulfur was present as sulfate at high supply compared to less than 10% at low external sulfate supply (Ende & Hüttl 1992). Such an impact of the storage capacity of plants for sulfate may also be responsible for the change in species composition in Polish woodlands. With an increase of the destruction of forest trees, acidophilous herbs, grasses and carpet mosses disappeared from the forest soil, and plant species characteristic of woodland clearings, *e.g. Senecio nemorensis* and *Rubus idaea* colonized the open forest patches in Polish woodlands (Medwecka-Kornaś & Gawroński 1990). One of the not yet investigated aspects is the reason for the disappearance of acidophilous plants. It seems unlikely that the well known aluminum resistance of *Luzula luzuloides* (= *L. albida*, Henrichfreise 1981) is too low to cope with an increasing aluminum availability with decreasing pH. It is more likely that the storage capacity of the vacuoles for sulfates (Ernst 1990a) may be insufficient, because the mean daily sulfur deposition in polluted areas can still amount up to 16 mg S m^{-2}, equivalent to 57 kg ha^{-1} $year^{-1}$. Due to the high amounts of fulvic acid in pine-oak forests, most of the sulfate may be adsorbed in sequioxide-rich subsurface soils (Johnson *et al.* 1982), thus disturbing the adsorption equilibrium with other nutrients in the soil solution (Ellenberg *et al.* 1986), and thus provoking disturbance of the nutrient balance in plants.

Table 2. Proven SO_2-resistance in populations of higher plants.

Plant species	Demonstrated or suspected (?) resistance mechanism	Reference
Pinus sylvestris	Exclusion	Rohmeder *et al.* (1962)
Geranium carolinianum	No exclusion, ?	Taylor & Murdy (1975)
	Improved repair system (?)	Taylor *et al.* (1986)
Lolium perenne	No exclusion	Bell & Mudd (1976)
	Increased sulfate accumulation (?)	Koziol *et al.* (1986)
	Greater stomatal resistance	
Rumex obtusifolium	?	Horsman & Wellburn (1977)
Lepidium virginicum	Glucosinolates (?)	Murdy (1979)
Glycine max	Increased sulfate accumulation (?)	Miller & Xerikos (1979)
Silene vulgaris	No exclusion, no increased SO_4 accumulation	Ernst *et al.* (1985)
	Balance between root and leaf uptake	

The advantage

Plant species which have an increased demand for sulfur, and a high capacity to store sulfur in inorganic and organic compounds, may benefit from increased anthropogenic sulfur emission. Although variation among and between plant species in resistance to air pollutants appears to be quite common in most species (Roose *et al.* 1982), only a few species have actually been tested for their tolerance to SO_2. From these few species, *i.e. Geranium carolinianum* (Taylor & Murdy 1975), *Lolium perenne* (Bell & Mudd 1976) and a few other grasses (Ayazloo & Bell 1981), *Silene vulgaris* (Ernst *et al.* 1985), *Rumex obtusifolium* (Horsman & Wellburn 1977), only one species, *i.e. Lepidium virginicum* (Murdy 1979) is a cruciferous one (Table 2). One condition responsible for the selection of SO_2-resistant genotypes is a relatively constant exposure of plant populations to increased SO_2-levels. Resistance may be realized by a diminished uptake of SO_2, as obviously found in coniferous trees with a slightly decreased SO_2 concentration of their needles (Rohmeder *et al.* 1962; Caput *et al.* 1978) or by an increased metabolic incorporation of SO_2 into cellular compounds. SO_2 is rapidly converted into HSO_3^- and SO_3^{2-} once it has entered the cytoplasm and consequently these anions will be either reduced to the sulfide level or oxidized to sulfate (Rennenberg 1984). Excess sulfate will be stored in the vacuole and aggravates the normal age-dependent sulfate storage in leaves either as sulfate (Ernst 1990a) or as flavonoid sulfates (Barron *et al.* 1988). This latter option will be achieved with difficulty because it demands an increased storage capacity of vacuoles, although such increased sulfate accumulation has been postulated (Miller & Xerikos 1979; Koziol *et al.* 1986). The former possibility of an enhanced incorporation of SO_2 into organic sulfur compounds may give more evolutionary potential for plant species with specific sulfur compounds.

Plants with an increasing sulfur demand may be those which use additional sulfur not only for growth (Marquard *et al.* 1968), but also for protection of their biomass by chemical defense compounds against pathogens and herbivores or by metal de-

Table 3. Endangered species of *Pteridophyta* and *Spermatophyta*, with emphasis on the *Brassicaceae*, in the flora of the Federal Republic of Germany (FRG) and the Netherlands (NL) in the period before and after 1950. The data sets are taken from Korneck (1984) and Haeupler & Schönfelder (1988) for the FRG and from van der Meijden (1990) for NL.

	All plants				*Brassicaceae*			
	FRG	%	NL	%	FRG	%	NL	%
Class frequency								
0 extinct	61	2.5	44	3.1	4	3.2	3	4.5
1 nearly extinct	101	4.1	161	11.4	2	1.6	6	9.1
2 strongly endangered	255	10.3	61	4.3	4	3.2	2	3.0
3 endangered	281	11.3	142	10.1	3	2.4	5	7.6
4 potentially endangered	165	6.7	102	7.2	5	4.0	3	4.5
subtotal no. of species	863	34.9	510	36.1	18	14.4	19	28.7
total no. of species	2477	100.0	1412	100.0	125	100.0	66	100.0

toxifying substances (Ernst *et al.* 1992; Rauser, this volume). There is growing evidence that plants with a high sulfur demand are synthesizing higher amounts of organic sulfur compounds with an increase in sulfur supply (glucosinolates: Marquard *et al.* 1968; dimethylsulfoniopropionate: van Diggelen *et al.* 1986; thiophenes: A.F. Croes, personal communication). A positive impact of SO_2 may be developed the best in ecosystems where the additional sulfur supply may be proportionally high compared to the natural background. Therefore, plant species synthesizing glucosinolates, alliin and allicin are of special interest in this context. Other taxa with specialized sulfur compounds are mostly restricted to the coastal zones, where the impact of SO_2 is low compared to the influence of sea salt sulfur, *e.g.*, the dimethylsulfoniopropionate accumulating *Spartina anglica* (van Diggelen *et al.* 1986), the choline-*O*-sulfate and/or flavonoid sulfate accumulating species of *Armeria maritima* and *Limonium vulgare* (Barron *et al.* 1988; Hanson & Gage 1991).

Members of the *Brassicaceae* are able to synthesize high amounts of glucosinolates, even at sulfate concentrations where plant growth is not further stimulated (Marquard *et al.* 1968). This high detoxifying mechanism and the potential role of glucosinolates and other organic sulfur compounds as repellants of herbivores and pathogens may give such plant species an advantage in SO_2 polluted areas, by either increased SO_2 resistance and/or by optimalization of their sulfur supply. To test this hypothesis, an analysis of the change of the distribution of *Brassicaceae* before and after 1950, *i.e.* before and after strong anthropogenic SO_2 emission was made for the Netherlands and Germany. Before 1950 the mean deposition rate of sulfur increased from 4 mg S m^{-2} d^{-1} in 1930 to 10 mg S m^{-2} d^{-1} in 1950 (Stuyfzand 1984). From 1950 to 1968 the deposition rate of sulfur jumped to 42 mg S m^{-2} d^{-1} in the Netherlands and other highly industrialized regions of Europe (Table 1). From 1969 onwards the substitution of coal by natural gas was the main factor diminishing the deposition rate of sulfur in the Netherlands to 10-15 mg S m^{-2} d^{-1} from 1980 onwards (Stuyfzand 1984; CBS 1983; CBS 1989). With regard to grasslands and arable fields having no special SO_2 interception, the above mentioned deposition rates have resulted in a total anthropogenic sulfur load of 7.5 kg S m^{-2} for the past 30 years

Table 4. Increase or decrease in the occurrence of *Brassicaceae* in the inventory quadrats (each 25 km^2) in the flora of the Netherlands in the period before and after 1950. The data set is based on the maps published by Mennema *et al.* (1980, 1985) and van der Meijden *et al.* (1989). The species are divided into (semi)natural ecosystems and man-made ecosystems.

Ecosystem	Species no.	decrease (-) or increase (+) of inventory quadrats	change of inventory quadrats per species
(Semi) natural ecosystems			
Coastal floodline	2	+ 45	+ 23
heavy metal vegetation	1	− 3	− 3
semi-natural grassland	6	− 84	− 14
wells	2	+ 467	+ 234
wetlands	9	+ 1140	+ 127
Man-made ecosystems			
walls	1	− 18	− 18
arable fields	16	+ 2309	+ 144
ruderal sites	26	+ 3158	+ 121

(equivalent to 75 t ha^{-1}). *Cruciferous* species were very successful in maintaining and extending their populations in the Netherlands and Germany during the last 40 years (Table 3). In both countries around 35% of all higher plant species were extinct or are endangered. With regard to the *Brassicaceae*, making up 5.0% (FRG) and 4,7% (NL) of the registered plant species in both countries, only 14.2 and 24.2% are endangered in FRG and in NL, respectively. This overall data set has to be corrected for other entirely human factors, which have reduced their distribution, such as the disappearance of flax growing with its accompanying specific weeds (*Camelina* spec.), the removal of spoil heaps and mine tailings (*Thlaspi*, *Cardaminopsis*) or disappearance of municipal walls (*Cheiranthus cheiri*). Independent of the nature of the ecosystems the increase of *Brassicaceae* in the inventory quadrats of the Netherlands varies between 121 and 234 on a mean per species, except at the coastal floodline with its fixed nature (Table 4). One of the reasons for this rapid and positive reaction to a changing environment may be the life history of the crucifers, with a high percentage of annuals (44.4% in F.R.G.) so that the selection can operate on the short term. The obviously positive correlation between sulfate supply and glucosinolate synthesis may have enhanced the vitality of the populations. If these correlations could be experimentally substantiated, they would explain an essential part of the change of the species composition in Central European ecosystems, because the greatest expansion of the area in this case study occurred in the heavily polluted parts of Germany and the Netherlands (Fig. 1). Enhanced sulfur deposition will not only improve growth but also seed quantity and seed quality, and thus increase the survival of seeds in the seed bank.

As a consequence of the increase of cruciferous plant species, the frequency of herbivores with a strong affinity for these plants should follow this trend. Such a plant-animal relationship exists for several butterflies. The cabbage butterfly, *Pieris rapae*, do oviposit on plant species such as *Brassica* spp., *Cardamine pratensis*, *Diplotaxis* spp. and some other members of the *Brassicaceae*. From 1910 onwards, the area of

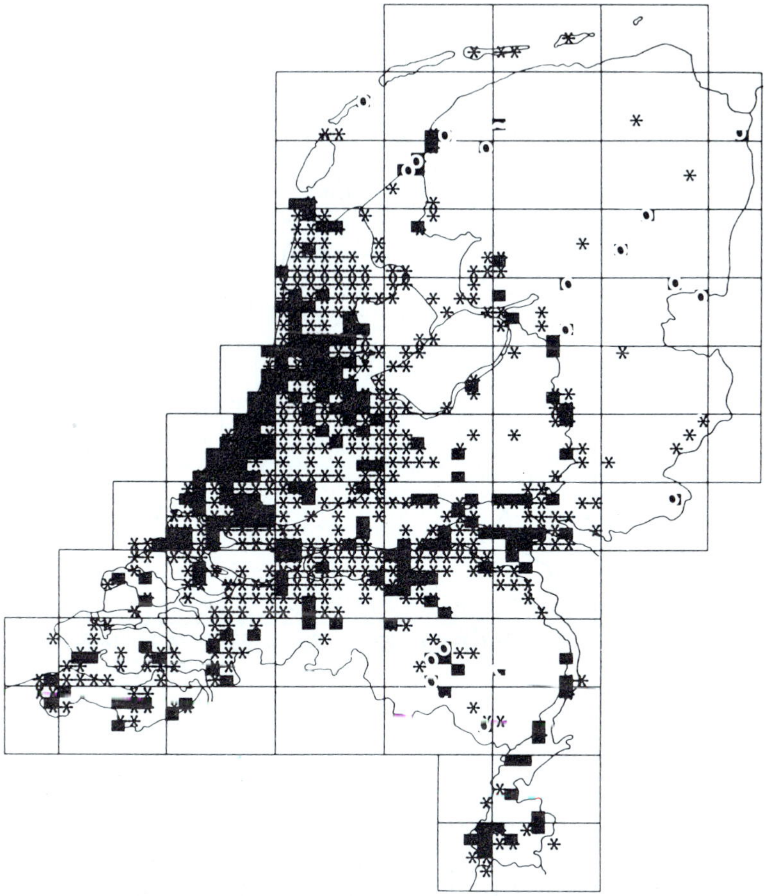

Fig. 1. The expansion of the occurrence of *Diplotaxis tenuifolia* from 1950 onwards in the Netherlands. The black quadrats are occurrences before 1950, the asterisks give the increase from 1950 onwards, the closed white circle in filled quadrats the disappearance of the population in the inventory quadrat after 1950 (after Van der Meijden *et al.* 1989).

occurrence of the cabbage butterfly in the Netherlands has increased (Tax 1989), from 1950 onwards by more than a factor of two (Fig. 2). The data set for other animals, obligately dependent on *Brassicaceae* in their larval stage, are too incomplete for a further evaluation. This advantage of the population increase of *Brassicaceae* and their hosts, however, should not be overestimated, because it can not outweigh the loss of non-cruciferous plant taxa.

Allium species may be a further test case for the selective advantage of SO_2 for sulfur-demanding plants. In Germany and the Netherlands there is a discrepancy of the reaction patterns. The population area of 13 from the 15 species in F.R.G. decreased from 3.3 up to 39.9%, whereas in NL five of the seven species increased varying from 25.9 to 400%, only two decreased by 28 to 32.8%. The number of plant species in both countries (n = 15, FRG; n = 7, NL), however, is too small to make any decisive statements for this taxon. The perennial life history and the first seed

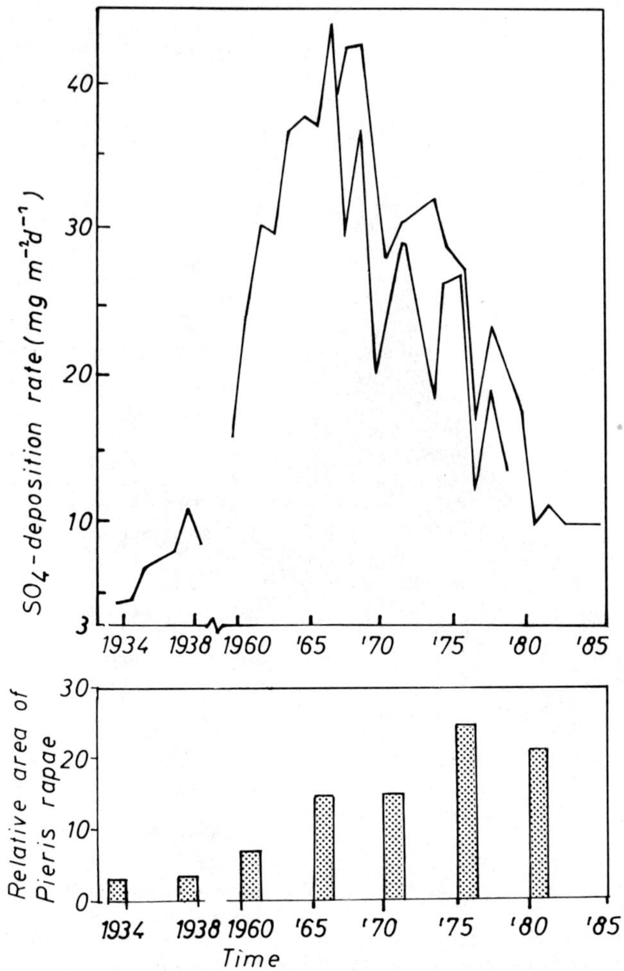

Fig. 2. The change of SO_4 deposition in the western part of the Netherlands from 1930 to 1980 (after Stuyfzand 1984) and the increase of the white cabbage butterfly (after Tax 1989).

production after 3-5 years of growth will also contribute to the delayed reaction of *Allium* species to increased sulfur supply (Ernst 1979).

In conclusion, for the understanding of disadvantages and advantages of SO_2 for plants, an interpopulation comparison of SO_2-sensitive and SO_2-tolerant genotypes (populations) will be the best approach to elucidate the triggers of sulfate uptake rate by roots and the SO_2-intake by leaves and to get insight into metabolic sulfur balances.

The evolution of secondary organic sulfur compounds

The evolution of secondary organic sulfur compounds has occurred quite independently during evolution of plants, sometimes the same compound class in many taxa,

Table 5. The occurrence of secondary organic sulfur compounds in higher plants.

Compound	Taxa	Reference
Dimethylsulfonio-propionate	*Chlorophyta*	Dickson *et al.* (1980)
	Rhodophyta	Challenger & Simpson (1948)
	Spermatophyta	
	Zostera	Gorham *et al.* (1980)
	Spartina	Van Diggelen *et al.* (1986)
Choline-*O*-sulfate	*Eubacteria*	Fitzgerald & Luschinski (1977)
	Chlorophyta	Blunden & Gordon (1986)
	Ascomycetes	Gravel (1976)
	Spermatophyta	
	Avicennia, Aegialitis	Benson & Atkinson (1967)
	Limonium, Armeria	Hanson & Gage (1991)
Phytochelatins	*Chlorophyta*	
	Ascomycetes	Kondo *et al.* (1984)
	Spermatophyta	Grill *et al.* (1985)
Flavonoid sulfates	*Pteridophyta*	Imperato (1982)
	Spermatophyta (32 families)	Barron *et al.* (1988)
Glucosinolates	*Spermatophyta* (15 families)	Rodman (1991a,b)
Alliin, allicin	*Spermatophyta*	
	Allium	Sendl & Wagner (1991)
Thiophenes	*Spermatophyta*	
	Asteraceae	
	Tagetes	Sütfeld (1982)
	Echinops	Lam *et al.* (1991)
Thionins	*Spermatophyta*	Bohlmann & Apel (1991)
Purothionins	*Poaceae*	
Viscotoxins	*Viscaceae*	
Crambin	*Brassicaceae* (Crambe)	
Raphanusanin	*Spermatophyta*	
	Raphanus sativus	Hasegawa *et al.* (1982)

but frequently related to environments enriched by sulfur such as oceans and coastal waters. Examples for this type of evolution are dimethylsulfoniopropionate in *Chlorophyta*, *Rhodophyta* and *Spermatophyta* (Table 5), cholin-*O*-sulfate from bacteria to higher plants, flavonoid sulfates restricted to *Pteridophyta* and *Spermatophyta*. Phytochelatins, being evolved in *Chlorophyta*, *Ascomycetes* and *Spermatophyta*, are examples for a short-term reaction of the metabolism to a surplus of heavy metals (Grill *et al.* 1985) without any impact on metal resistance (Ernst *et al.* 1992). The biosynthesis of some other compounds has only been evolved by some taxa within the higher plants, *e.g.* the independent evolution of glucosinolates with an emphasis on *Capparidales* (Rodman 1991a,b). Thionins have evolved in four completely unrelated families within the *Spermatophyta* (Bohlmann & Apel 1991). Raphanusanin and raphanusamide (Hasegawa *et al.* 1982) occur only in one *Raphanus* species, the alliins and allicins are restricted to the genus *Allium*, the thiophenes to several taxa of the *Asteraceae* (Table 5).

What is the function of these sulfur compounds in the ecology of species and plant communities? The suggestions are manifold (Table 6), ranging for the same compound, *i.e.* choline-*O*-sulfate, from sulfur storage to cytoplasmic osmoregulation,

Table 6. Suggested functional roles of secondary organic sulfur-compounds.

Compound	Function	Reference
Glucosinolate	Sulfur storage	Schnug, this volume
	fungicide	Ref. 40, 77, 78 in Ernst (1990a)
		Ref. 10, 14, 23, 60 in Schnug (1990)
	insecticide	Ref. 42, 59 in Schnug (1990)
	molluscide	Ref. 75 in Ernst (1990a)
	feeding repellent	
	to insects	Ref. 2, 18 in Schnug (1990)
	and mammals	Ref. 84 in Schnug (1990)
	attractant to insects	Ref. 33, 35, 37, 43, 76 in Ernst (1990a)
		Ref. 18, 53-55 in Schnug (1990)
Cholin-*O*-sulfate	osmoregulation	Hanson & Gage (1991)
Flavonoid sulfate	S-detoxification	Barron *et al.* (1988)
Alliin, allicin	attractant to insects	Lecomte & Thibout (1984)
	repellant to mammals?	
Thiophenes	bactericide	Chan *et al.* (1975)
	nematicide	Uhlenbroek & Bijloo (1958)
Thionins	bactericide	Fernandez de Caleya *et al.* (1972)
	fungicide	Ebrahim-Nesbat *et al.* (1989)
Phytochelatins	metal-binding	Kondo *et al.* (1984)
		Grill *et al.* (1985)
		Rauser (this volume)
Raphanusanins	growth inhibitor	Harada *et al.* (1991)

and for glucosinolates from sulfur storage, feeding repellent, bactericide, fungicide, nematicide, insecticide, to oviposition stimulant.

In an ecological context the reaction to pests may be species-specific and depends on the life history of the plant. The various developmental stages of a plant may adopt different procedures for protection. I will take the glucosinolates as an example to demonstrate the various aspects. In advance, it is necessary to state that the so-called coevolution of resistant pest organisms (Chew 1988), *i.e.* the function of glucosinolates as attractants for pests (see Table 6), is a taxa-independent reaction of a heterotrophic species to its host.

The genetic background

If a character is under selection, one to several genes have to be involved. The best investigated system is that of the cyanogenic glucoside in clovers, where at least two genes are involved, one for the enzyme linamarase and one for the product linamarine (Kakes 1990). In the glucosinolate system, however, at least three genes have to operate: the genes for the synthesis of the alkenyl- and indole-glucosinolate and one for the thioglucosidase. Up to now the elucidation of both systems is poor or even disappointing. More than 20 years ago, it was demonstrated that a high glucosinolate concentration was based on a dominant allele and that the glucosinolate content of the embryo is determined maternally (Josefsson 1970; Kondra & Stephansson 1970; Lein 1972; Love *et al.* 1990). Recently Haughn and coworkers (1991) have confirmed that the concentration of many of the alkenyl glucosinolates found in the wild type of *Arabidopsis thaliana* are due to a dominant allele Gms. The maternal involvement

Table 7. Possible genetic complexes operating in the glucosinalate-thioglucosidase system in *Brassicaceae*. [1]Bodnaryk & Palaniswamy (1990); [2]Bodnaryk (1991).

Developmental stage of plant	Fate of glucosinolates	Thioglucosidase activity	Production genes	Localization genes
Seedling	Decreasing concentration of alkenyl glucosinolates in cotyledons[1,2]	decreasing in cotyledons	none	none
	Roots?	roots?	none	none
Vegetative growth	Synthesis of indole glucosinolate	unknown	indole glucosinolates	shoot
			thioglucosidase	shoot
			indole glucosinolate	root
			thioglucosidase	root
Flowering plants				
flowers	Unknown	low	unknown	flowers?
leaves	Acropetal increase of glucosinolates (indole + alkenyl)	low	alkenyl glucosinolates (GSM) indole glucosinolates (?) epithiospecific proteins ESP thioglucosidase	increasing expression (?)
roots	Unknown	high	indole glucosinolates thioglucosidase	translocation genes? high expression in roots
Seed production				
pod	High production of alkenyl glucosinolates	unknown	alkenylglucosinolates GSM	fruit tissue
			glucosinolate transporter	fruit tissue
seed	Accumulation of alkenyl glucosinolates		thioglucosidase transporter	fruit tissue
		unknown	none	none

in seed glucosinolates demands a further gene, which is responsible for glucosinolate transport to the embryo. The biosynthesis of indole glucosinolates, usually absent from seeds (Chew 1988) and only present in vegetative plant organs, however, may be controlled by another gene. Nothing is known about the genetics of thioglucosidase; obviously its biosynthesis may be regulated by another gene. The occurrence of thioglucosidase in *Aspergillus* not able to synthesize glucosinolates indicates an independent inheritance of this enzyme. Furthermore, it is uncertain how the thioglucosidase activity is regulated in seeds where it is present in idioblasts (Maheshwari *et al.* 1981). It may be transported from the maternal tissue to the embryo, or the embryo itself has the capacity to synthesize the enzyme. Defense against the seed pathogenic yeast *Nematospora sinecauda* is based on an activation of the thioglucosidase-glucosinolate system in the cotyledons (Holley & Jones 1985). Due to the high dynamics of the various glucosinolate compounds and the thioglucosidase during a plant's life history (Falk *et al.* 1992) a lot of genes may be involved in the evolution and expression of the glucosinolate-thioglucosidase-system (Table 7).

Plant-pest specific resistance

The evidence of glucosinolates as defense substances will be critically evaluated. As mentioned above, seeds of *Brassica* are characterized by a predominance of alkenyl glucosinolates (cf. Schnug 1990). Selection of double low lines of *Brassica* species may be interesting for the feeding aspects of cattle, but it will increase the susceptibility of seeds to pathogens. It was already demonstrated that seed homogenates of low-glucosinolate lines of *Brassica campestris* were not toxic toward the seed pathogenic yeast *Nematospora sinecauda* (Holley & Jones 1985). In addition, the feeding activity of the seed weevil *Ceutorhynchus assimilis* is stimulated by the longer chain alkenyl glucosinolates, whereas the leaf and stem feeding flea beetles *Phyllotreta cruciferae* and other *Phyllotreta* species were stimulated by indolylmethyl glucosinolates (Nielsen 1978; Kjaer-Pedersen 1992; Pivnick *et al.* 1992). This differentiation of the microhabitat preference is a good example of ecological niche differentiation and may, dependent on the trigger concentration of alkenyl glucosinolates, influence seed attack in double-low lines.

Once the seed germinates, other glucosinolates may be the target for evolutionary adaptation. During growth of *Sinapis alba* seedlings, the concentration of p-hydroxybutyl glucosinolate (sinalbin) decreases within a few days by nearly 50% (Bodnaryk 1991). Per cotyledon pair the total amount, however, remains constant, indicating a lack of a *de novo* synthesis of glucosinolates in the cotyledons of mustard seedlings and questioning the proposed role of glucosinolates as sulfur storage (Schnug, this volume). This phenomenon is perhaps a general physiological behaviour of cotyledons towards the biosynthesis of secondary compounds (*cf.* cyanogenic glucosides, Frehner *et al.* 1990; pyrrolizidine alkaloids, Hartmann *et al.* 1989). In rape seedlings, labeled sulfate was only weakly incorporated into glucosinolates, even 40 days after germination (Qinzheng *et al.* 1991a). In addition, the difference in the dominance of glucosinolate species between seeds and young and mature plants, *i.e.* alkenyl versus indole glucosinolates demands the activity of a new enzyme for the biosynthesis of the latter group. The decrease of the alkenyl glucosinolate sinigrin in developing *Brassica* seedlings and the start of the de novo synthesis of 3-indolylmethyl glucosinolate (glucobrassicin) seven days after germination (Bodnaryk & Palaniswamy 1990) support this aspect. The functional role of glucosinolate as defense systems is especially weak just a week after germination, because both the glucosinolate concentration and the thioglucosidase activity decreased at that time (Bones 1990). The high attack of cotyledons of *B. napus* and *B. juncea* by insects (Bodnaryk & Palaniswamy 1990) emphasizes the weakness of the seedling defense system. A similar situation was found for the interaction between seedlings of *Sinapis alba* and the flea beetle *Phyllotreta cruciferae* (Bodnaryk 1991). Two possibilities may explain the lack of sufficient chemical defense in one to two-weeks old mustard seedlings: (1) under natural conditions the germination, mostly in early spring, occurs in the absence of a predator, due to delayed insect development at low temperatures; (2) another secondary product may be synthesized *e.g.* cardenolides and sapogenins, which deter herbivorous insects.

The growing seedling has to exploit the surrounding soil for nutrients and thus its roots are exposed to soil pathogens, *e.g. Plasmodiophora brassicae*, the cause of clubroot disease. Differences in sensitivity of *Brassica napus* to this pathogen are related to the concentration of indole-3-methyl glucosinolate, an auxin precursor (Rausch *et al.* 1981). In resistant plants the level of the precursor is low (Chong *et al.* 1985). Obviously, variability in differential gene activity between roots and shoots for the biosynthesis of indolmethyl glucosinolate may be present in populations of the various crucifer species. However, the information of its occurrence in wild and cultivated *Brassicaceae* is very low.

The attack of crucifer plants later in the growing season by various herbivores is not only based on glucosinolates and their degradation products, as often suggested, but also on three selection components. The most crucial stimulation seems to be (1) color to induce landing (Kolb & Scherer 1982), (2) tactile stimuli (Städler 1986) and (3) chemical stimuli. Recent research has shown, that the tarsal contact chemoreceptor in the cabbage butterfly is not sensitive to allyl glucosinolates but to other water-soluble compounds (Renwick & Radke 1988). The former overemphasize paid to glucosinolates finds a parallel situation on the chemoreceptors for *Ipidae*, which were believed for a long time to be pinene but electrophysiological measurements have detected completely other compounds (Masson & Mustaparta 1990). Furthermore, female *Pieris rapae* prefer to lay eggs on fertilized plants, thus enhancing the fitness for the offspring (Myers 1985). The importance of visual stimuli as cues of herbivorous insects during host finding in contrast or in addition to olfactory ones (Miller & Strickler 1984) may explain why in certain insects, as in *Pieris brassicae*, plant density is not related to search success, whereas in others *e.g.* the flea beetles (*Phyllotreta cruciferae, P. striolata*) odour plumes are the guide (Cromartie 1975). This diversity in reactions and cues supports the necessity for a reanalysis of the importance of glucosinolates in the crucifer-herbivore relationship as attractants and repellants of herbivores. In this reanalysis the occurrence and activity of thioglucosidase has to be taken into account, as well as other sulfur-containing compounds. As a first indication, the damage of animals by a high consumption of double low oilseed rate (hares: Seifert & Röbbelen 1988; roe deer: Doag *et al.* 1990; deer: Onderschenka *et al.* 1987) was suggested to be caused by *S*-methyl-cysteine sulfoxide, and not by glucosinolates (Schnug 1990).

Physiologically and ecologically very interesting is the aspect of the change of biosynthesis of glucosinolates during seed formation. Much evidence suggests that the biosynthesis of seed glucosinolates takes place in the fruit tissue (Lein 1972; Qinzheng *et al.* 1991b). The question, however, is not yet answered, if the preference of the accumulation of alkenyl glucosinolates in crucifer seeds is due to a differential gene activity for their biosynthesis in the fruit tissue or due to a preferred transport of these glucosinolates in comparison to indolyl glucosinolates. The evolutionary necessity of such a change in a major "defense" substance may be understood, if indolyl glucosinolates are less suitable for storage than alkenyl glucosinolates; the experimental evidence, however, has not been tested.

Concluding remarks

The obvious lack of glucosinolate and thioglucosidase synthesis in green cotyledons, the increase of the glucosinolate concentration in acropetal direction during plant growth (Glover *et al.* 1988), the change in the glucosinolate profile in a plant's life history from seedling to seed, the identical sulfate uptake of double and low wild type (Schnug 1990) with obviously strong effects on the biochemical compartmentation of sulfur, and the lack of full information about the thioglucosidase activity during the life history and its relation to the effectivity of the defense system, all these various aspects demonstrate that the understanding of the glucosinolate-thioglucosidase system is far from adequate. Recent investigations support ideas of the sulfur storage role of glucosinolates and of hormonal control of flowering (Clossais-Besnard & Larher 1991; Schnug, this volume). As a consequence many results, especially those on the plant-animal relationships in *Brassicaceae*, seem often to be interpreted as wishful thinking. A good documented example of the glucosinolate-myrosinase system as feeding deterrent to caddiflies, snails and amphipods is the hydrolysis of phenylethyl glucosinolate after damage of watercress (Newman *et al.* 1992).

The same speculative approach concerns the role of thiophenes as nematicides, because they are only effective if the plant root with the herbivore is exposed to light (Sütfeld 1982; Arnason *et al.* 1986). In such an ecological situation, most of the root system is already damaged by nematodes, and the remnant roots, if exposed to light, will have no advantage in the presence of thiophenes. From an ecological point of view, combined investigations of the biosynthesis of the intermediate and final products (*e.g.* glucosinolates) and the enzyme activities (*e.g.* thioglucosidases) are necessary to make progress in the understanding of the physiological and ecological aspects of organic sulfur compounds.

Acknowledgements

I thank Mrs. Thea Laan for typing the manuscript and Dr. R. van der Meijden, Rijksherbarium Leiden for the permission to reproduce the map of *Diplotaxis tenuifolia*.

References

Arnason, J.T., Philogène, B.J.R., Berg, C., MacEachern, A., Kaminski, J., Leitch, L.C., Morand, P. & Lam, J. 1986. Photoxicity of naturally occurring and synthetic thiophene and acetylene analogues to mosquito larvae. Phytochemistry 25: 1609-1611.

Autri, A.R. & Fitzgerald, J.W. 1991. Potential for organic sulfur accumulation in a variety of forest soils at saturating sulfate concentrations. Biol. Fert. Soils 10: 281-284.

Ayazloo, M. & Bell, J.N.B. 1981. Studies on the tolerance to sulphur dioxide of grass populations in polluted areas. I. Identification of tolerant populations. New Phytol. 88: 203-222.

Ayers, G.P., Ivey, J.P. & Gillett, R.W. 1991. Coherence between seasonal cycles of dimethyl sulphide, methanesulphonate and sulphate in marine air. Nature 349: 404-406.

Barron, D., Varin, L., Ibrahim, R.K. & Harborne, J.B., Williams, Ch.A. 1988. Sulphated flavanoids – an update. Phytochemistry 27: 2375-2395.

Bell, J.N.B. & Mudd, C.H. 1976. Sulphur dioxide resistance in plants: a case study of *Lolium perenne*. In: T.A. Mansfield (ed.), Effects of Air Pollutants on Plants. Cambridge University Press, Cambridge, pp. 87-103.

Benson, A.A. & Atkinson, M.R. 1967. Cholin sulfate and phosphate in salt secreting plants. Fed. Proceed. 26: 394.

Bentley, C.F., Gareau, L., Renner, R. & McElroy, L.W. 1956. Fertilizers and nutritive values of hays. I. Sulfur deficient grey wooded soils. Can. J. Agr. Sci. 36: 315-325.

Blunden, G. & Gordon, S.M. 1986. Betaines and their sulfonio analogues in marine algae. Progr. Phycol. Res. 4: 35-80.

Boag, B., Macfarlane Smith, W.H. & Griffiths, D.W. 1990. Observations on the grazing of double low oilseed rape and other crops by roe deer. Appl. Anim. Behav. Sci. 28: 213-220.

Bodnaryk, R.P. 1991. Development profile of sinalbin (p-hydroxybenzylene glucosinolate) in mustard seedlings. *Sinapis alba* L., and its relationship to insect resistance. J.Chem. Ecol., 17, 1543-1556.

Bodnaryk, R.P. & Palaniswamy, P. 1990. Glucosinolate levels in cotyledons of mustard, *Brassica juncea* L. and rape, *B. napus* L. do not determine feeding rates of flea beetle, *Phyllotreta cruciferae* (Goeze). J. Chem. Ecol. 16: 2735-2746.

Bohlmann, H. & Apel, K. 1991. Thionins. Annu. Rev. Plant Physiol. Plant Mol. Biol. 42: 227-240.

Bones, A.M. 1990. Distribution of β-thioglucosidase activity in intact plants, cell and tissue cultures and regenerant plants of *Brassica napus* L. J. Exp. Bot. 41: 737-744.

Caput, C., Belot, Y., Auclair, D., Decourt, N. 1978. Absorption of sulfur dioxide by pine seedlings leading to acute injury. Environ. Pollut. 16: 23-28.

CBS – Centraalbureau voor de Statistiek 1983. Algemene milieustatistiek 1979-1982. Staatsuitgeverij, 's-Gravenhage.

CBS – Centraalbureau voor de Statistiek 1989. Algemene milieustatistiek 1989. SDU, 's-Gravenhage.

Challenger, F. & Simpson, M.K. 1948. Studies on biological methylation. XII. A precursor of dimethylsulphide evolved by *Polysiphonia fastigiata*. Dimethyl & carboxyethylsulphonium hydroxide and its salt. J. Chem. Soc. 1948: 1591-1598.

Chan, G.F.C., Towers, G.H.N. & Mitchell, J.C. 1975. Ultraviolet-mediated antibiotic activity of thiophene compounds of *Tagetes*. Phytochemistry 14: 2295-2296.

Charlson, R.J. Lovestock, J.E., Andrea, M.O. & Warren, S.G. 1987. Oceanic phytoplankton, atmospheric sulphur, cloud albedo and climate. Nature 326: 655-661.

Chew, F.S. 1988. Searching for defense chemistry in the *Cruciferae*, or, do glucosinolates always control interactions of *Cruciferae* with these potential herbivores and symbionts? In: Spencer, K.C. (ed.), Chemical Mediation of Coevolution. Academic Press, Toronto, pp. 81-112.

Chong, C., Chiang, M.S. & Crete, R. 1985. Studies on glucosinolates in clubroot resistant selection and susceptible commercial cultivars of cabbages (*Brassica oleracea* ssp. *capitata*). Euphytica 34: 65-73.

Clossais Besnard, N. & Larher, F. 1991. Physiological role of glucosinolates in *Brassica napus*: Concentration and distribution pattern of glucosinolates among plant organs during a complete life cycle. J. Sci. Food Agric. 56: 25-38.

Cromartie, W.J. 1975. The effect of stand size and vegetational background on the colonization of cruciferous plants by hervivorous insects. J. Appl. Ecol. 12: 517-533.

De Kok, L.J. 1990. Sulfur metabolism in plants exposed to atmospheric sulfur. In: Rennenberg, H., Brunold, C., De Kok, L.J. & Stulen, I. (eds.), Sulfur Nutrition and Sulfur Assimilation in Higher

Plants. Fundamental, Environmental and Agricultural Aspects. SPB Academic Publishing, the Hague, pp. 111-130.
Dickson, D.M.J., Wyn Jones, R.G. & Davenport, J. 1980. Steady state osmotic adaptation in *Ulva lactuca*. Planta 150: 158-165.
Ebrahim-Nesbat, F., Behnke, S., Kleinhofs, A. & Apel, K. 1989. Cultivar-related differences in the distribution of cell-wall-bounds thionins in compatible and incompatible interactions between barley and powdery mildew. Planta 179: 203-210.
Ellenberg, H., Mayer, R. & Schauermann, J. (eds.) 1986. Ökosystemforschung. Ergebnisse des Sollingprojektes 1966-1986. Ulmer Verlag, Stuttgart.
Ende, H.P. & Hüttl, R.F.,1992. Insufficient sulfur nutrition of yellowed spruce and beech in the Southern Black Forest. Phyton 32: 45-48.
Ernst, W.H.O. 1979. Population biology of *Allium* ursinum in Northern Germany. J. Ecol. 67: 347-362.
Ernst, W.H.O. 1990a. Ecological aspects of sulfur metabolism. In: Rennenberg, H., Brunold, C., De Kok, L.J. & Stulen, I. (eds.), Sulfur Nutrition and Sulfur Assimilation in Higher Plants. Fundamental, Environmental and Agricultural Aspects. SPB Academic Publishing, the Hague. pp. 131-144.
Ernst, W.H.O. 1990b. Mine vegetation in Europe. In: A.J. Shaw (ed.) Heavy Metal Tolerance in Plants: Evolutionary Aspects. CRC Press, Boca Raton, Florida, pp. 21-37.
Ernst, W.H.O., Tonneijck, A.E.C. & Pasman, F.J.M. 1985. Ecotypic response of *Silene cucubalus* to air pollutants (SO_2, O_3). J. Plant Physiol. 118: 439-450.
Ernst, W.H.O., Verkleij, J.A.C. & Schat, H. 1992. Metal tolerance in plants. Acta Bot. Neerl. 41: *in press*.
Falk, A., Xue, J., Lenman, M. & Rask, L. 1992. Sequence of a cDNA clone encoding the enzyme myrosinase and expression of myrosinase in different tissues of *Brassica napus*. Plant Sci. 83: 181-186.
Feeny, P. 1977. Defense ecology of the *Cruciferae*. Ann. Missouri Bot. Gard. 64: 221-234.
Fernandez de Caleya, R., Gonzalez-Pascual, B., Garcia-Olmedo & F., Carbonero, P. 1972. Susceptibility of phytophatogenic bacteria to wheat purothionins in vitro. Appl. Microbiol. 23: 998-1000.
Fitzgerald, J.W. & Luschinski, P.C. 1977. Further studies on the formation of cholin sulphate by bacteria. Can. J. Microbiol. 23: 483-390.
Frehner, M., Scalet, M. & Conn, E.E. 1990. Pattern of the cyanide-potential in developing fruits. Plant Physiol. 94: 28-34.
Gibson, J.A.E., Garrick, R.C., Burton, H.R. & McTaggart, A.R. 1990. Dimethylsulfide and the alga *Phaeocystis pouchetii* in arctic coastal waters. Mar. Biol. 104: 339-346.
Glover, J.R., Chapple, C.C.S., Rothwell, S., Tober, J. & Ellis, B.R. 1988. Allylglucosinulate biosynthesis in *Brassica carinata*. Phytochemistry 27: 1345-1348.
Gorham, J., Hughes, L.J. & Wyn Jones, R.G. 1980. Chemical composition of salt-marsh plants from Ynys môn (Anglesey): the concept of physiotypes. Plant Cell Environ. 3: 309-318.
Gravel, R.A. 1976. Choline-*O*-sulphate utilization in *Aspergillus niger*. Genet. Res. 28: 269-276.
Grill, E., Winnacker, E.L. & Zenk, M.H. 1985. Phytochelatins: the principal heavy-metal complexing peptides of higher plants. Science 230: 674-676.
Guderian, R. 1977. Air Pollution. Springer Verlag, Berlin.
Haeupler, H. & Schönfelder, P. 1988. Atlas der Farn- und Blütenpflanzen der Bundesrepublik Deutschland. Ulmer Verlag, Stuttgart.
Hanson, A.D. & Gage, D.A. 1991. Identification and determination by Fast Atom Bombardement Mass Spectrometry of the compatible solute cholin-O-sulfate in *Limonium* species and other halophytes. Aust. J. Plant Plysiol. 18: 317-327.
Harada, N., Hagiwara, H., Ono, H., Uda, H., Ohba, S., Kubota, M., Nishiyama, S., Yamamura, S., Hasegawa, K. & Sakuda, M. 1991. Revision of the structure of raphanusanins, phototropism-regulating substances of radish hypocotyls. Tetrahedron Lett. 32: 6757-6770.
Hartmann, Th., Ehmke, A., Eilert, U., von Borstel, K. & Theuring, C. 1989. Sites of synthesis, translocation and accumulation of pyrrolizidine alkaloid N-oxides in *Senecio vulgaris*. Planta 177: 98-107.
Hasegawa, K., Shiihara, S., Iwagawa, T. & Hase, T. 1982. Isolation and identification of a new growth inhibitor, raphanusanin, from radish seedlings and its role in light inhibition of hypocotyl growth. Plant Physiol. 70: 626-628.
Haughn, G.W., Davin, L., Giblin, M. & Underhill, E.W. 1991. Biochemical genetics of plant secondary metabolites in *Arabidopsis thaliana*. Plant Physiol. 97: 217-226.
Heinze, M., Höhne, H. & Fiedler, H.J. 1982. Vergleichende Untersuchungen zum Elementgehalt von Bodenpflanzen auf Gipsstandorten. Flora 172: 493-510.

Henrichfreise, A. 1981. Aluminiumtoleranz von *Luzula albida* und *Milium effusum*, Pflanzen saurer und basischer Laubwaldböden. Acta Oecol. Plant. 2: 87-100.

Holley, R.A. & Jones, J.D. 1985. The role of myrosinase in the development of toxicity toward *Nematospora* in mustard seed. Can. J. Bot. 63: 521-526.

Horsman, D.C. & Wellburn, A.R. 1977. Effect of SO_2 polluted air upon enzymic activity in plants originating from areas with different annual mean atmospheric SO_2-concentrations. Environ. Pollut. 13: 33-39.

Imperato, F. 1982. Sulfate esters of hydroxycinnamic acid-sugar derivation from *Adiantum capillus veneris*. Phytochemistry 21: 2717-2718.

Johnson, D.W., Henderson, G.S., Huff, D.D., Lindberg, S.E., Richter, D.D., Shriner, D.S., Todd, D.E. & Turner, J. 1982. Cycling of organic and inorganic sulphur in a chestnut oak forest. Oecologia 54: 141-145.

Josefsson, E. 1970. Content of p-hydroxybenzylglucosinolate in seed meals of *Sinapis alba* as affected by heridity, environment and seed part. J. Sci. Food Aric. 21: 94-97.

Kakes, P. 1990. Properties and functions of the cyanogenic system in higher plants. Euphytica 48: 25-43.

Karsten, U., Wiencke, C., Kirst, G.O. 1991. Growth pattern and β-dimethylsulphoniopropionate (DMSP) content of green macroalgae at different irradiances. Mar. Biol. 108: 151-155.

Kiene, R.P., Bates, T.S. 1990. Biological removal of dimethyl sulphide from sea water. Nature 345: 702-705.

Kjaer-Pederson, C. 1992. Flight behaviour of the cabbage seedpod weevil. Entomol. Exp. Appl. 62: 61-66.

Kolb, G. & Scherer, C. 1982. Experiments on wavelength specific behaviour of *Pieris brassicae* L. during drumming and egg-laying. J. Comp. Physiol. 149: 325-332.

Kondo, N., Imai, K., Isobe, M., Goto, T., Murasugi, A., Wade-Nakagawa, C. & Hayashi, Y. 1984. Cadystin A and B, major unit peptide comprising cadmium binding peptides induced in a fission yeast saperation, revision of structures and synthesis. Tetrahedron Lett. 25: 3869-3872.

Kondra, Z.P. & Stephansson, B.R. 1970. Inheritance of the major glucosinolates of rapeseed meal. Can. J. Plant. Sci. 60: 643-647.

Korneck, D. 1984. Rote Liste der Farn-und Blütenpflanzen (*Pteridophyta et Spermatophyta*) 3. Fassung. Stand 21.12.1982. In: Blab, J., Nowak, E., Trautmann, W., Sukopp, H. (eds.), Rote Liste der gefährdeten Tiere und Pflanzen in der Bundesrepublik Deutschland. 4th ed., Kilda-Verlag, Greven, pp. 128-148,

Koziol, M.J., Shelvey, J.D., Lockyer, D.R. & Lohatley, F.R. 1986. Response of SO_2-sensitive and resistant genotypes of ryegrass (*Lolium perenne* L.) to prolonged exposure to SO_2. New Phytol. 102: 345-357.

Kozlowski, T.T., Kramer, P.J. & Pallardy, S.G. 1991. The Physiological Ecology of Woody Plants. Academic Press, San Diego.

Lam, J., Christensen, L.P. & Thomasen, T. 1991. Thiophene derivatives from Echinops species. Phytochemistry 30: 1157-1159.

Lecomte, C. & Thibout, E. 1984. Le pouvoir attractif chez la teigne du poiveau, *Acrolepiopsis assectella* Zell (*Lep. Yponomeutoïdea*) de quelques *Allium* du complexe *ampeloprassum* consommés par l'homme. Acta Oecol., Oecol. Appl. 5: 259-270.

Legrand, M., Feniet-Saigne, C., Saltzman, E.S., Germain, C., Barkov, N.I. & Petrov, V.N. 1991. Ice-core record of oceanic emissions of dimethylsulphide during the last climate cycle. Nature 350: 144-146.

Lein, K.A. 1972. Genetische und physiologische Untersuchungen zur Bildung von Glucosinolaten in Rapssamen: Lokalisierung des Hauptbiosyntheseortes durch Pfropfungen. Z. Pflanzenphysiol. 67: 333-342.

Liener, I.E. 1980. Toxic Constituents of Plant Food Stuffs. Academic Press, New York.

Likens, G.E., Bormann, F.H., Pierce, R.S., Eaton, J.S. & Johnson, N.M. 1977. Biogeochemistry of a Forested Ecosystem. Springer Verlag, Berlin.

Love, H.K., Rakow, G., Raney, J.P. & Downey, R.K. 1990. Genetic control of 2-propenyl and 3-butanyl glucosinolate synthesis in mustard. Can. J. Plant Sci. 70: 425-429.

Maheshwari, P.N., Stanley, D.W., Beveridge, T.J. & van de Voort, F.R. 1981. Localization of myrosinase (thioglucoside glucohydrolase, EC 3.2.3.1) in cotyledon cells of rapeseed. J. Food Biochem. 5: 39-61.

Marquard, R., Kühn, H. & Linser, H. 1968. Der Einfluß der Schwefelernährung auf die Senfölbildung. Z. Pflanzenernaehr. Bodenk. 121: 221-230.

Masson, C., Mustaparta, H. 1990. Chemical information processing in the olfactory system of insects. Physiol. Rev. 70: 199-245.

Medwecka-Kornás, A. & Gawróski, S. 1990. The dieback of fir *Abies alba* Mill. and changes in the *Pino-Quercetum* stands in the Ojców National Park (Southern Poland). Vegetatio 87: 175-186.

Mennema, J., Quené-Boterenbrod, A.J., Plate & C.L. 1980. Atlas van de Nederlandse Flora. 1. Uitgestorven en zeer zeldzame planten. Uitgeverij Kosmos, Amsterdam.

Mennema, J., Quené-Boterenbrood, A.J. & Plate, C.L. 1985. Atlas van de Nederlandse Flora. 2. Zeldzame en vrij zeldzame planten. Bohn, Scheltema & Holkema, Utrecht.

Meyer, S.E. 1986. The ecology of gypsophile endemisms in the Eastern Mojave Desert. Ecology 67: 1303-1313.

Miller, J.E. & Xerikos, P.B. 1979. Residence time of sulphite in SO_2-'sensitive' and 'tolerant' soybean cultivars. Environ. Pollut. 18: 259-264.

Miller, J.R. & Strickler, K.L. 1984. Finding and accepting host plants. In: Bell, W.J., Cardé, R.T. (eds.), Chemical Ecology of Insects. Chapman and Hall, London, pp. 127-157.

Minderman, G. & Leeflang, K.W.F. 1968. The amounts of drainage water and solutes from lysimeters planted with either oak, pine or natural dune vegetation, or without any vegetation cover. Plant Soil 28: 61-80.

Murdy, W.H. 1979. Effect of SO_2 on sexual reproduction in *Lepidium virginicum* L. originating from regions with different SO_2 concentrations. Bot. Gaz. 140: 299-303.

Myers, J.H. 1985. Effect of physiological conditions of the host plant on the ovipositional choice of the cabbage white butterfly, *Pieris rapae*. J. Anim. Ecol. 54: 193-204.

Newman, R.M., Hanscom, Z. & Kerfoot, W.C. 1992. The watercress glucosinolate-myrosinase system: a feeding deterrent to caddiflies, snails and amphipods. Oecologia 92: 1-7.

Nielsen, J.K. 1978. Host plant discrimination within *Cruciferae*: Feeding responses to four leaf beetles (*Coleptera*: *Chrysomelidae*) to glucosinolates, cucurbitacins and cardenoloides. Entomol. Exp. Appl. 24: 41-54.

Onderschenka, K., Tataruch, F., Steinbeck, T., Klansek, E., Vodnaresky, M. & Wagner, J. 1987. Investigations on the deer-losses caused by rape-intake. Wiener Tierärtzl. Monatsschr. 74: 369-377.

Pivnick, K.A., Lamb, R.J. & Reed, D. 1992. Response of flea beetles, *Phyllotreta* spp., to mustard oils and nitriles in field trapping experiments. J. Chem. Ecol. 18: 863-873.

Qinzheng, Z., Xiafua, L., Haifu, X., Zuegun, Z. & Yinhua, L. 1981. Conversion of SO_4^{2-} and changes in glucosinolate content in leaves of rape seedlings transplanted. J. Agron. Crop Sci. 166: 96-100.

Qinzheng, Z., Haifu, X., Xiafua, L., Zuegun, Z. & Yinhua, L. 1991b. Changes in content of glucosinolates and its accumulation in siliquae and seeds of oilrape (*Brassica napus* L). J. Agron. Crop Sci. 166, 152-156.

Rausch, T., Butcher, D.N. & Hilgenberg, W. 1981. Nitrilase activity in clubroot diseased plants. Physiol. Plant. 52: 467-470.

Rennenberg, E. 1984. The fate of excess sulfur in higher plants. Annu. Rev. Plant Physiol. 35: 121-153.

Renwick, J.A.A. & Radke, C.D. 1988. Sensory cues in host selection for oviposition by the cabbage butterfly, *Pieris brassicae*. J. Insect Physiol. 34, 251-257.

Rodman, J.E. 1991a. A taxonomic analysis of glucosinolate-producing plants, part 1: Phenetics. Syst. Bot. 16: 598-618.

Rodman, J.E. 1991b. A taxonomic analysis of glucosinolate-producing plants, part 2: Cladistics. Syst. Bot., 16: 619-629.

Rohmeder, E., Merz, W., von Schönborn, A. 1962. Züchtung von gegen Industrieabgase relativ resistenten Fichten- und Kiefernsorten. Forstw. Cbl. 81: 321-332.

Roose, M.L., Bradshaw, A.D., Roberts, T.M. 1982. Evolution of resistance to gaseous air pollutants. In: Unsworth, M.H., Omrod, D.P. (eds.), Effects of Gaseous Air Pollution in Agriculture and Horticulture. Butterworth Scientific, London, pp. 379-409.

Schnug, E. 1992. Glucosinolates. In: De Kok, L.J., Stulen, J., Schulze, E.D., Lange, O.L. & Oren, R. (eds.) 1989. Forest Decline and Air Pollution. Springer Verlag, Berlin.

Seifert, H.S.H. & Röbbelen, G. 1988. Production of OO-rapeseed varieties and its effect on wildlife. Double low rapeseed forum, London. NIAB Cambridge, pp. 62-65.

Sendl, A., Wagner, H. 1991. Isolation and identification of homologues of ajoene and alliin from bulb extracts of *Allium* ursinum. Planta Med. 57: 361-362.

Smith, W.H. 1981. Air Pollution and Forests. Interactions between Air Contaminants and Forest Ecosystems. Springer Verlag, New York.

Städler, E. 1986. Oviposition and feeding stimuli in leaf surface waxes. In: Juniper, B. & Southwood, R. (eds.), Insects and the Plant Surface. Edward Arnold, London, pp. 105-111.

Strickland, T.C., Fitzgerald, J.W., Ash, J.T. & Swank, W.T. 1987. Organic sulfur transformations and sulfur pool sizes in soil and litter from a Southern Appalachian Hardwood forest. Soil Sci. 143: 453-458.

Stuyfzand, P.J. 1984. Effecten van vegetatie en luchtverontreiniging op de grondwaterkwaliteit in kalkrijke duinen bij Castricum: lysimeterwaarnemingen. H_2O 17: 152-159.

Sütfeld, R. 1982. Distribution of thiophene derivatives in different organs of *Tagetes patula* seedlings grown under various conditions. Planta 156: 536-540.

Tax, M.H. 1989. Atlas van de Nederlandse dagvlinders. Vereniging tot Behoud van Natuurmonumenten in Nederland, 's-Graveland.

Taylor, G.E. Jr. & Murdy, W.H. 1975. Population differentiation of an annual species, *Geranium carolinianium*, in response to sulphur dioxide. Bot. Gaz. 136: 212-215.

Taylor, G.E. Jr., Tingey, D.T. & Gunderson, L.A. 1986. Photosynthesis, carbon allocation, and growth of sulfur dioxide ecotypes of *Geranium carolinium* L. Oecologia 68: 350-357.

Uhlenbroek, J.H. & Bijloo, J.D. 1958. Investigations on nematicides. I. Isolation and structure of a nematicidal principle occurring in *Tagetes* roots. Rec. Trav. Chim. Pays Bas 77: 1004-1008.

Unsworth, M.H. & Omrod, D.P. 1982. Effects of Gaseous Air Pollution in Agriculture and Horticulture. Butterworth Scientific, London.

Van der Meijden, R. 1990. Heukels Flora van Nederland, 21 ed. Wolters-Noordhoff, Groningen.

Van der Meijden, R., Plate, C.L. & Weeda, E.J. 1989. Atlas van de Nederlandse Flora. 3. Minder zeldzame en algemene soorten. Rijksherbarium/Hortus Botanicus, Leiden.

Van Diggelen, J., Rozema, J., Dickson, D.M.J. & Broekman, R. 1986. β-3-dimethylsulphoniopropionate, proline and quaternary ammonium compounds in *Spartina anglica* in relation to sodium chloride, nitrogen and sulphur. New Phytol. 103: 573-586.

Wookey, P.A. & Ineson, P. 1991. Chemical changes in decomposing forest litter in response to atmospheric sulphur dioxide. J. Soil Sci. 42: 615-628.

ACKNOWLEDGEMENTS

The workshop was sponsored by

Bayerisches Staatsministerium für Wirtschaft und Verkehr
Deutsche Forschungsgemeinschaft
Federation of European Societies of Plant Physiologists
Fraunhofer-Gesellschaft
Swiss National Academy of Sciences
The Sulphur Institute
University of Groningen – Faculty of Science

Beckmann Instruments GmbH
B.E.S.T. GmbH
Blackwell Scientific Publications Ltd.
Perkin-Elmer

Reviewers of the contributions in this volume

R. Bachhofen, Zürich, Switzerland
E.G. Beauchamp, Guelph, Canada
B.B. Buchanan, Berkeley, U.S.A.
D.T. Clarkson, Bristol, U.K.
W.J. Cram, Newcastle upon Tyne, U.K.
E.F. Elstner, München, F.R.G.
P.D. Goldan, Boulder, U.S.A.
T.C. Hutchinson, Peterborough, Canada
P.B. Jackson, Los Alamos, U.S.A.
P.J. Lea, Lancaster, U.K.
H.J.M. Linthorst, Leiden, the Netherlands
D.L. Messick, Washington, U.S.A.
J.B. Mudd, Riverside, U.S.A.
H. Sandemann, Neuherberg, F.R.G.
A. Schmidt, Hannover, F.R.G.
J.D. Schwenn, Bochum, F.R.G.
A. Trebst, Bochum, F.R.G.

Index

Mrs. C. Herschbach, Freiburg, F.R.G.

Technical assistance

Mrs. M.L. Cambridge, Maarssen, the Netherlands
J.M.A.M. Clement, Haren, the Netherlands
E. Leewinga, Haren, the Netherlands

PHYTON

Volume 32 (Fasc. 3), Special Edition

PROGRESS IN SULFUR METABOLISM OF HIGHER PLANTS

L.J. De Kok, I. Stulen, H. Rennenberg, D. Grill & H. Guttenberger
(Issue Editors)

Contents

Pfanz, H., Würth, G., Oppmann, B. & Schultz, G. Sulfite oxidation in, and sulfate uptake from the cell wall of leaves. *In muro* studies. pp. 95-98.

Polle, A., Morawe, B. & Rennenberg, H. Seasonal variations of glutathione reductase in leaves and leaf buds of beech trees (*Fagus sylvatica* L.). pp. 99-102.

Rao, M.V. & Dubey, P.S. Glutathione metabolism in four Indian wheat cultivars exposed to long-term sulfur dioxide. pp. 103-108.

Rüegsegger, A & Brunold, Ch. Effect of cadmium and/or removal of kernels or shoots on the levels of cysteine, γ-glutamyl-cysteine, glutathione, and TCA-soluble thiols in maize seedlings. pp. 109-112.

Schneider, A. & Rennenberg, H. Degradation of glutathione (GSH) in heterotrophic tobacco cells. pp. 113-117.

Schnug, E. & Evans, E.J. Monitoring of the sulfur supply of agricultural crops in northern Europe. pp. 119-122.

Schnug, E. & Haneklaus, S. Sulfur and light element determination in plant material by X-ray fluorescence spectroscopy. pp. 123-126.

Schröder, P., Nathaus, F.J., Rusness, D.G. & Lamoureux, G.L. The induction of glutathione *S*-transferase and *C-S* lyase in the needles of spruce trees. pp. 127-132.

Shahak, Y., Arieli, B., Hauska, G., Herrmann, I. & Padan, E. Isolation of sulfide-quinone reductase (SQR) from prokaryotes. pp. 133-137.

Stuiver, C.E.E., De Kok, L.J. & Kuiper, P.J.C. Does glutathione play a role in freezing tolerance of plants? pp. 139-142.

Turner, W.L., Lea, P.J. & Pallett, K.E. The purification and characterization of cystathionine-β-lyase from *Echinochloa colonum*. pp. 143-146.

Wingsle, G., Strand, M., Karpinski, S & Hällgren, J.-E. Influence of SO_2 and NO_2 exposure on antioxidants and photosynthesis in needles of *Pinus sylvestris*. pp. 147-153.

INDEX OF AUTHORS

SUBJECT INDEX